International
Association
of Fire Chiefs

International
Association of
Arson Investigators

T0276453

Fire Investigator

Principles and Practice

SIXTH EDITION

International Association of Fire Chiefs

International Association of Arson Investigators

National Fire Protection Association

Fire Investigator

Principles and Practice

SIXTH EDITION

JONES & BARTLETT
LEARNING

World Headquarters
Jones & Bartlett Learning
25 Mall Road
Burlington, MA 01803
978-443-5000
info@jblearning.com
www.jblearning.com
www.psglearning.com

1 Batterymarch Park
Quincy, MA 02169
www.NFPA.org

2111 Baldwin Avenue, Suite 203
Crofton, MD 21114
www.firearson.com

4025 Fair Ridge Drive
Fairfax, VA 22033
www.IAFC.org

Jones & Bartlett Learning books and products are available through most bookstores and online booksellers. To contact the Jones & Bartlett Learning Public Safety Group directly, call 800-832-0034, fax 978-443-8000, or visit our website, www.psglearning.com.

Substantial discounts on bulk quantities of Jones & Bartlett Learning publications are available to corporations, professional associations, and other qualified organizations. For details and specific discount information, contact the special sales department at Jones & Bartlett Learning via the above contact information or send an email to specialsales@jblearning.com.

Copyright © 2023 by Jones & Bartlett Learning, LLC, an Ascend Learning Company

All rights reserved. No part of the material protected by this copyright may be reproduced or utilized in any form, electronic or mechanical, including photocopying, recording, or by any information storage and retrieval system, without written permission from the copyright owner.

The content, statements, views, and opinions herein are the sole expression of the respective authors and not that of Jones & Bartlett Learning, LLC. Reference herein to any specific commercial product, process, or service by trade name, trademark, manufacturer, or otherwise does not constitute or imply its endorsement or recommendation by Jones & Bartlett Learning, LLC and such reference shall not be used for advertising or product endorsement purposes. All trademarks displayed are the trademarks of the parties noted herein. *Fire Investigator: Principles and Practice, Sixth Edition* is an independent publication and has not been authorized, sponsored, or otherwise approved by the owners of the trademarks or service marks referenced in this product.

There may be images in this book that feature models; these models do not necessarily endorse, represent, or participate in the activities represented in the images. Any screenshots in this product are for educational and instructive purposes only. Any individuals and scenarios featured in the case studies throughout this product may be real or fictitious but are used for instructional purposes only.

The International Association of Arson Investigators and the publisher have made every effort to ensure that contributors to *Fire Investigator: Principles and Practice, Sixth Edition* materials are knowledgeable authorities in their fields. Readers are nevertheless advised that the statements and opinions are provided as guidelines and should not be construed as official International Association of Arson Investigators policy. The recommendations in this publication or the accompanying resources do not indicate an exclusive course of action. Variations, taking into account the individual circumstances and local protocols, may be appropriate. The International Association of Arson Investigators the publisher disclaim any liability or responsibility for the consequences of any action taken in reliance on these statements or opinions.

24705-3

Production Credits

Vice President, Product Management: Marisa R. Urbano
Vice President, Content Strategy and Implementation: Christine Emerton
Director, Product Management: Laura Carney
Director, Content Management: Donna Gridley
Manager, Content Management: Kim Crowley
Content Strategist: Jennifer Deforge-Kling
Director, Project Management and Content Services: Karen Scott
Manager, Project Management: Kristen Rogers
Program Manager: Kathryn Leeber
Senior Digital Project Specialist: Angela Dooley
Director, Marketing: Brian Rooney

Director, Sales, Public Safety Group: Brian Hendrickson
Content Services Manager: Colleen Lamy
VP, Manufacturing and Inventory Control: Therese Connell
Composition: S4Carlisle Publishing Services
Cover and Text Design: Scott Moden
Media Development Editor: Faith Brosnan
Rights & Permissions Manager: John Rusk
Rights Specialist: Liz Kincaid
Cover Image (Title Page, Part Opener, Chapter Opener): © Jones & Bartlett Learning. Photographed by Glen E. Ellman.
Printing and Binding: LSC Communications

Library of Congress Cataloging-in-Publication Data

Names: International Association of Arson Investigators, editor. | International Association of Fire Chiefs. | National Fire Protection Association.
Title: Fire investigator : principles and practice / edited by International Association of Arson Investigators, (IAAI).
Description: Sixth edition. | Burlington, MA : Jones & Bartlett Learning, [2023] | Includes bibliographical references and index.
Identifiers: LCCN 2021047336 | ISBN 9781284247053 (paperback)
Subjects: LCSH: Fire investigation--Standards. | Fire investigation--Examinations--Study guides.
Classification: LCC TH9180 .F487 2023 | DDC 363.37/65--dc23
LC record available at https://lccn.loc.gov/2021047336

6048

Printed in the United States of America
26 25 24 23   10 9 8 7 6 5 4 3

Brief Contents

Header image: © Jones & Bartlett Learning. Photographed by Glen E. Ellman.

Contents

CHAPTER 3
Building Construction and Systems for Fire Investigators   41

CHAPTER **8**
Examining the Fire Scene　219

CHAPTER 15
Analyzing the Incident   359

CHAPTER 16
Explosions   389

CHAPTER 17
Automobile, Marine, and Equipment Fires   409

CHAPTER 18
Wildland Fires   453

Acknowledgments

Lead Editor and Author

George Codding, IAAI-CFI
Fire Investigator
Louisville Fire Department
Louisville, Colorado

Reviewers

Jones & Bartlett Learning would like to thank the experienced subject matter experts from the International Association of Fire Chiefs, the National Fire Protection Association, and the International Association of Arson Investigators for their review and feedback of the *Sixth Edition*.

Dale Anderson
Director of Fire Science Casper College
Casper, Wyoming

Jeffrey R. Barlow, BS, AA, EFO, CFEI, CFII, CVFI
Burlington Fire Protection District
Burlington, Kentucky

Staff Sergeant Brian K. Blomstrom, MPA, CFO, CTO, FIT
Greenville Department of Public Safety
Greenville, Michigan

Richard A. Boisvert
Brighton Area Fire Authority
Brighton, Michigan

Edward P. Bonollo, CFI, CFEI, CVFI
Principal/Forensic Fire Investigator
Paradise Fire Consultants
Ewa Beach, Hawaii

Andrew Buterbaugh, MSc, CFEI
International Association of Arson Investigators, New Jersey
    Chapter
Union County College
Cranford, New Jersey

Matthew Claflin, BAS, OFO
University of Akron
Akron Fire Department
Akron, Ohio

Donny Collins, BAS
Mississippi State Fire Academy, Instructor Advanced
Jackson, Mississippi

William A. Cooke
Assistant Chief/Fire Marshal
Rochester Hills Fire Department
Rochester Hills, Michigan

Peter Cutrer, CFI-IAAI, CFPS, CFPE, NFPA CFI-II
7CS Consulting, LLC
Shapleigh, Maine

Bradley D. Drury, IAAI-CFI
Assistant Chief
DeWitt, Michigan

Robert S. Goldenberg, MPA, BA, EMT-P
Florida State Fire College
Ocala, Florida

Dr. Rusty Nolan Graham, EdD, MS
Guilford Technical Community College
Ellenboro, North Carolina

Kristopher Grod, BS
Fire Chief
Sussex Fire Department
Sussex, Wisconsin

Margueritte Hickman, BA, MS, IAAI-CFI
Rogue Community College
Sage Fire Solutions, LLC
Medford, Oregon

John A. High, Sr.
Champaign, Illinois

Gary S. Hudson, BS, IAAI-CFI
EFI Global, Inc.
Sandy, Utah

James Iammatteo, IAAI-CFI
Minnesota State Fire Marshal Division
Keewatin, Minnesota

Matthew W. Knott, MS, CFO, FM, CEM
Rockford Fire Department
Rockford, Illinois

Daniel J. Kramer, MPA, LP
Windcrest Fire Department
Windcrest, Texas

Stephen S. Malley, MPA
Weatherford College
Public Safety Professions
Weatherford, Texas

David J. Marcarelli
North Haven Fire Department
North Haven, Connecticut

Jerry Marrison, MBA, FM, CFPS
North Port, Florida

Jamie T. Meece, EMD, AS, AAS
Firefighter and EMT Chaplain
Fort Knox Fire Department (Retired)
Pulaski County, Kentucky

Michael Moscatello, NAFI-CFEI, CVFI, IAAI-FIT, NJCFI
NJ Fire Official
West Milford Township Fire Marshal's Office
International Association of Arson Investigators New Jersey
    Chapter
Atlantic Professional Services
West Milford, New Jersey

Edward Nunn, Jr., BS, IAAI-FIT, CVFI, NAFI RFEI, NJ CFI
NJ Fire Official and Fire Officer
Madison Fire Department
Madison, New Jersey

Sean F. Peck, MEd, CFO, FM, NRP
Arizona Western College
Yuma, Arizona

Jeff Pricher, NAFI-CFEI, CFII, IAAI-FIT(V)
Fire Chief/Fire Marshal
Scappoose Fire District
Scappoose, Oregon

James Pulito, BSME, MFO, CFEI
Columbia River Fire & Rescue
St. Helens, Oregon

Frank C. Schaper, BS, MA
St. Charles, Missouri

Leigh H. Shapiro, MS
Deputy Fire Chief/Senior Tour Commander (Ret.)
Hartford, Connecticut Fire Department
Hartford, Connecticut

Nicholas Steker, MPA
Franklin Park Fire Department
Triton College
Northeastern Illinois Public Safety Training Academy (NIPSTA)
River Grove, Illinois

Mark J. Sylvester, CFEI, IAAI FIT(V), ECT, CI
Bernards Township Fire Prevention Bureau
Roxbury Township Fire Prevention Bureau
International Association of Arson Investigators New Jersey
    Chapter
Netcong, New Jersey

David VanBeek, AAS, BAEd, BS, EFO
Everett Community College
Everett, Washington

Kevin S. Walker, JD, MBA
Eastern Oregon University
LaGrande, Oregon

Jeremy Williams
Abilene Fire Department
Abilene, Texas

Scott Wilson, CFEI, CVFI, IAAI-FIT(V)
O'Neill Associates
Paterson Fire Department
International Association of Arson Investigators New Jersey
    Chapter
Franklin Lakes, New Jersey

Douglas J. Zordan
State of Connecticut – Office of Education and Data
    Management
Cromwell, Connecticut

Header image: © Jones & Bartlett Learning. Photographed by Glen E. Ellman.

CHAPTER 1

Introduction to Fire Investigation

KNOWLEDGE OBJECTIVES

After studying this chapter, you should be able to:

- State the purpose and objectives of a fire investigation.
- Identify the subjects that are considered requisite knowledge for fire investigators in National Fire Protection Association (NFPA) 1033, *Standard for Professional Qualifications for Fire Investigator*. (**NFPA 1033: 4.1.7, 4.1.7.1**, p. 3)
- Describe how fire investigators apply the job performance requirements listed in NFPA 1033. (**NFPA 1033: 4.1.1**, p. 3)
- Identify the methods of remaining current in requisite knowledge for fire investigators. (**NFPA 1033: 4.1.7.2, 4.1.7.3**, p. 4)
- Identify the continuing education requirements for fire investigators found in NFPA 1033. (**NFPA 1033: 4.1.7.3**, p. 4)
- Describe the ethical responsibilities of a fire investigator.
- Describe the types and roles of fire investigators.
- Describe how NFPA documents are developed.
- Describe the purpose and development of NFPA 1033, *Standard for Professional Qualifications for Fire Investigator*. (**NFPA 1033: 4.1.1**, p. 8)
- Describe the purpose and development of NFPA 921, *Guide for Fire and Explosion Investigations*.
- Describe the steps of the scientific method.
- Explain application of the scientific method to fire investigation. (**NFPA 1033: 4.1.2**, pp. 10–14)
- Identify the basic steps of the scientific method in the context of a fire investigation. (**NFPA 1033: 4.1.2, 4.6.5**, pp. 10–14)

SKILLS OBJECTIVES

After studying this chapter, you should be able to:

- Use the scientific method when conducting a fire investigation. (**NFPA 1033: 4.1.2; 4.6.5**, pp. 10–14)
- Use the scientific method in origin determination. (**NFPA 1033: 4.1.2, 4.6.5**, pp. 10–14)
- Use the scientific method in cause determination. (**NFPA 1033: 4.1.2**, pp. 10–14)

You are the Fire Investigator

You are a public agency investigator, and you have just been notified that a fire investigation you assisted in has resulted in arson charges against an individual. You receive a subpoena to testify in a preliminary hearing in the criminal matter.

From your basic fire investigation training, you remember that NFPA 921, *Guide for Fire and Explosion Investigations,* governs the methodology to be followed in a fire investigation and that this document is considered authoritative in many courts. You are confident that your investigation followed NFPA 921 and that your report clearly articulated that your investigation and hypothesis testing followed the scientific method. However, you also recall that NFPA 1033, *Standard for Professional Qualifications for Fire Investigator,* is the standard dictating the knowledge and skills required to be a fire investigator, and you wonder how a court would view your training and experience under the framework provided by this document.

1. How do NFPA 921 and NFPA 1033 differ from each other?
2. How does each document figure into your work as a fire investigator?
3. What could have been the consequences had you not followed the methodology provided in NFPA 921 during your investigation?

 *Access Navigate for more practice activities.*

Introduction

The purpose of any fire investigation is to evaluate and determine the facts presented and, ultimately, to arrive at the truth of what occurred in a specific incident. The objectives are to determine the fire's origin and cause, the factors that contributed to the fire's start or spread, the acts or omissions of persons involved, and who may be responsible for the fire. The overall goals of fire investigation are protection of lives and property, and the prevention of future incidents.

The National Fire Protection Association (NFPA) has developed NFPA 921, *Guide for Fire and Explosion Investigations*, to aid fire investigators from all sectors in the analysis of fire and explosion incidents and in the formulation of opinions regarding origin, cause, fire development, and fire prevention. NFPA 1033, *Standard for Professional Qualifications for Fire Investigator*, was developed by the NFPA to use as the minimum job performance requirements (JPRs) for the fire investigator. At the beginning of each chapter in this text, the applicable JPRs of NFPA 1033, 2022 Edition, will be cited. It is important for the reader to understand how the information in this text is related to the JPRs of fire investigation.

NFPA 921 and NFPA 1033 are in widespread use in the United States and Canada by fire investigators from the public and private sectors, and are frequently cited in court cases in which fire investigators are called upon to testify and provide opinions.

This text discusses these documents throughout, as well as other sources of information. Candidates to become fire investigators and current fire investigators are encouraged to review these NFPA documents thoroughly.

Becoming a Fire Investigator

Fire investigations are conducted by a variety of persons, including fire marshals, fire department personnel, police officers and detectives, investigators on a state or provincial level, federal agents, insurance company representatives, engineers, and private investigators and firms. Other persons involved in the field directly or indirectly include criminal prosecutors and defense attorneys, attorneys involved in civil litigation, scientists, insurance adjusters, and code enforcement personnel.

According to NFPA 1033, a fire investigator is a person "who has demonstrated the necessary skills and knowledge to conduct, coordinate, and complete a fire investigation." Fire investigation is a complex endeavor requiring substantial knowledge, training, and skill, as well as a high level of professionalism and ethics. The fire investigator strives to maintain and improve knowledge by seeking training and learning opportunities, and by learning from other investigators and professionals whenever possible.

Qualifications

NFPA 1033, *Standard for Professional Qualifications for Fire Investigator*, requires a fire investigator to be at least 18 years of age and to possess a high school diploma or equivalent level of education. Because the fire investigator will be in a position to examine the personal property of others, a complete and thorough background and character investigation must be completed by the **authority having jurisdiction (AHJ)**.

As with any profession, fire investigators must meet a basic level of education and training. They must meet the JPRs established in NFPA 1033 and by their employer, as well as any requirements established by law, to carry out their duties successfully. This requires that the fire investigator be evaluated periodically by qualified personnel to ensure that the investigator is meeting those requirements **FIGURE 1-1**. Employers often offer or require technical reviews and/or peer reviews of the investigator's written products, which can provide important feedback and ideas for improvement.

Knowledge

In the introductory paragraphs to its JPR sections, 4.1.7 in NFPA 1033 states that for investigators to successfully complete the JPR tasks, they must "remain current" on requisite knowledge subjects, which include the following four categories and 16 subcategories:

- **Fire Science**
  - Fire chemistry
  - Thermodynamics
  - Fire dynamics
  - Explosion dynamics

- **Fire Investigation**
  - Fire analysis
  - Fire investigation methodology
  - Fire investigation technology
  - Evidence documentation, collection, and preservation
  - Failure analysis and analytical tools

- **Fire Scene Safety**
  - Hazard recognition, evaluation, and basic mitigation procedures
  - Hazardous materials
  - Safety regulations

- **Building Systems**
  - Types of construction
  - Fire protection systems
  - Electricity and electrical systems
  - Fuel gas systems

Note that the categories themselves are important topics and may be broader than the subcategories below them—meaning that, in effect, the fire investigator must remain current on 20 areas of knowledge. Also, a comparison of the 2022 version of NFPA 1033 to the 2014 version shows that two requisite knowledge areas have been removed: thermometry and computer fire modeling (removed from 4.1.7, but included in the subtopics in Annex D1.2[3]). Importantly, several new areas have been added:

- **Fire Scene Safety** (category)
  - Hazard recognition, evaluation, and basic mitigation procedures
  - Safety regulations
- **Building Systems** (category)
  - Types of construction
  - Fuel gas systems

All fire investigators, including those who have already been working in the field, should review these new categories and ascertain whether they have the requisite knowledge in these subjects. If they feel that their knowledge base needs to be supplemented, they should seek training and education immediately to close any gaps and remain current with NFPA 1033.

FIGURE 1-1 Fire investigators must be evaluated periodically to ensure that they are meeting all requirements established by NFPA 1033 and the laws in their jurisdiction.
© Jones & Bartlett Learning. Photographed by Glen E. Ellman.

///////////////////////

TIP

The Annexes are not part of the standard, but rather add supplemental information to assist the investigator in a more thorough understanding of the standard.

Continuing Education and Training

Continuing education and training are necessary for all fire investigators. A critical part of the fire investigator's professionalism is the ability to identify and recognize areas in which further training and knowledge is needed, and to obtain education in those areas. Training programs should be designed to maximize the instructional process to prepare investigators to meet their roles and responsibilities. Although fire investigator training need not be received in any particular order, fundamental fire investigation courses and academies can be very useful in establishing baseline knowledge. Fire investigation training can be obtained from local agencies, state or provincial governments, commercial training companies, fire academies, professional associations, and online programs.

In the United States, the U.S. Fire Administration's National Fire Academy (NFA) provides investigation and courtroom training for the public investigator. Scientific and technical journals and articles provide further opportunities for learning about topics related to fire and fire investigation, and the International Association of Arson Investigators (IAAI) provides an extensive online training resource, CFI Trainer (www.cfitrainer.net). Even after receiving fundamental training, the fire investigator must obtain continuing education on relevant topics on an ongoing basis; NFPA 1033 specifies that the investigator should earn 40 hours of such training in a five-year span through seminars, workshops, and courses.

The investigator must also stay current with new investigative methodologies, fire protection technology, and code requirements. Also, the investigator should strongly consider joining organizations relating to the field, such as the IAAI, IAAI state and regional chapters, the National Association of Fire Investigators (NAFI), the NFPA, state associations, local task forces, and others. Networking with others in the field allows the investigator to stay current on technology and developments, and to maintain contact information for others whom the investigator may need to call or consult during an investigation.

Certification

Holding a certification in fire investigation is different from receiving training, although training is a prerequisite to obtain most certifications. As a fire investigator's career develops, it may be necessary to apply for relevant certifications, as these help demonstrate knowledge and skills in the field. Certification requirements for fire investigators vary by state or provincial jurisdiction, and may be dictated by the investigator's employer. Numerous states in the United States have certification programs, and certifications are also available from national and international organizations. The IAAI offers the Certified Fire Investigator (CFI) certificate and the Fire Investigation Technician (FIT) designation. The NAFI offers the Certified Fire and Explosion Investigator (CFEI) credential. These programs are designed to assess and document the investigator's professional qualifications and knowledge and adherence to relevant NFPA documents.

In many states, non-public-sector fire investigators must be licensed as private investigators or private fire investigators to conduct investigations in the state. Some statutes have exemptions for insurance adjusters or employees. Further, states generally require the licensing of professional engineers. All investigators and engineers should understand the relevant statutes in each state where they intend to work, determine the legal requirements applicable to them, and identify how the state defines the respective roles of investigator and engineer.

Ethics

Fire investigators work in a field that requires a high degree of professionalism and ethics. Fire investigation associations normally require their members to adhere to written Codes of Ethics. The IAAI's Code of Ethics contains 11 distinct provisions governing the fire investigator's personal and professional conduct, integrity, confidentiality, and honesty **TABLE 1-1**.

A fire investigator who fails to adhere to the Code of Ethics may be investigated, censured, or even dismissed as an IAAI member, and may be subject to the revocation of certification(s) or designation(s). Further, being discredited as a fire investigator can cause the investigator to lose a court case, and can permanently damage the investigator's career and credibility when testifying in the future. In modern times when most people have Internet access, even a small ethical issue can follow an investigator permanently.

Types and Roles of Fire Investigators

A fire investigator may occupy a variety of roles depending on employment, assignment, and other factors. A public-sector investigator may be employed by a police or fire department, and have a role defined by law. A private fire investigator may be an

TABLE 1-1 IAAI's Code of Ethics
■ I will, as a fire/arson investigator, regard myself as a member of an important and honorable profession.
■ I will conduct both my personal and official life so as to inspire the confidence of the public. I will exhibit professionalism and integrity in all aspects of the performance of my duties.
■ I will not use my profession and my position of trust for personal advantage or profit.
■ I will regard my fellow investigators with the same standards as a hold for myself. I will never betray a confidence nor otherwise jeopardize their investigation.
■ I will regard it my duty to know my work thoroughly. It is my further duty to avail myself of every opportunity to learn more about my profession.
■ I will avoid alliances with those whose goals are inconsistent with an honest and unbiased investigation.
■ I will make no claim to professional qualifications which I do not possess.
■ I will share all publicity equally with my fellow investigators, whether such publicity is favorable or unfavorable.
■ I will be dutiful to my superiors, to my subordinates, and to the organization I represent.
■ I will utilize electronic media and other communication technologies in a professional manner that does not exhibit, dishonor, or demean my profession or the International Association of Arson Investigators.
■ As a fire/arson investigator, I am first and foremost, a truth seeker.

Reproduced from International Association of Arson Investigators. 2020. "Code of Ethics." Accessed October 2020. https://www.firearson.com/Member-Network/Code-Of-Ethics.aspx

engineer or an insurance employee whose work is limited to a particular scope of employment. Some fire investigators focus on wildland or vehicle fires. The work performed by any of these investigators may also be limited or restricted by the law pertaining to their profession or assignment. Further, fire investigators may be retained or assigned for limited purposes; for example, an engineer may be assigned to evaluate an appliance or device, an insurance employee may have a role that is limited to insurance coverage determinations, and a private investigation firm may be retained solely to review and evaluate the data from an earlier investigation or to represent a contractor, installer, tenant, or manufacturer of a specific appliance.

NFPA 921 and NFPA 1033 recognize that a fire investigator's work and role may be limited by assignment, time availability, resource limitations, and relevant policies. Investigators should never exceed their level of training and expertise, nor attempt to perform tasks for which they are not qualified.

The National Fire Protection Association

The National Fire Protection Association (NFPA) is the publisher of almost 300 codes, standards, recommended practices, and guides related to fire and life safety. All NFPA documents are voluntary documents, which means that they do not have the power of law unless an AHJ adopts them. A number of NFPA's codes and standards have been adopted into law in various jurisdictions around the world, including the following:

- NFPA 1, *Fire Code*
- NFPA 54, *National Fuel Gas Code*
- NFPA 70, *National Electrical Code*
- NFPA 101, *Life Safety Code*

One of the distinctive features of all NFPA documents is that they are developed through a consensus process, which brings together technical committee volunteers representing varied viewpoints and interests to achieve consensus on the document content **FIGURE 1-2**. The breadth of expertise of technical committee members helps ensure that NFPA standards and other documents are well rounded and accurate. More than 200 of these technical committees currently exist, and they are governed by a 13-member Standards Council, which is also composed of volunteers.

Public Input to NFPA Documents

All of the committee meetings and the NFPA annual meeting are open to the public, and anyone, whether an NFPA member or not, can submit public inputs (proposals) to change a document or make public comments on actions taken by the committee.

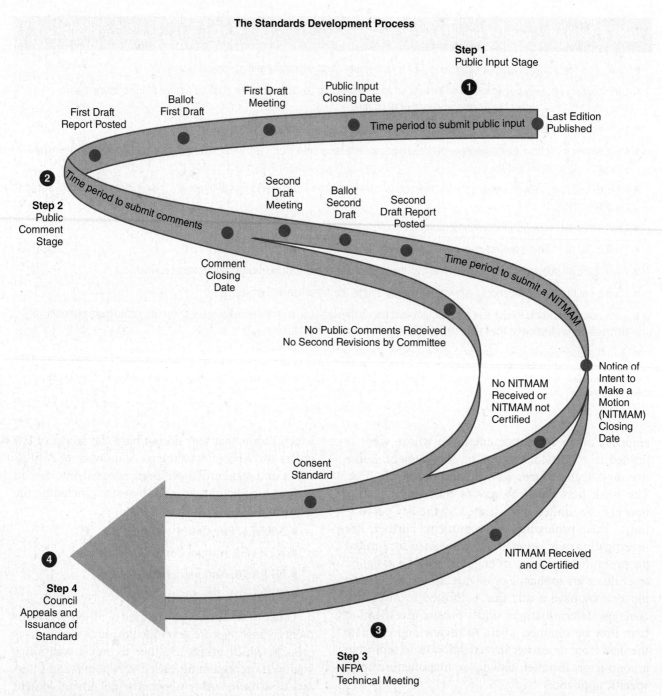

The Standards Development Process

FIGURE 1-2 NFPA standards development process.
Courtesy of National Fire Protection Association (NFPA) 2017.

The only step in the process that is limited to NFPA members is the voting that takes place at the annual technical session. The schedule for public input on NFPA documents varies according to the document. Public input deadlines are posted on the NFPA website, and public input is submitted through a web-based electronic submission system.

Tools and techniques are constantly being updated and modified between editions of NFPA 921. Consequently, newer methods that are not covered in the current edition may be used at a fire scene. For this reason, it is important that fire investigation professionals be diligent in making proposals and comments to the document during its revision cycles.

Revision Cycles

All NFPA documents are updated on regular cycles of 3 to 5 years, with anywhere from 20 to 45 documents reporting in a given revision cycle. NFPA 921 has been on a 3-year cycle, although a 4-year cycle sometimes applies (including for the issuance of the 2021 edition). NFPA 1033 has been on a 5-year cycle, but the newest edition issued in 2022.

Sequence of Events in Standards Development

TABLE 1-2 describes the current sequence of events for the standards development process (in this context, the term *standard* includes guides such as NFPA 921).

Corrections to Documents

Between revisions of a document published by the NFPA, errata and **Temporary Interim Amendments (TIAs)** may be issued. For example, if a publishing error is found in some part of the document, the NFPA can issue an erratum correcting the error. In the next

TABLE 1-2 Standards Development Process	
Step 1: Input Stage	■ Input accepted from the public or other committees for consideration to develop the First Draft ■ Committee holds the First Draft Meeting to revise the Standard (23 weeks) ■ Committee(s) with Correlating Committee (10 weeks) ■ Committee ballots on the First Draft (12 weeks) ■ Committee(s) with Correlating Committee (11 weeks) ■ Correlating Committee First Draft Meeting (9 weeks) ■ Correlating Committee ballots on the First Draft (5 weeks) ■ First Draft Report posted
Step 2: Comment Stage	■ Public Comments accepted on the First Draft (10 weeks) ■ If the Standard does not receive Public Comments and the Committee does not wish to further revise the Standard, the Standard becomes a Consent Standard and is sent directly to the Standards Council for issuance ■ Committee holds the Second Draft Meeting (21 weeks) ■ Committee(s) with Correlating Committee (7 weeks) ■ Committee ballots on the Second Draft (11 weeks) ■ Committee(s) with Correlating Committee (10 weeks) ■ Correlating Committee Second Draft Meeting (9 weeks) ■ Correlating Committee ballots on Second Draft (8 weeks) ■ Second Draft Report posted
Step 3: Association Technical Meeting	■ Notice of Intent to Make a Motion (NITMAM) accepted (5 weeks) ■ NITMAMs are reviewed and valid motions are certified for presentation at the Association Technical Meeting ■ A Consent Standard bypasses the Association Technical Meeting and proceeds directly to the Standards Council for issuance ■ NFPA membership meets each June at the Association Technical Meeting and acts on Standards with "Certified Amending Motions" (certified NITMAMs) ■ Committee(s) and Panel(s) vote on any successful amendments to the Technical Committee Reports made by the NFPA membership at the Association Technical Meeting
Step 4: Council Appeals and Issuance of Standard	■ Notification of intent to file an appeal to the Standards Council on an Association action must be filed within 20 days of the Association Technical Meeting ■ Standards Council decides, based on all evidence, whether to issue the Standard or to take other action

Reproduced with permission from NFPA 921-2021. 2020. Guide for Fire and Explosion Investigations. National Fire Protection Association. This reprinted material is not the complete and official position of the NFPA on the referenced subject, which is represented only by the standard in its entirety.

revision of the document, the error is corrected. If emergency changes are proposed that must occur between revisions, a TIA is placed before the committee to address the issue. The committee must conduct a letter ballot on proposed TIAs and a notice for public comments is sent out.

A recommendation for approval is established if three-fourths of the voting members have voted in favor of the TIA. With this approval and the review of public comments, the Standards Council may issue the TIA. The TIA then becomes a public input item during the next cycle of the document. Several errata and TIAs have been issued for NFPA 921 since its inception.

Types of NFPA Documents

The NFPA publishes four types of documents: standards, codes, guides, and recommended practices.

- A **standard** contains only mandatory provisions, using the word "shall" to indicate requirements, and is in a form generally suitable for adoption into law, or for mandatory reference by another standard or code. NFPA 1033 is a standard.

- A **code** is a standard that is an extensive compilation of provisions covering a broad subject matter or that is suitable for adoption into law independently of other codes and standards. An example is NFPA 70, *National Electrical Code (NEC)*.

- A **guide** is a document that is advisory or informative in nature and that contains only nonmandatory provisions. NFPA 921 is a guide.

- A **recommended practice** is similar in content or structure to a standard or code, but contains only nonmandatory provisions.

NFPA 1033 and NFPA 921 Overview

NFPA 1033

As previously discussed, NFPA 1033 is the national professional qualification standard for anyone serving as a fire investigator. It sets forth the minimum **job performance requirements (JPRs)** to be a qualified fire investigator in either the public or private sector. As such, it serves as a benchmark for all fire investigators. It does this in four ways:

- By specifying 20 broad-ranging categories and subcategories of requisite knowledge on which investigators must remain current

- By identifying the main responsibilities of fire investigators

- By pinpointing the job performance requirements that encompass the job of a fire investigator

- By listing the knowledge and skills necessary to do each task involved in an investigator's job

In 1972, the Joint Council of National Fire Service Organizations (JCNFSO) created the National Professional Qualifications Board (NPQB) to facilitate the development of nationally applicable performance standards for uniformed fire service personnel. As a result of that process, in May 1977, the NFPA adopted NFPA 1031, *Standard for Professional Qualifications for Fire Inspector and Plan Examiner*. In the years that followed, it was decided to create an independent standard to apply to fire investigators only. In June 1987, NFPA adopted NFPA 1033. Since the original release of this document, it has been updated on a regular basis.

Relevance of NFPA 1033 to Fire Investigators

NFPA 1033 specifies the minimum knowledge and skills required for a fire investigator to evaluate a fire scene, conduct the investigation, and document the scene safely. NFPA 1033 also addresses skills related to evidence collection, interviewing of witnesses, report writing, and final presentation of investigative findings. This standard is intended to apply to all fire investigations, including wildland, vehicle, and structural fires. Jurisdictions may also choose to enforce their own requirements exceeding those set forth in NFPA 1033.

NFPA 921

NFPA 921 was developed by the NFPA and approved by the American National Standards Institute (ANSI). Although NFPA classifies NFPA 921 as a guide and not a standard, it provides the appropriate and established principles and methodology for the field. Specifically, NFPA 921 "establishes guidelines and recommendations for the safe and systematic investigation or analysis of fire and explosion incidents." It also serves as a model for the advancement and practice of fire and explosion investigation, fire science, technology, and methodology. The content of NFPA 921 is "not intended to be a comprehensive scientific or engineering text;" the investigator may have to utilize additional resources in the course of an investigation.

NFPA 921 advocates a systematic approach, known as the scientific method, that provides the process and framework to ensure a successful fire or explosion investigation. Deviations from these procedures are not necessarily wrong or inferior, but do need justification. The scientific method must be followed in every instance. The scientific method is discussed later in this chapter.

Because NFPA 921 is a guide, not every portion of it needs to be applied to each fire or explosion investigation. For example, an investigator seeking to determine the origin and cause in a residential fire likely will not need to rely on the principles and information contained in this document's Marine Fires chapter. Instead, investigators are expected to apply the appropriate recommended procedures to a given incident, based on the nature of the incident and the scope of the investigation.

Although NFPA 921 was never originally intended as a document to define legal issues involving civil or criminal litigation of fire or explosion incidents, it has come to fill that role in the minds and practice of many fire investigation professionals, including attorneys and trial courts. Despite its official status as a guide, many courts have recognized NFPA 921 as the "standard of care" in the profession and have cited NFPA 921 or the principles of methodology utilized within it. If a court of appeals declares NFPA 921 to be a "standard of care" in a particular case, then depending on the language of the decision, the document may become tantamount to a standard in the jurisdiction covered by the court. When a trial court declares NFPA 921 to be a standard, it becomes the controlling "law of the case."

As the courts' rulings in numerous fire cases make clear, expert opinion must be supported by established and tested principles, and not by assumptions, speculation, or years of experience. When a fire investigator testifies as an expert, the court will apply accepted standards to examine the fire investigator's qualifications and the investigation performed, to determine whether the expert should be permitted to testify on a particular matter. NFPA 921 and NFPA 1033 will figure prominently into this determination, referred to in many courts as a "Daubert" analysis. Further information on this process is found in Chapter 5, *Legal Considerations for Fire Investigators.*

Relevance of NFPA 921 to Fire Investigators

Although investigators may point to the fact that NFPA 921 is a guide when their actions are questioned, the methodology and practices defined by this document are generally accepted within the fire investigation community. Likewise, many federal and state courts in the United States have stated that they consider NFPA 921 to represent the industry standard and/or a best practice document for fire investigation. Thus, a fire investigator who fails to follow the principles and practices defined by this document will likely be challenged in court, and will be required to explain clearly what method was used in the investigation and why it was chosen in place of the methods prescribed in NFPA 921. Different jurisdictions have varying interpretations of the importance of NFPA 921, so investigators should be aware of rulings related to this issue in their local jurisdiction.

Since the first edition of NFPA 921 was introduced in 1992, there have been significant advances in fire science as applied to fire investigations. Although every effort is made to ensure NFPA 921 is as complete as possible, inevitably some areas are not covered in each edition. Given the infinite number of items that could be involved in a fire, such as different types of fuels, objects burned, or individual sources of ignition, this guide cannot possibly address each and every potential fire scenario. However, through the NFPA's revision process, the guide remains a living document, updated regularly to include new material, as described in the "Revision Cycles" section in this chapter.

NFPA 921 and NFPA 1033 are closely related, and it is useful to read them together. Importantly, NFPA 1033 specifically refers the reader to NFPA 921 as a source of the knowledge areas it requires investigators to have, and further refers the reader to NFPA 921 for basic methodology and procedures, which are relevant to the skills listed in NFPA 1033 as mandatory for fire investigators. Although NFPA 921 is defined as a "guide" and NFPA 1033 is defined as a "standard," it is NFPA 921 that is most frequently cited in court cases as a source when evaluating the qualifications of investigators as expert witnesses and assessing the reliability of their opinions.

JPRs and Their Application

As previously mentioned, the NFPA 1033 standard contains the JPRs for fire investigators working in any aspect of the field. The fire investigator must meet these JPRs along with any requirements established by the investigator's employer. In addition to identifying the areas in which investigators are expected to maintain up-to-date knowledge, the JPRs identify the tasks that fire investigators are expected to be able to perform in the completion of their duties as well as measurable or observable outcomes. They also identify the knowledge and skills that are required to be able to properly complete these tasks.

NFPA 1033 does not provide a specific framework for how an investigator is to be evaluated under the JPRs, and does not require or expect that an investigator should perform each and every JPR during a given investigation. Investigators should be prepared to demonstrate their ability to perform each of the JPRs in NFPA 1033 and should be able to demonstrate training and knowledge in each of the knowledge areas. As such, the JPRs and knowledge areas provide an excellent template for fire investigator training and development. The investigator is encouraged to do a self-evaluation under these standards and to obtain further practice or training in any area that might be needed.

The Methodology of Fire Investigation

The investigation of a fire or explosion most often involves the identification, collection, and analysis of a variety of data. The ability to recognize these data, or facts, and then to analyze them properly and objectively is paramount in fire investigation. Each investigation must be carried out in a consistent manner to ensure that all aspects of any given fire scene are addressed. A systematic approach based on the scientific method must be employed to perform a thorough physical evaluation of the scene, careful collection of data, and complete documentation of the scene and physical evidence, all of which are necessary to ensure that the analysis has been done correctly and that the results are supported by scientific principles. (Chapter 4 of NFPA 921 covers basic methodology.)

Scientific Method

The **scientific method**, as defined by NFPA 921, is "the systematic pursuit of knowledge involving the recognition and definition of a problem; the collection of data through observation and experimentation; analysis of the data; the formulation, evaluation, and testing of hypotheses; and, where possible, the selection of a final hypothesis" **FIGURE 1-3**.

In an origin and cause investigation, the scientific method should be followed first to determine the fire's origin and then to determine its cause. In other words, the origin of a fire must be identified prior to the determination of cause. NFPA 921 states that if the origin cannot be appropriately determined, then the cause normally cannot be determined; additionally, if the origin is not correctly identified, the determination of cause is likely to be incorrect.

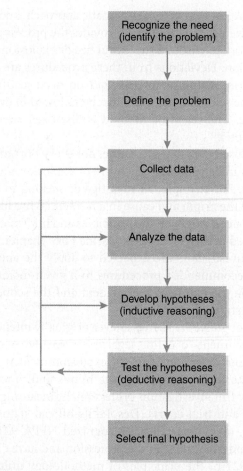

FIGURE 1-3 The scientific method.

Reproduced with permission from NFPA 921-2021. 2020. Guide for Fire and Explosion Investigations. National Fire Protection Association. This reprinted material is not the complete and official position of the NFPA on the referenced subject, which is represented only by the standard in its entirety.

Fire Investigation and the Scientific Method

Fire investigators have often followed the scientific method but sometimes have not defined it by name. Performing an investigation systematically can be beneficial, and investigators often examine fire scenes from outside to inside, for example, or from least damaged to most damaged areas. But a systematic approach means far more than performing the same steps in each investigation, and it implies more than just gathering facts. The following steps are used to apply the scientific method to fire investigations.

Step 1: Recognize the Need to Use the Scientific Method

The first step is to realize that there is a problem to be resolved—something that is often self-evident because an investigator is notified of an incident and asked to determine its origin and cause. The exact nature of the problem that each investigator needs to address depends on the role and responsibility that the investigator fulfills during the investigation. An example would be receiving a call to investigate a house fire in a single-family dwelling on Green Street.

Step 2: Define the Problem

Upon arriving at the scene or when the assignment is given, an investigator must define the problem to be resolved. This may depend on the role, employment, or authority of the investigator. Defining the problem may include determining the origin and cause of the fire, determining responsibility for an incident, determining whether an appliance contributed to a fire cause, or determining the circumstances surrounding the death of a victim. It is important to define the problem so that the investigator can apply resources in the most effective and efficient manner. Defining the problem includes defining the manner in which the problem will be resolved, such as the steps to be taken and the tasks to be performed. As an example, an investigator might be tasked to conduct a comprehensive examination of the scene to determine the area of origin, point of origin, and cause of a fire.

Step 3: Collect Data

The investigator should begin by conducting a systematic evaluation of the scene and gathering all of the data needed to find the answer to the problem that has been defined. Several chapters in NFPA 921 provide information on data collection. These data can include (but are not limited to) the following:

- Recognition of physical evidence, such as fire patterns, fuel loads, ventilation, building construction, and other facts
- Collection of materials, such as debris samples, for laboratory evaluation
- Results of laboratory examinations
- Documentation of personal observations, such as witness statements
- Documentation of the fire scene through photographs, sketches, and notes
- Official reports, such as those of fire and police departments
- The documentation or results of prior scene investigations
- Other information such as weather reports, property records, and prefire photos

An example of data collection is as follows: An investigator notes that the presence of a smoldering mattress, without bed linens, on the front lawn of the dwelling on Green Street. The investigator locates the homeowner and begins to take notes on his observations and the events leading up to the fire. Also noted is the condition of the homeowner, who appears to be intoxicated. After speaking to a few other witnesses at the scene and taking notes on their observations, the investigator examines the mattress on the lawn more closely and begins to collect and document any additional evidence found at the home. The investigator evaluates and examines the rooms where the fire and smoke damage are located, and sifts through the debris in these rooms. As part of the reconstruction of the scene, the mattress is replaced into the room of origin for comparison to fire patterns found in the room and on the mattress.

According to NFPA 921, "the data collected using the scientific method that will be later analyzed are [referred to as] **empirical data**." Empirical data are factual data that are based on actual measurements, observations, or direct sensory experiences that can be verified or are known to be true.

It is important that an investigator properly document and collect the data that will ultimately be used to test and verify the hypotheses. Because memories fade over time, proper documentation of interviews must occur. Also, once the fire scene has been demolished or repaired, the opportunity to collect additional data from the scene may be lost.

Collecting data may also include literature reviews, pattern analysis, scene documentation, photography and diagramming, evidence recognition and preservation, and review and analysis of others' investigations. Documentation of the fire scene is a topic that is addressed elsewhere in this text.

Step 4: Analyze the Data

At this point, the evidence is examined as objectively as possible to evaluate its meaning and to determine possible hypotheses for the events being investigated. During data analysis, the investigator uses only data that are known to be true (empirical data). Data that are speculative, unconfirmed, or not related to the fire scene should not be considered. The analysis is based on the expertise, knowledge, experience, and training of the person doing the analysis.

To continue with the example, during the reconstruction of the scene, the investigator begins to analyze the data found, including the burn patterns around the mattress. To understand the burning properties of the mattress better, the investigator reviews manufacturing labels and researches listed standards. Reviewing the investigator's own knowledge of fire dynamics and spread allows for a comparison with the patterns that are observed in the building and on the mattress.

Step 5: Develop Hypotheses

Through the analysis of the data, the investigator will develop **hypotheses**, based on the empirical data found, to explain the events of the incident. This process, which is referred to as **inductive reasoning**, is necessary both in the determination of origin and later in the determination of fire cause. The hypotheses represent potential solutions for the question(s) or problem(s) identified in Step 2, such as determining the origin, cause, fire spread, responsibility, or some combination of these as directed by the investigator's assignment.

Whether considering the origin, or cause, or another question, the investigator should develop all possible hypotheses that might be supported by the analysis of the data. If a hypothesis under consideration does not have factual (data) support, it should not remain under consideration, and the investigator may need to gather further data to formulate more hypotheses. The viable hypotheses are then brought forward to Step 6 for testing.

For example, based on the physical evidence found at the Green Street scene and the interviews conducted, the remains of a candle recovered from the mattress are determined to be the origin of the fire. The interviews confirm that a candle was accidentally knocked over onto the mattress by the occupant. Thus, the investigator develops a cause hypothesis that the open flame from the candle ignited a portion of the mattress.

Step 6: Test the Hypotheses

Testing each viable hypothesis will help in the investigator's quest to arrive at the final hypothesis or conclusions. Testing a hypothesis can be done in different ways, but in the fire investigation field, it most often relies on **deductive reasoning**. Full-scale laboratory or physical testing of the entire fire scene and fire dynamics is not economically or practically feasible in most fire investigations, although experimental and physical tests may be done in many cases. In virtually all cases, however, investigators use their knowledge and skills, along with the facts of the case, to analytically challenge or test the hypothesis.

Cognitive testing is the use of a person's thinking skills and judgment to evaluate the empirical data and challenge a hypothesis based on data and scientific knowledge. Other investigators may also provide a forum to test or challenge a hypothesis. Essentially, the goal is to disprove (falsify) the hypothesis by finding reasons why it may not be true. Through deductive reasoning, the hypotheses at issue are supported, unsupported, or refuted by the complete body of evidence and data. Hypothesis testing includes evaluating other

potential origins or causes against the hypothesis being tested.

Any hypothesis that is incapable of being tested (either physically or cognitively) is an invalid hypothesis. A hypothesis that is developed based on the absence of data is one such example. The inability to refute a hypothesis does not mean that it is true.

A critical step in determining whether a hypothesis is valid is seeing whether it can withstand reasonable challenge. Whenever possible, hypothesis testing should include critical examination and challenge by others in the field through, for example, administrative or technical review by other knowledgeable persons who can review the process and conclusions of the investigator and, when appropriate, pose challenges to the conclusions **FIGURE 1-4**.

In all cases, testing a hypothesis requires critical self-challenge, in which the investigator asks questions regarding the hypothesis under consideration. A key question to determine whether a hypothesis is valid is: What other hypotheses could be supported by the same set of facts? If alternative hypotheses exist that might be valid based on the currently known facts, it is likely that an investigator has not gathered enough data. In such case, the investigator should obtain more information, reapply the steps in the scientific method, and attempt to disprove all hypotheses. Ultimately, for the investigator, the goal after full hypothesis testing is to find the hypothesis that is valid to a probability (more likely than not), and that is determined to be uniquely consistent with the facts and with the principles of science.

Other types of questions to use in testing the hypothesis include the following:

- Is there another way to interpret the facts (data)?
- If yes, why is the investigator's interpretation more likely true?

FIGURE 1-4 Enlist the knowledge of an experienced administrator or investigator to review your process or conclusions.
© Jones & Bartlett Learning. Photographed by Glen E. Ellman.

- What are the weaknesses in the hypothesis (analysis of the data)?
- What arguments would someone else (an opposing expert) use to refute the hypothesis?
- Are there facts that contradict the hypothesis?
- Is there research that supports the hypothesis?
- Can the hypothesis be proven to someone else?
- Does the hypothesis make sense?

There are three situations in which the investigator will be required to seek further data, engage in further analysis, or conclude that the matter is undetermined: (1) A valid hypothesis cannot be developed; (2) one or more hypotheses have been developed but are not supportable within an acceptable level of certainty; or (3) multiple possible hypotheses remain that have not been falsified. A hypothesis that is *possible* is one that has not been falsified, but is not supportable within an acceptable level of certainty.

TABLE 1-3 shows the levels of hypothesis certainty. These levels relate to how strongly an investigator holds their conclusion or final hypothesis, and are based on the investigator's confidence in the collected, analyzed, and tested data. In turn, the level of certainty may impact how the findings are used in court or other proceedings. It is recommended that only hypotheses that rise to the level of "probable" be expressed as expert opinions with the certainty required in the legal system. In most jurisdictions in the United States, an expert opinion must meet a level or standard of certainty to be admissible in a legal proceeding—often stated as "within a reasonable degree of scientific certainty in the fire investigation profession."

Because every fire is unique and destroys data and evidence of its own existence, the certainty may not be an "absolute" or "without any doubts." Thus, the fire investigator is not required to eliminate all possible doubt before expressing an expert opinion. Nevertheless, fire investigators should always be willing to state that they do not have an opinion on the matter at issue, if that is the case, and if no further data are available, to state that the matter is undetermined.

In the mattress fire scenario, the hypothesis is compared with the known and proven data: The fire originated on the mattress. Manufacturing data and lab tests show that the mattress will ignite only with the application of an open flame. The evaluation of the scene included consideration of other possible ignition sources that were eliminated during the testing of the hypothesis process. Physical evidence was discovered, in the form of a candle on the mattress. An interview with the owner revealed that a candle was burning and was knocked onto the mattress, causing it to ignite.

Chapter 15, *Analyzing the Incident*, describes additional methods of testing a hypothesis.

Bias

The investigator must approach every incident with an open mind and without presumption or **expectation bias**, which is the making of premature conclusions without having examined, or considered, all of the relevant data. Any preconceived determinations or conclusions will, consciously or unconsciously, influence an investigator's efforts and should be avoided to maintain a proper level of objectivity during the investigation.

TABLE 1-3 Hypothesis Certainty	
Level	**Explanation**
Probable	More likely true than not (more than 50 percent likely to be true). This level of certainty is usually associated with opinions involving the formation of a final hypothesis that cannot be eliminated during analysis and testing and when all other reasonable competing hypotheses have been evaluated and eliminated.
Possible	Feasible (often used when two hypotheses have the same level of certainty). This can be the result of the formation of two or more rival or competing hypotheses formed from a critical review of available data, but neither hypothesis can be eliminated during the analysis and testing process. A common example of this is open flame ignition of available combustibles versus smoldering ignition caused by carelessly discarded smoking materials. At times, the evidence and data may support the formation of both hypotheses, but neither can be eliminated without additional information or testing.
Suspected	When the opinion does not qualify as an expert opinion; a term that should not normally be used in fire investigation reports or conclusions. This can be the result of an inability to collect enough data to formulate a particular hypothesis or an inability to analyze or test a hypothesis properly.

It is also important to avoid **confirmation bias**. Confirmation bias means relying only on the data that support a preferred hypothesis and ignoring data that are not supportive or are contrary. Indeed, available data often can support multiple hypotheses or even contradictory ones. To safeguard against confirmation bias, consider all theories that are supported by the data and work to disprove (falsify) the hypothesis under consideration by comparing it to all of the data and evidence in an effort to show that it is not true. Confirmation bias can lead to a failure to properly formulate and test alternative hypotheses or to prematurely discount hypotheses that seem to contradict data without proper assessment.

The data from the fire or explosion scene should stand on their own and be appropriately evaluated and tested in relation to the fire or explosion scene. If a hypothesis cannot withstand a reasonable challenge or test, it cannot stand. If no hypotheses are supported, you will need to return to earlier steps in the scientific method, such as collecting additional data, analyzing those data, or redeveloping the hypothesis.

Step 7: Select the Final Hypothesis

The last step in using the scientific method is to select the final hypothesis. This step occurs only after all the previous steps have been performed fully and carefully. If all other reasonable hypotheses have been reliably tested and disproven, an investigator may be able to conclude that the remaining hypothesis is the one to adopt as the appropriate final hypothesis. Reliable testing and consideration of all hypotheses are critical to this process and cannot be ignored.

The final hypothesis must be reliably and fully tested just as well as the other hypotheses are tested; it must not be selected simply because it remains after others are disproven. The selected hypothesis must be independently supported by evidence and consistent with the facts, and it must be challenged as rigorously as all of the other proposed explanations. The investigator should never take a shortcut in this process. Document the facts and information supporting the final hypothesis and the reasons why it is the final hypothesis. Further, review the entire process up to this point to confirm that all data are accounted for, and all of the feasible alternative hypotheses have been ruled out.

In some instances, more than one hypothesis may be consistent with all the known facts, and none can be excluded so as to allow one to be more probable. In other instances, the hypothesis testing may leave no hypotheses that are supported and consistent with the facts. In either event, further facts, information, and testing may be needed prior to a final hypothesis being reached. In these situations, an investigator may be left with no option except to report the issue in question as undetermined.

Negative Corpus

Fire investigators are cautioned not to rely on *negative corpus* analysis when determining the cause of a fire. This term is used by some investigators to refer to the identification of an ignition source that is not supported by data, based on the supposed elimination of all of the other ignition sources that were found, known, or suspected in the area of origin. In the past, negative corpus theories were used to support the determination of an incendiary fire, by stating that because known sources were ruled out, the fire "must" have been intentionally started with an open flame. This process is inconsistent with the scientific method and is not deemed acceptable in NFPA 921. All hypotheses must be supported by facts and data, and the logical inferences that come from them; observations; calculations; experiments; and the laws of science.

The Basic Methods of Fire Investigation

The scientific method is the analytical framework that investigators must apply to ensure that their conclusions are supported. As a practical matter, follow these general steps to ensure that the scientific method is applied properly in each case.

1. *Receive the assignment.* You are notified of the fire loss and goal of the requested investigation. This step can be equated to "recognizing the need" or "identifying the problem" within the framework of the scientific method.
2. *Prepare for the investigation.* For the given assignment and the nature of the fire scene, determine the tools required, safety concerns to be addressed, and personnel needs to complete the task. Additionally, evaluate and ultimately document what legal authority will give you the right to conduct the investigation. This step can be equated to "defining the problem" within the framework of the scientific method. (For additional information about legal authority, see Chapter 5, *Legal Considerations for Fire Investigators.*)
3. *Conduct the investigation.* Begin a systematic process of information gathering to include examination and documentation of the scene, interviews, and information research. This is the "collecting data" step within the framework of the scientific method.

4. *Collect and preserve evidence.* Identify possible evidence, and document and preserve it for future testing or legal presentation. This is also the "collecting data" step within the framework of the scientific method.
5. *Analyze the incident.* Using the scientific method, analyze the collected and available data, and develop and test hypotheses. These are the "analyze the data," "develop a hypothesis," and "test the hypothesis" steps within the framework of the scientific method.
6. *Draw conclusions.* Determine the final tested hypothesis. This is the "select a final hypothesis" step of the scientific method.
7. *Report.* As determined by your responsibility as the investigator, the process and conclusions are reported in written or oral forms.

Reviewing the Procedure

As the investigator, your work products, such as reports, will often be reviewed by others, including your supervisor. These reviews are generally categorized into three types: administrative, technical, and peer reviews. Note that these are not the same type of reviews performed by opposing parties during litigation.

The **administrative review** is usually carried out within your organization to ensure the product meets the organization's procedures, such as for content as outlined in the organization's procedures manual, grammar, and spelling. This type of review is not a true critique of the investigation, the conclusions, or the testing of the hypotheses. The reviewer may not have the necessary skills to perform a technical review and may not review all of the data that you have developed or relied on.

The **technical review**, in contrast, is a critique of your work and findings. As such, the reviewer must have the skills, knowledge, and background to perform such an evaluation. This type of evaluation requires the reviewer to have access to all of the data on which you relied. The reviewer will challenge and critique the work product and investigative findings, as well as testing of the hypotheses based on the findings. This type of review is beneficial if it adds a second serious challenge to the investigation, but bias may be present because of working relationships, and reviewers may be perceived to have an interest in the outcome of the investigation.

The **peer review** is a formal procedure that is generally applied to scientific or technical papers. The peer review is considered independent and objective because the reviewers do not have interest in the outcome of the report. The author does not select the reviewers, and the reviewers should not be coworkers of or part of the same organization as the author; indeed, they often remain anonymous to the writer. Peer reviewers have sufficient knowledge to evaluate flaws in the process or procedures employed, but likely do not have the resources to question the investigator's data. For this reason, the peer reviewers may not be able to confirm the validity of your conclusions.

A proper and unbiased technical review usually represents the best means to assess the validity of your conclusion. During litigation, you may encounter other parties who challenge your methods and findings, and you will want to ensure your work stands up to their scrutiny.

After-Action REVIEW

IN SUMMARY

- NFPA 921 can be used by anyone who is responsible for investigating and analyzing fire and explosion incidents.
- NFPA 921 establishes guidelines and recommendations for the safe and systematic investigation or analysis of fire and explosion incidents.
- The distinction between a standard and a guide is important because different jurisdictions have varying interpretations of the importance of NFPA 921.
- NFPA 1033 is a standard that establishes the minimum job performance requirements (JPRs) for service as a fire investigator. It establishes the minimum knowledge and skills required for a fire investigator to evaluate a fire scene, conduct an investigation, and document the scene safely.
- Because the fire investigator will be in a position to examine the personal property of others, a complete and thorough background and character investigation of each person filling this role must be completed by the authority having jurisdiction (AHJ).
- After the investigator's initial training, they must remain current with new investigative methodologies, fire protection technologies, and code requirements.

- All NFPA documents are voluntary documents, which means that they do not have the power of law unless an AHJ adopts them.
- More than 6000 volunteers serve on NFPA's technical committees, with selection of these volunteers based on their background and expertise.
- The definitions contained within NFPA 921 are critically important because they provide a consistent language for all fire investigators to use when conducting a fire investigation and preparing a case.
- The scientific method is a systematic method of problem solving.
- The steps in the scientific method include recognizing and identifying the problem, defining the problem, collecting information, analyzing the information, developing hypotheses, testing the hypotheses to determine whether the results are reliable and supported by facts and scientific principles, and determining the final hypothesis, if possible.
- Proper documentation will validate analysis of the data that are collected in relation to a fire incident.
- After all evidence and data are collected, the whole body of evidence and data is reviewed using inductive reasoning; evidence is looked at objectively to evaluate its meaning and determine a possible hypothesis.
- The hypothesis is challenged or tested through deductive reasoning.
- If a hypothesis cannot withstand a challenge or test, an investigator may need to return to earlier steps of the scientific method.
- The steps of the scientific method must be followed in the determination of the fire's origin, and again in the determination of the fire's cause.

KEY TERMS

Administrative review Review of an investigator's work carried out within the investigator's organization to ensure that policy has been followed and quality standards have been met.

Authority having jurisdiction (AHJ) An organization, office, or individual responsible for enforcing the requirements of a code or standard, or for approving equipment, materials, an installation, or a procedure. (NFPA 921)

Code A standard that is an extensive compilation of provisions covering broad subject matter or that is suitable for adoption into law independently of other codes and standards. (NFPA 921)

Cognitive testing The use of a person's thinking skills and judgment to evaluate the empirical data and challenge the conclusions of the final hypothesis.

Confirmation bias The attempt to prove a hypothesis instead of disproving a hypothesis, by relying only on data that support the hypothesis and ignoring or dismissing contradictory information, resulting in a failure to consider alternative hypotheses or the premature discounting of an alternative hypothesis.

Deductive reasoning The process by which conclusions are drawn by logical inference from given premises. (NFPA 921)

Empirical data Factual data that are based on actual measurements, observations, or direct sensory experiences, which can be verified or are known to be true. (NFPA 921)

Expectation bias Preconceived determination or premature conclusions as to what the cause of the fire was without having examined or considered all of the relevant data.

Fire investigation The process of determining the origin, cause, and development of a fire or explosion. (NFPA 921)

Fire investigator An individual who has demonstrated the skills and knowledge necessary to conduct, coordinate, and complete a fire investigation. (NFPA 1033)

Guide A document that is advisory or informative in nature and that contains only nonmandatory provisions. A guide may contain mandatory statements, such as when a guide can be used, but the document as a whole is not suitable for adoption into law. (NFPA 921)

Hypothesis Theory supported by the empirical data that the investigator has collected through observation and then developed into explanations for the event, which are based on the investigator's knowledge, training, experience, and expertise. (NFPA 921)

Inductive reasoning The process by which a person starts from a particular experience and proceeds to generalizations; the process by which hypotheses are developed based on observable or known facts and the training, experience, knowledge, and expertise of the observer. (NFPA 921)

Job performance requirements (JPRs) A statement that describes a specific job task, lists the items necessary to complete the task, and defines measurable or observable outcomes and evaluation areas for the specific task. (NFPA 1033)

Peer review Formal review of a professional's work performed by other qualified professionals not selected by the author of the work, and often conducted anonymously.

Recommended practice A document that is similar in content and structure to a code or standard but that contains only nonmandatory provisions using the word "should" to indicate recommendations in the body of the text. (NFPA 921)

Scientific method The systematic pursuit of knowledge involving the recognition and definition of a problem; the collection of data through observation and experimentation; analysis of the data; the formulation,

evaluation, and testing of a hypothesis; and, when possible, the selection of a final hypothesis. (NFPA 921)

Standard An NFPA standard, whose main text contains only mandatory provisions using the word "shall" to indicate requirements, and which is in a form generally suitable for mandatory reference by another standard or code or for adoption into law. Nonmandatory provisions are not to be considered a part of the requirements of a standard and shall be located in an appendix, annex, footnote, or other means as permitted in the *NFPA Manual of Style*. (NFPA 921)

Technical review Review of a professional's work by other professionals qualified in all aspects of the author's profession, with access to all relevant documentation.

Temporary Interim Amendment (TIA) An amendment to an NFPA standard processed according to NFPA regulations on an urgent basis after the publication of the standard, without the opportunity to publish in the first and second draft reports for review and comment.

REFERENCES

National Fire Protection Association, International Association of Arson Investigators. 2005. *User's Manual for NFPA 921: Guide for Fire and Explosion Investigations*, 2nd ed. Sudbury, MA: Jones & Bartlett.

National Fire Protection Association. 2020. *NFPA 2021: Guide for Fire and Explosion Investigations*, 2021 ed. Quincy, MA: National Fire Protection Association.

National Fire Protection Association. 2021. *NFPA 1033: Standard for Professional Qualifications for Fire Investigator*, 2022 ed. Quincy, MA: National Fire Protection Association.

On Scene

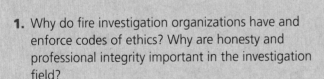

1. Why do fire investigation organizations have and enforce codes of ethics? Why are honesty and professional integrity important in the investigation field?

2. After review of the latest edition of NFPA 1033, are there knowledge areas in which you could use some further training or education? Where can you turn to receive this training and education?

3. What training and learning resources are available to you (either in your area, or elsewhere, or online) to increase your knowledge of fire investigation or of the knowledge areas in fire investigation?

4. What are the steps of the scientific method, and how would you follow them in the investigation of a fire?

5. Can you recall an investigation (or other situation) in which two viable theories or hypotheses existed to explain an event? How would scientific methodology assist in resolving such a situation?

Chapter Opener: © Jones & Bartlett Learning. Photographed by Glen E. Ellman; On Scene siren: © Bildgigant/Shutterstock.

CHAPTER

2

Fire Science for Fire Investigators

KNOWLEDGE OBJECTIVES

After studying this chapter, you should be able to:

- Explain the basic chemistry of fire. (**NFPA 1033: 4.2.4; 4.2.5**, pp. 20–22)
- Describe phase changes and thermal decomposition. (**NFPA 1033: 4.2.4; 4.2.5**, p. 21)
- Identify and describe products of combustion. (**NFPA 1033: 4.2.4; 4.2.5**, p. 21)
- Describe the components of fire dynamics. (**NFPA 1033: 4.2.4; 4.2.5**, pp. 22–26)
- Define heat flux. (**NFPA 1033: 4.2.4; 4.2.5**, p. 23)
- Describe the roles of fuel items and fuel packages in fire spread. (**NFPA 1033: 4.2.4; 4.2.5**, pp. 26–29)
- Define heat release rate. (**NFPA 1033: 4.2.4; 4.2.5**, p. 27)
- Explain the factors that affect ignition. (**NFPA 1033: 4.2.4; 4.2.5**, pp. 29–33)
- Describe the process of flame spread. (**NFPA 1033: 4.2.4; 4.2.5**, p. 33)
- Describe compartment fire behavior. (**NFPA 1033: 4.2.4; 4.2.5, 4.2.8**, pp. 33–36)

SKILLS OBJECTIVES

There are no skills objectives for this chapter.

ADDITIONAL NFPA REFERENCE

- NFPA 921, *Guide for Fire and Explosion Investigations*

You are the Fire Investigator

You are requested to investigate a "suspicious" residential fire. You are told by responding firefighters that the residence was heavily involved in fire when they arrived. They stated that on their arrival, the front door was open; they also indicated that they encountered heavy fire on both the ground floor and the upper level. A neighbor tells you it was a "fast-moving fire," spreading through several rooms during the time he waited for the arrival of the fire engines. For this reason, he suspects that someone must have used an accelerant to cause the fire. Your initial examination shows significant damage to the living room on the ground floor, with several areas of pronounced heavy damage. The living room has a sliding door that appears broken and burned, and a charcoal grill is visible on a deck outside along with staining materials.

1. Is the neighbor's report objective or subjective? What reasons might exist for a fire to be "fast moving" other than the presence of an accelerant?

2. What might be some reasons for two apparently unconnected fires in a room that has gone to flashover?

3. Should self-heating be considered during the examination of this fire, and, if so, what materials and conditions would have to be present?

NAVIGATE *Access Navigate for more practice activities.*

Introduction

Fire has been used by humans for thousands of years for many purposes, including heating homes, cooking food, and waging war. In each period of human existence, tragic events have occurred as a result of the destructive power of fire. Every fire investigator must fully understand the fundamental properties of fire and fire development to determine accurately the origin and cause of any fire or explosion event they may investigate.

Fire Chemistry

All matter is composed of elements; combinations of elements are known as **compounds**. Elements are substances that will not break down into simpler substances. There are 92 natural elements, such as carbon, sulfur, and hydrogen. All elements are composed of smaller units, called **atoms**. Atoms are the smallest unit of an element that can take part in a chemical reaction.

Matter may be present in one of three states: solid, liquid, or gas. Fuels must generally be in the gas (also called vapor) state to be involved in the combustion process, so a phase change or decomposition may be required before the fuel can burn. The application of heat to matter can effectively result in a phase change—from a solid to a liquid, for example—in which no atoms change. Alternatively, it may result in thermal decomposition, as in **pyrolysis**, where atoms are broken down and rearranged to form different matter.

In a fire, many different chemical reactions occur. Overall, though, fire is considered an **oxidative reaction**. That is, atoms from a fuel are joining with the oxygen in the air or being oxidized. As the atoms in a material are broken down and rearranged, heat energy is produced and released into the atmosphere.

Energy is a property of matter that is manifested as an ability to perform work, either by moving over a distance against a force or through transferring heat. It can be changed in form or transferred to other matter, but cannot be created or destroyed. **Power** is a property that describes energy emitted, transferred, or received over a unit of time.

Because it releases energy, fire is referred to as an **exothermic reaction**. In the fire environment, the heat that is generated helps to continue the reaction until it is interrupted (uninhibited chain reaction). The molecular rearrangement and breaking down of atoms during this process results in the formation of new substances and the presence of free radicals, atoms that have open bonds and are highly reactive. Free radicals are not as stable as regular atoms and easily react with other substances. Thus, the continued formation and reaction of free radicals is part of the combustion process. Other reactions, such as some phase changes (in which the matter is not altered), may be **endothermic reactions**, in which heat is absorbed to continue the reaction.

Although many compounds may be found at a fire scene, the most important class for fire investigators comprises the carbon-based or organic compounds. Among the organic compounds, some of the best fuels are formed of hydrocarbons, compounds that are composed solely of carbon and hydrogen.

Fire is defined as a rapid oxidation process—that is, an exothermic chemical reaction resulting in the release of heat and light energy in varying intensities. **Oxidation** generally involves loss of electrons in a material. It can happen rapidly, as in fire, or slowly, as in the rusting of iron.

In general, fire occurs only when materials are in their gas phase. Solids, for example, must be heated, causing the material to decay and produce fire gases. This process of decomposition of a material into simpler molecular compounds by the effects of heat alone is known as pyrolysis; it is evident when wood, heated sufficiently, breaks down into vapors and char. In contrast, for liquids to burn, they must be heated to produce ignitable mixtures in air through a process known as **vaporization**.

For a fire or **combustion** to occur, four components must be present: fuel, oxidizing agent, heat, and an uninhibited chemical chain reaction. By removing or eliminating one of the four components, the fire can be extinguished or prevented. The four components of fire are often depicted as the fire tetrahedron **FIGURE 2-1**.

Fuel

The first component of the fire tetrahedron, **fuel**, comprises any substance that can undergo combustion. The most common fuels fire investigators will encounter are organic fuels, which contain carbon. These fuels include wood, plastics, and petroleum products. Inorganic fuels, such as combustible metals, do not contain carbon. When a solid fuel is heated,

the compounds and elements that make up the material begin to break down. This process of pyrolysis produces ignitable gases that will burn on the surface of the material. Similarly, vaporization of liquids leads to the production of ignitable vapors and gases, which will also burn on the surface of the liquid.

Gaseous fuels are probably the most dangerous of the three physical states because they need only an ignition source for combustion to occur.

Oxidizing Agent

Fires require an **oxidizing agent** to support the combustion process. Most fires use the oxygen in air for this purpose, as it is the most readily available source of an oxidizer. However, certain other materials, such as hydrogen peroxide, can also provide the oxidizer needed for a fire.

Heat

Heat provides the energy needed to create and ignite vapors produced from the fuel source. Heat is a form of energy, which is measured in joules, calories, or British thermal units (Btu). The rate at which heat is released, known as the heat release rate (HRR), is measured in joules per second or watts.

Uninhibited Chemical Chain Reaction

The last component of the fire tetrahedron, an **uninhibited chemical chain reaction**, provides a self-sustaining event that continues to develop fuel vapors and sustain flames even after the removal of the ignition source. As a fire continues to burn, this reaction radiates heat back to the surface of the fuel, producing more vapors and continuing the combustion process.

Phase Changes and Thermal Decomposition

The response of materials to heat varies dramatically. During a fire, materials may change their physical state as a result of being heated. These physical changes or **phase changes** include melting (in which solids turn into liquids) and vaporization (in which liquids turn into gases). Both melting and vaporization are reversible upon cooling because the chemical structure of the material does not change. When an irreversible change occurs in either a solid or a liquid, the process is referred to as **thermal decomposition**.

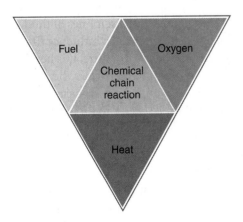

FIGURE 2-1 The fire tetrahedron.

Combustion: Premixed and Diffusion Flames

Flames produced during the combustion process can be categorized as either premixed or diffused. **Premixed burning** (flame) can occur when fuel vapors such as natural gas are dispersed and mixed with air prior to ignition. These fuels can produce both fires and explosion events; the fuel and air are already combined and need only a competent ignition source to start the chemical reaction. For each fuel, there is an optimal ratio of fuel to air at which the most efficient means of combustion exists. This optimal mixture is referred to as the **stoichiometric ratio**. The stoichiometric ratio is a concentration above the **lower explosive limit (LEL)** and below the **upper explosive limit (UEL)**. Below the LEL, there is not enough fuel to sustain combustion; above the UEL, there is not enough air to sustain combustion.

Most fuels are actually consumed by **diffusion flame** burning, in which the fuel and air mix together at the region of combustion. This includes the flame from solid materials that are being pyrolyzed. This process produces vapors that are oxidized, usually above the surface of the material. One example of a diffusion flame is the flame from a candle, in which the fuel and the air meet and mix in the luminous zone. Like a premixed flame, the diffusion flame can exist only when a sufficient concentration of oxygen is present. While normal room air contains approximately 21% oxygen, depending on the conditions, combustion may be sustained at much lower oxygen concentrations.

Flames may transition from a premixed to a diffusion flame. An example is when the premixed vapors above a gasoline pan are ignited and consumed. The flame then becomes a diffusion flame, relying on the burning of the vapors from the pan.

Products of Combustion

When a fuel burns, it can produce chemical compounds that are released as both visible and invisible products of combustion. Hundreds of different compounds can be produced during the combustion process, including carbon monoxide, carbon dioxide, and water vapor. If a fire has a limited amount of air available for combustion, an increase in the amount of visible products of combustion (such as soot and smoke) and carbon monoxide will occur.

Smoke is created by the combination of the various products of combustion, dependent on the particular fuel being consumed. These products include solids such as soot or ash, small liquid droplets, and various other fire gases. Some fuels, such as alcohols and propane, burn without the production of smoke, whereas plastics and other hydrocarbons tend to create large amounts of dense black smoke. As these products of combustion begin to migrate away from the fire and their temperature cools, they accumulate on both horizontal and vertical surfaces. Surfaces that are normally cooler, such as ceramic tile, tend to collect more of these combustion products than warmer surfaces do.

> **TIP**
>
> Smoke and flame color are not valid indicators of the burning material. Many factors, including suppression activities, can alter the appearance of smoke and flames.

Smoke and Flame Color

The color of smoke is not a reliable indicator of the material burning. Although certain fuels may produce a particular color and density of smoke, other factors (including decreasing oxygen levels and ventilation-controlled fires) can produce heavy, dark smoke that resembles the smoke normally associated with oil and other hydrocarbons burning. As a fire goes through its various phases, smoke production will generally increase. Firefighting operations can also change the color of smoke by mixing condensing vapors with black smoke, producing a white to gray color. This phenomenon can cause witnesses and firefighters alike to believe that other fuels may be burning.

Although fuels may burn at a particular temperature and produce a certain color flame, neither temperature nor flame should be relied on as an indicator of what is burning.

Fire Dynamics

Fire dynamics is the study of how chemistry, fire science, and the engineering disciplines of fluid mechanics and heat transfer interact to influence fire behavior. To understand the behavior of fire, it is helpful to have some understanding of fluid flows, heat transfer, ignition and flame spread, fuel packages, heat flux, and the distinction between fuel-controlled fires and ventilation-controlled fires.

Fluid Flows

Fluids include both liquids and gases. In fires, however, the primary focus is usually gas flows. Although these flows can be generated by mechanical forces such as fans, in most cases the fire actually creates its own buoyant flow. A buoyant flow occurs because hot gases are less dense than cool gases, so they rise. The primary engine that creates flows in fires is the fire itself. The hot gases created by the fire rise above the fire in a **plume**. As these gases rise, they *entrain* (draw in) cool air, which causes the velocity of the gas flow to increase. At the same time, the temperature of the plume decreases. The **air entrainment** also causes the diameter of the plume to increase, resulting in a cone shape.

When the plume reaches the ceiling of a room, the movement of the fire gases parallel to the ceiling is known as a **ceiling jet**. The ceiling jet flows along the ceiling until it encounters a vertical obstruction, such as a wall, that impedes its flow.

In compartment fires, the buoyant gases flow in and out through multiple vents. If there is only one vent in the compartment, the gases flow out the upper portions of the opening, and the fresh air flows into the compartment to feed the combustion through the lower portions of the vent opening.

Heat Transfer

Heat energy naturally moves from areas of higher temperature to lower temperature. The exchange of thermal energy between materials, known as **heat transfer**, is measured as energy flow per unit of time. The rate of transfer increases when the temperature difference between the objects is greater. The measure of the rate of transfer between objects, called **heat flux**, is commonly expressed as kilowatts per square meter (kW/m^2) or watts per centimeter squared (W/cm^2). As a fire progresses through its various phases or stages, the effects of heat transfer create the fire patterns and physical evidence that you will rely on to establish the origin and cause of the fire accurately.

Most people use the terms *heat* and *temperature* interchangeably; however, they have different meanings. **Temperature** is a measurement of the amount of molecular activity when compared with a reference or standard, whereas *heat* refers to a form of energy. An increase or decrease in the amount of energy transferred to an object can affect the molecular activity within an object and result in a corresponding change in an object's temperature. Heat may be transferred by conduction, convection, or radiation. A fire investigator must be aware of how each mode of transfer can influence the development and spread of a fire.

Conduction

Conduction occurs when solid objects are heated, and energy is transferred from hotter to cooler areas

through direct contact. An example of this can be seen when a spoon is left in a hot liquid and the end of the handle begins to get warmer. The rate of this transfer is influenced by the difference in temperature and the thermal conductivity of an object. **Thermal conductivity (k)** is a measure of the amount of heat per unit of time that will flow across an area with a temperature difference (gradient) of one degree per unit of length (W/m-K). Thermal conductivity can be an important property when assessing how easily a fuel package can be ignited.

Heat capacity is a measure of the amount of energy required to raise the temperature of an object by one degree of a unit mass (J/kg-K). Materials with higher heat capacities will need more energy to raise their temperature compared to materials with lower heat capacities. The more conductive a material is, the faster heat will transfer to it. Denser objects tend to be better conductors than lower-density objects.

According to NFPA 921, **thermal inertia** reflects "[t]he properties of a material that characterize its rate of surface temperature rise when exposed to heat;" simply put, it is a measure of how easily the surface temperature of the material will increase when heat flows into it. The thermal inertia is the product of the thermal conductivity (k), the density (ρ), and heat capacity (*c*). This mathematical expression, $k\rho c$, is pronounced *kay-row-see*.

Plastic foams have a low density and a low thermal conductivity, so when they are exposed to heat, their surface temperature increases quickly. Metals, by contrast, have a high density and a high thermal conductivity, so the surface temperature of a metal object increases more slowly when exposed to the same heat source.

As materials are heated, the surfaces of an object begin to absorb heat, which is then transferred through the remainder of the object via conduction. The speed at which an object's surface absorbs heat is determined by its thermal inertia and is most influential during the initial heating of the object. For instance, objects that are thin tend to heat and ignite faster than objects of like materials that are thicker. This effect has a direct impact on ignitability and flame spread.

Convection

When heat is absorbed from heated gases or liquids by cooler objects or materials, the process is referred to as **convection**. In the early stages of a fire, convection plays a major role in the spread of heat and flames. Convection may occur by one of two mechanisms: natural or forced. With natural ventilation, the buoyant effect of the heated gases causes the heat to move away from the source, typically by rising, then spreading laterally through a space. Forced convection involves the influence of external sources, including the use of fire streams and ventilation fans.

Radiation

The transfer of heat via electromagnetic waves is known as **radiation**. Electromagnetic waves are longer than visible light waves but shorter than radio waves. They increase the sensible temperature of any substance or object that is capable of absorbing the radiation. Heat from the fire is transferred in a direct line away from the source object and absorbed by cooler objects, including liquids and gases. Most exposure fires result from radiant heat transfer.

The rate of heat transfer from a radiating material is proportional to that material's absolute temperature (the temperature of a material measured in degrees Kelvin or on the Rankine scale) raised to the fourth power. As an example, a radiating material whose absolute temperature doubles will have a 16-fold increase in the radiation emanating from the material.

TIP

In heat transfer, heat energy moves from areas of higher temperature to areas of lower temperature. This heat transfer impacts the fire patterns and other physical evidence at a scene. There are three primary types of heat transfer:

- Conduction: Heat transfer through direct contact of solid objects
- Convection: Heat transfer from heated gases or liquids to cooler objects or materials
- Radiation: Heat transfer through electromagnetic waves

FIGURE 2-2

The rate of heat transfer between two objects is affected mainly by the distance between the objects, but other factors come into play as well. For example, the relative temperatures of the two objects and the view (i.e., the angle between the radiator and the target) affect the heat transfer rate. Radiation is transferred more readily between two parallel surfaces than between two perpendicular surfaces. When the target surface is not parallel to the radiator, the energy is spread out over a broader area, reducing the number of watts per unit area.

As a fire develops within a compartment, radiation becomes the dominant means of heat transfer. Given this reality, an understanding of radiation is critical to your ability to comprehend fire growth and spread. In addition, it is important to be able to relate the concept of heat flux to everyday experience. The data in **TABLE 2-1** describe the effects of radiant heat flux.

Thermometry

Thermometry is "[t]he study of the science, methodology, and practice of temperature measurement," as defined in NFPA 921. Although thermometry is seldom used directly at the fire scene, it can play an important role in the investigation and analysis of a fire event. For example, during the investigation of suspected spontaneous combustion fires in decomposing haystacks, core temperature measurements may be taken at the scene as part of the hypothesis testing process to determine whether other sections of material are at critical temperatures.

Additionally, an understanding of the appropriate application of thermometry is necessary to use fire dynamics equations and formulae correctly. Results from thermometry studies are frequently used in fire safety or code compliance cases as well. Several systems are available for measuring degrees of temperature, which can broadly be divided into empirical temperature scales and thermodynamic temperature scales.

The Fahrenheit and Celsius (Centigrade) scales are the most common empirical temperature scales. They are based on the relative temperatures at which water boils and freezes, along with other empirical comparisons.

Thermodynamic temperature scales are based on the lowest possible temperature of absolute zero; hence, they are called absolute temperature scales.

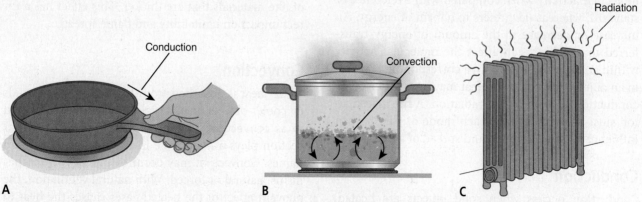

FIGURE 2-2 The primary types of heat transfer. **A.** Conduction. **B.** Convection. **C.** Radiation.

TABLE 2-1 Effect of Radiant Heat Flux (NFPA 921 Table 5.5.4.2)

Approximate Radiant Heat Flux (kW/m²)	Comment or Observed Effect
170	Maximum heat flux as currently measured in a post-flashover fire compartment.
80	Heat flux for a Thermal Protective Performance (TPP) Test of protective clothing.[a]
52	Fiberboard ignites spontaneously after 5 seconds.[b]
29	Wood ignites spontaneously after prolonged exposure.[b]
20	Heat flux on a residential family room floor at the beginning of flashover.[c]
20	Human skin experiences pain with a 2-second exposure and blisters in 4 seconds with second-degree burn injury.[d]
15	Human skin experiences pain with a 3-second exposure and blisters in 6 seconds with second-degree burn injury.[d]
12.5	Wood volatiles ignite with extended exposure[e] and piloted ignition.
10	Human skin experiences pain with a 5-second exposure and blisters in 10 seconds with second-degree burn injury.[d]
5	Human skin experiences pain with a 13-second exposure and blisters in 29 seconds with second-degree burn injury.[d]
2.5	Human skin experiences pain with a 33-second exposure and blisters in 79 seconds with second-degree burn injury.[d]
2.5	Common thermal radiation exposure while firefighting.[f] This energy level may cause burn injuries with prolonged exposure.
1.0	Nominal solar constant on a clear summer day.[g]

The unit kW/m² defines the amount of heat energy or flux that strikes a known surface area of an object. The unit kW represents 1000 watts of energy, and the unit m² represents the surface area of a square measuring 1 m long and 1 m wide. For example, 1.4 kW/m² represents 1.4 multiplied by 1000 and equals 1400 watts of energy. This surface area may be that of the human skin or any other material.
[a]From NFPA 1971, *Standard on Protective Ensembles for Structural Fire Fighting and Proximity Fire Fighting.*
[b]From Lawson, "Fire and the Atomic Bomb."
[c]From Fang and Breese, "Fire Development in Residential Basement Rooms."
[d]From SFPE Engineering Guide, "Predicting 1st and 2nd Degree Skin Burns from Thermal Radiation," March 2000.
[e]From Lawson and Simms, "The Ignition of Wood by Radiation," pp. 288–292.
[f]From U.S. Fire Administration, "Minimum Standards on Structural Fire Fighting Protective Clothing and Equipment," 1997.
[g]From the *SFPE Handbook of Fire Protection Engineering*, 2nd ed. Quincy, MA: NFPA.
Reproduced with permission from *NFPA 921-2021, Guide for Fire and Explosion Investigations,* Copyright © 2020. National Fire Protection Association. This reprinted material is not the complete and official position of the NFPA on the referenced subject, which is represented only by the standard in its entirety.

These scales are based on the fundamental laws of thermodynamics or statistical mechanics, where temperature is measured in kelvins (K) or using the Rankine (R) scale **FIGURE 2-3**.

To obtain proper results when using equations involving measures of temperature, you must both understand the equations and use the correct units of temperature in the equations. Frequently, equations involving temperature require that a specific unit of temperature be used, and the use of the wrong unit (e.g., degrees Fahrenheit when degrees Celsius is required) will result in incorrect results.

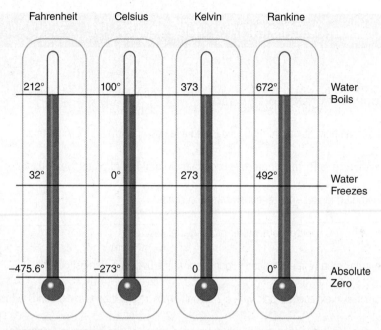

FIGURE 2-3 Relationship of temperature measurement scales.

Fuel Items and Fuel Packages

The amount of fuel present is referred to as the **fuel load**. In any fire scenario, it is important to consider the fuel load at the scene. The total fuel load of a given fire will be a reflection of the potential energy; it does not determine how fast the fire develops once ignition occurs.

The materials consumed during a fire are referred to as the **fuel items**. When fuel items are assembled and placed close to each other, their collection creates a **fuel package**. If only one specific fuel item exists, it may also be considered a fuel package. If other fuel packages are nearby, ignition via radiant heat can occur. Examples of commonly found fuel packages include the following:

- Furniture and other contents in a dorm or bedroom
- Personal items in a commercial storage space
- Combustible raw materials stored in a paper manufacturing facility

The specific fuel items within a fuel package influence the speed and intensity of a particular fire, as well as the best method of fire extinguishment.

The location and position of a fuel package are also important to consider. For example, fuel packages that are located next to or near a wall or corner will restrict the amount of air entering the plume and create an imbalance. This imbalance will bend the flame and thermal plumes toward the restricting surface or

object. Once the flame and thermal plumes begin to bend toward the wall, they may attach themselves to the wall, restricting the entrainment of air into the base of the fire. The geometry of the fuel package and its proximity to the wall will determine the extent to which the flame and thermal plume will bend. This decrease in the entrainment of air will also cause an increase in plume and upper-layer temperatures because of the smaller amount of cooler air entraining into the fire, which dilutes the higher temperatures of the plume. As the plume temperature increases and releases energy to the upper layer, the temperature of the upper layer will naturally increase.

When entrainment is decreased, less air is available to mix with fuel vapor, and the vapor must be transported over a longer distance to mix completely with this reduced quantity of air. This can result in increased flame heights. Whether the flame height will increase depends on numerous variables. Flame height is also affected by positioning of the fuel package against the wall—in such a case, heat will radiate back from the wall into the fire, increasing its energy. Increased flame heights have been reported in experiments involving fire against a wall, although NFPA 921 states that experiments predict no significant increase in flame length for such fires.

When the fuel package is positioned in a corner, air entrainment is further reduced and reflected radiant heat increases, causing an increase in the absolute temperature of the upper layer relative to the absolute temperature that would have been measured if the fire was positioned away from a corner. Similarly, a marked increase in the flame height is seen when flames become attached to the walls in a corner configuration.

Thus, the position of the fire in relation to walls and corners must be considered when interpreting damage patterns. These effects may also be observed with outdoor fires when similar conditions are present.

Heat Release Rate

The power of a fire is determined by calculating the energy being released by the individual fuels being consumed. Known as the **heat release rate (HRR)**, this value is measured in either watts or kilowatts. The HRR of a fuel is calculated by multiplying the mass burning rate by the heat of combustion of the fuel. It also depends on the total area involved in the combustion process. The mass burning rate or mass loss rate typically reflects how much material is consumed during combustion, measured in grams of material per second per area burning. The heat of combustion measures how much energy is being released as a result of this reaction and is typically expressed in kilojoules per gram of material.

The HRR of a fuel is often illustrated on a generalized curve to indicate the energy being released during the various stages of a fire. **FIGURE 2-4** displays an example of an ideal HRR curve for the stages in a fire in which the oxygen is not limited (fuel-controlled fire).

When a fuel item is burned, the highest HRR value that is measured is termed the *peak HRR*. **TABLE 2-2** provides general peak HRR values for fuel packages, which can assist fire investigators in explaining both fire development and spread issues. Although these values may be helpful, they should not be considered infallible—fuel items that may have a similar use can produce varying HRR values. Testing and experimentation are the most accurate methods of determining a particular fuel item's peak HRR, but such testing data are usually not available. When using HRR estimates to test fire-spread hypotheses, it is a best practice to estimate at the high end of the ranges shown in Table 2-2.

When a fire is unconfined and allowed to grow, the HRR will generally increase as the fuel item becomes more involved. Once flames have spread across the entire surface of the fuel item, the peak HRR will be reached, and the rate will begin to decrease as the fire moves toward the decay phase. In the instance of a compartment fire, the HRRs of all fuel items and packages that are involved are added together to produce an HRR for the entire compartment. The HRR may also increase in the compartment fire as heat is contained and radiated back on the burning items.

Flame Height and Heat Release Rate

Generally, larger flame heights correspond to higher rates of heat release, but this holds true only when the fuel packages being compared are in the same relative location in a compartment. Some care is necessary in estimating flame height because people tend to focus on the *highest* portion of the flame, when the property of interest is actually the *average* flame height.

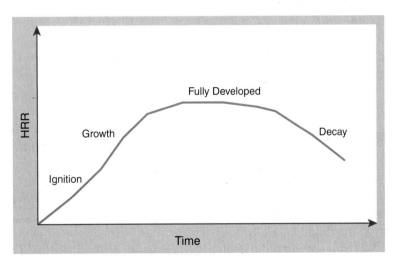

FIGURE 2-4 Idealized HRR curve.

Reproduced with permission from *NFPA 921-2021, Guide for Fire and Explosion Investigations*, Copyright © 2020. National Fire Protection Association. This reprinted material is not the complete and official position of the NFPA on the referenced subject, which is represented only by the standard in its entirety.

TABLE 2-2 Representative Peak HRRs (Unconfined Burning) (NFPA 921 Table 5.6.3.1)

Fuel	Weight		Peak HRR (kW)
	Kilograms	**Pounds**	
Wastebasket, small	0.7–1.4	1.5–3	4–50
Trash bags, 42 L (11 gal) with mixed plastic and paper trash	2.5	7.5	140–350
Cotton mattress	12–13	26–29	40–970
TV sets	31–33	69–72	120 to >1500
Plastic trash bags/paper trash	1.2–14	2.6–31	120–350
PVC waiting room chair, metal frame	15	34	270
Cotton easy chair	18–32	39–70	290–370
Gasoline/kerosene in 0.2 m² (2 ft²) pool	19	42	400
Christmas trees, dry	6–20	13–44	3000–5000
Polyurethane mattress	3–14	7–31	810–2630
Polyurethane easy chair	12–28	27–61	1350–1990
Polyurethane sofa	51	113	3120
Wardrobe, wood construction	70–121	154–267	1900–6400

Values are from the following publications: Babrauskas, V. "Heat Release Rates," in *SFPE Handbook of Fire Protection Engineering*, 3rd ed. Quincy, MA: NFPA, 2002; Babrauskas, V., and Krasny, J. *Fire Behavior of Upholstered Furniture*. NBS Monograph 173: Fire Behavior of Upholstered Furniture. Gaithersburg, MD: National Bureau of Standards, 1985; Lee, B. T. *Heat Release Rate Characteristics of Some Combustible Fuel Sources in Nuclear Power Plants*. NBSIR 85-3195. Gaithersburg, MD: National Bureau of Standards, 1985; NFPA 72, *National Fire Alarm Code*, 1999 ed., Annex B.
Reproduced with permission from *NFPA 921-2021, Guide for Fire and Explosion Investigations,* Copyright © 2020. National Fire Protection Association. This reprinted material is not the complete and official position of the NFPA on the referenced subject, which is represented only by the standard in its entirety.

Flame heights are defined for three regions of the fire:

- Plume region: the area above the visible flame
- Intermittently flaming region: the upper region of the flame
- Continuously flaming region: the lower visible flame

Correlations of flame height with HRR were determined experimentally using liquid pools, so they may not apply in all cases of burning solid fuels. Moreover, the diameter of the flame is an important consideration.

Of more importance to the fire investigator is the location of the fuel package within a room. A fuel package burning in the center of the room will have a plume that is able to entrain air from all sides. If that package is moved against the wall, less air is available for entrainment and heat radiates back to the fire, so the flame height may increase. If that package is located in the corner of the room, significantly less air is available, resulting in a much taller flame. This so-called wall effect or corner effect can lead to misinterpretation of the fire's origin because heavier fire damage will be located above fuel packages located against walls or in corners. The effect is illustrated in **FIGURE 2-5**, which shows three identical sofa cushions 17 seconds after ignition. Also, keep in mind that ceiling height and distance from a plume greatly affect how quickly fire protection devices react to the fire.

FIGURE 2-5 The effect of location in a compartment on flame height. **A.** A couch cushion burning in the center of a room. **B.** A couch cushion burning against a wall. **C.** A couch cushion burning in a corner.
© Jones & Bartlett Learning. Photographed by Glen E. Ellman.

Ignition

With any fire investigation, the source of ignition is critical in determining the cause of a fire. An ignition source can be defined as either smoldering (as is the case with hot coals and glowing embers) or flaming (for example, ignition when a match or lighter is the source). A source of ignition may also be characterized as either piloted or autoignition. Sparks, arcs, and open flames are common sources of piloted ignition, in which an external source ignites the fuel item. Autoignition, by contrast, occurs when the fuel item is heated to a temperature at which fuel gases being released from the object ignite without a piloted ignition source.

To heat a fuel sufficiently to generate ignitable vapors, a number of interrelated factors must come together:

- Form of the fuel: Is it a solid, liquid, or gas? Does it have a lot of surface area that can absorb heat and a relatively small amount of mass?

- Amount or mass of the fuel: How much fuel is present that needs to be heated? The greater the mass, the more heat that is needed.

- Proximity of the fuel to the heat source: Generally, the closer the fuel is to the heat source, the faster the fuel's temperature is raised.

- Amount of heat being generated: How much heat is the heat source generating? If it is a relatively small heat source, the fuel needs to be either closer or exposed longer for its temperature to increase.

- Duration of exposure: The longer the duration, the more the fuel's temperature will increase. However, some fuels may not need long exposures to have their temperature raised sufficiently to generate combustible vapors.

For example, a single match likely cannot heat a large block of wood sufficiently to ignite it, even though the temperature of the heat source may be above the minimum ignition temperature of the target fuel. In this case, there is too much mass in the wood, and the match generates insufficient heat. Similarly, a solid piece of wood, such as a 2-inch by 4-inch piece of lumber, that is exposed to a brief electrical arc would probably not ignite because the duration of heat exposure is not sufficient.

In the initial stages of a fire, an upholstered chair does not ignite from the radiant heat of a fire in a wastebasket that is several feet away. The fire is too small and the distance too great to raise the temperature of the upholstery sufficiently to generate ignitable vapors. However, if the same wastebasket fire is moved to within a foot of the chair, the chair may easily ignite.

Most fuels need to liberate fuel gases or vapors for ignition to occur. When the gases and vapors are released into the atmosphere, they must mix with air to produce an ignitable mixture. Some fuels, especially liquids, have flash points that produce ignitable vapors at ambient temperatures, releasing vapors into the atmosphere that will ignite if the vapors are heated to their ignition temperature and an ignition source is present.

To heat a fuel item to its ignition temperature, the rate of heat transfer must overcome the rate of heat loss due to convection, conduction, and radiation, as well as the loss of energy as a result of pyrolysis or vaporization. Thus, the source of heat energy needs to be greater than the ignition temperature of the fuel item. An exception to this is instances where spontaneous ignition occurs. **TABLE 2-3** lists ignition temperatures for various materials.

TABLE 2-3 Reported Burning and Sparking Temperatures of Selected Ignition Sources (NFPA 921 Table 5.7.1.1)

Source	Temperature, °C	Temperature, °F
Flames		
Benzene[a]	920	1690
Gasoline[a]	1026	1879
JP-4[b]	927	1700
Kerosene[a]	990	1814
Methanol[a]	1200	2190
Wood[c]	1027	1880
Embers[d]		
Cigarette (puffing)	830–910	1520–1670
Cigarette (free burn)	500–700	930–1300
Mechanical sparks[e]		
Steel tool	1400	2550
Copper–nickel alloy	300	570

[a]From Drysdale, D. *An Introduction to Fire Dynamics.* Wiley, 1999.
[b]From Hagglund, B., and Persson, L. E. *Heat Radiation from Petroleum Fires.* FOA Report C20126-D6(A3). Stockholm, Sweden: National Defence Research Institute, 1976.
[c]From Hagglund, B., and Persson, L. E. *An Experimental Study of the Radiation from Wood Flames.* FOA Report C4589-D6(A3). Stockholm, Sweden: National Defence Research Institute, 1974.
[d]From Krasny, J. *Cigarette Ignition of Soft Furnishings: A Literature Review with Commentary.* NBSIR 87-3509. Washington, DC: Center for Fire Research, National Bureau of Standards, June 1987.
[e]From NFPA *Fire Protection Handbook*, 15th ed., Section 4, p. 167.

Reproduced with permission from *NFPA 921-2021, Guide for Fire and Explosion Investigations,* Copyright © 2020. National Fire Protection Association. This reprinted material is not the complete and official position of the NFPA on the referenced subject, which is represented only by the standard in its entirety.

Some materials, such as cigarettes, cellulose insulation, and sawdust, are permeable, allowing better air infiltration—which, in turn, supports a smoldering fire or solid-phase combustion. In smoldering fires, flames are absent; however, more products of combustion are produced per unit of mass burned. Although these materials may not produce flames when they burn, they can be a competent heat source to other fuels that will produce flaming combustion.

Ignition of Flammable Gases

For a flammable gas to ignite, it must be present in a concentration that will allow for a piloted ignition from either a spark or a flame. This concentration is referred to as the **flammable range** of a gas and has both lower and upper limits. Below the lower flammability limit (LFL), the mixture is too "lean" to burn (not enough fuel); above the upper flammability limit (UFL), the mixture is too "rich" to burn (not enough air). The terms *flammability limit* and *explosive limit* are used interchangeably. LFL and UFL are more commonly expressed as LEL and UEL, respectively.

Flammable gases may also ignite without a piloted ignition source if the gas–air mixture is heated to a specific temperature, referred to as the **autoignition temperature (AIT)**. The autoignition temperature is the lowest temperature at which a gas–air mixture will ignite in the absence of an ignition source. Factors such as the size and geometry of gas volume and concentration will influence the autoignition of a flammable gas. When the volume of a gas–air mixture increases, the ignition temperature may decrease; similarly, a gas–air mixture at its stoichiometric ratio may favor a lower ignition temperature.

Ignition of Liquids

Liquids are frequently encountered as fuel sources in fires. A fire investigator should understand the ignition characteristics of these types of fuel packages.

Flash Point

The **flash point** is the lowest temperature at which a liquid produces a flammable vapor. It is measured by determining the lowest temperature at which a liquid will produce enough vapor to support a small flame for a short period of time if a piloted ignition source is present. To sustain burning after the removal of an ignition source, the liquid must be heated to its **fire point**. The fire point is usually only a few degrees higher than the flash point and, in some instances, may be the same. In most cases, this is a distinction without a difference.

When a large pool of liquid is present, ignition may occur only to the portion of the liquid that is heated to its fire point. Flaming combustion may then spread to the remainder of the pool after ignition, or produce a wick effect in which the fuel is drawn toward the fire.

When liquids are dispersed in an atomized mist, the **surface-to-mass ratio** will increase and can allow for piloted ignition below the normal fire point of a bulk liquid. Liquids such as hydraulic fluids have been shown to be ignitable when in the form of a spray, despite having a very high flash point.

Liquids may also ignite when the liquid itself begins to oxidize, creating an exothermic reaction. This process may occur when certain liquids, such as linseed oil, are suspended in porous material, such as cloth rags.

Ignition of Solids

Solids may be ignited by either a smoldering ignition, piloted flaming ignition, or flaming autoignition.

Smoldering Ignition and Initiation of Solid-Phase Burning

For their ignition to occur, most cellulosic products, including wood and paper, must be pyrolyzed first to produce char. This char will then allow for smoldering or solid-phase burning to consume the remaining material. Some materials, such as carbon and magnesium, are capable of solid-phase burning without being reduced or converted to char.

Certain materials, such as thermoplastics, do not produce a char when burned and are not capable of solid-phase burning. Similarly, materials that change to a liquid when heated will also not support solid-phase burning.

Note that smoldering is a result of an oxidation process. It is not caused by an external heat source impinging on the fuel.

Self-Heating and Self-Ignition. Some materials are capable of initiating a chemical chain reaction and increasing in temperature just by coming in contact with air. This process, termed **self-heating**, usually involves solid fuels.

Even though some materials are capable of self-heating, a condition of **thermal runaway** needs to occur for self-ignition to occur. In thermal runaway, the heat generated exceeds the amount of heat lost from the material. This condition most often arises in the best-insulated regions of the fuel package, mainly the center of the package.

Self-heating may also result from materials, mostly organics and certain metals, reacting with oxygen. Metal powders may rapidly oxidize and self-ignite, producing metal oxides as by-products. Additionally, certain chemical reactions, such as those involving polymers, resins, and adhesives, produce heat while they are forming solids during the curing process. Spontaneous combustion is possible when organic materials that contain fatty acids begin to react with oxygen, generating heat.

Ignition by self-heating of a fuel typically requires a lower minimum surrounding or exposure temperature than ignition without self-heating. For example, rags soaked in linseed oil can ignite at a surrounding temperature of 68°F (20°C), but the pure liquid form of linseed oil has a flash point of 428°F (220°C).

Other materials, such as unsaturated molecules that contain double bonds of carbon to carbon, may also begin to generate heat. Oils and other saturated hydrocarbons that contain carbon-to-carbon single bonds will self-heat or spontaneously combust only when the temperature is elevated, or they are present in very large piles of materials. These materials, such as hydraulic and motor vehicle oils, generally do not produce sufficient heat energy to self-ignite.

To self-heat to the point of ignition, materials must be porous and permeable and capable of being oxidized. These materials must also be capable of producing a char without melting.

As with any fire, a self-heating material requires sufficient air (oxygen) to support the combustion process. Thus, if linseed oil–soaked rags are placed into a sealed container, the air supply is limited, greatly reducing the risk of an accidental fire.

Other factors to consider when evaluating a material's ability to self-heat include the size and shape of the fuel package and the environmental conditions present. Fuels with smaller surface areas are more likely to self-heat than fuels with larger surface areas, due to the ability of a smaller surface area fuel to insulate itself and increase the interior temperature of the fuel. Objects that are round or square will sustain self-heating more efficiently than large, flat surfaces of similar mass. **FIGURE 2-6** illustrates the critical conditions that must exist for spontaneous ignition to occur. The five numbered points of the "star" in the center of the figure show situations in which just one condition is not met and ignition does not occur.

The typical situations in which you are likely to encounter self-heating fires involve commercial clothes dryers with oily laundry, painting and staining operations, and piles of vegetable materials such as cotton bales or haystacks. Wood, like other cellulosic materials, is subject to self-heating when exposed to elevated temperatures that are below its normal ignition temperature of 482°F (250°C). The scientific community has not reached a consensus on the self-ignition

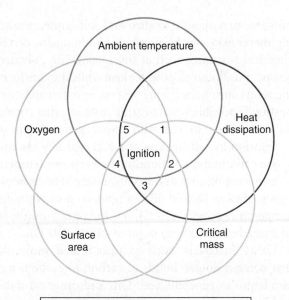

(1) Insufficient surface area
(2) Insufficient oxygen
(3) Ambient temperature too low
(4) Insufficient insulation–heat radiated away
(5) Insufficient material

FIGURE 2-6 Conditions required for spontaneous ignition to occur in materials capable of self-heating.

Reproduced with permission from *NFPA 921-2021, Guide for Fire and Explosion Investigations,* Copyright © 2020. National Fire Protection Association. This reprinted material is not the complete and official position of the NFPA on the referenced subject, which is represented only by the standard in its entirety.

of wood. Charcoal briquettes have been suspected of self-heating to ignition; however, laboratory testing has shown that commonly sized bags of briquettes do not reach sufficient temperatures to self-ignite.

Certain elements may spontaneously ignite when exposed to air. White phosphorus, sodium, potassium, and some finely divided metals such as zirconium may react in this manner. Such elements are known as pyrophoric materials.

Oxidizer Fires. A material that is not combustible but that can increase the rate of combustion or produce spontaneous combustion when combined with other substances is considered to be an oxidizing agent. Commonly encountered oxidizers include chlorine tablets for swimming pools and certain fertilizers that may begin to self-heat when combined with organic materials such as hydraulic fluids.

Transition to Flaming Combustion. When sufficient heat and vapors are created, smoldering materials may produce flaming combustion upon the introduction of a piloted ignition source. The amount of time required for this transition is difficult to predict and depends on numerous factors.

When the ignition source for flaming combustion is a smoldering material such as a cigarette, a long period of time may lapse before the first flames are observed. This smoldering heat can also begin

preheating the fuel, permitting a more vigorous flaming combustion once it occurs. In other words, a "slow smoldering fire" may burn quite rapidly once the transition occurs. Therefore, the investigator should not automatically eliminate a smoldering ignition source as an option just because the fire was perceived to burn rapidly after its discovery.

Piloted Flaming Ignition of Solid Fuels

To create flaming combustion of a solid fuel item, heat must be applied to the fuel to generate flammable vapors through either pyrolysis or vaporization. These vapors may then be ignited by a piloted ignition source such as an arc, spark, or flame once their ignition temperature is reached.

The ignition temperature is an engineering approximation, but in general, the temperature of ignition of solids ranges from 518°F to 842°F (270°C to 466°C). Ignition temperatures for non–fire-retardant plastic range from 518°F to 680°F (270°C to 360°C), and those for wood-based products range from 626°F to 707°F (330°C to 375°C).

Additionally, unlike smoldering fires (which require only a minimum radiant flux of approximately 7 to 8 kW/m²), piloted ignition requires a critical radiant heat flux near 10 to 15 kW/m² to sustain flaming combustion. In experiments to determine the minimum heat flux, results have been shown to depend on the duration of the testing. Piloted ignition has been observed in some cases after approximately 1 hour of radiant heating.

Finally, the thickness of the material plays a role in the ignition of the fuel. Items that are thicker are generally more difficult to ignite than thinner ones. Further, thinner fuels such as papers and thin wood pieces or shavings allow for heating of more than one side, reducing the time required to ignite the fuel item. The greater the surface-to-mass ratio a fuel has, the less energy that is required for its ignition. For example, it takes far more energy to ignite a log than it does to ignite the same log after it has been reduced to chips or sawdust.

Autoignition

In the absence of a smoldering ignition (including self-heating) or a piloted ignition source, ignition of solid materials can only occur from autoignition of the flammable gases that are generated when the solid material is heated. Wood will autoignite at approximately 750°F (450 °C). If the autoignition occurs because the material is subjected to radiant heat transfer, the autoignition temperature will likely need to be higher than if the heat is coming from a convective source, because with the latter, the air surrounding the gases is already at a higher temperature.

Flame Spread

As a fire grows, flames begin to move across the surface of the fuel at varying rates. The rate of flame spread depends on not only the individual fuel properties but also the position and orientation of the fuel surfaces. Flame spread can either be **concurrent** (with the movement of the gases or wind direction) or **counterflow** (counter to or opposed to gas flow). Concurrent flame spread occurs faster than counterflow flame spread does. On vertical surfaces, concurrent flame spread is upward, and counterflow flame spread is downward. On horizontal surfaces, flame spread is normally counterflow.

On liquids, flame spread depends on the temperature of the liquid in relation to its flash point. If the liquid is below its flash point, flames spread across its surface through liquid flow. When the liquid is above the flash point, flame spread results from the ignition of gases.

For flames to move across the surface of a solid fuel, the material must be heated to produce fuel gases that ignite and facilitate the flame spread in a manner similar to that seen in the ignition process. Rates of flame spread depend on the mechanism of spread (concurrent versus counterflow) as well as the particular properties of the fuel (e.g., thickness, thermal inertia). With thin fuels like a matchstick or piece of paper, downward (counterflow) spreading flame will attach to the fuel from both sides, with the burning region being fairly short. Similarly, upward (concurrent) flame spread on these fuels will often attach on all sides.

Role of Melting and Dripping in Flame Spread

Flame spread may also occur as the result of melting or dripping materials from a fuel package. When a material melts, it flows with gravity, usually collecting on a horizontal surface below the object. This flow can allow the flames to spread away from the burning fuel, igniting other portions of the burning fuel, or to spread to other fuel packages not currently on fire. This may limit the ability of the flame front to spread across the surface of the fuel package. In addition, flame spread may be accelerated when certain fuels begin to melt and pool. For example, when urethane foams burn, such as from a cushion on a recliner, pools of burning liquid are created on the floor underneath the recliner and increase the rate of flame spread.

Role of External Heating on Flame Spread

The rate of flame spread for a given fuel package may also be accelerated when radiant heat produced by other burning objects, as well as radiant heat from the upper gas layers of a compartment fire, impacts the given fuel package.

Compartment Fire Spread

The way fire interacts with its surroundings in a compartment fire differs from the way fire acts in the open. Understanding what happens in a confined fire is critical to being able to understand its aftermath.

Whereas flame spread involves the movement of fire across the surface of a fuel item, fire spread refers to the ignition of other fuel items and packages that may be present or located nearby. Fire spread may result from either direct flame contact or remote ignition. Fire may spread to a nearby fuel package or to combustible wall and ceiling surfaces through direct flame contact. Remote ignition can occur through any of the three modes of heat transfer—conduction, convection, and radiation—discussed earlier.

During a compartment fire, superheated gases and smoke rise and are confined by the ceiling. As the temperature at ceiling level rises, the radiant heat created begins to accelerate the rate of fire spread as well as the rate of heat release of the burning fuel items. As the flames or fire plume begin to grow and impinge on the ceiling, the plume starts to spread across the ceiling in all directions as a ceiling jet. This jet continues to spread laterally until it is confined by vertical surfaces such as a wall or partition. A heated layer of gas and smoke is then created that begins to bank downward into the room. This heat radiates back on the burning fuel package as well as other "target" fuels located within the compartment **FIGURE 2-7**.

As the fire continues to burn, hot gases and energy continue to be added to the gas layer, which begins to descend until it either exits the room through an opening or reaches the base of the fire. The gas layer stops descending when an opening such as a window or door allows gases to flow out of the compartment at a rate equal to the hot gases being produced. Unless the fire is contained or extinguished, the hot gas layer continues to increase in temperature and to radiate heat back to other target fuels, ultimately igniting them. A distinct pattern indicating the flow of the hot gases and smoke is created by the gases exiting the room at the top of the doorway, with cooler air entraining to the base of the fire below.

Fuel-Controlled Fire

During the ignition and growth phases of a compartment fire, there is a sufficient amount of air available to allow the fire to continue to burn. When the size of the fire is controlled by how much fuel is burning, it is referred to as a **fuel-controlled fire**.

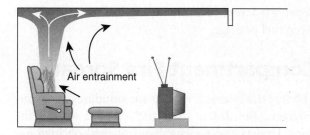
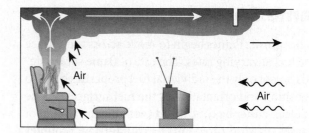
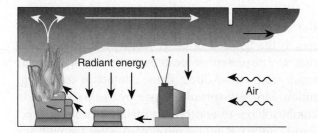
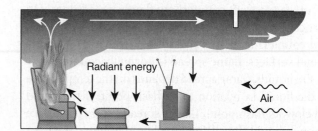
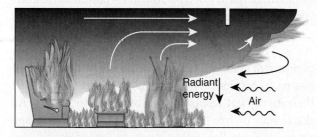
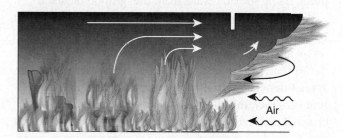

FIGURE 2-7 The progression of a typical compartment fire.

Flashover

Once floor-to-ceiling flames are produced, the fire begins to grow rapidly. The ceiling layer temperature continues to rise, reaching approximately 1100°F (600°C). At this point, convected and radiated heat energy impinges on the other combustible items within the room, producing fire gases from them. These items ignite at nearly the same time and, if oxygen is available, fire spreads rapidly throughout the space, causing full-room involvement. This event, known as a **flashover**, is a transition phase in which the fire progresses from being "a fire in a room" to "a room on fire." A heat flux of approximately 20 kW/m² on the floor level is generally considered sufficient energy to bring flashover. After flashover, the room is said to be fully involved. If the combustion rate is using up the oxygen in the space, as is commonly the case in compartment fires, the fire will progress from fuel-controlled burning to ventilation-controlled burning.

Not all compartment fires evolve to the flashover stage. For example, rooms that lack sufficient ventilation or very large compartments, such as warehouses, may not reach full involvement. The minimum size of a fuel package that can cause flashover in a given room is determined by the size of the ventilation provided. The time to flashover in residential room fires may be

as little as 3 to 5 minutes, or even shorter with non-accelerated fires. These times will be greatly affected by the HRR of the fuel, fuel package placement, and room geometry.

An approximation of the HRR required for flashover for a compartment with a single vent opening can be generated using the following equation:

$$Q_{fo} = 750\, A_o \sqrt{H_o}$$

where

Q_{fo} = heat release rate for flashover (kW)

A_o = area of opening (m²)

$\sqrt{H_o}$ = height of opening (m)

In a compartment with a single vent opening the size of a standard door passage (3 feet [0.9 m] wide by 7 feet [2 m] high), a minimum fuel package of about 2000 kW would be required to bring the fire to flashover.

Ventilation-Controlled Burning

If the rate of combustion begins to exceed the amount of air flow into the compartment, the fire transitions from fuel controlled to ventilation controlled. The hot gas layer begins to darken because of the unburned pyrolysis products as well as an increase in the amount of carbon monoxide. Unless the fire is extinguished or sufficient heat and smoke are removed from the room, the gas layer continues to extend to the floor, resulting in a flashover and full room involvement. Note, however, that the gas layer may not extend completely to the floor prior to full room involvement. Likewise, a flashover may not necessarily proceed to full room involvement.

One of the major factors determining whether flashover will occur is the presence of a hot gas layer with sufficient energy to radiate downward to involve exposed fuel packages. To counteract this risk, firefighters often ventilate the roof of a structure to allow the hot gas layer to escape before flashover occurs. **TABLE 2-4** lists several of the variables that determine whether and how fast flashover will occur.

For fire investigators, it is important to be able to recognize which rooms in a structure have progressed to full room involvement. Full room involvement can result in the production of low patterns, irregular patterns, and holes in the floor. Although these patterns might appear unusual in a compartment that did not become fully involved, they are common in a fully involved compartment. This topic is discussed further in Chapter 4, *Fire Effects and Patterns.*

Following flashover, a fire is likely to become significantly underventilated, resulting in a drop in oxygen levels. With the reduction in available oxygen, smoke production rates will increase noticeably. Much as with a diffusion flame in general, when oxygen concentrations drop in areas without ventilation sources, flaming combustion will be limited in those areas. By contrast, areas with an adequate oxygen supply will

TABLE 2-4 Factors That Influence Flashover	
Factor	**Effect**
Size (volume) of the compartment	The larger the compartment, the more time required to fill the upper layer with hot gases and to raise the temperature of the target fuels.
Height of the ceiling and distance between the fire and the ceiling	The greater the height, the longer it takes for the compartment to reach flashover conditions.
Ventilation (or lack thereof) in the compartment	The growth of the fire that can create flashover is closely tied to the ventilation available.
Amount of fuel in the compartment	The amount of fuel includes the initial fuel package, the target fuel packages, and the lining material in the compartment.
Type of fuel in the compartment	The type of fuel determines how much radiant heat will be required to raise the temperature of the target fuels to a point where sufficient combustible gases are being released to be ignited.
Layout of the room's contents	Target fuels located closer to the initial fire are heated sooner and release combustible gases sooner than those that are remote from the initial fire.
Location of the fire in the compartment	Fires that are located in the center of a compartment have more air entering the plume and cooling it by entrainment, shorter plume heights, and no restrictions to the geometry of the plume. If the same fire is located against a wall, there is less entrainment because of the wall barriers. Therefore, the following conditions may result: ■ Longer flames ■ A faster rise in upper-layer temperatures ■ Less time to flashover If the fire is located in a corner, the entrainment is reduced even more. This can result in even longer flames, a faster rise in upper-layer temperatures, and a reduced time to flashover than if the fire is located against a wall or in the center of the room. The location of the fire in relation to vertical barriers is important in interpreting the fire development and spread.

continue burning, which may result in significant fire patterns near ventilation points and in other places that are not near or related to the origin of the fire. Investigators must understand the role that ventilation plays in the creation of fire patterns, especially in compartments and areas that have proceeded to flashover and full room involvement. Patterns in these fires may be deceiving to investigators seeking to determine the fire's origin. This topic is discussed further in Chapter 13, *Determining the Origin*.

Vent Flows

The air flow required for a compartment fire may be provided by mechanical means such as the heating, ventilation, and air conditioning (HVAC) system or, more commonly, by natural openings. The hot gas layer in the room, being more buoyant, will flow out the upper regions of the vent opening, and air will flow in the lower region of the vent opening. The line where the flow of the two layers changes is called the **neutral plane FIGURE 2-8**. As the hot gases accumulate

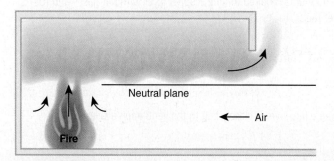

FIGURE 2-8 Neutral plane.

within the room, the height of the neutral plane moves downward. A fire in a compartment will have only one neutral plane even if multiple ventilation openings are present. In fires that have multiple vents, those vents may potentially serve only as a vent for the hot gases or only as an inflow, depending on the level of the vent relative to the fire's neutral plane.

TIP

Be cautious of statements made by witnesses as they relate to the rate of fire growth. Many times, witnesses describe the rate of fire growth from the time they discovered the fire, which may be difficult to compare with the actual time of ignition. Factors such as the compartment size, fuel configuration, ventilation, and fuel load present all influence the rate of growth. For this reason, rapid fire growth in itself is not a reliable indicator of an incendiary fire.

With a single ventilation opening, there is a proportional relationship between the inflow of air and the ventilation factor, which is the area of the opening (A_o) multiplied by the square root of the height of the opening (H_o). This ventilation factor is an integral component in evaluating the peak HRR of a fire in a compartment. For example, a standard door opening (3 feet [0.9 m] wide by 7 feet [2 m] high) can support an HRR of about 4000 kW, but if the lower half of the opening is blocked, the opening can support an HRR of only about 1600 kW.

As the fire progresses, the ventilation may change. Windows may fail as the glass cracks out, and additional vents may emerge as the fire destroys doors or creates new openings. As these new vents appear, the neutral plane height will change. Openings that were at first venting may transition to inflow.

After-Action REVIEW

IN SUMMARY

- Fire investigators must fully understand the fundamental properties of fire and fire development to determine accurately the origin and cause of any fire or explosion event they may investigate.
- For a fire to occur, four components must be present: fuel, an oxidizing agent, heat, and an uninhibited chemical chain reaction. Collectively, they are referred to as the fire tetrahedron.
- The most common fuels fire investigators will encounter are organic fuels, which contain carbon (e.g., wood, plastics, and petroleum products).
- Most fires require an oxidizing agent, such as oxygen from the air, to support the combustion process.
- The rate at which heat is released, known as the heat release rate (HRR), is measured in joules per second or watts.
- An uninhibited chemical chain reaction creates a self-sustaining event that continues to develop fuel vapors and sustain flames even after the removal of the ignition source.
- During a fire, materials may change their physical state as a result of being heated. These phase changes include melting and vaporization.

- Flames produced during the combustion process can be categorized as either premixed or diffused.
- If a fire has a limited amount of air available for combustion, the amount of visible products of combustion, such as soot, smoke, and carbon monoxide, will increase.
- The color of smoke should not be relied on as an indicator of the material burning.
- Fluid flows, heat transfer, ignition and flame spread, fuel packages, heat flux, and the distinction between fuel-controlled fires and ventilation-controlled fires are all components of fire dynamics.
- Fluids include both liquids and gases, but in fire investigation, the focus is generally on gas flows.
- Heat energy naturally moves from areas of higher temperature to lower temperature through heat transfer, measured as energy flow per unit of time.
- Thermometry is the study of the science, methodology, and practice of temperature measurement.
- The total fuel load of a given fire will be a reflection of the potential energy; it does not determine how fast the fire develops once ignition occurs.
- The specific fuel items within a fuel package influence the speed and intensity of a particular fire, as well as the best method of fire extinguishment.
- The power of a fire is determined by calculating its HRR, the energy being released by the individual fuels being consumed. It is measured in either watts or kilowatts.
- Generally, larger flame heights correspond to higher rates of heat release, but this holds true only when the fuel packages being compared are in the same relative location in a compartment.
- An ignition source can be defined as either smoldering or flaming or as either piloted or autoignition.
- Flammable gases may ignite without a piloted ignition source if the gas–air mixture is heated to its autoignition temperature.
- Rates of flame spread depend on not only the individual fuel properties but also the position and orientation of the fuel surfaces.
- When a material melts, it flows with gravity, usually collecting on a horizontal surface below the object and allowing the flames to spread away from the burning fuel to other fuel packages.
- The rate of flame spread may be accelerated through radiant heat produced by other burning objects and from the upper gas layers of a compartment fire.
- Whereas flame spread involves the movement of fire across the surface of a fuel item, fire spread refers to the ignition of other fuel items and packages that may be present or located nearby.
- In a fuel-controlled fire, the size of the fire is controlled by how much fuel is burning.
- After flashover, the room is said to be fully involved and often progresses from fuel-controlled burning to ventilation-controlled burning.
- One of the major factors determining whether flashover will occur is the presence of a hot gas layer with sufficient energy to radiate downward to involve exposed fuel packages.

KEY TERMS

Air entrainment The process of air or gases being drawn into a fire, plume, or jet. (NFPA 921)

Atoms The smallest unit of an element that can take part in a chemical reaction.

Autoignition temperature (AIT) The lowest temperature at which a combustible material ignites in air without a spark or flame. (NFPA 921)

Ceiling jet A relatively thin layer of flowing hot gases that develops under a horizontal surface (e.g., ceiling) as a result of plume impingement and the flowing gas being forced to move horizontally. (NFPA 921)

Combustion A chemical process of oxidation that occurs at a rate fast enough to produce heat and usually light, in the form of either a glow or flames. (NFPA 921)

Compounds Combinations of elements.

Concurrent Flame spread that is occurring in the same direction as the gas flow from the fire, or in the wind direction.

Conduction Heat transfer to another body or within a body by direct contact. (NFPA 921)

Convection Heat transfer by circulation within a medium such as a gas or a liquid. (NFPA 921)

Counterflow Flame spread that is counter to, or opposed to, the gas flow from the fire.

Diffusion flame A flame in which the fuel and air mix or diffuse together at the region of combustion. (NFPA 921)

Endothermic reaction A reaction or process that absorbs or uses energy.

Energy A property of matter manifested as an ability to perform work, either by moving an object against a force or by transferring heat. (NFPA 921)

Exothermic reaction A reaction or process that releases energy in the form of heat.

Fire A rapid oxidation process; an exothermic chemical reaction resulting in the evolution of light and heat in varying intensities. (NFPA 921)

Fire dynamics The detailed study of how chemistry, fire science, and the engineering disciplines of fluid mechanics and heat transfer interact to influence fire behavior. (NFPA 921)

Fire point The lowest temperature at which a volatile combustible substance continues to burn in air after its vapors have been ignited (as when heating is continued after the flash point has been reached).

Flammable range The range of concentration of a gas or vapor to air between the upper explosive limit (UEL) and the lower explosive limit (LEL), at which combustion can occur in air.

Flash point The lowest temperature of a liquid, as determined by specific laboratory tests, at which it gives off vapors at a sufficient rate to support a momentary flame across its surface. (NFPA 921)

Flashover A transition phase in the development of a compartment fire in which surfaces exposed to thermal radiation reach ignition temperature more or less simultaneously, and given sufficient availability of oxygen, fire spreads rapidly throughout the space, resulting in full room involvement or total involvement of the compartment or enclosed space. (NFPA 921)

Fuel A material that will maintain combustion under specified environmental conditions. (NFPA 921)

Fuel-controlled fire A fire in which the heat release rate and growth rate are controlled by the characteristics of the fuel, such as quantity and geometry, and in which adequate air for combustion is available. (NFPA 921)

Fuel items Any articles that are capable of burning.

Fuel load The total quantity of combustible contents of a building, space, or fire area, including interior finish and trim, expressed in heat units or the equivalent weight in wood. (NFPA 921)

Fuel package A collection or array of fuel items in close proximity with one another such that flames can spread throughout the array.

Heat A form of energy characterized by vibration of molecules that is capable of initiating and supporting chemical changes and changes of state. (NFPA 921)

Heat capacity The amount of heat necessary to raise the temperature of a unit mass by one degree, under specified conditions; it is measured in units such as J/kg-K or Btu/lb-°F.

Heat flux The measure of the rate of heat transfer to a surface or an area, typically expressed in kW/m^2 or W/cm^2. (NFPA 921)

Heat release rate (HRR) The rate at which heat energy is generated by burning. (NFPA 921)

Heat transfer The exchange of thermal energy between materials through conduction, convection, and/or radiation. (NFPA 921)

Lower explosive limit (LEL) The minimum concentration of a gas or vapor to air at which the gas or vapor will burn in air.

Neutral plane The line where the flow of the hot gas and cooler air changes.

Oxidation A chemical reaction in which an element combines with oxygen, resulting in loss of electrons.

Oxidative reaction A reaction in which the atoms in a material are broken down and rearranged, producing heat energy.

Oxidizing agent A substance that promotes oxidation during the combustion process.

Phase change The conversion of a material from one state of matter to another; it is reversible and does not change the chemical composition of the material.

Plume The column of hot gases, flames, and smoke rising above a fire; also called *convection column, thermal updraft*, or *thermal column*. (NFPA 921)

Power A property of a process, such as fire, which describes the amount of energy that is emitted, transferred, or received per unit of time. (NFPA 921)

Premixed burning Burning in which the fuel and oxidizer are mixed prior to combustion, as in a laboratory Bunsen burner or a gas cooking range; propagation of the flame is governed by the interaction between flow rate, transport processes, and chemical reaction. (NFPA 921)

Pyrolysis Process in which material is decomposed, or broken down, into simpler molecular compounds by the effects of heat alone; pyrolysis often precedes combustion. (NFPA 921)

Radiation Heat transfer by way of electromagnetic energy.

Self-heating A result of exothermic reactions that occur spontaneously in some materials, whereby heat is generated at a sufficient rate to raise the temperature of the material. (NFPA 921)

Stoichiometric ratio The optimal ratio in a fuel–air mixture at which point combustion will be most efficient (above the lower explosive limit and below the upper explosive limit).

Surface-to-mass ratio Ratio of the surface area of a solid or gas to its mass. The higher the surface-to-mass ratio, the more surface area of the material that is exposed to air.

Temperature The degree of sensible heat of a body as measured by a thermometer or similar instrument. (NFPA 921)

Thermal conductivity (k) The measure of the amount of heat that will flow across a unit area with a temperature gradient of one degree per unit of length; it is measured in units such as W/m-K and Btu/hr-ft-°F.

Thermal decomposition An irreversible change in chemical composition as a result of pyrolysis.

Thermal inertia The properties of a material that characterize its rate of surface temperature rise when exposed to heat; related to the product of the material's thermal conductivity (κ), density (ρ), and heat capacity (c). (NFPA 921)

Thermal runaway A condition in which the heat generated exceeds the amount of heat lost from a material.

Thermometry The study of the science, methodology, and practice of temperature measurement. (NFPA 921)

Uninhibited chemical chain reaction One of the elements of the fire tetrahedron; it provides for the combination and interaction of the other elements.

Upper explosive limit (UEL) The maximum concentration of a gas or vapor to air at which the gas or vapor will burn in air.

Vaporization A phase transition from a liquid or a solid phase to a gas phase.

REFERENCE

National Fire Protection Association. 2020. *NFPA 2021: Guide for Fire and Explosion Investigations*, 2021 ed. Quincy, MA: National Fire Protection Association.

On Scene

1. If you were asked to fully describe in fire chemistry terms the ignition process of a match igniting newspaper, how would you describe it?

2. If an explosive gas escapes from an appliance into the basement of a residence, what are some reasons why ignition might or might not occur, assuming the presence of at least one ignition source, such as a pilot light, in the basement?

3. What are some of the attributes of a compartment that would increase the likelihood of the compartment to reach flashover? That would decrease it?

4. Explain the difference between a fuel-controlled fire and a ventilation-controlled fire. What are some conditions in which the same fire could transition from one of these states to the other?

CHAPTER 3

Building Construction and Systems for Fire Investigators

KNOWLEDGE OBJECTIVES

After studying this chapter, you should be able to:

- Identify the design, construction, and structural elements of buildings and describe their effect on fire development, spread, and control. (**NFPA 1033: 4.2.2**; **4.2.3**; **4.2.5**; **4.2.8**, pp. 43–46)
- Identify and describe types of building construction. (**NFPA 1033: 4.2.2**; **4.2.3**, pp. 46–52)
- Assess the structural integrity of construction assemblies during a fire. (**NFPA 1033: 4.2.3**, pp. 52–54)
- Identify the components of heating, ventilation, and air conditioning (HVAC) systems and describe their potential involvement in fire. (**NFPA 1033: 4.2.3**; **4.2.8**, pp. 54–56)
- Explain the impact of passive fire protection systems on fire investigation. (**NFPA 1033: 4.2.3**, pp. 56–58)
- Describe the considerations when documenting a passive fire protection system. (**NFPA 1033: 4.2.8**, p. 58)
- List the types of active fire protection systems.
- Identify and describe the components of common fire alarm and detection systems. (**NFPA 1033: 4.2.8**, pp. 58–66)
- Identify and describe the operational characteristics of common fire alarm and detection systems. (**NFPA 1033: 4.2.8**, pp. 58–66)
- Identify and describe the operational characteristics of water-based fire suppression systems. (**NFPA 1033: 4.2.8**, pp. 69–79)
- Identify and describe the components of fire suppression systems. (**NFPA 1033: 4.2.8**, pp. 69–84)

- Identify and describe the operational characteristics of non–water-based fire suppression systems. (**NFPA 1033: 4.2.8**, pp. 79–84)
- Describe how to document fire protection systems. (**NFPA 1033: 4.2.8**, p. 84)

SKILLS OBJECTIVES

There are no skills in this chapter.

ADDITIONAL NFPA REFERENCES

- **NFPA 13**, *Standard for the Installation of Sprinkler Systems*
- **NFPA 80**, *Standard for Fire Doors and Other Opening Protectives*
- **NFPA 220**, *Standard on Types of Building Construction*
- **NFPA 921**, *Guide for Fire and Explosion Investigations*
- **NFPA 101**, *Life Safety Code*

You are the Fire Investigator

You have been assigned to investigate a fire in a commercial building. The responding firefighters told you that when they arrived, light smoke was visible from the eaves of the building. They did not find an active fire inside the building. The firefighters opened the attic access and found heavy smoke throughout the attic and flames in one end of the attic. Shortly thereafter, the fire vented through the roof. The firefighters reported that a large portion of the roof collapsed into the building.

During your investigation, you learn the building was constructed in the late 1950s with a flat roof. Approximately 20 years later, a pitched roof was added during a building remodel.

1. Does the addition of the pitched roof affect your investigation?

2. What is the type of building construction?

3. Who needs to be interviewed, and what information are you hoping to obtain?

4. Is it possible to determine the area of origin of the fire?

Access Navigate for more practice activities.

Introduction

The design, materials, and construction type of a structure influence the development and movement of any fire that may occur within it. This influence can be either positive, containing the fire within a compartment, or negative, allowing the fire to spread from the area of origin.

To analyze fire growth and movement properly, you must have an understanding of building construction, building systems, and fire protection systems, as well as the definitive fire patterns that remain after a fire.

Building Systems Overview

Some building techniques are a direct result of the analysis of past fires—often catastrophic fires—including some conflagrations that occurred at the turn of the 20th century. Today, more than half of modern code requirements are related to fire protection.

One primary cause of severe fire damage to structures is failure to contain or confine the fire. Basic principles of construction provide that fire should be contained to the room of origin, area of origin, or structure of origin. When a fire does move from the room of origin, it is most often through unprotected or improperly protected openings or construction defects rather than as a result of system failures. Open doors and penetrations of fire resistance–rated assemblies, such as walls and ceilings for building or utility services, provide ample corridors for fire movement.

Fires originating in interstitial spaces are not afforded the same level of protection as a fire that develops within a compartment (see Section 7.2.2.5 of NFPA 921, *Guide for Fire and Explosion Investigations*). As noted earlier, interstitial spaces are the spaces "between the building frame and interior walls and the exterior façade [as well as] spaces between ceilings and the bottom face to the floor or deck above" (e.g., a cockloft). Fires that originate in these concealed locations may develop undetected and move freely and rapidly without any barriers to stop their spread.

Fire investigators need to understand active and passive building systems, including manual and automatic fire detection and fire suppression; heating, ventilating, and air conditioning (HVAC) systems; utilities; and building compartmentation, including the use and location of fire resistance–rated assemblies. Knowing where these systems are installed and how they operate can help the investigator determine a system's overall performance and its effect on the fire and will allow for identification of any alterations and/or failure indicators. It is important to remember that the system failure may not have occurred through system faults or tampering, but rather the system simply may have been overcome by the fire development.

An HVAC system operating at the time of a fire can facilitate the spread of smoke and fire throughout a structure. As a fire investigator, you will need to examine such systems to determine their status at the time of the fire. Consider using outside resources to help examine systems with which you are not familiar. Mechanical and electrical inspectors or engineers are valuable assets and often available to consult with during an investigation.

Design, Construction, and Structural Elements

A building's characteristics affect the development, spread, and control of a fire. These characteristics include the type of construction; the integrity and performance of its structural elements under a fire load; and its building systems, including active and passive fire protection systems. Everything from the construction type of the building, to the number of doors, windows, and other openings it contains, can affect how the structure responds to a fire or explosion and the level of damage the building sustains.

Once a window breaks or a door is opened, the introduction of additional oxygen can greatly increase the fire's size and speed. Exterior winds can rapidly move a fire down unprotected corridors. During a fire investigation, you must examine and consider the environment and mechanical conditions at the time of the fire that may have impacted the size, speed, and spread of the fire. Ventilation effects can have a tremendous impact on the fire's growth. Additional factors that influence the origin, development, and spread of a fire include the interior layout, interior finish materials, and building services and utilities. **FIGURE 3-1** shows the structural components in a typical single-family dwelling.

Building Design

In a building fire, fire spread and development are largely effects of radiant and/or convective heat transfer. In compartment fires, the following factors significantly affect fire spread:

- Room size
- Room shape
- Ceiling height
- Placement and area of doors and windows
- Interior finishes and furnishings
- HVAC systems
- Fuel packages and location

Limiting the fire spread to a specific area through compartmentation is a primary objective of fire safety design. Fire investigators must be prepared to encounter a variety of construction techniques and building materials that may have an impact on the fire spread.

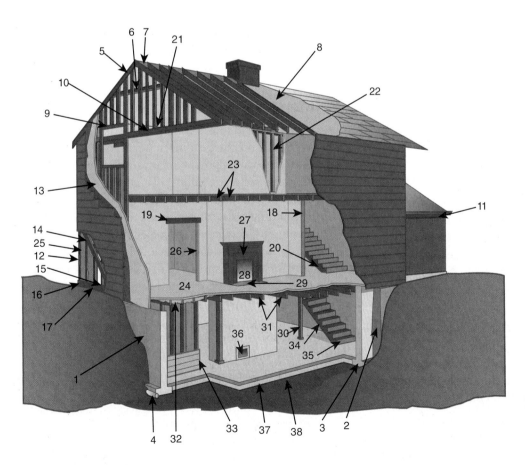

1. Foundation wall
2. Frost wall
3. Wall footing
4. Perimeter drain
5. Rafter
6. Collar beam
7. Ridge board
8. Roof sheathing
9. Window header
10. Attic joist
11. Box beam
12. Exterior wall stud
13. Wall sheathing
14. Corner bracing
15. Exterior wall plate
16. Box sill
17. Sill
18. Wall stringer
19. Header
20. Stair partition casing
21. Attic insulation
22. Partition studs
23. Second floor joists
24. Finish flooring
25. Wall insulation
26. Cripple stud
27. Damper control
28. Ash door
29. Hearth
30. Post or column
31. First floor joist
32. Subfloor
33. Basement partition
34. Stair stringer
35. Tread and riser
36. Cleanout door
37. Concrete floor slab
38. Granular fill

FIGURE 3-1 Typical single-family dwelling.

TIP

Become familiar with not only the most common building designs and construction, but also less-common styles of buildings or compartments, such as atriums, stadiums, and tunnels, that could have different impacts on fire spread. Special designs may also incorporate special materials.

Building Loads

Building loads, or forces acting on a structure, are classified as either live loads or dead loads. A **live load** comprises the weight of movable, temporary loads that need to be designed into the weight-carrying capacity of the structure, such as people, furniture, furnishings, equipment, machinery, wind, snow, and rainwater. In contrast, a **dead load** is constant and immobile. It consists principally of the weight of materials that are part of the building, such as the structural components, roof coverings, and mechanical equipment. Building designs and the construction techniques used take into account both of these loads.

Under certain conditions, a building may no longer be capable of supporting its loads. Loads applied above and beyond the structure's design parameters may create instability and ultimately failure. Examples of such conditions include the following:

- Extreme wind or snow loads
- Additional contents, such as stock
- Additional mechanical components, such as HVAC and elevator equipment
- Large congregation of people in a limited area
- Water from firefighting operations

Structural Changes

Structural changes can occur either deliberately—during building renovations, for example—or during a fire. As a fire grows and progresses, it may damage structural support elements so that they are no longer capable of supporting their designed loads. An exterior examination of the building should always be performed during an investigation. There are several reasons for doing such a survey:

- To determine whether there is any potential structural compromise that may have occurred during the fire and that could be a hazard if collapse were to take place
- To preserve any existing evidence
- To interpret the fire-damaged areas noted from the exterior

- To identify all potential means of entry and egress, which helps to ensure that there are always a minimum of two ways out of any area being processed
- To identify the utilities used in the structure

All of these items are also vital when performing the initial interior survey prior to commencing scene processing.

Room Size

The geometry of a room—its height, width, and especially the distances of the walls from each other—is an important factor in fire development there. Given a similar fuel package and similar ventilation points, a smaller room will reach flashover more quickly than a larger room does. Flashover is an important mechanism for fire spread beyond the room of origin. In an extremely large room, a fire may never have sufficient heat energy transfer to cause flashover.

Compartmentation

The **compartmentation** of a structure refers to its subdivision into separate sections or units. A common mechanism of fire spread outside the compartment of origin is horizontal movement through openings such as doors, windows, and unprotected openings and penetrations in walls. Vertical openings such as stairways, utility chases, and shafts also contribute to fire spread outside the compartment of origin.

Interstitial and Concealed Spaces

Concealed and **interstitial spaces** exist in most buildings and can provide a mechanism for fire spread beyond the compartment of origin. Interstitial spaces generally lack fire stops. They can contribute to spread of fire in buildings through openings between the building frame and interior walls and the exterior façade, and in spaces between ceilings and the underside of the floor or deck above. A lack of fire stops may allow for vertical spread of fire. If fire stops were provided, inspect them to determine whether they maintained their integrity to stop or limit the fire spread.

Concealed spaces are another major concern. Failure to account for concealed spaces can lead to erroneous interpretation of fire patterns. In some buildings, concealed spaces are equipped with fire sprinklers, early detection devices, or fire stops. Building codes sometimes allow concealed spaces if they are constructed of noncombustible material. Concealed spaces can contribute to fire spread when they are used for the general storage of combustible items—a use that is often prohibited by code.

Planned Design Versus "As-Built" Conditions

Where possible, obtain a copy of the structure's floor plan from the owner, fire department, or building department with jurisdiction over the area. The building department's file will show whether the proper permits were obtained for any alterations made to the interior or exterior portions of the structure. In many cases, however, walls, wiring, and mechanical systems are changed without the required permits.

A building's original design may not—and frequently does not—reflect what was finally constructed. Consequently, the building plans available to you may or may not reflect the true "as-built" structure. It is important to determine the accuracy of any plans in relation to the actual structure because conditions may exist that vary from the original building plans. These changes may potentially be identified through examination of the fire scene, examination of similar houses built by the same contractor in the same project, or witness interviews and occupant photographs.

Building Materials

Building materials can greatly influence a fire's ignition and growth. The types of materials and their chemical composition, thermal conductivity, and density all have a definitive impact on the fire's growth and speed. Items commonly used in today's construction, such as plastics and synthetic materials, can greatly elevate heat release rates. Thermoplastics can be transformed from solids to liquids to ignitable gases and can deform into flaming drips and pools when they burn, causing drop-down damage and patterns. Thermoset plastics pyrolyze directly to ignitable gases and do not tend to drip and flow. It is important to determine the type of plastic material in the area of origin so as not to confuse a drop-down pool of thermoplastic material with a suspected ignitible liquid. **TABLE 3-1** identifies some of the factors to consider in relation to building materials.

Orientation, position, and placement of materials make a difference in how the materials react under fire conditions. For instance, carpet is generally placed horizontally on the floor; when carpet is placed vertically on the wall, the flame spread related to the carpet is greatly increased.

Orientation, Position, and Placement of Loads

The fuel load is "[t]he total quantity of combustible contents of a building, space, or fire area, including interior finishes and trims, expressed in heat units

TABLE 3-1 Building Materials

Characteristics	Influencing Factors
Ignitability	■ Minimum ignition temperature ■ Minimum ignition energy ■ Time/temperature relationship for ignition
Flammability	■ Heat of combustion ■ Average and peak heat release rate ■ Time to peak heat release rate ■ Mass loss rate ■ Air entrainment
Thermal inertia	■ Reaction to heating ■ Ease of ignition ■ Fuel load
Thermal conductivity	■ Good conduction versus poor conduction
Toxicity	■ Quantity and types of gases produced by a material while burning
Physical state and heat resistance	■ Temperature at which a material changes phase ■ Amount of heat required to ignite the material in its different phases
Orientation, position, and placement	■ Different burning characteristics exhibited by the materials, depending on whether they are vertical or horizontal ■ Flame spread ratings, which can be obtained through a Steiner Tunnel Test (see NFPA 921, Section 7.2.3.7.2)

or the equivalent of weight in wood," according to NFPA 921. One thing that fire investigators may overlook is the orientation, position, or placement of fuel loads within a building or area. Where an item is located, or how it is installed, can have an impact on how materials burn and how fast a fire progresses. For example, a trash can that is located 3 feet (0.9 m) from a desk that catches fire will have a different potential for fire spread as compared with a trash can located immediately adjacent to the edge of the desk.

Interior finishes and the composition of furnishings also play major roles in the speed and intensity of a fire. A piece of 2-inch (5-cm) by 4-inch (10-cm) oak wood will burn much slower and create less char compared to a piece of wood with less density, such as a 2-inch (5-cm) by 4-inch (1-cm) piece of balsa wood.

The Steiner Tunnel Test. The Steiner Tunnel Test (ASTM E84) is used to establish the flame spread indexes commonly used in fire and building codes. Many of the materials tested in the tunnel are not designed for or intended to be applied in building designs as wall or ceiling covers. If not installed as designed (e.g., if their orientation is different from planned), the actual flame spread of the material may be different and bear no real relationship to its flame spread index classification. It would not, therefore, be an acceptable interior finish material per the fire and building codes. The interior finish requirements in fire and building codes are determined according to occupancy type and size. When an assembly is not rated, no information will be available as to the reaction of the assembly under the standard fire tests.

Devastating consequences are possible when interior finishes do not meet the code requirements. In many instances, multiple fire deaths have resulted from violations of these codes. An awareness of this possibility can help explain the speed of a fire or the reason for the loss of life.

Occupancy

The fire investigator should determine whether the use or occupancy classification changed at any point during the life of the building. The occupancy is the intended use of a structure, for example, commercial or residential. A change in use or occupancy classification can result in the introduction of fuel loads for which the building's fire protection systems were not designed. A change of occupancy is often accompanied by changes to structure and systems, putting the "as-built" structure at even more variance from the original plans. In these cases, egress, active and passive fire protection, and other safeguards may not be adequate.

For example, consider a situation in which a large home-improvement retail occupancy has extra rack storage sprinkler protection for the paint section of the building. The owners decide to move the paint section to another portion of the store but do not add the additional sprinkler protection. If a fire were then to occur in this area, the standard sprinkler protection may be overwhelmed by the fuel load presented by the aerosols, paints, thinners, and strippers.

(See Chapter 15, *Analyzing the Incident,* for additional discussion of failure analysis and how it can affect fire spread in a building.)

> **TIP**
>
> It is important to determine whether a building's occupancy classification has changed since it was built or renovated. Different occupancies require different fire protection systems depending on their load, contents, and storage configuration.

Types of Building Construction

Many types of construction are employed throughout the world. This text addresses the most common types in the United States. A fire investigator should determine and document the type of construction based on the structural elements of the building. Note any structural elements, breaches, structural changes, or other factors that may influence the integrity or fire spread of the structure. The following discussion of construction types includes a list of features typical of each.

> **TIP**
>
> A wood wall stud is commonly called a 2 by 4, reflecting the nominal lumber measurement. The actual dimension of the 2 by 4 [2 × 4] is 1¾ (44.5 mm) by 3½ inches (88.9 mm), which is the dimensional lumber measurement.

Ordinary Construction

In **ordinary construction**, according to NFPA 921, "exterior walls [consist] of masonry or other noncombustible materials" **FIGURE 3-2**. This type of construction is referred to as "ordinary" because it is used in a wide variety of buildings. The floor, roof, and partition framing are wood assemblies and use braced or platform framing methods. NFPA 220, *Standard on Types of Building Construction*, classifies ordinary construction as Type III construction. Open vertical shafts, combustible materials, and multiple ceilings are a few examples of factors that affect fire spread in this type of construction.

Wood-Frame Construction

In **wood-frame construction**, exterior walls and load-bearing components are constructed from wood

FIGURE 3-2 Ordinary construction is used in a wide variety of buildings.
Courtesy of Ken Hammon/USDA.

TABLE 3-2 Ordinary Construction Versus Frame Construction

	Ordinary Construction	Frame Construction
Exterior walls	▪ Masonry or noncombustible material	▪ Wood
Interior walls	▪ Wood assemblies ▪ Platform or braced frame assembly	▪ Wood assemblies ▪ Platform or braced frame assembly

or other combustible materials. Wood-frame construction is generally associated with residential or commercial construction. Buildings utilizing wood frames are of limited size, with their floor joists and vertical supports generally spaced 16 inches (0.4 m) on center. Vertical supports can be 2-inch (5-cm) by 4-inch (10-cm), or 2-inch (5 cm) by 6-inch (15-cm) nominal wood members. These components alone offer little fire resistance; flames and hot gas can readily penetrate the spaces between joists or studs, allowing the fire to spread. They may be sheathed with a fire-resistive material such as gypsum board, lath and plaster, or mineral tiles. Even non–fire-rated sheathing can provide some fire resistance. However, a fire-rated wall may not necessarily stop heat from being conducted through the wall to underlying members or to combustible materials on the other side of the wall. If a building has a masonry veneer exterior over wood framing, it is still considered wood-frame construction.

A downside to wood-frame construction is the gaps that can be present between the joists or studs. As just noted, these spaces can allow hot gases to penetrate and fire to spread outside the area of origin. To compensate for this risk, wood-frame construction can be sheathed with a fire-resistive membrane, which can provide up to a 2-hour fire resistance when applied according to ASTM testing standards.

Wood-frame construction is classified as Type V construction, as defined in NFPA 220. The characteristics of ordinary construction versus frame construction are shown in **TABLE 3-2**.

Platform-Frame Construction

Platform-frame construction is used for a majority of modern wood-frame construction **FIGURE 3-3**. In this method, walls are placed on top of platforms or floors.

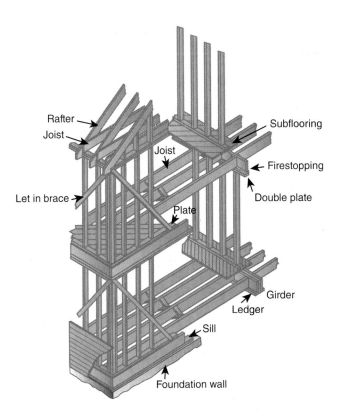

FIGURE 3-3 Platform-frame construction.

Thus, the platforms act as a fire stop for floor-to-floor vertical fire spread. Nevertheless, barriers that are combustible can be overcome in time. Areas of concern with this type of construction include concealed spaces in the **soffits** (the horizontal undersides of the eaves or cornice) and vertical openings for utilities.

TIP

Checking the floor or roof to see whether it is spongy does not ensure safety with modern lightweight construction.

Balloon-Frame Construction

In **balloon-frame construction**, the exterior wall studs extend from the foundation wall to the roofline **FIGURE 3-4**. Because of this feature, building codes have for many years required the installation of fire stops in all vertical channels created by the studs. Fire stops can take the form of wood boards or filling of the void space with noncombustible materials, such as insulation. If fire stops are properly installed, the fire performance of the building can be similar to that of platform-frame construction; however, a lack of fire stops can allow uninhibited vertical fire spread or fire ignition by fall down from the attic area. Some forms of fire stops may be removed during the installation of utilities such as building wiring, HVAC, or other services.

Balloon-frame construction can cause rapid fire extension through unprotected vertical channels, and horizontal extension is probable due to the open connections of the floor joists to the vertical channels. Open channels also allow for fall down of burning debris and for combustion gases to convect downward.

More extensive burning may occur at the upper level than where the fire originated. An investigator must identify all avenues of fire travel in this type of construction. Fire can break out in an area remote from the point of origin or bypass one floor and reach another because of the ability to spread through vertical channels.

Plank-and-Beam Construction

In **plank-and-beam construction**, larger beams—for example, 4 inches (10 cm) by 10 inches (25 cm), or 5 inches (12.5 cm) by 12 inches (30 cm)—are used **FIGURE 3-5**. These beams are more widely spaced at 4 or 6 feet (1.2 or 1.8 m) on center, and are supported by posts. The floor decking has a minimal thickness of 2 inches (5 cm) and is generally of the tongue-and-groove type, which helps limit fire spread. Plank-and-beam construction type has a limited number of concealed spaces, and the exterior finish has no structural value. It allows for large spans of unsupported finish material, which may result in failure of structural sections, with large frame members left standing. Large areas on the interior often have exposed combustibles that may allow for flame spread.

Post-and-Frame Construction

Post-and-frame construction is similar to plank-and-beam construction. This type of structure uses larger elements with an identifiable frame or skeleton

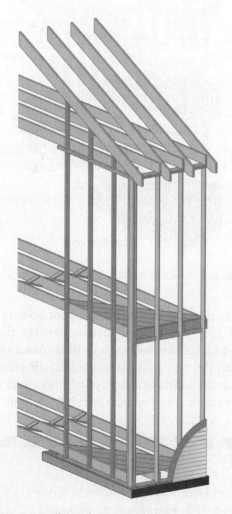

FIGURE 3-4 Balloon-frame construction.

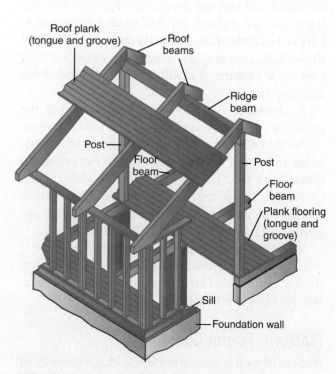

FIGURE 3-5 Plank-and-beam construction has a limited number of concealed spaces.

FIGURE 3-6 Post-and-frame construction.
Courtesy of Glenn Corbett.

of timber that is fitted together **FIGURE 3-6**. An example is a typical barn construction, where the posts provide a majority of the support, and the frame provides a place for the exterior finish to be applied.

> **TIP**
>
> Fires that have burned in balloon-frame walls destroy the structural integrity of the building. Collapse is a serious threat.

Heavy Timber Construction

Heavy timber construction employs structural members made of unprotected wood, with the smallest dimension being 6 or 8 inches (15 or 20 cm) **FIGURE 3-7**. Floor assemblies are constructed of 2-inch (5-cm) thick, tongue-and-groove end-matched lumber, and no concealed spaces are permitted. When heavy timber construction is used, building codes require the bearing walls to have a 2-hour rating. Fire spread may be rapid due to the wood-frame components, open spaces, and large areas of interior combustibles.

Contemporary log homes use specially milled logs for their exterior walls and for many of their structural elements. The interior construction is generally 2 by 4 wood frame, but the interior finishes are typically made to look like the milled log exterior walls. This inconsistency between construction and materials

FIGURE 3-7 Heavy timber construction has exterior walls that consist of masonry construction and interior walls, columns, beams, floor assemblies, and roof structures that are made of wood.
© solkanar/Shutterstock.

may cause confusion in cases of rapid fire spread and failure. Another hazard with contemporary log homes is that they often contain large, open spaces, which facilitate fire spread.

Mill construction is an early form of heavy timber construction. Mill construction was influenced by and developed largely through the efforts of insurance companies, which recognized a need to reduce large fire losses occurring in factories. The exterior walls are of masonry construction, with the columns and beams constructed of heavy timber. Walls were generally assigned a fire rating of at least 2 hours, with interior walls having rated doors between compartments. The lack of concealed spaces in this type of construction enhances the fire-resistive nature of the building. Protection of vertical openings plays a critical role in limiting the passage of fire. Fire sprinkler systems are included in this type of construction, and scuppers are provided in the walls for drainage to reduce water damage.

Alternative Residential Construction

In addition to wood-frame, site-built construction, other forms and materials are being used for residential construction, such as manufactured housing and steel-framed construction.

Manufactured housing is produced through a construction technique whereby the structure is built in one or more sections. In traveling mode, the sections are 8 feet (2.4 m) or more in width and 40 feet (12.2 m) or more in length. These sections are transported to the building site and assembled there, where they can be 320 ft^2 (29.7 m^2) or larger. Another type of manufactured housing builds the structure on a steel frame equipped with wheels, which allows it to be

FIGURE 3-8 Manufactured home.
© Leifr/Shutterstock.

transported easily to the site. This type of manufactured housing is often referred to as a mobile home.

Manufactured homes are the least expensive structures to build. Some incorporate house-like features such as steeper pitch roofs, residential-size windows, and covered decks. Manufactured homes consist of four major components or subassemblies: chassis, floor system, wall system, and roof system **FIGURE 3-8**. The chassis generally consists of two longitudinal steel beams, braced by steel cross members, that receive all of the vertical loads from the wall, roof, and floor. When placed onsite, these loads are transferred to stability devices, which may consist of piers or footings, or to a foundation. Steel outriggers cantilevered from the outside of the main beams bring the width of the chassis to the approximate width of the overall structure.

Since the mid-1970s, U.S. Department of Housing and Urban Development (HUD) Standard mobile homes have offered more substantial construction than homes built before the implementation of this standard. Construction of HUD-approved structures includes wood studs with interior finishes of (most often) gypsum wallboard. Exterior sidings can be metal, vinyl, or wood. Roof systems in HUD-approved structures are stronger and more solidly built, and often feature gable or hip-style rooflines. Roof construction can be similar to that of a site-built home, consisting of an asphalt paper/shingle roof covering over oriented strand board (OSB) or plywood attached to roof rafters or wood trusses. The wall system is bonded into a complete unit between the roof and floor system by steel ties.

Older (pre-HUD) units are usually constructed of 2-inch (5-cm) by 2-inch (5-cm) wood exterior studs covered with asphalt paper as a vapor barrier. This paper is then covered with an aluminum siding skin that is attached directly to the studs. The interior walls are generally wood paneling with no insulation in the wall

voids. Roof construction is most often a minimal flat wood frame with a steel exterior skin.

The older type of construction allows for rapid spread of fire with greater intensities due to the minimalist construction and the lack of interior finishes that resist fire. By contrast, the newer manufactured homes use gypsum wallboard on the walls and ceilings and incorporate code-required smoke alarms. Comparatively speaking, fires in these structures are similar to those found in site-built homes, and fire spread is not as rapid as in pre-HUD homes. The use of early detection devices wired to the house power reduces the chance of a large fire and lowers the risk for death and injury.

A **modular home** is sometimes referred to as a system-built home or prefabricated home. Modular homes are similar to manufactured homes in that they are built in a controlled environment, utilizing technology similar to that found in newer site-built homes. These prefabricated houses consist of multiple sections that are manufactured in an offsite facility and then delivered to the homeowner's location of choice. There, the modules are assembled into a single residential building using either a crane or trucks. Modular homes do not have axles or a frame, but rather are built on a contractor-provided foundation. They can be multilevel or single floor in design and can be customized to suit the buyer's preferences. Because of the type of construction, fires in modular homes will have burn tendencies that are similar to those of site-built wood-frame homes.

Steel-framed residential construction is becoming more common. Site-built, panelized, and pre-engineered systems use steel framing methods. Steel-framed construction has characteristics similar to those of wood-frame construction but is noncombustible; however, steel framing can lose its structural capacity during exposure to extreme heat **FIGURE 3-9**.

Manufactured Wood and Laminated Beams

Laminated beams are structural elements that have characteristics similar to solid wood beams. They are composed of many wood planks that are glued or *laminated* together to form one solid beam and are generally for interior use only. Commonly referred to as glulam beams, they behave like heavy timber until failure occurs. Effects of weathering decrease their load-bearing ability. During an investigation, it is important to document the size of individual members as well as the overall size of the beam.

Wood I-beams have smaller dimensions than floor joists; consequently, they can burn through and fail

FIGURE 3-9 Steel will begin to lose its strength at temperatures of 1000°F (537°C).
© DedMityay/Shutterstock.

sooner than dimensional lumber. Openings in the web (the vertical center portion) for utilities may reduce the structural integrity of the web. I-beams must be protected by gypsum board to help delay their collapse under fire conditions.

Lightweight wood trusses are similar to other trusses in design. Individual members are fastened using nails, staples, glue, or metal gusset plates (gang nail plates) or wooden gusset plates. Truss failure can occur from gusset plates failing even before wood members are burned through. When one member of the truss fails, the other members take on additional loads and may become stressed, which may cause the entire truss to fail.

Rating of Wood-Frame Assemblies

A fire rating is accomplished by covering the wall with a noncombustible finish, commonly gypsum board—though a fire-rated assembly may still conduct heat through to the underlying combustible members. Field installations can vary from those used in testing and evaluating the fire integrity. Therefore, a fire-rated wall in the field may fail sooner than expected depending on the quality of installation.

Noncombustible Construction

Noncombustible construction is used primarily in commercial, industrial, and high-rise construction. The building materials used in this type of construction do not add to the fuel load. Examples of these materials include brick, stone, metal, and non-reinforced concrete.

Metal Construction

Metal is commonly found exposed in unfinished spaces and may fail in a period as short as 3 minutes

during flashover. Fire-rating tests are not necessarily indicative of how the member will perform in fire scenes, which are unpredictable and can generate conditions that differ from those used in laboratory tests. As a fire investigator, you should familiarize yourself with the conditions of the building and with its interior finishes and ventilation effects, and determine whether there may have been exposed metal that could have contributed to an earlier-than-expected failure.

Ductile materials such as steel will deform before they fail during fire conditions, and this deformation can be elastic or plastic. In the elastic range, a material deforms and then assumes its original shape with no loss of strength after the load is removed. If the material is subject to stress beyond its elastic limit, it enters the plastic range, in which it is permanently deformed but may continue to bear the load. In either event, elongation or deformation of structural materials can produce building collapse or damage.

TABLE 3-3 describes some of the features of metal construction.

Concrete and Masonry Construction

Concrete and masonry constructions have inherent fire resistance because of their mass, high density, and low thermal conductivity **FIGURE 3-10**. They are strong under compression but weak under tension. Structural failure often occurs at connection points. Failures in these types of constructions may result

TABLE 3-3 Steel Versus Masonry/Concrete	
Material	**Factors**
Steel	Can conduct electricity. Good conductor of heat. Loses its ability to carry a load well below the maximum temperatures encountered in a fire. Can distort, buckle, or collapse as a result of fire exposure. The amount of distortion depends on factors such as the heat of the fire, duration of exposure, physical configuration, and composition of the steel.
Masonry and concrete	Will generally absorb more heat than steel, due to its mass. Good thermal insulator. Does not heat up quickly, and does not transfer heat through itself as easily as steel. Masonry and concrete will carry a load much longer at a given temperature when compared with steel.

FIGURE 3-10 Concrete is noncombustible and provides thermal protection around steel reinforcing rods.
© Marek Pawluczuk/Shutterstock.

from heat transfer through the concrete or masonry or surface spacing, which exposes the reinforcement to fire temperatures. Failure in the steel connections between components can occur at temperatures well within the range found in structure fires.

Construction Assemblies

Assemblies consist of manufactured parts put together to make a completed product and may or may not be fire-resistance rated. A collection of components, such as structural elements, forms an assembly such as a wall, floor, or ceiling. Doors form a part of a larger unit, so they are considered a component. Most non-rated assemblies do provide some resistance to fire or smoke. Failure of one assembly, however, may lead to failure of another. The way that assemblies react in a fire often influences how a fire grows and spreads. Because assemblies are designed as a complete unit, the integrity of the overall unit and its ability to perform during a fire depend on the unit being manufactured, installed, and maintained as intended.

Variables can compromise the assembly and cause it not to function according to its design and testing. Examples include fire doors that are blocked open between one compartmentalized fire area to another, and a fire-rated wall that has unprotected holes to pass utilities through from one side to the other. Holes left in the walls, ceiling tiles missing or not clipped down, ductwork that passes through areas without proper fire and/or smoke dampers, and so forth can allow for passage of fire.

Fire-rated assemblies are rated for a specific fire test criterion and under specific test conditions. The actual fire conditions found in a structure may be more severe and may cause the assembly to fail in less time than indicated by the hourly rating assigned as a result of the fire test. Fire investigators should examine rated assemblies after a fire to identify any flaws in a component in the overall assembly.

Floor, Ceiling, and Roof Assemblies

Floor, ceiling, and roof assemblies are of particular concern to the firefighter as well as the fire investigator. These assemblies are among the first to fail when structural elements are exposed to fire conditions. Types of failures include collapse, deflection, distortion, heat transmission, and fire penetration.

Factors that can affect the failure of floor, ceiling, and roof assemblies include the type of structural element, protection from the elements, span, load, and beam spacing. Added loads, such as water injected into the structure during firefighting operations, can also contribute to failure. Penetrations in assemblies for utilities are common. Although such penetrations are required to be sealed in a fire-rated assembly, often they are not. Floor assemblies are tested for fire spread from below, not from above. Roof stability can be a critical factor during firefighting operations and for fire dynamics.

The potential for collapse should be a vital concern at the fire scene. Structural elements are not intended to maintain their strength when subjected to fire conditions. The protection afforded to structural elements, such as gypsum wallboard, is intended to prevent the immediate involvement of the structural elements; however, once this protection fails and the structural elements are affected, their ability to maintain their load-bearing capabilities is diminished.

Walls

Walls serve as barriers to fire and smoke spread. They can be made to a wide variety of standards and in a wide variety of types. They may or may not be fire-resistance rated or load bearing **FIGURE 3-11**.

Penetrations in these assemblies for utilities are common and are required to be sealed in a fire-rated assembly. A **fire wall** separates buildings or compartmentalizes interior areas of large buildings to prevent the spread of fire, while creating a fire-resistance rating and structural stability. A **fire barrier** also resists the passage of fire and smoke. Fire barrier walls using gypsum board will use a Type X gypsum wallboard. Fire walls and fire barriers normally do not need to meet the same requirements as smoke barriers but

During a fire investigation, it is necessary to examine the building's interior walls for several reasons. First, attempt to determine which walls are load-bearing, and if so, whether any of their structural elements were removed, as this could create undue stress on the building, leading to a potential collapse. Second, examine the walls to determine whether they were constructed to resist passage of fire and/or smoke. Fire walls and smoke barriers are constructed of specific materials that have been tested and proven to prevent the passage of smoke or fire. If the fire has spread through the wall, examine the wall for any penetrations that may have allowed for the passage of the fire. It is not uncommon for contractors to install lines or other components and penetrate the compartmented separations without sufficiently closing off the openings in conformance with fire/building codes.

Doors

Doors can serve as a critical factor to limiting fire spread throughout the structure. Doors may be made of a variety of materials and may be fire rated or non–fire rated.

Any opening in a fire-rated wall or partition is required to have a fire-rated door and associated door assembly installed. Fire-rated assemblies include rated frames, hinges, closures, latching devices, and, if provided (and allowed), glazing. Fire doors are built using a variety of methods and materials and have a fire protection rating. Fire doors may be constructed of solid wood, steel, or steel with an insulated core of wood or mineral material. The insulating value of fire doors aids egress, particularly in multistory buildings, and provides some protection against autoignition of combustibles near the opening's unexposed side. They must be closed to provide an effective barrier to smoke, heat, and fire. Any door installed in a fire resistance–rated wall assembly should be installed as a fire-protection-rated door assembly **FIGURE 3-12**.

The hourly rating of the door depends on the rating of the fire wall and is usually less than the wall system rating. A fire-rated door assembly must have

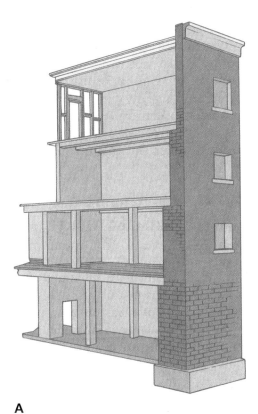

A

B

FIGURE 3-11 A. A load-bearing wall provides structural support. **B.** A non-load-bearing wall supports only its own weight.

may have to, should they also be constructed to act as a smoke barrier. A **smoke barrier** is a "continuous membrane, either vertical or horizontal"—such as a wall, floor, or ceiling assembly—"designed and constructed to restrict the movement of smoke," according to NFPA 921.

Non–fire-rated walls can provide varying levels of fire resistance. An assembly without a rating means that no test was done for that type of wall, ceiling, or floor component. As such, no time frame for when this component may fail has been established, although it may provide some resistance to the spread of fire within the building.

Assemblies with smoke damper systems are designed to restrict the passage of smoke. Such barriers may or may not have protected openings to block the passage of fire.

FIGURE 3-12 A fire door has a label indicating its classification and rating.
Courtesy of SecurallDoors.com.

the following properly operating components rated as part of that system:

- Hinges
- Closures
- Latching devices
- Glazing

See NFPA 80, *Standard for Fire Doors and Other Opening Protectives*, for more information on fire doors.

During a fire investigation, fire doors should be examined to see whether they had been propped open with doorstops or other objects prior to the fire. If the doors had approved magnetic devices to hold them open, the automatic closers should be inspected to determine whether they worked on activation of the fire alarm system and, if they did, whether they closed and latched securely. If the doors were propped open, determine whether this was a normal circumstance for this business or an unusual situation. It is also possible that firefighters may have propped these doors open during firefighting activities. Inspect fire doors to determine whether they had holes or penetrations or had been modified in a way that they no longer conformed to the intended design.

Concealed Spaces

Concealed spaces commonly have areas where penetrations are used to provide access for HVAC systems; plumbing; electrical, computer, and/or telephone lines; and other functions. These penetrations are required to be sealed to meet the rating of the wall or floor through which they pass. Many times, contractors or others create openings in these rated separations after construction without properly sealing them. Such openings may have allowed passage of fire or smoke from one protected area to another.

Construction Materials

Unprotected structural steel loses strength at high temperatures and should be protected from exposure to the heat produced by building fires. A variety of methods may be employed to provide the appropriate insulation needed to protect structural steel. One such method is encasement with materials such as poured concrete or board systems that provide a barrier. Examples of board systems include calcium silicate and gypsum. Surface treatments such as spray-on fireproofing or intumescent coatings may also be used.

A less-common form of protection is filling hollow members, which acts as a heat sink to reduce the temperature increase in the steel element. Reinforced concrete can also be used as a protective coating but can affect the fire resistance of the concrete, including the concrete's density, aggregates, and moisture content. Wood encasement may or may not ensure reasonable structural integrity during a fire. The size and moisture content of the wood are other factors to be considered, as is the presence of fire retardant in the wood to delay ignition and reduce combustion.

Heating, Ventilation, and Air Conditioning Overview

Residential and commercial buildings commonly contain HVAC systems that may provide heat, cooling, or both. Often, both heat and cold may be provided by the same system, depending on the settings and on the ambient air temperature. The fire investigator should become familiar with the types of HVAC systems found in these occupancies, as well as their components and how they perform, to better understand their potential role in a fire incident.

Heating Systems

By definition, all heating systems produce heat, though they may take a variety of forms. Thus, fire investigators often need to analyze whether these systems were involved as ignition sources in an incident, whether they contributed to the incident in some other way, or whether they are uninvolved in the ignition of a fire.

Components of Heating Systems
Fuels

A variety of fuel types are used in heating systems. Some use natural gas, which is typically piped into the building; others use propane gas, which normally is delivered into a tank that can be found onsite. A further discussion of gas systems appears later in this chapter. Other heating systems are designed to use fuel oil, wood, coal, or other unique types of fuels. Systems that are all electric, such as electric baseboard heaters, do not rely on fuels.

Devices

Heat-producing devices in heating systems include the following types:

- **Furnaces.** These devices may use electricity or gas, oil, or solid fuel. Most fuel-consuming (non-electric) furnaces have a vent or chimney.

Furnaces can be central air furnaces, or they can be mounted in the structure's floor or on the wall.

- **Boilers.** Normally constructed of cast iron or steel, boilers run on either hot water or steam, and are powered by propane, natural gas, fuel oil, or solid fuel. They contain safety devices that monitor pressure, low water, and temperature to prevent fire and explosion conditions.

- Radiant or convective heaters. Flat panels installed in walls, floors, or ceilings, and generate heat via electric elements or via water passing through pipes.

- Stoves. These devices use wood, wood pellets, or coal. They consist of a fire chamber surrounded by an enclosure of metal or soapstone, and have vents or chimneys.

- Fireplaces. Built from masonry, fireplaces typically burn solid fuel or gas, and normally do not have any control on the supply of incoming air. Factory-built fireplaces are usually of metal construction and use liners, refractories, and insulation to protect the wooden enclosures in which they are installed. Some may have a gas igniter and/or a mechanism to circulate air.

- Electric heating units. These devices can include baseboard heaters, permanently installed room heaters in a wall or floor, and central forced air.

Chimneys and Vents

Chimneys and vents exhaust heat-producing devices to the exterior of a building. Masonry chimneys and unlisted metal smokestacks are not tested prior to installation. Factory-built chimneys come as full assemblies and have been tested to safety standards. Vents are factory-built units with specific types and designations for use with specific furnaces or appliances.

Controls and Safety Devices

A variety of controls and safety devices are required in heating systems to enhance safety by preventing improper and unexpected operation of the device:

- Pressure switches are used for several purposes, including monitoring draft or air flow, fuel pressure on oil and gas systems, and water or steam pressure.

- The high-temperature limit, a heat sensing switch, is closed during normal device operation, but will open if a preset temperature is sensed.

- Door switches ensure that necessary doors to furnaces and other systems are in place to allow these systems to operate safely.

- Flame sensors ensure that a flame is present when fuel is being fed to a heating device, to avoid accumulation of unburned fuel gas.

- Controls often exist that start or stop a burner in response to changes in demand.

- Flame rollout detectors shut a gas-fired heater down when they sense flames rolling out of the combustion chamber—a condition that may happen if the device is improperly drafted or if soot has built up in the combustion chamber.

- **Control thermostats** control the operating temperature of devices by, for example, turning on a circulating fan when the heat exchanger reaches a certain temperature.

Installation, Use, and Maintenance

Numerous codes and standards address the installation of heating devices and their fuel supplies, although the investigator may need to refer to the manufacturer's instructions for specific installation requirements. Topics that must be considered in installation include appliance placement, venting and controls, fuel supply, and air for combustion and cooling.

Depending on the situation, the fire investigator may need to inquire as to the installation history and details for a particular heat-producing appliance. Further, issues may arise concerning the use and maintenance of these appliances, as they must be run and maintained according to the manufacturer's instructions to ensure safety.

Heating Systems as Potential Causes of Fire

Although heating systems can certainly serve as the ignition source for a fire event, the mere presence of a heating device within the area of origin does not automatically mean it is the fire's ignition source. As required by NFPA 921, a hypothesized fire cause involving a heating device must be tested along with any other applicable hypothesis. At times, investigators have determined that a given heating device did not produce sufficient heat to ignite nearby combustibles, or was not functioning, or ruled out a heating device for other reasons.

The fire investigator should be aware of the various ways in which a heating device might be a heat source, but should also remember that qualified specialists may be needed to examine a heating device to determine the specific failure mode that might have led to a fire. The investigator who finds a heating device at the origin of a fire must consider which type of scenario could have resulted in a fire. For example, the following scenarios might be examined:

- Whether the heating appliance was operating at normal temperatures but was located against or near a fuel within a distance that could have allowed sufficient heat to cause the device to ignite (and, if so, whether the appliance was capable of producing enough heat at normal temperatures to cause ignition)

- Whether a failure or fault caused the device to exceed normal temperature, causing heating or ignition of materials that were near the device

- Whether the device was installed or used in a manner outside of the uses intended by the manufacturer and applicable codes—for example, a chimney installed without proper clearances, or a stove or furnace supplied with improper fuel

Under normal circumstances, the fire investigator should not attempt to disassemble a heating appliance to determine whether a failure mode occurred within the appliance. See the spoliation discussion in Chapter 5, *Legal Considerations for Fire Investigators.* Information on the collection of heating devices for evidentiary purposes is provided in Chapter 9, *Identification, Collection, and Preservation of Physical Evidence.* The various potential fire causes from heating devices are discussed in Chapter 14, *Cause Determination.*

Air Conditioning and Air Handling

Components

Air-conditioning systems within buildings typically share components with heating systems, if they exist. A condenser coil, often located outside the building, removes heat from refrigerant, which is then pumped to the circulation system. If the system is combined with the furnace, an evaporator coil uses the cooled refrigerant to cool the air exiting from the furnace to the building.

In both furnaces and air conditioners, an air handling unit includes the blower, the evaporator coil if it exists, a filter or filters, and other components. The unit blows air into air supply ducts leading into the occupied areas of the building. Air is supplied to the furnace through air return vent(s) that draw air from the conditioned areas into a return plenum, or via a plenum space between a drop ceiling and upper floor or under a raised floor area.

In commercial buildings, air movement systems can be considerably more elaborate than the simple scheme just described. Air handlers may be located in an equipment room, on the roof, or on the ground. The duct work may contain smoke detectors linked to the building's alarm system and may have automatic dampers to prevent the spread of fire and smoke. Larger systems may be required by code to shut down upon the activation of the fire alarm. Some systems provide smoke exhaust fans, and some provide pressurization of areas of the building upon alarm activation to prevent products of combustion from entering areas of refuge and egress.

Investigation

In any building fire, the fire investigator must be aware of the ways in which HVAC systems can be involved in the spread of fire and smoke. Heat, smoke, and toxic combustion products can enter them for a variety of reasons and be circulated to other parts of the building, including plenum spaces, whether or not the blower is operating at the time of the fire. The ducts can also serve to supply fresh air—that is, an oxygen supply—to a fire, thereby supporting continuing combustion.

The presence and layout of HVAC systems and ductwork should be considered when analyzing fire and smoke spread scenarios. The investigator should consider whether the supply or return ductwork facilitated the oxygen supply to the fire, or facilitated fire or smoke spread impacting other compartments or occupants. Consider how each of these factors may have impacted the fire patterns found on scene. In addition, the fire investigator should determine whether system components such as dampers were maintained and functional.

Other issues to consider in larger buildings include whether occupant escape was prevented or slowed by air pressure acting on egress doors, and whether smoke handling systems properly protected occupants in their effort to evacuate.

Passive Fire Protection Systems

Part of your job as the fire investigator is to assess whether a building is provided with a **passive fire protection system**. Passive fire protection can best be defined as fire resistance–rated wall and floor assemblies used to create fire-rated compartmentalized areas meant to control the spread of fire. These assemblies can take the form of occupancy separations, fire partitions, or fire walls used to keep fires, high temperatures, and flue gases within the fire compartment of origin, and to aid firefighting and evacuation processes. Determine whether any of these provided systems failed and, if so, how and why. Pay attention to how damage to the systems may have aided in the development of the fire's growth. Problems associated with failure may include the following issues:

- Improper design
- Inadequate installation
- Change in occupancy and associated hazards
- Breaches in compartment walls or damage to applied coatings after the initial system installation

Design and Installation Parameters of the System

When a passive fire protection system is identified, evaluate whether each system and/or component had a role in the evolution of the fire, taking into account which codes, guides, standards, and manufacturers' instructions were in effect at the time of construction, or whether an update might have been required to bring the system up to the most current code. In addition, analyze the protection of openings and doors, including the fire rating of the door, door frame, door hardware, and construction surrounding the door. Likewise, evaluate windows in the protected openings, including the window glass (type, thickness, and number of panes), opening mechanisms/hardware, and frame and construction surrounding the windows.

Ductwork penetrations in a fire-rated wall/partition/floor and ceiling must be examined to determine any impact on the fire evolution. If smoke and/or fire dampers were required, did they operate as required? Examine all penetrations to determine whether they complied with applicable codes, standards, and manufacturers' instructions.

Examination at Fire Scene

During the origin and cause determination phase, the fire investigator's examination of passive fire protection systems should include whether each system was assembled properly, whether it impeded fire and smoke as expected, and whether it failed prematurely. This examination should include the components identified in **TABLE 3-4**.

TABLE 3-4 Considerations for Passive Fire Protection Systems	
Component	**Considerations**
Penetrations	■ Were they properly sealed?
Joint systems	■ Did the system fail by not resisting the passage of fire as expected by its fire rating? ■ Was joint material securely in or near the joint for the entire length of the component? ■ Did the building code contain exceptions for requiring fire-rated joint systems?
Fire doors	■ Were all the necessary components included in the assembly? ■ Did the closing device perform properly? ■ An examination of the door label should be recorded if available.
Fire windows	■ Were the glass and glazing the proper thickness? ■ Was wire glass required or not? ■ Was the framing appropriate for the window rating?
Duct and transfer openings	■ Were any fire dampers, smoke dampers, fire/smoke dampers, and/or ceiling radiation dampers required? If so, did they operate as required when activated by fire, smoke, or automatic activation? ■ Examine the actuating device/method to ensure that it operated as needed.
Fireblocking and draftstopping	■ Did the structure require fireblocking or draftstopping in combustible concealed spaces? ■ Was the fireblocking or draftstopping installed in the correct locations? ■ Was the material used appropriate for the rating, and was it the proper thickness?

Documentation and Data Collection

When a passive fire protection system is present at a fire scene, documentation of the system can be very important. To assist in documentation, as well as to increase understanding of the system and any role it might have played in the fire, the investigator might review design plans, design specifications, as-built drawings, equivalencies or alternative levels of protection approved by the authority having jurisdiction, building and fire permits, invoices of work to the system, measurements, diagrams and/or photographs, and maintenance, inspection, and testing records, as each of these helps document the systems found at a fire scene. The fire investigator should document any occupancy changes that may have occurred since the original installation to ensure that the passive fire protection system met the current occupancy and fuel load. Note that an installation can vary from permit plans in significant ways if the occupant made changes to the installation that are not reflected in the permit materials. Check with local code officials if modifications or installations found in the structure do not appear to conform to code or to the permitted work.

Code Analysis

As a fire investigator, it is necessary to be familiar with codes and to know where to find information on the installation of passive fire protection systems. Some of the codes an investigator may need to be familiar with are listed here:

- Building codes
- Fire codes
- Property maintenance codes
- NFPA 101, *Life Safety Code* (if adopted by the authority having jurisdiction)
- Localized code amendments

In addition, it is a good idea to consult with the local building official and fire prevention staff.

Permit applications and/or building permits can provide dates indicating which codes may have been in effect at the time of the installation of the passive fire protection system. Both the name of each code and the specific edition are important in determining whether the installation met the code at that time. Building codes will address the following points:

- Type of construction
- Fire resistance–rated construction

- Requirements for glass and glazing, gypsum board and plaster, and other materials
- References to ASTM and Underwriters Laboratories

Design Analysis

Items that may be found in the approved construction documents include:

- Drawings
- Specifications
- Design calculations
- Floor plans
- Details
- Cross-section elevations
- Schedules for the work

If computer modeling or calculations are used to assess the performance of the passive fire protection system, data on the thermal properties of the walls, ceilings, and floors will likely be needed for the evaluation.

Testing and Maintenance

The fire investigator should assess whether there were any requirements for periodic testing or examination of the passive fire protection system, including fire prevention inspections. The building owner may be required to maintain the systems by ensuring that they are not damaged or breached.

Active Fire Protection Systems

Active fire protection systems are categorized into three general areas: fire alarm systems, including detection and notification appliances; water-based systems; and fire suppression systems. The systems can provide valuable analytical information that can help the fire investigator determine when and where a fire started, how the fire progressed, and how the activation and operation of the system affected the fire. The performance of fire protection systems may figure into the formation and testing of origin and cause hypotheses, as well as hypotheses concerning fire spread. Thus, it is important to understand, document, and preserve these systems. Note that documentation and analysis of these systems often require the assistance of technical specialists to avoid destruction of critical data.

Fire Alarm Systems

Surviving a structure fire depends on the amount of time between the fire starting, the occupants realizing there is a fire, and the occupants getting out of the building. A properly designed, installed, tested, and maintained fire alarm system increases building occupants' chances of survival by detecting fires early and alerting the occupants to take action. In addition, a fire alarm system can provide direct notification to supervising stations, which then contact the public safety answering points (PSAPs) and initiate the appropriate response, all without the need for an individual to call in the fire emergency. Some alarms are monitored directly by the PSAP. This alert will expedite the emergency response that in many instances may help limit property damage and potentially save a life.

System Components

A **fire alarm system** is an assortment of interconnected components and devices that, once activated, outputs audio and visual signals to alert building occupants of a fire emergency to initiate a response. Three basic component groups make up a fire alarm system: the control unit, the **initiating devices**, and the **notification appliances**. Within these groups are a number of different devices and components that perform a variety of functions. Associated with the control unit are primary and secondary power supplies that power the unit, the initiating device circuits, the notification device circuits, and the alarm reporting services. Initiating devices such as smoke and heat detectors may automatically respond to a fire condition and through the control unit initiate an alarm. In contrast, a **manual fire alarm box** requires a person to operate the device. Notification appliances such as bells, horns, and strobe lights provide audible and visual warnings to prompt the occupants to react to the emergency. Alarm reporting services are signals sent via phone or radio signals to the supervising stations.

General System Operation

The operation of a fire alarm system starts when an initiating device automatically detects or reacts to a fire condition or when a building occupant discovers a fire and manually initiates the alarm. Once the initiating device is activated, a signal is transmitted to the fire alarm control unit, which then activates the notification appliances. In addition to notifying the occupants, many systems send a signal to an onsite or offsite monitoring location, where operators receive the alarm signal information and take the appropriate action. Modern fire alarm systems in commercial buildings often interface with other building control and management systems and fire protection systems to initiate additional life-safety operations.

These operations might include starting fans to pressurize stairways, elevator shafts, and vestibules; closing smoke dampers and doors to compartmentalize a building; capturing elevators and moving them to predetermined floors; and unlocking certain doors to ensure safe ingress and egress are possible.

Although some system technology dates back 50 years or more, even the oldest and most basic systems can provide information that is useful during an investigation. When an alarm activates, the fire alarm control unit can provide information that might be as simple as a light marked with a location identifier, such as the floor number, and the type of device that was activated. Modern fire alarm control units can provide even more comprehensive information, including graphic and touch screen units that display the building footprint and/or alphanumeric information readouts that show the device type, type of alarm signal, floor, and exact location of the device on the floor. A basic understanding of system operation will greatly assist you in your role as fire investigator.

Key Components of Systems
Fire Alarm Control Unit

Two types of **fire alarm control units (FACUs)** are still in use today. The **conventional technology** unit has been around for more than 50 years, but even with technological advances it still provides only basic information, performance, and monitoring capabilities. This type of unit has some ongoing uses, especially in small buildings where there are few alarm system devices and it does not take much time to move through the building and locate the event. In contrast, **addressable technology** units use state-of-the-art computer and circuit technology to provide first responders with extensive information concerning the fire event. They can also monitor the system's integrity for conditions that could inhibit the system and components from operating properly and provide an interface to other building fire protection and life-safety systems.

Fire alarm control units generate three different types of output signals, and each has a specific purpose. The **alarm signal** alerts building occupants to a fire emergency with the expectation that the occupants will leave the building, notifies the supervising station, and takes any other necessary predetermined action to ensure life safety. The **trouble alarm signal** alerts a responsible party (usually the building engineer or maintenance personnel) that there is a problem with the fire alarm system's integrity. System integrity problems may include ground faults, breaks in system wiring, power or component failure, communications problems, and device removal. The **supervisory alarm signal** alerts a responsible party (usually the building engineer or maintenance personnel) that the system's normal (ready) status has changed. Changes in normal status include closing electronically supervised fire suppression system control valves, high or low air pressure levels for dry-pipe sprinkler systems, fire pump failures or abnormalities, and air and water temperatures that are outside the appropriate range. Other changes in the system status may include water tanks with a low water level, water pressures that are too low or high, and any operation of associated fire protection equipment during nonfire conditions, such as starting a fire pump to exercise the pump and associated components.

The trouble and supervisory signals may be transmitted to the supervising station, which will then notify the building's responsible party. If a predetermined time has lapsed without communications with the responsible party, the supervising station may notify the fire department. To clear any of the three signals—and in most cases reset the fire alarm control unit—the condition that caused the signal must be investigated and resolved.

Power Supplies

The power supply is an integral part of the fire alarm unit and supplies the electrical power to the unit, system components, and system devices. Two power sources are required: the primary power source, which usually comes from the local commercial power utility, and the secondary or backup power source, which usually consists of rechargeable batteries. In some installations associated with factories or industry, natural gas, diesel fuel, or steam engine–driven generators act as both the primary and secondary power sources. The design capacity of both the primary and secondary power supplies takes into account the current draw from all the system circuits, components, and devices, plus some additional capacity for a margin of safety. Having additional capacity is especially important for the secondary source, because code requirements establish the minimum time that a system must be able to sustain and operate on secondary power without failure. Codes do not permit the electric power circuit supplying power to the fire alarm control unit to power any other appliances.

Initiating Devices

Initiating devices interact with the fire alarm control unit through automatic or manual means to activate the fire alarm system. Automatic initiating devices activate the system when changes are detected in the monitored environment; they include various types of fire detectors, water flow detectors, and component monitoring devices. Fire detection devices can be spot-type or line-type units, or they can use video images. A **spot-type detector** is a single-element device, such as a smoke or heat detector, that is usually installed on the ceiling and is designed to protect a certain square footage area as determined by the manufacturer. Some spot-type detectors are installed in HVAC units and ductwork. A **line-type detector** runs along a continuous route throughout an area of a building to cover a hazard by using some type of electrically or thermally sensitive cable or light beam. A **video detector** uses high-definition video cameras to analyze digital images for changes in the pixels due to smoke and flame generation. Water flow and monitoring initiating devices are discussed later in this chapter.

A **manual initiating device** signals the fire alarm system when a person operates the device. Manual initiating devices are limited to manual fire alarm boxes and **electronic valve supervisory devices**. Manual fire alarm boxes, also known as manual pull stations, all operate in the same basic way, albeit with some slight differences related to the type of signal that is generated and how some boxes must be operated. These differences include coded versus noncoded signaling, discussed later in the section, and the need for one action (single action) versus two actions (double action) to operate the alarm box. Operation of an electronic valve supervisory device, also known as a tamper switch, requires turning a wheel, handle, or other lever that, once moved a certain distance, initiates a supervisory alarm.

Smoke Detection

A **smoke detector** relies on detection of the smoke particles generated from a fire, rather than heat, to initiate an alarm signal. Because sufficient smoke must develop to trigger a smoke detector before other detection systems will operate, smoke detectors are considered life-safety devices. These devices detect smoke by one of two operating principles: by using a radioactive element or by obscuring or scattering light. Both types of detectors are reliable, but depending on the type of fire and conditions, one type of detector might respond faster than the other. The choice of a detector is based on a number of factors, including the environment, the construction and design characteristics of the building, the building contents and their burning characteristics, the shape of the room and the ceiling height, and airflow and ventilation conditions. These factors determine which operating principle and device would be best for the conditions.

Smoke Detectors Versus Smoke Alarms

To many people, smoke detectors and smoke alarms are the same device. In reality, one is installed primarily in commercial occupancies and the other primarily in residential occupancies. A smoke detector is a component of a fire alarm system that is usually installed in commercial occupancies. It initiates an alarm through the fire alarm control unit and relies on the fire alarm control unit to supply its power. A **smoke alarm** is a detector that receives power from a house circuit, has a battery for backup power, has an integrated sounding device, and is usually interconnected with the other installed smoke alarms. This arrangement, called multi-station smoke alarms, relies on the idea that if one device sounds an alarm, they will all sound an alarm. Some smoke alarms, called single-station smoke alarms, are stand-alone units that are powered only by a battery.

In some situations, the smoke alarm is part of a combination system in which security, medical, gas detection, and fire signals all report to a monitoring panel, which notifies a supervising station of the alarm to take the appropriate action. In other situations, the smoke alarm is a part of a complete fire alarm system that operates in a similar manner as a commercial system.

TIP

Smoke detectors and alarms remain some of the best early-warning life-safety devices, but they do have limitations. For example, spot-type smoke detectors cover only a certain area reliably and, over time, may become overly sensitive or undersensitive compared to their original design. This could be problematic because the level of sensitivity must be appropriate for the environment in which detectors are installed. If the devices are too sensitive, false alarms will happen; if they are not sensitive enough, detectors may not react rapidly enough to provide an appropriate warning.

Ionization Smoke Detectors

An **ionization smoke detector** uses a small amount of radioactive material and two electrically charged plates to detect a fire. The radioactive material charges the air particles within the detector's sensing chamber to make the air conductive. Because the air is conductive, a small but measurable current is created by the

ionized particles that attach to the oppositely charged plate. When they enter the sensing chamber, smoke particles attach to the charged air particles and reduce the rate of current flow between the plates. The detector activates once the current flow decreases to a predetermined level **FIGURE 3-13**. Ionization detectors are typically spot-type detectors that are best suited for fires that rapidly develop more flame than smoke, such as a burning piece of paper. They are the most common type of smoke alarms in residential occupancies.

Photoelectric Detectors

A **photoelectric detector** uses a light source and photoelectric cell receiver to initiate an alarm. Within the sensing chamber, a light source and a receiver are installed at an angle to each other. When smoke or other particulate matter enters the sensing chamber, the light

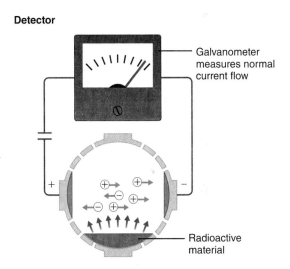

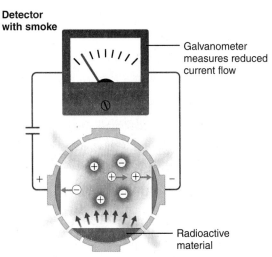

FIGURE 3-13 Principle of operation for an ionization smoke detector.
© Jones & Bartlett Learning.

source refracts off the particles into the receiver and initiates an alarm **FIGURE 3-14**. Photoelectric smoke detectors are typically spot-type detectors that are able to respond more quickly to smoldering fires because the smoke particles tend to be larger than those produced by rapidly developing flaming fires. Flaming fires do not refract as much light because the particles tend to be smaller, resulting in delayed alarm initiation. Photoelectric detectors are commonly found as components in fire alarm systems.

Air Sampling (Aspirating) Detectors

Air sampling smoke detection, also known as aspirating detection, is one of the more sophisticated types of smoke detection in use. Such systems constantly draw air for analysis into a detection chamber,

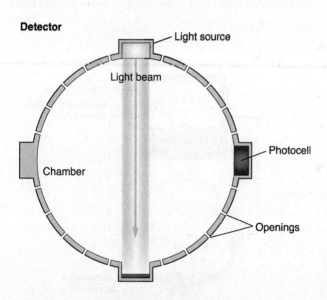

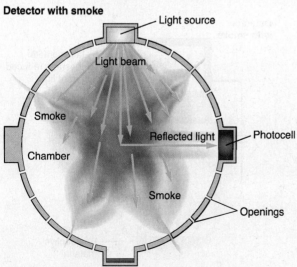

FIGURE 3-14 Principle of operation for a light-scattering photoelectric smoke detector.
© Jones & Bartlett Learning.

through piping with sampling holes that are located in the area of the building being protected by the system **FIGURE 3-15**. By constantly drawing air for analysis, the air sampling smoke detection system identifies undetectable by-products that form when a material breaks down during the precombustion stage of a fire and does so long before an occupant would notice. The detection chamber may be located inside or outside the room or area being protected and is capable of sampling air from several tubes; thus, it is more capable of covering a larger area than spot-type detectors.

As the air is drawn into the sensing chamber, it is analyzed by highly sensitive light refractions. Other methods, including chemical gas analysis, may also be used. This type of air sampling system is typically known as very early warning smoke detection and can sense products of combustion well before typical spot-type smoke detectors can. One of the major advantages of air sampling smoke detection is that once the detector analyzes the particles and determines that there may be a problem, the system can activate different levels of alarms. The first-level alarm indicates that the system has detected an abnormal air characteristic, the second level indicates there is potentially a fire, and the final level indicates that there is a fire in progress.

Air sampling detection systems can be stand-alone units, or they can be part of an overall fire protection strategy. These systems are typically installed in high-value operations where concealed detection is important, maintenance access is limited, or critical processes cannot be disrupted. Historic buildings, telecommunications facilities, computer/server operations centers, production facilities, electrical substations, semiconductor manufacturing facility clean rooms, warehouses, aircraft hangars, textile mills, and atriums are some of the occupancies where this type of detection may be found.

Projected Beam Detectors

A **projected beam detector** sends a light source to a receiver that spans a protected area. When the receiver senses a reduction in light intensity due to smoke or some other obscuration of the light source, an alarm is initiated **FIGURE 3-16**. The projected range can be more than 300 feet (91 m) between the transmitter and the receiver, covering a width of approximately 60 feet (18 m). This type of smoke detector is suitable for locations where spot detection would be impractical or not effective—for example, the very high ceilings found in atriums, concert halls, warehouses, gymnasiums, and factories.

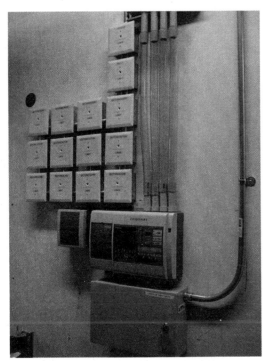

A   **B**

FIGURE 3-15 A. Air sampling smoke detection system control panel/analyzer. **B.** The ends of these tubes have small holes to capture air samples, which are then analyzed by the control panel/analyzer.
© A. Maurice Jones, Jr./Jones & Bartlett Learning.

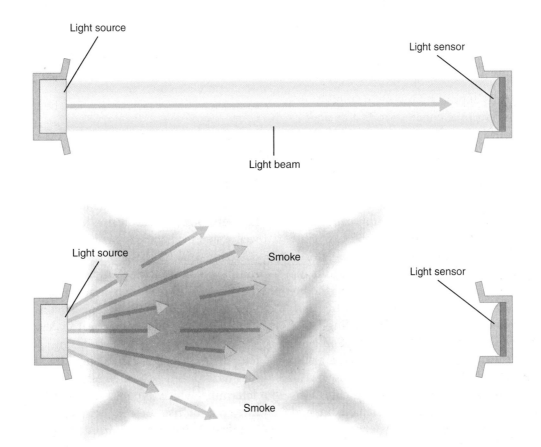

FIGURE 3-16 Principle of operation for a light-obscuration photoelectric smoke detector.
© Jones & Bartlett Learning.

FIGURE 3-17 Duct detector air sampling tubes are located inside the ductwork to capture air samples.
© A. Maurice Jones, Jr./Jones & Bartlett Learning.

Duct Smoke Detectors

A **duct smoke detector** is installed on many commercial air distribution systems to sample the air as it moves through the system ductwork in a building **FIGURE 3-17**. Depending on the decision of the code authority having jurisdiction, the detector could initiate an alarm signal or a supervisory signal.

When the detector senses some type of particulate matter in the airflow, the duct detector shuts down the associated air distribution unit. Without the assistance provided by this unit, smoke and toxic gases cannot move as easily from one side of a building or floor to another. Thus, this approach prevents or inhibits exposure to the products of combustion that create significant and potentially life-threatening problems for occupants.

Duct smoke detectors are not required on all air distribution systems. Installation requirements depend on airflow capacity based on the amount of cubic feet per minute of airflow the unit delivers. In addition, duct smoke detectors must be able to operate over the complete range of air velocities, temperatures, and humidity expected for the installation conditions.

Duct smoke detectors can be mounted on the inside of the ductwork or on the outside of the unit, where the detector connects to air sampling tubes located inside the ductwork **FIGURE 3-18**. Some are ganged together and installed in front of an air duct intake. Many duct detectors are mounted above ceilings or are too high to see, making it difficult to investigate their activation; inspect, test, maintain, and service the detectors; and, in some instances, reset the detector. NFPA 72, *National Fire Alarm and Signaling Code*, requires that any detector installed more than 10 feet (3 m) above the finished floor or in a location

FIGURE 3-18 Duct detector mounted on an air duct line.
© A. Maurice Jones, Jr./Jones & Bartlett Learning.

not readily visible has some type of remote indicator that identifies the HVAC unit with which it is associated. Remote lamps or annunciation devices should be mounted on a wall, ceiling, or as close as possible to the detector so the detector can be quickly located by first responders or building maintenance personnel **FIGURE 3-19**.

Heat Detection

A **heat detector** is one of the most reliable and oldest fire detection devices in use. This type of detector is well suited to challenging environments where smoke detection may not be applicable, such as dusty environments where temperatures may fluctuate, where the fire will have a high heat output, or where fire detection speed is not a primary concern. Heat detectors identify changes in heat by sensing predetermined fixed temperatures or sensing specified rates of temperature change. Three operating principles—the fuel, hazard, and associated rate of fire development—determine which type of detector—fixed temperature, rate of rise, or rate compensation—should be installed. Fixed-temperature detectors operate when the sensing element temperature rating is reached, rate-of-rise detectors measure temperature changes over a fixed amount of time, and rate-compensation

FIGURE 3-19 Remote indicator for a difficult-to-locate or concealed duct detector.
© A. Maurice Jones, Jr./Jones & Bartlett Learning.

detectors react to predetermined air temperatures. Most heat detectors are spot-type devices, but line-type detectors could be more appropriate for certain hazards and environments.

Because heat detectors rely on the heat generated by the fire for their activation, rather than the formation of smoke, these devices typically react more slowly to fires than other types of detectors do, especially smoke detectors. For this reason, heat detectors are not considered life-safety devices and should not replace smoke detectors unless approved by the reviewing authority having jurisdiction. When a heat detector is substituted for smoke detectors, that decision is usually based on ambient conditions and environments that could make a smoke detector malfunction or inappropriately alarm.

Radiant Energy–Sensing Fire Detection

A **radiant energy–sensing detector** does not depend on smoke or heat plumes to operate, but instead detects specific portions of the visible and invisible light spectrum produced by flames, sparks, or embers. These detectors are well suited to environments where exposure to non-fire-related sources of radiant energy is generally greater, such as in manufacturing and industrial settings. Radiant energy–sensing detectors are typically installed to protect high-risk and high-value facilities where rapid and sensitive detection is critical and where remote detection of small fires is crucial.

Selecting the appropriate type of device depends on careful evaluation of the environment and fuel type present, with attention directed to other sources of nonfire radiant energy that could cause false alarms or, in some cases, no alarm at all. These detectors are found at high-risk locations, including petroleum processing, storage, and loading facilities; print shops; paint application facilities; airplane hangars; woodworking facilities; textile mills; museums; ordnance facilities; and computer data rooms.

Flame Detectors

A **flame detector** is designed to detect a certain part of the light spectrum. Detectors operate in the ultraviolet (UV), infrared (IR), ultraviolet/infrared (UV/IR), or multiple-wavelength infrared range. The dual UV/IR and multiple-wavelength IR detectors have two sensing elements and require that both elements detect a specific light spectrum band before an alarm will be triggered. Dual sensing improves detection of an actual fire and reduces the possibility of detecting a nonfire radiant energy source. Radiant energy detectors must have an unobstructed view of the protected area and be close enough to detect the energy to initiate an alarm. When this is a problem, additional detectors are required to ensure complete coverage of the hazard area. Flame detectors may be designed and adjusted for different flames and/or flame intensities.

Spark/Ember Detectors

A **spark/ember detector** is not fuel specific, but simply looks for sparks or embers that may be present during a manufacturing process. Usually, spark and ember detectors are installed on, in, or near conveyor belts or duct lines where solid materials move through a manufacturing process and the possibility of a spark and ember starting a fire is high. These detectors operate in the IR range and are typically installed in a closed, dark environment. Once a spark or ember is detected, some type of fire suppression—typically a water spray nozzle connected to an ultra-high-speed water spray system—is activated.

Other Types of Detectors

Gas-sensing fire detection may be designed to detect a specific gas, any type of toxic gas, or a variety of gases and vapors associated with processes utilizing hydrocarbons. Several different sensing technologies are available, including infrared, semiconductor, electrochemical, and catalytic beads. These detection systems are installed to protect commercial, industrial, and residential occupancies, including high-hazard environments; off-shore oil and gas rigs; oil, gas, and petrochemical facilities; gas turbines; HVAC air intakes; oil and gas wells; and enclosed air handling ductwork.

Gas-sensing fire detectors can be manufactured only for sensing gas or as a combination gas/smoke detector. As with any device, inspection, testing, and maintenance (i.e., calibration) are necessary to ensure the device operates properly.

Notification Appliances

Occupant notification is the most important function of a fire alarm system, with the objective being to prompt occupant evacuation or relocation in the event of a fire. When an initiating device sends an activation signal to the fire alarm control unit, the unit then sends a signal to the notification appliances to produce audible and visual building alarms. Several types of notification appliances are available to provide the audible and visual alerts to the occupants, including horns, bells, speakers, chimes, sirens, buzzers, various types of lamps, strobe lights, and any number of these devices in combination.

The type, number, and location of notification appliances will depend on many factors, including the code requirements; use and occupancy condition of the building; the building layout; room sizes; the environment; and how much, what type, and to whom the audible and visual information needs to be communicated. Notification signals may be transmitted throughout the building, known as **public mode notification**, or only to a certain monitored location of a building, known as **private mode notification**. Public mode is appropriate for occupancies where people are capable of taking action upon alarm activation, including office buildings, stores, and movie theaters. Private mode requires that the alarm be transmitted to a location where trained individuals will receive the alarm, interpret it, and then take the appropriate action. Private mode is appropriate for occupancies where movement is restricted or where people are mentally or physically unable to respond, such as hospitals, nursing homes, and correctional facilities.

Once operating, notification appliances may transmit a coded signal, a noncoded signal, or a textual signal. A **coded signal** transmits a predetermined number of audible or visual patterns over a specific interval that in some cases helps to identify the location of the initiating device. A **noncoded signal** transmits a consistent audible and/or visual signal that continues until the initiating device and system are reset. Typically, the required audible signal is a three-pulse temporal signal. In some older and very specific installations,

the signal may be different. A **textual signal** can take the form of a prerecorded voice message heard from speakers, alphanumeric information on a fire alarm or annunciation panel, or a symbol or picture that flashes on a display screen at a constantly attended location where trained individuals monitor incoming detection, trouble, and supervisory signals.

An **annunciation panel** is a type of notification or indicating panel that provides critical information to emergency responders via a light panel, LCD/alphanumeric text, graphic display, or touch-screen video monitor. This panel is usually located at the building entrance and interfaces with the fire alarm control unit to provide critical information. Information can include the type of signal initiated, the system or device that is activated, and the location of the system or device, including floor level and zone. An annunciation panel will echo portions of the information provided by the FACU. It is extremely valuable because the information provided by the panel can help isolate the incident to specific systems, devices, and areas of a building and can identify the types and numbers of signals generated during the incident.

> **TIP**
>
> No matter the severity of the incident, the amount of damage to the fire alarm system and components, or whether the primary or secondary power has been disconnected or compromised, you should photograph the fire control panel and annunciation panel to capture the indications and the events that took place **FIGURE 3-20**. If primary power is lost, some secondary power sources will sustain the system for only a limited amount of time. Thus, once the secondary source expires, system data could be lost—making timely recording of this information critical.

Operation and Installation Parameters of the System

FACU Features

Frequently described as the "brains" of the alarm system, the FACU is where the system components and devices interface and interact. Its function is to receive the initiating device and other component input signals; to interpret and manage those input signals; and then to deliver the appropriate alarm, trouble, or supervisory output signal. This includes providing power to support all system devices, processing automatic and manual input signals from the initiating devices, monitoring the integrity of the system circuits and power supplies, powering and activating the

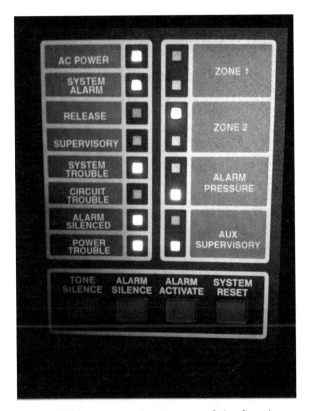

FIGURE 3-20 Always take pictures of the fire alarm control panel and annunciation panel displays to ensure that valuable event information is preserved.
Courtesy of A. Maurice Jones. Jr.

notification appliances, and interfacing with and activating other building and control systems during a fire emergency. The FACU must have a secondary power source, such as batteries or a building generator.

TIP

You can obtain valuable information by interviewing the first on-scene responders who viewed the information communicated by the annunciation panel. That information establishes the status of the fire protection systems when the firefighters first arrived.

The capabilities of the system will depend on its type, and on the dates when it was manufactured and installed. Some systems perform only basic detection and notification functions, whereas others interact and control different fire protection and building systems. Modern FACUs are capable of storing information in the panel memory circuits; sending information to a printer; or sending information about the system status and the signal input and output events to a proprietary or supervising station monitoring company, acknowledging the alarm condition, silencing the alarm condition, and resetting the system.

Location and Spacing of Devices

NFPA 72, manufacturers' guidelines, and, in some cases, local codes govern the installation of initiation devices and notification appliances. NFPA 72 provides information on performance, reliability, and quality of fire alarm and detection systems, as well as significant information regarding the application, design, installation, inspection, testing, and maintenance requirements for these systems. This code also addresses components, hardware, system types, calculations, power requirements, and supplemental integration and interface guidance. However, NFPA 72 is not an installation specification or approval guide, nor does it provide the methods necessary to achieve the requirements. Thus, during the evaluation of a building, it is critical for the fire investigator to consider the environment, design, features, and construction characteristics, including room layout and ceiling height, to identify conditions that could create operational problems for the fire alarm and detection system.

Internal System Communication

Most fire alarm and detection system components are connected together with wire or cable designed specifically for the purpose and installation parameters unique to fire alarm systems. The wiring is the transmission medium used to communicate, initiate, activate, and power the system components, circuits, and devices. The use of wire to connect the system components has been the standard since the first fire alarm systems were designed and installed. Now, however, wireless systems and devices are meeting the same standards and requirements as hard-wired systems, leading to increased versatility of systems along with their approval and acceptance throughout the fire protection and building industries.

TIP

Code requirements, installation conditions, and manufacturers' guidelines determine the type of wire and cable that must be installed and where it can be installed so it is protected from damage and does not contribute to a fire.

Means of Transmission

There are many approved systems and means to communicate between a fire alarm system and third-party monitoring center. To ensure the signal is received at the monitoring facility, two methods of transmission are typically required; this dual transmission method helps ensure that if the primary transmitting link fails, the system will automatically switch to the backup. The most frequently used and approved primary and backup methods include hard-wired and fiber-optic communication, one-way radio transmission, two-way radio frequency multiplex, digital cellular technology, and Internet protocol. Both the primary and backup communication links require supervision and must report any failure. At no more than 24-hour intervals, both links should undergo automatic testing to ensure the integrity and continuity of the link. Given the increased reliability of wireless technologies, some systems used today may have only one transmission method.

Systems Monitoring and Control Supervising Stations

The supervising station is the location where the system signals are received and processed. Three main categories of supervising stations are distinguished: central, proprietary, and remote. Each has the same basic duties of receiving the signals from a protected property and then processing the signal and transmitting the information to the appropriate person or emergency response agency.

Central Supervising Station. A **central supervising station** typically handles any fire protection system–related issues at a protected property it serves. A central station service provider provides services both at the protected property and at an offsite monitoring facility. Its responsibilities include installation, testing, and responding to the site service; signal monitoring; retransmission of the signals; and record keeping. Central station service providers must meet strict performance requirements and be audited by certifying agencies to verify that standards and requirements are met.

Once the central station receives and recognizes a signal, the station personnel must take the appropriate action. If the signal is an alarm, the central station notifies the fire department and then dispatches a responder, who must arrive within 2 hours of the alarm to silence and/or reset any systems. If the signal is supervisory, a responder must arrive within 2 hours. In contrast, for trouble signals, arrival is required within 4 hours. Supervisory signals from a security guard tour (fire watch) require the responder to investigate within 30 minutes of notification.

Proprietary Supervising Station. With a **proprietary supervising station**, building and property owners are responsible for monitoring the fire

alarm system signals at their facilities. Typically, the proprietary system consists of a number of fire alarm systems on a property or spread throughout an area or region in buildings that are under the control of a single owner. All signals are transmitted to a central monitoring location on the property, to a regional monitoring location, or to a central monitoring location elsewhere. Staffing of the monitoring facility is provided 24 hours a day by trained personnel, who monitor and transmit fire alarm signals to the fire department and, if established, the facility's emergency response personnel. Monitoring facility personnel also notify the emergency response personnel or service personnel of any trouble or supervisory signals, regardless of whether they are located at the site or at a remote location.

Proprietary systems also can be set up to handle all types of building control functions, including elevator control, smoke control, stair tower pressurization, stair tower and exit door lock release, and shutdown of manufacturing or other processes. The building or room housing the monitoring equipment must be constructed of fire-resistant materials and be separated from any hazardous processes to ensure survival if there is a fire or other facility emergency.

Remote Supervising Station. A **remote supervising station** reports alarm, trouble, and supervisory signals generated by a protected property fire alarm system that signals an offsite location that is distant from the protected property. Some fire departments permit signals to go directly to a fire or public safety answering point via approved fire department radio frequencies, a dedicated one-way (outgoing only) telephone line, or dedicated communication circuits. When the signals do not go to a fire or public safety answering point, they must go to an approved location where trained individuals monitor the signals 24 hours a day. When personnel monitoring the system receive fire alarm signals, they must immediately contact the fire department. In contrast, a trouble or supervisory signal is reported to the owner or the owner's representative for immediate investigation and resolution of the problem.

Postfire Analysis of Fire Alarm Systems

If the investigation reveals that the building is equipped with any fire protection system, the fire investigator must thoroughly document and analyze the system to determine its role, if any, during the fire. The collection of data, as referenced in NFPA 921 and covered in detail in Chapter 10, *Data Collection Outside of the*

Fire Scene, should include gathering information on building systems and contents, reviewing passive and active fire protections, obtaining data from systems, and interviewing occupants and first responders.

Fire Alarm Effectiveness

Fire alarm systems, when properly designed, installed, inspected, tested, serviced, and maintained, are highly effective in providing early occupant notification when fire occurs. However, even when these systems are in proper working order, an investigator may need to consider scenarios in which occupants did not respond to alarm signals as expected. Many factors can influence occupant reactions, including a person's response to sounds when sleeping, their ability to hear, whether they are impaired, and whether they have a disability. Thus, as part of the assessment of a fire scenario, the fire investigator should investigate occupant reactions and evaluate any reasons why they did not react as expected.

Impact on Human Behavior

The presence of a fire protection system could have substantial impacts on human behavior, both good and bad. Some individuals will always respond to a fire alarm signal, whereas the presence of these systems might give others a false sense of safety, resulting in a delayed response or even no response from the individual. In addition, some individuals might respond better to a certain type of audible notification, be it a horn, bell, or voice message. Sometimes building occupants fail to respond to fire alarm activations due to repeated past alarms that were found to be false or if they did not receive clear directions regarding what they should do when an alarm occurs, including where and when to evacuate.

Water-Based Fire Suppression Systems

Purpose of Systems

The purpose of a water-based fire suppression system is to protect property and provide life safety by delivering sufficient amounts of water over the fire area at the appropriate rate in an effort to control, and in some cases extinguish, the fire. This is accomplished by the automatic activation of a sprinkler head, a device that reacts to changes in temperature; activation of a fire alarm and detection system's automatic initiating device; or, for some types of systems, activation by manual fire alarm box.

General System Operation

The main extinguishing mechanism for water-based fire suppression systems is reduction of the fire's heat by applying sufficient amounts of water to cool the fire area and prevent the fire from spreading. Application of a sufficient amount of water depends on a number of factors, because different materials release different amounts of energy when burning. In essence, to control a fire, there needs to be an adequate and reliable water source that can deliver water at a sufficient rate over a certain area.

Another extinguishing mechanism involves mixing a **foam concentrate** with water to create a **foam solution** blanket that both cools the fuel and isolates it from the oxygen. In this case, the amount of water must be adequate for the hazard, but the foam and water must also be properly mixed: Too much water may result in a weak solution that cannot control or extinguish a fire, and too much foam concentrate may result in the system running out of foam before the fire is controlled or extinguished.

Types of Water-Based Systems

Water-based sprinkler systems can be categorized into four types: wet, dry, preaction, and deluge **TABLE 3-5**. These four types are the most commonly installed systems in commercial and industrial settings, and most of the other water-based fire sprinkler systems are variants of them. Water mist and foam–water systems are discussed in the next section.

All of these systems use NFPA 13, *Standard for the Installation of Sprinkler Systems*, as their design and installation standard. The installation of a particular type of system mainly depends on two factors: the hazard it must protect and the operating environment. In many instances, all four types of systems will be appropriate for the hazard and conditions, but other NFPA standards may need to be referenced to deal with special hazard situations.

Specialized Sprinkler Systems

In some situations, the use of a standard automatic fire sprinkler system is not appropriate for the applications and conditions of a particular building, process, or hazard. In these instances, specialized sprinkler systems—including low-, medium-, and high-expansion foam systems; compressed-air foam systems; foam–water sprinkler/foam–water spray systems; and water-spray fixed systems—may be installed to protect the hazard **TABLE 3-6**. These systems are used where the conditions and applications of standard automatic fire sprinkler systems are not

TABLE 3-5 Types of Water-Based Systems

Type	Description	Application
Wet-pipe sprinkler system	■ Has water in the piping at all times. ■ This system is preferred because upon activation of a sprinkler head exposed to sufficient heat, water immediately flows onto the fire.	■ Environment must be able to maintain a temperature of 40°F (4°C) or higher so water does not freeze and damage or break the fittings, piping, and components ■ Commonly installed in stores, offices, and homes
Dry-pipe sprinkler system	■ The system pipes hold pressurized air or, in some rare cases, a gas such as nitrogen. ■ The air or gas holds pressure on the dry-pipe valve clapper so it does not release the water into the system until necessary. ■ A fire sprinkler head must be activated to release the pressurized air or gas in the piping. ■ Once sufficient pressure is lost, the valve opens and floods the piping with water. ■ Water must reach open sprinkler head(s) within a specified time frame (typically 60 seconds) to be effective.	■ Mostly used where the environment is not able to maintain a temperature of at least 40°F (4°C) ■ Commonly installed in unheated warehouses, loading docks, attics, and parking garages
Preaction sprinkler system	■ The sprinkler system interfaces with a dedicated fire alarm control unit and initiating devices to release a preaction valve. ■ Initiating devices are located in a protected area. ■ Three operational configurations are available, each requiring one or more activation events to take place before water enters the piping (e.g., activation of a fire sprinkler head, activation of a smoke or heat detector, or operation by manual means). The need for multiple activation events to take place provides a higher level of certainty that the system is responding to an actual fire.	■ Protect high-value occupancies, including telecommunication centers, server farms, computer operation centers, freezers, document storage centers, museums, and libraries
Deluge sprinkler system	■ All sprinkler heads or nozzles are open to the atmosphere. ■ The sprinkler system interfaces with electric, hydraulic, or pneumatic initiating devices to release the deluge valve and flow water from all the open sprinkler heads or nozzles onto the hazard. ■ This kind of system provides large amounts of water over the hazard to prevent a rapidly growing fire from spreading. ■ An adequate water supply is required, and frequently a fire pump is necessary to deliver the required water volume and pressure to meet the system design demand. ■ A foam concentrate is commonly mixed with water to protect certain hazards.	■ Protect high-hazard occupancies where it is necessary to flood an entire area to prevent the fire from spreading (e.g., aircraft hangars, power plants, woodworking facilities, cooling towers, refineries, explosive and ordnance factories, and chemical storage or processing facilities)

TABLE 3-6 Specialized Water-Based Systems			
Type	**Description**	**Application**	**NFPA Standard**
Water mist system	■ Specialized sprinkler heads and spray nozzles discharge a very fine spray mist of water droplets, where 99 percent of the droplets discharged from these systems must be 1000 microns (1 mm) or less in size. ■ Fire is extinguished by cooling, displacing oxygen, and blocking the radiant heat. ■ Systems are classified based on the working pressure of the system (low, intermediate, or high) and whether one pipe (single fluid water only) or two pipes (twin fluid water and gas) supply the head or nozzle. ■ Advantages include discharging less water than a conventional sprinkler system and the fact that some systems are self-contained and are designed for complete fire extinguishment. ■ Disadvantages include higher design and installation costs compared to other sprinkler or suppression systems, the self-contained tank systems only having a finite amount of water, the need for reserve tanks, the need for each specific application to be tested and approved, and the possibility of nozzles clogging when water quality is not exceptional.	■ Land, sea, and air applications, including computer rooms, data centers, laboratories, archive storage, museums, historic buildings, electrical switchgear rooms, telecommunication facilities, tunnels, underground mass-transit facilities, boats, yachts, naval ships, passenger ships, offshore operations, aircraft cargo spaces, and aircraft hangers	■ NFPA 750, *Standard on Water Mist Fire Protection Systems*
Low-, medium-, and high-expansion foams	■ Foam systems are used to protect hazards that involve flammable and combustible liquids. ■ Foam is less dense than fuel or water and is able to flow freely over the liquid surface. ■ Foam is produced by mixing a *foam concentrate* with water to produce a *foam solution*. ■ Foam concentrate is available in 1 percent, 2 percent, 3 percent, and 6 percent mix ratios. ■ Each percentage represents the amount of concentrate to mix with water, where the resulting mixture equals 100 percent. ■ Mixing the foam concentrate and water creates the foam solution, but air must be introduced by mechanical means to form the foam bubbles that are discharged to protect the hazard. ■ Foam products fall into three categories defined based on the expansion ratio of the foam: low, medium, and high expansion. ■ Expansion is a function of properly mixing the manufacturer's defined percentage of foam concentrate and water. ■ Low-expansion foam increases from a 2:1 ratio of foam to water up to a 20:1 ratio. Medium-expansion foam increases from a 20:1 ratio up to a 200:1 ratio. High-expansion foam increases from a 200:1 ratio up to a 1000:1 ratio.	■ Flammable and combustible liquid outdoor storage tanks, containment dikes, interior flammable and combustible liquid storage, and fuel loading racks	■ NFPA 11, *Standard for Low-, Medium-, and High-Expansion Foam*

(continues)

TABLE 3-6 Specialized Water-Based Systems *(continued)*

Type	Description	Application	NFPA Standard
	■ Different types of foam concentrates are available, including aqueous film-forming foam, alcohol-resistant aqueous film-forming foam, film-forming fluoroprotein foam, protein foam, and fluoroprotein foam. ■ Some foams are designed for specific applications, whereas other foams have characteristics that make them suitable for use with many different hazards.		
Compressed-air foam systems	■ Air or nitrogen, water, and foam concentrate are combined to create high-momentum, greater-expansion-ratio foam. ■ The foam concentrate is normally stored in nonpressurized tanks that range from 5 to 500 gallons (19 to 1893 L). ■ High-pressure air cylinders supply the air or nitrogen coupled with a high-pressure manifold and pressure regulation devices. ■ Compressed-air foam systems use mixing chambers to combine the solution that feeds the piping and flows to the specialized nozzles that deliver the foam.	■ Flammable and combustible liquids, liquid spill fires, three-dimensional fires, ships, wildland fires, and some structural firefighting	■ NFPA 11, *Standard for Low-, Medium-, and High-Expansion Foam*
Foam–water sprinkler/foam–water spray systems	■ Foam and water are used as companion suppression agents; the foam–water solution serves as the primary agent, and the water is used as the secondary agent. ■ When the foam supply from these systems becomes depleted, the water supply continues until it is shut off. ■ The primary design goal is to extinguish fire, but these systems are also appropriate for prevention, control, and exposure protection. ■ Foam–water sprinkler systems protect areas of a building or structure and equipment. ■ Foam–water spray systems use directional foam application to protect a specific hazard or piece of equipment within a building or structure.	■ Class B flammable and combustible liquid hazards, including aircraft hangars and petroleum dispensing and storage facilities; acceptable for use with certain Class A hazards	■ NFPA 16, *Standard for the Installation of Foam–Water Sprinkler and Foam–Water Spray Systems (withdrawn in 2020 and incorporated into NFPA 11)*
Water-spray fixed systems	■ These systems are designed to deliver a concentrated and directed water spray pattern onto the surface of, into an area within, or around the hazard for the purpose of fire control, prevention, extinguishment, or exposure protection. ■ One adequate water supply is needed, which is released by automatic or manual activation. ■ Nozzles are either automatic or open configurations. ■ Automatic nozzles use the same basic technology as automatic sprinkler heads (heat-sensitive glass bulb/fusible link) and operate when a predetermined design temperature rating is achieved. ■ The open nozzle is open to the atmosphere.	■ Equipment or structural members surrounding or supporting the equipment, electrical transformers, oil switches, motors, cable trays, paper, wood, textiles, flammable liquid and gas materials, and certain hazardous solids	■ NFPA 15, *Standard for Water Spray Fixed Systems for Fire Protection*

appropriate, where a supplemental suppression agent is needed, or where specialized equipment and components are required for application. They can also be found in enclosed areas and areas with flammable liquids. Low-expansion foam covers burning materials with a foam blanket, whereas medium- and high-expansion foams are used to fill enclosures.

The four most prominent types of automatic fire sprinkler systems and these specialized systems do share some similarities: Water is the primary suppression agent, and it must come from an automatic supply; the components are the same or similar; and all components must be automatically activated and use the same types of system valves. However, specialized systems have some key differences, including specialized spray nozzles instead of the typical fire sprinkler heads, the use of foam agents with water in certain types of systems, the need for specialized components to mix the foam solution and water appropriately, and the design and application methods.

Residential Sprinkler System

The benefits of **residential fire sprinkler systems** derive from a change in the overall goal from property protection, which is the motivation for commercial fire sprinkler systems, to life safety. This difference in motivation is manifested by reducing the system's water supply requirements, creating a special type of sprinkler head, and placing the fire sprinkler heads in those locations where a fire is most likely to start. Each of these changes contributes to achieving the design goals of residential fire sprinkler systems, which are to prevent flashover in rooms with sprinklers where the fire started and to provide a better chance for the occupants to escape or be rescued.

Commercial fire sprinkler systems and residential systems have many similarities, but also many differences. First, two NFPA standards deal specifically with residential occupancies: NFPA 13D, *Standard for the Installation of Sprinkler Systems in One- and Two-Family Dwellings and Manufactured Homes*, and NFPA 13R, *Standard for the Installation of Sprinkler Systems in Low-Rise Residential Occupancies*. These standards establish requirements that are different from NFPA 13 for water supply, design, installation, components, and testing. The majority of residential sprinkler systems are wet-pipe systems and use similar types of components as commercial systems do. However, their components tend to be smaller because less water is needed based on the design requirements to supply the system and make it work properly.

Some NFPA 13D and NFPA 13R systems share the water supplied by an appropriately sized water line fed from a reliable community water system, but most water suppliers require separate domestic and fire service water lines. In some instances, the water supply consists of an adequately sized stored water source that has an automatically operated pump. This arrangement is common in communities where there is no water system or no reliable automatic water source or well water is the main source of water. As part of the fire sprinkler system, a storage tank of adequate capacity with an automatic fill line holds the water until needed. This arrangement is predominantly associated with NFPA 13D systems, rather than NFPA 13R systems, due to the size and capacity of the storage tank that would be needed for an NFPA 13R system. Both NFPA 13D and NFPA 13R outline a number of approved water sources—but no matter the source, it must be automatic and supply an adequate amount of water for a specific time to the system.

Two other important distinctions between NFPA 13 and NFPA 13D/NFPA 13R systems are the use of a residential sprinkler head and the locations of sprinkler heads within a residence or dwelling unit. First, residential sprinkler heads must meet different testing criteria than standard commercial fire sprinkler heads do. The residential sprinkler head must prevent the harmful fire by-products and the associated temperatures from reaching levels that could inhibit safe passage or rescue of the occupants. Second, NFPA 13D and 13R systems exempt sprinkler head installation in certain locations within a residence. These locations are deemed less likely to be the room of fire origin; statistically, the likelihood of a fire in these spaces is low. Both standards list a number of locations that do not require fire sprinkler heads, including bathrooms that are less than 55 ft² (5 m²), closets and pantries that are no greater than 24 ft² (2.2 m²) and where the least dimension does not exceed 3 ft (0.9 m), and areas where the wall and ceiling construction is of limited or noncombustible materials. Also included are garages, open attached porches, carports, attics, equipment rooms, and concealed spaces.

Key Components of Water-Based Systems

Although some specialized water-based systems employ components specific to the type of system, all water-based fire protection uses the same basic components, including automatic sprinkler heads or nozzles, piping, pipe support and stabilization assemblies, fittings, gauges, and various types of valves. These components form the building blocks of the various systems. In most cases, the major difference between systems is the type of system valve—that is, whether an alarm valve is for a wet system, a dry valve, a preaction valve, or a deluge valve.

All components of the system must be listed, approved, or labeled by a nationally recognized evaluation agency to ensure they perform as required when tested under the conditions in which the components must operate. Without this oversight, components could fail, leaving the building without a fire sprinkler system. The fire investigator examining a water-based system should determine whether all components in the system were appropriately listed, both individually and as a combined system.

Sprinklers and Spray Nozzles

Automatic sprinkler heads and nozzles are spray devices that distribute water over a limited area at a designated flow rate. Their purpose is to distribute water to reduce the heat from a fire by controlling and limiting fire growth beyond the early stage of fire development.

Sprinkler heads are heat-activated devices that operate at predetermined temperatures. As the heat generated by a fire rises to the ceiling, everything, including the sprinkler head, is subjected to the higher temperatures. When the sprinkler head reaches the predetermined activation temperature for its heat-sensitive element, it releases so that the disk holding back the water falls away from the head, allowing the water to

flow from the head and spray onto the fire. Each head operates independently; therefore, only the heads that reach the predetermined activation temperature will operate and flow water.

Sprinkler heads are defined by characteristics such as temperature rating, orientation, **K-factor**, and presence of a coating on the head that allows it to be used in a particular environment. Temperature ratings range from 135°F to 650°F (57°C to 343°C). Sprinkler heads' temperature ratings can be ascertained by examining the color of the liquid in the bulb or factory-applied paint or coating on the frame or deflector, or by noting the rating engraved on the frame or deflector **TABLE 3-7**. Although there are many types of sprinkler heads that fall into different categories and subcategories, generally all heads are oriented in either a pendent, upright, or sidewall position. When performing calculations, the K-factor is a value assigned by the manufacturer to the head that represents the orifice size. The larger the K-factor number, the larger the orifice, meaning the more water that will flow from the head.

Each sprinkler head has specific characteristics and considerable information applied to it to inform the end user of the type of head, temperature rating, listing or approval, model number, K-factor, year manufactured, and sprinkler identification number **FIGURE 3-21**. Nozzles also must carry a listing or

---

TABLE 3-7 Temperature Ratings, Classifications, and Color Codings (NFPA 13, Table 7.2.4.1)

Maximum Ceiling Temperature		Temperature Rating		Temperature Classification	Color Coded	Glass Bulb Colors
°F	°C	°F	°C			
100	38	135–170	57–77	Ordinary	Uncolored or black	Orange or red
150	66	175–225	79–107	Intermediate	White	Yellow or green
225	107	250–300	121–149	High	Blue	Blue
300	149	325–375	163–191	Extra high	Red	Purple
375	191	400–475	204–246	Very extra high	Green	Black
475	246	500–575	260–302	Ultra high	Orange	Black
625	329	650	343	Ultra high	Orange	Black

Reproduced with permission from *NFPA 13-2022, Standard for the Installation of Sprinkler Systems,* Copyright © 2021. National Fire Protection Association. This reprinted material is not the complete and official position of the NFPA on the referenced subject, which is represented only by the standard in its entirety.

FIGURE 3-21 With close observation, a person can determine a number of sprinkler head characteristics, including the manufacturer, the model, the temperature rating, the K-factor, the year of manufacture, the sprinkler identification number, and the listing or approval by a third-party testing agency.
© A. Maurice Jones, Jr./Jones & Bartlett Learning.

approval by a nationally recognized testing and certification agency. Nozzles are generally used for special hazard applications and are designed to apply a special water spray pattern, but most do not have a heat-sensitive element. Depending on what is being protected, nozzles surround the hazard or equipment and apply large amounts of water directly onto the hazard or equipment.

Piping and Fittings

Joined pipe and fittings form a conduit to move the water from the source to and out of the fire sprinkler heads. To be acceptable for use as part of a fire sprinkler system, these pipe and fittings are required to meet or exceed certain standards from a number of nationally recognized organizations, including ASTM International, the American Society of Mechanical Engineers, the American National Standards Institute, and the American Welding Society. In addition, the pipe and fittings must be able to handle the maximum permitted system working pressure of 175 psi or carry an appropriate rating for the anticipated pressure greater than 175 psi. Many different materials are approved for use in pipes, including ferrous, black steel, galvanized steel, copper, certain alloy materials, and chlorinated polyvinyl chloride (CPVC).

Fittings connect piping, valves, sprinkler heads, and other components through use of elbows, tees, flanges, crosses, unions, couplings, bushings, plugs, and reducers. These fittings may be made of cast or malleable iron, steel or wrought carbon steel, alloy steel, copper, or CPVC.

Joining methods include screw thread, grooved, flanged, welded, soldered, heat fusion, brazed, pressure, or glue.

TIP

Installation of a sprinkler head with a temperature rating that is too low relative to the normal-high ambient temperature could result in unwanted sprinkler head activation. However, if the temperature rating of the head is too high, delayed activation could allow the fire to spread to the point where it would be difficult to control.

System Valves

At least one **water control valve** is required to be installed on all automatic fire sprinkler systems. The purpose of this valve is to permit or prevent water flow into and through the system. Under normal conditions, the control valve or valves should be fully open and remain open to permit water to flow freely into and through the system **FIGURE 3-22**. However, in the event the system needs modification, maintenance, service, or repair, the system can be shut down. Once the work has been completed, any valve that was closed must be reopened to place the system back in service.

To show whether the water control valve is open or closed, all automatic fire sprinkler systems' water control valves are required to be indicating-type valves. Indicating-type valves allow a person to look at

FIGURE 3-22 Many performance failures of water-based fire suppression systems are due to a closed control valve. This valve was closed, and the electronic monitoring device was disabled so that no supervisory alarm would sound.
Courtesy of A. Maurice Jones, Jr.

the valve and quickly determine whether the valve is open, partially open, or shut. In addition, all indicating valves must be identified with a permanent sign and indicate the area of the facility the valve is controlling. Indicating valves for fire sprinkler systems are usually 2 inches (5 cm) or larger and include outside screw and yoke (OS & Y), butterfly indicator, wall post indicator (WPIV), and post indicator (PIV) types.

////////////////////////////////////

TIP

Water under pressure can be very dangerous, so do not attempt to operate any fire suppression system water control valve before analyzing its location and understanding the purpose of the valve's position relative to the event being inspected or investigated. Opening a closed valve could cause considerable water damage if the valve is closed due to a broken sprinkler line or an open/activated sprinkler head. Closing a previously open valve could shut down part or all of the system, preventing water from flowing if the system should be activated. If you do not possess the knowledge needed to address the situation, contact a technical expert or seek the supervision of a trained individual to avoid flooding, damage, injury, or a fire that grows out of control due to no water supply.

Proportioners

Expansion foam systems require a mechanical device called a proportioner to mix the foam concentrate with the water to achieve the proper ratio. Several different proportioning devices are available, including Venturi, pressure, and balanced pressure proportioners. A **Venturi proportioner** flows water over an open orifice within the proportioning device, which creates a lower pressure that draws the foam concentrate out of the storage arrangement and into the water stream. A **pressure proportioner** directs the incoming water supply into a storage tank, where it exerts pressure on a collapsible bladder inside the tank that holds the foam concentrate, thereby pushing the concentrate out of the tank and to the proportioner for mixing. A **balanced pressure proportioner** uses a pump and an atmospheric tank to flow the foam concentrate into the water supply. The proportioner achieves the correct ratio by balancing the foam and water volumes.

Water Supply

One of the most important elements of any water-based fire protection system is access to at least one automatic water supply. Automatic water supplies do not require a person to interact with any part of the system for water to flow. The water can come from one or a number of different sources, including wells, rivers, lakes, municipal water systems, storage tanks, reservoirs, and cisterns. However, one or more pumps may be required to support the system with adequate pressure and flow for the duration necessary to control or extinguish a fire. If any of these elements is deficient, the system will not perform as required and may fail to protect the intended property or hazard. Water supplies should be reevaluated periodically to ensure their pressure, flow, and duration have not changed.

Operation and Installation Parameters of the System

Essentially, all water-based fire suppression systems are some type of fire sprinkler system. The appropriate system design ultimately depends on many factors, including the hazard level, occupancy and use of the building or structure, storage arrangement, the amount and type of fuel load, the potential rate of heat release, processes, and construction materials. These factors play a major role in selecting the design approach, type of fire sprinkler system, and types of heads that will be installed to protect the hazard.

Location and Spacing of Sprinklers

Automatic fire sprinkler heads are not randomly placed throughout buildings. Instead, they are carefully located and spaced so that upon activation, they will discharge a sufficient amount of water over a defined area to control a fire within minutes of the fire starting. However, before any head is installed, the design professional must undertake an evaluation of the hazard. The evaluation for most designs starts with classifying the occupancy, commodities, or storage arrangement that the system must protect. Any one or a combination of occupancy, commodity, and storage arrangements will produce certain combustible characteristics that will output energy through fire development. Based on the level of energy output, the system design must deliver enough water to absorb the energy from the fire, ensuring that it does not grow any larger. Evaluating the combustibility, amount of combustibles, and rate of heat release of a commodity, material, or product is critical to determining the hazard classification and design approach.

Head location and spacing are functions of the physical environment and the requirements outlined in the adopted codes and standards that describe how much area a head must cover to protect the hazard. When the hazard is determined to be high, the sprinkler heads are placed closer together. When the hazard is determined to be low, the sprinkler heads can be placed farther apart. Spacing can range from 90 ft² (8.4 m²) per head for high-hazard environments to 225 ft² (20.9 m²) per head for low-hazard environments. These spacing requirements apply to commercial or industrial occupancies with full sprinkler systems. The adopted codes and standards do include a number of exceptions for commercial and industrial occupancies, but as previously discussed, certain rooms and areas in residential occupancies and dwelling units can automatically omit sprinkler heads.

Another factor in determining sprinkler head location and spacing is the choice of sprinkler head to protect the hazard. Sprinkler heads must undergo the same evaluation and approval process as any other piece of fire protection system equipment, and, when specified for use, their use and installation must follow the manufacturer's specifications and the requirements outlined in the adopted codes and standards. Because a large variety of sprinkler head types are available, certain conditions and situations may allow for the use of a head that can exceed 225 ft² (20.9 m²) spacing as long as it meets the manufacturer's specifications and is within the requirements of the adopted codes and standards.

As important as proper head spacing is for protecting the hazard, the location of the head relative to any component of the structure is critical to ensure successful activation and water discharge. The codes and standards include installation requirements that specify how far away from walls, ceilings, obstructions, and equipment sprinkler heads should be to work properly, albeit with some exceptions that provide relief for specific conditions. Generally, sprinkler head deflectors can be located between 1 and 12 inches (2.5 and 30.4 cm) below the ceiling and at least 4 inches (10.2 cm) away from the wall. Some special storage situations may require spacing between 18 and 36 inches (45.7 and 91.4 cm).

Pipe Size and Arrangement

The pipe size and arrangement of a fire sprinkler system are determined by evaluation of the inherent hazard level of the occupancy, as outlined by the adopted design standard. Once the level of hazard is determined, a decision must be made whether to use a pipe schedule for the design or the hydraulic design method.

Pipe schedules list the pipe sizes and the number of fire sprinkler heads that the pipe can support based on the hazard classification. Specifically, a pipe schedule spells out that a certain size pipe can have a certain number of total heads attached. Large pipes can have more heads than small pipes, but the total number of heads attached to a system is fixed by the largest pipe size and hazard classification. As the pipe size decreases, the number of heads that can be on the system decreases. These systems were common until the 1970s. Since the advent of hydraulic design, new pipe schedule systems have been rarely installed because of the size limitations imposed by the sprinkler standards and the limits to expanding existing pipe schedule systems.

Today, the requirements for the vast majority of systems are calculated based on **hydraulic design**. Unlike the pipe schedule, in which the number of heads on a pipe is based on pipe size, this method establishes pipe size throughout a system by calculating the **friction loss** at certain points in the system—typically the hydraulically most remote—based on the available water pressure and required water flow at those points. These calculations often allow for pipes throughout a system to be of similar size. If the calculations indicate that smaller pipes are able to support the pressure and flow demand to meet the system water requirements, then smaller pipes will be used because they reduce the cost of installation.

Sprinkler Coverage and Distribution

Many different types of sprinkler heads are manufactured to handle almost any design challenge, hazard, or environmental condition. As noted earlier, heads are installed in one of three basic installation orientations: pendent, upright, and sidewall. Pendent heads discharge water downward, upright heads discharge water upward into a deflector, and sidewall heads discharge water horizontally. Sprinkler heads are designed and manufactured to provide unique water discharge patterns and water droplets sized for the specific environmental and installation conditions, including heads made for residential occupancies, rack storage, attics, and fur vaults. The variety provides the design professional with many options to meet the protection goal, but the head that is chosen must always be able to provide sufficient flow and cover a certain area to control a fire.

When a sprinkler head activates, it discharges water in a preset pattern over a certain area and at a specific flow rate to meet the design objective. Meeting the design objective requires that an adequate water supply be available. If the water supply is insufficient, the fire is unlikely to be controlled. The discharge pattern, coverage, and flow rate are a function of the sprinkler head design. Two sprinkler head components—the orifice and the deflector—are critical to developing the discharge pattern and flow rate. The size of the orifice determines how much water flows out of the head, whereas the deflector determines the direction of water flow, water reach, and size of the droplets. When the hazard level is high, fire development is expected to be rapid and intense. Therefore, specifying a head that can discharge large water droplets will provide the protection necessary to penetrate the fire plume and control the fire. Critical to achieving the desired water delivery is sufficient pressure, measured at the most remote head, for proper flow and discharge to take place.

Water Flow Rate and Pressure

When designing a fire sprinkler system, one of the first things that must be determined is the hazard level of the occupancy. Based on this determination, the minimum amount of water flow and required pressure needed to reach the hydraulically most **remote area** can be established. If the water supplied to this area is adequate, it should be better than adequate in other parts of the system because they are closer to the water source. Pressure is critical because sprinkler heads require a certain amount of pressure to operate properly, as defined by the manufacturer's specifications. If the operating pressure is too low, the head will not flow enough water to control the fire. The typical sprinkler head requires a minimum of 7 psi to function properly; however, other hazardous conditions may require more pressure and water flow.

The amount of water required to flow from the head within the remote area is called the **design density**. The design density value is based on years of testing, in which fires were set in controlled conditions and water was applied to determine how much water it would take to control a certain size and type of fire. These values are depicted on the **density/area curves** published in NFPA 13; these curves provide a graphic representation of the minimum amount of water that needs to flow over a specific area. Density/area curves have been developed for light, ordinary, high-hazard, and other challenging situations **FIGURE 3-23**. There are a number of points on each curve where the design professional can select a density, measured in gallons of water per minute per square foot (gpm/ft^2), over a given area. The curves are designed to balance the water flow and area covered by establishing minimum and maximum density and area parameters. Most design professionals will flow as much water as the curves will permit, and properly performed hydraulic calculations will provide a more accurate indication of pressure and flow throughout the system, allowing for more design flexibility.

Activation Mechanisms and Criteria

Fire sprinkler systems may be activated to flow water onto a fire in several different ways. Many fire sprinkler systems, such as wet and dry sprinkler systems, use heat-activated sprinkler heads. Once the predetermined temperature is reached at these heads, the heat-activated element releases, allowing the disk covering the head orifice to fall away and flow water. Typically, the heat-sensitive element is a glass or frangible bulb, a fusible solder link, or a chemical pellet.

The choice of a sprinkler head with a heat-sensitive element is based on a number of factors, and it is essential to use the appropriate head for the situation. Pertinent factors include the occupancy condition,

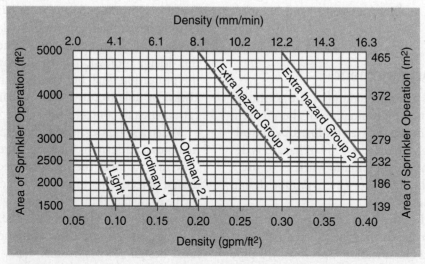

FIGURE 3-23 NFPA 13 density/area curve (NFPA 921, fig. 8.4.4.4.1).

Reproduced with permission from *NFPA 921-2021, Guide for Fire and Explosion Investigations*, Copyright © 2020. National Fire Protection Association. This reprinted material is not the complete and official position of the NFPA on the referenced subject, which is represented only by the standard in its entirety.

the environment, the water supply, and the protection goal, all of which could have an impact on how rapidly a head activates. The amount of time a head takes to activate once it is exposed to temperatures above the predetermined activation temperature is called the **response time index (RTI)**. A lower RTI indicates that the head should operate more rapidly. A head with a low RTI is a must for certain occupancy conditions, because the system is designed to protect the occupants (e.g., in residential occupancies), or where the sprinkler system relies on a certain type of head to protect the property (e.g., high-piled storage warehouses).

Other systems, such as deluge systems, use open sprinkler heads that are dependent on either human actions or interaction with a heat-activated device such as a heat detector to initiate water flow. Still other systems, such as preaction systems, utilize closed sprinkler heads and fire or smoke detectors. The detectors will open a valve to release the water to the sprinkler heads, where the heat-sensitive elements in the sprinkler head will become activated if necessary and release the water onto the fire.

Systems Monitored and Controlled

Fire sprinkler systems are monitored and controlled by various devices and systems, which notify those responsible when the normal status of the system has changed. Status changes could relate to a water flow condition, a change in air pressure in a dry system, a valve closure, a temperature change, or a water level change. Usually, this monitoring is performed in conjunction with a fire alarm system. These notifications can be local or sent to an offsite entity that will notify the responsible parties.

All sprinkler systems have at least one **water flow detection device**. If there are multiple sprinkler risers, the system may have multiple water flow devices, depending on the size of the building and the number of systems within the building.

Wet-pipe sprinklers are equipped with a **paddle-type water flow device**, which utilizes a flexible round plastic paddle or vane that inserts into the piping and is slightly smaller than the inside diameter of the pipe. The paddle moves with sustained water flow to activate a switch connected to the exterior of the pipe. Once activated, a signal is sent to the fire alarm panel to sound a fire alarm. The device returns to the normal, ready position once the water flow stops, leaving only the fire alarm system to be reset and put back in service.

Because the velocity and force of the water rushing through the pipe could break the paddle arm and create a blockage somewhere in the pipe, paddle-type water flow devices are not installed on dry, preaction, or deluge systems. These devices have an adjustable delay mechanism to compensate for fluctuations in water pressure surges between the sprinkler system and the water supply. The delay mechanism device is usually set to between 30 and 60 seconds, because setting it to no delay will result in multiple false alarms due to water surges.

Another type of water flow detection device, the **pressure switch**, is used to detect water flow in dry, deluge, preaction, and older wet sprinkler systems that use mechanical bell alarms. Pressure switches operate when the preset threshold either rises above or drops below the pressure setting. There is no time delay mechanism; once the pressure threshold is reached, a signal is sent to the alarm panel. Unlike a paddle-type water flow device, a pressure switch may require a manual reset after it is activated.

Not only can pressure switches be used to monitor water flow, but they can also be used to monitor air pressure changes in dry and preaction systems and pressure tanks. When conditions change, a pressure switch can generate a supervisory signal that notifies the system of low- and high-air-pressure conditions.

Electronic supervision of water control valves through the fire alarm system is critical because one valve in the wrong position could render an entire fire suppression system out of service. Supervision occurs through a **tamper switch**, a device that is either integrated as part of the valve or externally attached to the valve. With as little as a few turns of the valve handle, a signal is sent to the fire alarm control panel; this signal, in turn, initiates a local supervisory signal and in some cases notifies an offsite party.

Other monitoring devices that are critical to ensure the system will perform as required include air and water temperature sensors and water level sensors. Temperature sensors detect drops or increases in temperature in buildings and systems where temperatures out of the norm could render a system useless. Water level sensors monitor the amount of water in a storage tank. If the water level falls or rises above the normal level, the sensor notifies the responsible party to take the appropriate action.

Non-Water-Based Fire Suppression Systems
General Information

Non-water-based fire suppression provides an alternative to water-based fire suppression when water may not be the best choice for the hazard involved.

This type of suppression is also referred to as special hazard fire suppression because it uses gaseous and chemical-based agents to protect special and specific hazards, processes, and equipment.

Purpose of Systems

When choosing a suppression agent, water is usually the first and obvious choice. In some conditions and situations, however, water would have little or no effect, would require an additive to work appropriately, would not be appropriate for the hazard, or would make the situation worse. Non-water-based fire suppression provides a means to handle and control these kinds of challenging fire conditions, which would otherwise be more problematic or inappropriate to mitigate with water-based suppression. The most common non-water-based systems use wet and dry chemicals, dry powders, and gaseous agents.

System Components

The design of many non-water-based fire suppression systems starts with extensive laboratory testing, where engineers study and analyze how the agent and associated systems handle certain fire conditions and scenarios. Based on these data, manufacturers can provide engineered and pre-engineered systems that incorporate the necessary component type, size, proper location, design criteria, and design limitations. Unlike in water-based fire suppression systems, in which the majority of the components are interchangeable, non-water-based system components are specific to the type of system and the agent in use and cannot be interchanged.

Suppression Agents

Gaseous suppression agents are divided into three basic categories: halon, halon replacements, and inert gases. Chemical agents are either dry or wet chemicals or chemical-based dry powders.

Halon and Halon Replacements

Halogenated hydrocarbons (halons) mix carbon and one or more elements from the halogen series in the periodic table: fluorine, chlorine, bromine, or iodine. The chemical makeup of a halon gas is based on a numbering system that lists the number of atoms for each element that make up the agent. For example, Halon 1211 is a mixture of one part carbon, two parts fluorine, one part chlorine, and one part bromine, while Halon 1301 is a mixture of one part carbon, three parts fluorine, zero parts chlorine, and one part bromine. Because neither of these halons uses iodine, no corresponding number for iodine is attached to the formula. Halon 1211 and 1301 are the only two halons still in use with fire protection systems or fire extinguishers.

Halon agents are odorless, colorless, noncorrosive, nonconductive, and nontoxic at low concentration levels. However, these products have been generally banned because of their effects on the environment. Reclaimed gas and existing reserves are still permitted to be used, so it may be a number of years before all halon gases are depleted. However, eventually all halon systems will be obsolete because manufacturers have stopped making replacement parts for these systems and offer little or no technician support for older systems.

To overcome this problem, manufacturers started to develop new inert extinguishing agents called **halocarbons (clean agents)**. The goal was to address the environmental concerns and potential toxicity by producing gases that were environmentally safe and not electrically conductive, vaporized readily, and left no residue upon evaporation. Although they were successful in addressing the environmental concerns, the clean agents were never a direct replacement for halon because it took more clean agents and different components to provide the same level of protection.

Clean agents fall into two categories: halocarbon-based agents or **inert gas**-based agents. Halocarbon-based agents extinguish fires by eliminating the heat from the flame's reaction zone or by interrupting the uninhibited chain reaction. They are grouped into five subcategories: FC, FK, FIC, HFC, and HCFC. These designations identify the chemical composition of the mixture, which may include carbon, chlorine, bromine, fluorine, iodine, and hydrogen. HFC-227ea (heptafluoropropane, C_3F_7H) is one of the most commonly used clean agents; it has the trade name FM-200.

Inert agents reduce the oxygen in the protected area to a level that will not sustain combustion. Inert gases (IG) are a mixture of helium, neon, argon, nitrogen, and small amounts of carbon dioxide. IG-541 (N_2ArCO_2) is a combination of nitrogen, argon, and about 8 percent carbon dioxide; it has the trade name Inergen.

///////////////////////

TIP

Timely investigation is important because some dry chemical agents leave corrosive residue that, when exposed to moisture, will cause damage, especially to electronic equipment and components, potentially contaminating evidence.

Inert Gases

With few exceptions, carbon dioxide (CO_2) gas is a very good extinguishing agent with a long history of use for almost all combustible materials, flammable and combustible liquids, and electrical fires. This odorless, colorless, noncombustible, nonconductive gas leaves no residue. What makes CO_2 such a good suppression agent is that it displaces and reduces the level of oxygen below the 15 percent ordinarily necessary to sustain combustion. Because CO_2 is approximately 50 percent heavier than air, it will separate the air from the fuel, making it an especially good choice when dealing with a deep-seated fire. However, this gas is dangerous insofar as a person in a confined space who is directly or indirectly exposed to discharged CO_2 risks asphyxiation due to its displacement of oxygen. Therefore, a person working in the area should leave once a predischarge warning is issued. Because of this hazard, the decision to install a CO_2 system requires a life-safety analysis.

The pressure and temperature of the environment determine whether CO_2 exists as a gas, a solid, or a liquid. When used as a suppression agent, it is stored in tanks as a highly pressurized liquid that expands to a gas and combines with moisture in the air to form a cloud that consumes the hazard.

TIP

Before entering any area where a gas-based suppression agent has discharged, especially in an area protected by a CO_2 system, all emergency responders and investigators should wear self-contained breathing apparatus and take air quality readings to determine the amount of oxygen, residual gas-based suppression agent, or other gases present to ensure the area is safe to enter.

Dry Chemical

Dry chemical extinguishing agents are small, solid, powdery particles that extinguish the flame by covering and smothering the protected surface area of a fire, cutting off the oxygen to the fuel to prevent reignition and flame spread to adjacent areas. Dry chemical extinguishing agents adhere to, coat, and insulate the surface area of the hazard, whereas other types of agents would run off the hazard and be ineffective. In addition, dry chemical agents do not react with flammable liquids and gases, and they are not conductive. The particles in these agents use pressurized nitrogen, carbon dioxide, or air as the transport medium. They are suspended in the gas to facilitate flow and distribution of the agent, which makes dry chemicals suitable for use with fire extinguishers and hose lines. Upon activation, the particle/gas mixture flows out of the storage container, into the piping network, and out of the nozzles. Although not considered dangerous or toxic, the discharging agent is under pressure and creates a cloud that limits visibility and can cause respiratory problems. The hazard and protection requirements determine the type of nozzle used with dry chemicals and the placement of the nozzle.

Dry chemical agents are used primarily to extinguish fires in flammable liquids. They should not be confused with "dry powder" agents, which are primarily used on Class D fires.

TIP

Manufacturers of dry chemical systems and agents have stopped making fixed dry chemical extinguishing systems and parts for commercial kitchen applications, as such agents have failed to meet the UL 300 test standard for control and extinguishment of commercial kitchen fires and, therefore, lost their approval for these applications.

Sodium Bicarbonate–Based Dry Chemicals

Sodium bicarbonate–based agents are also known as **regular or ordinary dry chemicals**. Similar to baking soda, sodium bicarbonate is one of the oldest and most widely used dry chemical agents. This compound has historically been used for fire suppression in cooking operations, as the agent chemically reacts with the oils and fats to create a soap-like liquid that covers the surface of the hazard (saponification). However, the use of dry chemical agents related to cooking has greatly decreased in recent times because today's larger, energy-efficient cooking equipment now heats cooking oils to higher temperatures, and the dry chemical agents cannot provide sufficient cooling and isolation to prevent reignition. Sodium bicarbonate is generally not recommended for use on Class A fires. To differentiate it from other dry chemical agents, the sodium bicarbonate is colored blue or white.

Dry Chemicals Based on Potassium Salts

Potassium bicarbonate, potassium chloride, and urea-based potassium agents are considered more effective than sodium bicarbonate because smaller amounts are needed to provide much higher fire extinguishing capabilities. These agents work well on Class B fires, except for those associated with cooking operations. Potassium salts are also effective on Class C fires but are not recommended for Class A fires. Potassium bicarbonate is purple in color to differentiate it from other dry chemicals.

Multipurpose Dry Chemical

Ammonium phosphate is a multipurpose dry chemical that works very well on Class A, B, and C fires. However, it is not effective on fires associated with cooking operations, especially those involving deep-fat fryers. When heated, this agent decomposes and forms a molten residue that sticks to the heated material to isolate the oxygen from the material. Ammonium phosphate is colored yellow to differentiate it from other dry chemicals.

Wet Chemicals

Wet chemical extinguishing agents are proprietary water-based solutions. In these chemicals, water is mixed with potassium acetate, potassium carbonate, potassium citrate, or, in some instances, a mixture of agents and other additives to form an alkaline solution. Just like the ordinary dry chemical agents, wet chemical agents react with the cooking oil or fat to form a soapy foam blanket (saponification) that separates the oxygen and the fuel for a fire. The soapy foam blanket decreases or eliminates the fuel vapors by separating, smothering, and cooling the fuel to prevent reignition. Whereas industrial kitchens use a number of different suppression agents, commercial kitchens use wet chemical extinguishing agents because they are the most effective agent for Class K fires that could start in appliances such as deep-fat fryers; ranges; griddles; grills; woks; or char, chain, and upright broilers. Wet chemical systems are designed as local application systems.

Condensed and Dispersed Aerosols

Condensed and dispersed aerosols are used in either total flooding or local application systems. A condensed aerosol consists of finely divided solid particles and gaseous matter generated by a combustion process of a solid aerosol-forming compound. In contrast, a dispersed aerosol consists of finely divided particles of chemicals that are present in a pressurized agent storage container and suspended in a halocarbon or inert gas.

Key System Components
Suppression Agent Supply

The containers that store the suppression agent must be made of materials (usually metals) that will not react with the agent, and must be of an appropriate size to store the amount of agent needed to protect the hazard. They must also be able to handle the pressures that will be encountered; this is especially true for gas-based agents, which can be pressurized to high levels in their containers. A limited supply of agent is available in any fire suppression system, and the container will define the amount available. As with all fire protection components and systems, containers must be listed, approved, or labeled by a nationally recognized testing and certifying agency.

Pressure Sources

Depending on the type of system and agent, some containers may simply store the agent. In that case, a secondary pressure cartridge or pressure container, upon activation, will either push the agent out of the container or activate a valve to release the agent from the container. Usually, the pressure cartridge or container is filled with nitrogen or carbon dioxide gas as the propellant. This design is common with dry chemical systems and some wet chemical systems.

Distribution Piping

The material used for piping must be noncombustible and compatible with the agent that will flow through the pipe. In addition, the piping must be able to handle the anticipated storage and operating pressures and the environment that the piping is protecting. Depending on the type of system and agent, the materials used for this purpose are likely to be some type of metal—for example, black iron, chrome-plated or stainless steel, galvanized steel, copper, or brass.

Valves, Hoses, and Fittings

Any valve, hose, fitting, or component that is part of a fire suppression system must be listed, approved, or labeled for use with the system by a nationally recognized testing and certification agency. Valves, hoses, fittings, or other components that are installed but not listed, approved, or labeled for use with the system could cause the system not to work properly or to fail completely, resulting in an out-of-control fire.

Distribution Nozzles

A nozzle distributes the extinguishing agent onto the hazard. Each system manufacturer makes different types of nozzles for its specific suppression system, because each nozzle is designed for an intended application that requires listing, approval, or labeling for the particular application. Nozzles must be made of noncombustible, corrosion-resistant materials that will not deform or fail if exposed to fire. In addition, nozzles must be permanently marked for proper identification and installed to ensure the correct nozzle will operate without failure at the anticipated discharge pressures and to protect the hazard.

Actuation System

Suppression systems may use either automatic or manual actuation methods to release the extinguishing agent onto the hazard. Actuation systems include fusible links, heat detectors, smoke detectors, fire sprinkler heads, and manual fire alarm boxes. Usually, these actuation devices tie into a fire alarm and detection system that initiates an alarm signal. Fusible links are common in wet and dry chemical systems, whereas heat and smoke detectors are often found in gas-based systems. Heat detectors, smoke detectors, and fire sprinkler heads are usually associated with expansion foam systems.

System Monitoring and Control

Just like fire sprinkler systems, fire suppression systems are monitored and controlled by various devices and systems that notify those responsible when the status of the system has changed. Status changes could involve system activation, a change in container pressure, a valve closure, a temperature change, or a system leak. Typically, the monitoring takes place in conjunction with the building fire alarm and detection system, with notifications being sent locally or to an offsite organization that will notify the responsible parties. Most gas-based systems require a secondary fire alarm panel that handles special safety functions such as prewarning alarms to ensure occupants are able to exit the hazard area before the agent discharges. The system may also be required to shut off fuel or heat sources.

Operation and Installation Parameters of the System
Location and Spacing of Nozzles

Nozzle location and spacing are determined by the manufacturer's installation guidelines and the applicable codes and standards, but the type of nozzle and the hazard will also influence the placement. Since many non-water-based fire suppression systems are engineered or pre-engineered systems, it is important to precisely follow the system design and nozzle installation requirements to ensure appropriate coverage of the hazard. Usually, the higher the hazard level, the larger the number of nozzles, spaced closer together, that are required to protect the hazard.

Pipe Size and Arrangement

Depending on the suppression system, the pipe size and arrangement may be pre-engineered based on laboratory testing by the manufacturer or, if not pre-engineered, hydraulically calculated. Pre-engineered systems have the advantage of protecting known hazards that generally do not change, making it easy to follow the manufacturer's guidelines. However, when a hazard changes—for example, in terms of the size of the protected area or the size of the hazard, or due to the environmental conditions—hydraulic calculations can be performed to determine the appropriate pipe size and location. This determination is based on the flow rate and friction losses associated with the pipe size and length, fittings, and other system devices and components as the agent flows through the pipe and out of the nozzles.

Nozzle Coverage and Distribution

Nozzles are designed to deliver the agent in a manner that provides the appropriate coverage for the hazard based on the nozzle's discharge characteristics and the pressure available. Many different nozzles are available to protect most hazards. These nozzles can be broadly categorized as those for "local application" systems versus those for "total flooding" systems. Local application systems use nozzles that are directed onto or into equipment, processes, operations, or specific areas to smother, cool, and extinguish the fire. Nozzles used with total flooding systems deliver the agent to an enclosed hazard or area within a structure. To be effective, the area must be enclosed tightly enough to hold the agent there for a sufficient period of time to extinguish the fire and prevent reignition. The numbers of nozzles and the amount of agent needed to protect a hazard or area depend on the volume of the area, the manufacturer's design and installation requirements, and the type of hazard.

Activation Mechanisms and Criteria

Non-water-based fire suppression systems use automatic and manual actuation methods to release the extinguishing agent onto the hazard. Different types of fire detectors can be used that tie into a fire alarm and detection system; activation can also occur through detection devices integrated with the suppression system. Although some systems provide a manual means to activate the extinguishing system, most activations are automatic.

Systems Monitored and Controlled

Fire suppression systems are monitored and controlled by various devices and systems, which notify those responsible that the system's status has changed. Status changes could indicate system activation, a change in container pressure, a valve closure, a temperature change, or a system leak. Again, the monitoring is carried out in conjunction with the building fire alarm and detection system, with notifications being sent locally or to an offsite organization. Most gas-based systems require a secondary fire alarm panel that handles special safety functions such as prewarning alarms to ensure occupants are able to exit the hazard area before the agent discharges.

Analysis of Suppression Systems

The fire investigator can gather important data from fire protection and fire suppression systems after a fire. When these systems are involved, it is important to begin gathering this information promptly. Postfire analysis of fire protection and suppression system is discussed in detail in Chapter 10, *Data Collection Outside the Fire Scene.*

System Documentation

As a fire investigator, you should document the fire suppression system, including its various components, their locations, and their condition. Additionally, documentation should address the system's power supply and the expected and actual activation of the detectors and suppression. Also document how the system interacted with other fire systems, including alarms and notifications, smoke control, and control of elevators, doors, barriers, and the HVAC system.

Thoroughly photograph all of the components of the system that are visible, along with any visible product from the suppression system, and sprinkler heads and other equipment that appear to have activated during the event. Check and document system parts and valves from the event back to the earliest component involved, such as where the water entered the building, if applicable. Document any displays or panels associated with the system, including their status and information on any displays.

Document any sites where any tampering with sprinklers or other systems is apparent. A person intending to burn a building may tamper with sprinkler systems in an effort to slow or disable their response, allowing the fire to grow larger and do more damage.

Codes and Standards for Fire Suppression Systems

Codes and standards adopted by a municipality establish the requirements for water-based and non-water-based fire suppression systems for certain hazards. Typically, the codes establish a requirement to install a system and the adopted standard that must be used to design, install, inspect, test, and maintain the system. Collectively, these codes and standards establish requirements for the built environment from design through construction and for the remainder of the system's life span. A majority of jurisdictions in the United States have codes that are based on NFPA 72, which may have been adopted in full or in part, or with local modifications or modifications from International Code Council (ICC) series codes. In any case, it is critical to determine which codes and standards were in effect both when the building was built and when the fire occurred. Consult the authority having jurisdiction to determine the code version, any local amendments to the code, whether the applicable code includes maintenance requirements, and whether variances were granted during the design of the building. The building's insurer may have imposed additional alarm requirements that were included in the design.

Determining which code was in place will be important to an analysis of the manner and timing in which water-based and non-water-based fire suppression systems reacted during the fire incident. Generally, if the building has not changed its use or occupancy condition from the time it was built, the codes and standards in place at the time of construction continue to apply, but performing research to find this information is critical to making the appropriate determinations.

Both water-based and non-water-based fire suppression systems are typically required to meet criteria established by NFPA codes and standards. NFPA standards match the type of system to the type of occupancy and the hazard needing protection. Recall that there are NFPA standards that deal with every type of fire protection system. Thus, use of the appropriate standard for the system installed is critical **TABLE 3-8**.

The ICC-established model code series provides for a platform of minimum requirements applicable to the built environment. These minimum requirements can be (and many times are) amended to reflect the emphasis in a particular municipality, so they often vary from locality to locality.

Design Analysis

To interpret the actions and effects that a fire suppression system had during and against a fire, you must

TABLE 3-8 NFPA Fire Protection System Standards

NFPA 10, *Standard for Portable Fire Extinguishers*
NFPA 11, *Standard for Low-, Medium-, and High-Expansion Foam*
NFPA 12, *Standard on Carbon Dioxide Extinguishing Systems*
NFPA 12A, *Standard on Halon 1301 Fire Extinguishing Systems*
NFPA 13, *Standard for the Installation of Sprinkler Systems*
NFPA 13D, *Standard for the Installation of Sprinkler Systems in One- and Two-Family Dwellings and Manufactured Homes*
NFPA 13R, *Standard for the Installation of Sprinkler Systems in Low-Rise Residential Occupancies*
NFPA 14, *Standard for the Installation of Standpipe and Hose Systems*
NFPA 15, *Standard for Water Spray Fixed Systems for Fire Protection*
NFPA 16, *Standard for the Installation of Foam–Water Sprinkler and Foam–Water Spray Systems (withdrawn in 2020 and incorporated into NFPA 11)*
NFPA 17, *Standard for Dry Chemical Extinguishing Systems*
NFPA 17A, *Standard for Wet Chemical Extinguishing Systems*
NFPA 20, *Standard for the Illustration of Stationary Pumps for Fire Protection*
NFPA 22, *Standard for Water Tanks for Private Fire Protection*
NFPA 24, *Standard for the Installation of Private Fire Service Mains and Their Appurtenances*
NFPA 25, *Standard for the Inspection, Testing, and Maintenance of Water-Based Fire Protection Systems*
NFPA 70, *National Electrical Code*
NFPA 72, *National Fire Alarm and Signaling Code*
NFPA 750, *Standard on Water Mist Fire Protection Systems*
NFPA 2001, *Standard on Clean Agent Fire Extinguishing Systems*

understand the basis for the system design, functionality, and capabilities.

The design of a system takes a number of factors into consideration relative to the hazard it must protect. Assuming that the installed system was approved and installed as designed, it is up to you as the investigator to evaluate whether the installed system was still able to protect the hazard. It is common for the hazard level to change but for systems to stay the same. If the hazard level increases, the system may not be designed to deliver the amount of suppression agent necessary to control or extinguish a fire. To make this determination, you may need to review manufacturer information, codes, and design standards, and may have to consult with a fire protection engineer or design professional.

Hazard Protection

The selection of a water-based or non-water-based fire suppression system is predicated on the inherent hazard(s) throughout a property. In addition, some requirements to install a particular type of system may be derived from a building code or an insurance company requirement. Once the hazard is determined, the design of the system will consider the appropriate delivery system, how the system detects fire and becomes activated, how much suppression agent is required, how it will be applied, the rate of application, and the concentration. Hazards change frequently; therefore, it is important to evaluate the hazard and conditions and then to examine and analyze the type of system installed relative to the hazard encountered to determine if the system was adequate for the task. Clearly, understanding the nature of the hazard is important to this process.

To illustrate this point, years of laboratory testing and analysis of fire scenes and fire scene data led to the development of tables that classify hazards in relation to the amount of fuel load and the potential output of the fuel load, which are published in NFPA 13. The resulting density/area curves list the amount of water, in gallons per square foot over a certain area in square feet, that is needed to control or extinguish a fire for the hazard condition. The hazard condition categories are as follows:

- Light Hazard
- Ordinary Hazard (Groups 1 and 2)
- Extra Hazard (Groups 1 and 2)
- Special Occupancy Hazard

The underlying idea is that one can evaluate the fuel load for the occupancy condition and correlate that condition to the hazard classification. For example, a school is designed as a Light Hazard building or occupancy, so it would not require as much water to discharge from the sprinkler heads as would a fire in a sawmill, which is classified as an Extra Hazard Group 1 building.

It is very common for a building to be designed and built for a specific use or occupancy condition but later change to a different use or occupancy condition. If this change increases the hazard level, and if the water-based fire suppression system is not updated to handle the new hazard, a fire could overwhelm the suppression system and lead to a catastrophic loss. For example, if a building was once a public laundry but was purchased and became a dry cleaner, the water-based fire suppression system would need to be changed to support the increased hazard level. Although it would appear that the use or occupancy condition is the same or very similar with both types of businesses, a dry cleaner has a greater fuel load because of the amount of clothes (and in some cases chemicals) present, whereas a laundry uses water and soaps that pose a lesser hazard.

Other aspects to consider when evaluating the system's design include the appropriateness of the equipment, including initiating devices, location and spacing of components, and system monitoring. Evaluate whether the system is appropriately powered and designed from an electrical standpoint and whether the notification appliances were appropriately audible, visible, and intelligible.

Placement

As with other types of building systems, water-based and non-water-based fire suppression systems must be properly designed, installed, inspected, tested, and maintained to ensure their optimal performance during a fire incident. Both water-based and non-water-based fire suppression systems are designed to control or fully suppress a fire. To achieve that design objective, sprinkler head and nozzle placement is critically important. During the course of the fire scene analysis, document sprinkler and nozzle locations and spacing relative to walls, ceilings (including suspended ceilings), storage arrangements, shelving, light fixtures, and ductwork, along with any other condition that may have affected the proper operation of the sprinkler head or nozzle. As part of any investigation, you must determine whether sprinkler heads or nozzles were obstructed, damaged, displaced, or improperly placed, allowing for inadequate control and uninhibited fire spread. Additionally, document and examine the hazards, fuel packages, or processes protected by the sprinkler heads or nozzles to determine these devices' effectiveness. Moving an expected fuel or target (e.g., relocation of cooking equipment) can diminish the effectiveness.

The location and positioning of detectors, initiating appliances, and warning appliances are also important, as they have an influence on system activation times and occupant responses.

Verify which design standard was used for the system, and review the approved plans, reference standards, and manufacturer's data sheets against the installed system. Of special importance is the review of the manufacturer's data sheets for the sprinkler heads or nozzles. You may need to establish whether the installed system was capable of handling the fire or if any system manipulation occurred prior to the fire that may have enabled fire propagation. Many issues can affect system performance, but sprinkler head and nozzle placement is critical to successful distribution of the suppression agent.

Capacity

Water-based and non-water-based fire suppression systems are designed to deliver predetermined quantities of water or some other agent, typically through sprinkler heads and nozzles over a given area, and to provide additional water capacity for fire hose streams. Because system design is based on occupancy and hazard classification, the investigator should understand the capacity for which the system was designed and how the installed system reacted during the fire incident.

For example, NFPA 13 states that an educational facility is classified as a Light Hazard and, based on the density/area curve for that hazard level, requires 0.10 gallon of water per square foot (4 liters per square meter) over 1500 ft^2 (139 m^2) of area. In other words, every sprinkler head within the designated 1500-ft^2 (139-m^2) area must provide at least 0.10 gallon per square foot (4 liters per square meter).

This information can be found on the **hydraulic data nameplate**, which states the important design characteristics of the system **FIGURE 3-24**. During system installation, a hydraulic data nameplate is required to be posted on or near the system valve to which the information applies. It can prove very valuable when determining whether the system that was installed was appropriate for the hazard condition at the time of the fire.

During the postincident examination, fire investigators will document the number of devices that operated during the fire. Depending on the number and location of the devices that were activated, they will determine whether the system design was appropriate for the hazard level or if other factors were present that compromised system performance. They will document all mitigating items and factors, including appropriate and allowable fuel loads, the possibility of system tampering (e.g., closed valves), water supply reductions, increased water flow due to more than anticipated sprinkler or nozzle activation, structural and nonstructural obstructions to device operation for proper water delivery, and interjection of ignitable liquids or other incendiary compounds to create a larger fire or retard fire suppression system performance.

Coverage

The amount of coverage that a fire suppression system provides depends on a number of factors, including when the building was built, the codes in force

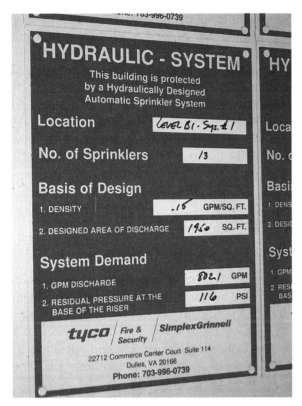

FIGURE 3-24 This hydraulic data nameplate provides information concerning the fire sprinkler system design and performance characteristics that can be valuable when investigating a fire.
© A. Maurice Jones, Jr./Jones & Bartlett Learning.

at the time of construction, and the design standards that applied to the type of system. Therefore, to analyze and assess flame spread and fire damage, you as the fire investigator must determine the extent of suppression system coverage within a structure. The type and extent of suppression system coverage will have a direct impact on flame spread and fire damage. For example, full coverage will minimize flame spread and fire damage, whereas a building with limited or no coverage will probably suffer extensive flame spread and heavy fire damage.

Total (Complete) Coverage

It can be very helpful to understand and remember some basic code and design standard terms and definitions. If a system is designed for total or complete coverage, the design requirement should encompass every space within a structure, including concealed areas and spaces, unless there is a specific allowable exception. Generally, any such exception will require passive protection, such as fire-rated construction, in lieu of fire suppression system coverage. Total coverage systems are common in many commercial properties that exceed certain code-defined thresholds. For assistance in determining the requirements for

coverage, contact the local code office that provided the review, approvals, and inspections for the systems.

Partial or Selective Coverage

Partial or selective coverage is based on use and occupancy conditions in a building that may not require full or any coverage. For example, coverage might be provided in portions of a building that have a higher hazard level, or sprinkler coverage might be omitted in areas in which such absence is permitted by code. The portions protected or covered are designated by specific code and standard requirements.

Local Coverage

Sometimes you may encounter a fire suppression system that protects only a piece of equipment or machine, but not any part of the building around the equipment or machine. Typically, these situations exist due to specific insurance company requirements or owner preference. Often this protection takes the form of a water spray system using open nozzles, combined with heat or flame detection. You should approach these systems designed for local coverage in the same way that you evaluate code-required coverage, assessing the design parameters for appropriateness and proper functionality.

Installation

Becoming familiar with the appropriate technical resources and system design and installation standards can help fire investigators who are examining fire suppression systems, because these systems may require complicated and detailed engineering analysis. Interview those occupants who were nearest to the system during its activation to understand any details regarding the fire and system functionality. Other people who can be interviewed to obtain information about the system include engineers, design professionals, manufacturers' representatives, personnel from the regulatory agency that reviewed and approved the design and inspected the system, and the installing contractor(s).

Examination, analysis, and documentation of a water-based fire suppression system installation should begin with the incoming water supply, whether it is from the local water purveyor, a tank, or another resource. Check whether the main water supply valve is open to confirm that there is an available water supply. The examination then progresses along the system supply line to confirm that system control valves are open. Once the water supply is confirmed, identify and examine components such as risers and the main supply piping, control and check valves, smaller branch or sprinkler line piping, and water delivery devices such as sprinkler heads or nozzles for proper type and orientation. Verify proper component installation and operation, and examine and document improper component installation and operation throughout the system. Examples of improperly installed components include incorrect sprinkler head orientation, delayed sprinkler operation due to the use of improper temperature-rated sprinkler heads, check valves installed backward, and fully or partially closed water control valves.

System Performance

As the fire investigator, you must determine and document how the system performed or if it performed at all. When considering the system's performance, examine whether it had appropriate power and was functional and the conditions under which it was operating. Evaluate the time required to activate initiating and notification appliances, and determine whether the interconnection with other systems was functional and appropriate. Examine occupant response, as well as any evidence of transmission of, and response to, alarm and trouble signals.

Evaluation of system performance is similar to installation analysis, requiring you to apply a thorough failure analysis process through examination, analysis, and documentation of the suppression system's role in the incident. Improper installation will affect system performance; therefore, you must verify whether the system operated correctly or, if the system failed to operate correctly, why.

Testing and Maintenance

The main reasons that fire suppression systems fail are improper inspection, testing, and maintenance. Systems must be properly tested and maintained at regular intervals in accordance with manufacturers' recommendations and adopted standards to ensure and verify their correct operation and full functionality.

As part of an investigation, consider requesting test records, maintenance records, and documentation for the system to determine if the minimum required maintenance, inspections, and tests were performed. These records can help demonstrate whether unusual conditions were found, whether corrections were needed and completed, and what the current status of the system(s) was. Document whether the required inspection tags were attached to the system. Additionally, ascertain whether a system was connected to an

offsite monitoring company, which might be able to provide additional data related to system operation, functionality, and notification.

Water-based systems should be inspected, tested, and maintained based on the requirements in NFPA 25, *Standard for the Inspection, Testing, and Maintenance of Water-Based Fire Protection Systems*, which establishes minimum intervals for the different systems and components. Other system-specific NFPA standards may provide guidelines for inspection, testing, and maintenance, as may local building codes.

Impact on Human Behavior

Human behavior information should be examined in every fire investigation, including people's interaction with the fire and the fire suppression system. It is very important to analyze occupant awareness and response prior to and during system operation. This information can assist you in determining the fire origin, growth, spread, and cause.

Buildings protected by fire suppression systems that are properly designed, installed, inspected, tested, and maintained offer the occupants a significantly better chance of survival than is present in a building without such protection. When interfaced with a fire alarm system, not only does activation of the system alert occupants to hostile fire conditions, but fire suppression systems can also control and/or suppress the fire, allowing for increased survivability and greater time to respond appropriately to the conditions and circumstances of the fire event. Having adequate time to react and respond is the single most important factor in survivability. Even so, occupants must not take for granted that a system is in place and that they have more time to respond, because these systems do not remove smoke and gases that could be in the air prior to system activation.

Exposure to some types of suppression agents may inhibit visibility or breathing, disorient occupants, or impair a person from seeking refuge, escaping, or exiting an area. Although such events are very rare, death and injury can occur even when systems operate correctly.

Building Electrical and Gas Systems

Modern buildings commonly have electrical systems, including service equipment, one or more fuse or circuit breaker panels, branch circuit wiring, and switches and outlets. In many building fires, the fire investigator will need to include electricity in the hypotheses considered as a potential heat (ignition) source.

The presence of electrical wiring or equipment in the area of origin does not, in itself, indicate that it was the ignition source. The investigator will need to identify whether a given electrical item acted as an ignition source, and what failure mode or other situation occurred to allow it to do so. This requires an understanding of how to determine whether an electrical item caused the fire, or was victim of it.

Gas distribution systems are also common in buildings and may be found within an area of origin. As with electrical systems, their presence does not mean they are involved in the fire cause, and the investigator must have an understanding of such systems to appropriately evaluate their role, if any, in a fire. Chapter 7, *Electricity and Fuel Gas Systems*, is devoted to the discussion of building electrical and gas systems.

After-Action REVIEW

IN SUMMARY

- The design, materials, and construction type of a structure influence the development and movement of any fire that may occur within it.
- Modern design considerations and construction features are a direct result of the analysis of (often catastrophic) fires.
- A building's characteristics affect the development, spread, and control of a fire. These characteristics include the type of construction; the integrity and performance of its structural elements under a fire load; and its building systems, including but not limited to active and passive fire protection systems.
- Determine and document the type of construction based on the structural elements of the building. Document structural elements, breaches, structural changes, or other factors that may influence the building's integrity or fire spread within the structure.
- The way that assemblies react in a fire often influences how a fire grows and spreads. Because assemblies are designed as a complete unit, the integrity of the unit and its ability to perform during a fire are dependent on the unit being manufactured, installed, and maintained in the form for which it was intended.
- Weather can impact systems by, for example, wind spreading the fire, rain and snow increasing the load, increased humidity raising the fuel moisture, increased temperature speeding evaporation, and more.
- Fire spread is affected by the materials used in the structure's construction, such as steel, reinforced concrete, and wood.
- Passive fire protection can best be defined as fire resistance–rated wall and floor assemblies used to form fire-rated compartmentalized areas meant to control the spread of fire. These assemblies can take the form of occupancy separations, fire partitions, or fire walls used to keep fires, high temperatures, and flue gases within the fire compartment of origin and to aid firefighting and evacuation processes.
- Active fire protection systems can be a valuable tool when analyzing a fire event. The type of system and the code and standard in force when the system was installed are factors that will determine how much information the system is designed to provide.
- A fire alarm system can provide direct notification to public safety answering points and to proprietary, central, and remote supervising stations that initiate the appropriate response without the need for an individual to call in the fire emergency. This alert will expedite the emergency response, which in turn may help limit property damage and save a life.
- A water-based fire suppression system is intended to protect property and life safety by delivering sufficient amounts of water over the fire area at the appropriate rate in an effort to control, and in some cases extinguish, the fire.
- Non-water-based fire suppression provides an alternative to water-based fire suppression when water may not be the best choice for the hazard involved. It is also referred to as special hazard fire suppression because it uses gaseous and chemical-based agents to protect special and specific hazards, processes, and equipment.

KEY TERMS

Active fire protection system A system that uses moving mechanical or electrical parts to achieve a fire protection goal. (NFPA 72)

Addressable technology Fire alarm and detection systems in which devices and system components are assigned a data address, are capable of two-way communication, and can perform certain activities based on signal input.

Air sampling smoke detection A method of smoke detection designed to draw air from the protected area into a detection chamber for analysis.

Alarm signal A warning signal that alerts occupants of a fire emergency.

Annunciation panel A device that uses indicating lamps, alphanumeric displays, or other means to provide first responders with status, condition, and location information concerning fire alarm system components and devices.

Assemblies Manufactured parts put together to make a completed product.

Balanced pressure proportioner A proportioner that uses an atmospheric tank and pump to introduce the foam concentrate into the system with the water supply.

Balloon-frame construction A construction type in which the exterior wall studs go from the foundation wall to the roofline. The floor joists are attached to the walls by the use of a ribbon board, which creates an open stud channel between floors, including the basement and attic.

Boiler A closed vessel in which water is heated, steam is generated, steam is superheated, or any combination thereof by the application of heat from combustible fuels in a self-contained or attached furnace. (NFPA 85)

Central supervising station A facility that receives fire alarm signals from properties that is staffed with trained, qualified, and proficient personnel who take the appropriate action to process, record, respond, and maintain monitored fire alarm systems.

Coded signal A signal that generates a predetermined visual or audible pattern to identify the location of the initiating device that is operating.

Compartmentation A concept in which fire is kept confined in its room of origin, minimizing smoke movement to other areas of a building.

Control thermostat A device that automatically regulates temperature, or that activates a device when the temperature reaches a certain point.

Conventional technology The technology used in fire alarm and detection systems that provides limited communication between the fire alarm control panel and system devices and, upon activation of an alarm, provides only general information concerning the device and its location.

Dead load The weight of materials that are part of a building, such as the structural components, roof coverings, and mechanical equipment.

Deluge sprinkler system A type of automatic fire sprinkler system equipped with open fire sprinkler heads, which simultaneously discharges water from all sprinkler heads.

Density/area curves Graphs that establish the relationship between the required amount of water flow from a sprinkler head (density) and the area that must be covered by the water for different hazard classifications.

Design density The minimum predetermined amount of water that must flow from the fire sprinkler heads in the most hydraulically demanding part of the fire sprinkler system (remote area) to control or extinguish a fire.

Dry chemical extinguishing agents Dry powder suppression agents made from sodium bicarbonate–based, potassium-based, or ammonium phosphate chemicals that cover and smother the burning material.

Dry-pipe sprinkler system A type of automatic fire sprinkler system equipped with automatic sprinkler heads; the pipes contain pressurized air or nitrogen. When a sprinkler head activates to reduce the air or nitrogen pressure, the dry pipe valve clapper opens, flooding the system piping with water.

Duct smoke detector A smoke detector that samples the air moving through the air distribution system ductwork or plenum; upon detecting smoke, the detector sends a signal to shut down the air distribution unit, close any associated smoke dampers, or initiate smoke control system operation.

Electronic valve supervisory devices Devices or switches that integrate with or attach to a water control valve; they detect movement or changes in the position of the valve and then send a signal that the normal open position has changed.

Fire alarm control units (FACUs) Equipment that monitors the integrity of the fire alarm system's circuits and devices, processes manual and automatic input signals from the initiating devices, drives the notification appliances, provides an interface to control or activate other fire protection and building systems in a fire emergency, and provides the power to support all of the system devices.

Fire alarm system A group of components assembled to monitor and annunciate the status of automatic and manual initiating devices so building occupants may respond appropriately.

Fire barrier A wall, other than a fire wall, that has a fire-resistance rating. Fire walls and fire barrier walls do not need to meet the same requirements as smoke barriers. (NFPA 101)

Fire wall A wall separating buildings or subdividing a building to prevent the spread of fire, with a fire-resistance rating and structural stability. (NFPA 221)

Flame detector A radiant energy detector that senses specific portions of the visible and invisible light spectrum.

Foam concentrate A condensed form of the foam product.

Foam solution A solution created by mixing foam concentrate and water in the correct proportions.

Friction loss As water flows away from the source through a water system, a progressive pressure drop due to changes in the direction of the pipe, length of pipe, type of pipe, size of the pipe, types and sizes of the fittings, and other components in line with the piping.

Furnace A device used to heat air for residential homes and commercial buildings.

Gas-sensing fire detection A type of detector designed to sense a specific gas, toxic gases, or a variety of gases and vapors produced when processing hydrocarbons.

Halocarbons (clean agents) Chemical compounds made up of carbon and hydrogen, chlorine, fluorine, bromine, or iodine.

Halogenated hydrocarbons (halons) Chemical compounds made up of carbon and one or more elements from the halogen series of elements (fluorine, chlorine, bromine, or iodine).

Heat detector An initiating device that operates after detecting a predetermined fixed temperature or sensing a specified rate of temperature change.

Heavy timber construction A construction type in which structural members (i.e., columns, beams, arches, floors, and roofs) are made of unprotected, solid, or laminated wood, with large cross-sectional areas (6 or 8 inches [15 or 20 cm] in the smallest dimension, depending on the reference).

Hydraulic data nameplate A permanent and rigid sign posted on or near a fire sprinkler system riser that provides information about the hazard, design density, size of the remote area, required gallons and pressure for the water supply, and location that the system is serving.

Hydraulic design A mathematical method of determining flow and pressure at any point along the sprinkler system piping for the purpose of determining pipe size throughout the system.

Inert gas A gas that could contain a mixture of helium, neon, argon, nitrogen, and small amounts of carbon dioxide, and that does not become contaminated, react, or become flammable.

Initiating devices Systems of components and devices that provide a manual or automatic means to activate fire alarm and supervisory signals.

Interstitial spaces The spaces between the building frame and interior walls and the exterior façade, and between ceilings and the bottom face to the floor or deck above. (NFPA 921)

Ionization smoke detector A smoke detector that uses a small amount of a radioactive material, which electrically charges air particles to produce a measurable current in the sensing chamber; once smoke enters the sensing chamber, a current drop below a predetermined level initiates an alarm.

K-factor A number assigned to represent the discharge coefficient for the orifice of a sprinkler head that is used when calculating the water flow or water pressure at a specific location in the fire sprinkler system.

Laminated beams Structural elements that have the same characteristics as solid wood beams. They are composed of many wood planks that are glued or laminated together to form one solid beam, and are designed for interior use; the effects of weathering decrease their load-bearing ability.

Lightweight wood trusses Similar to other trusses in design; individual members are fastened using nails, staples, glue, metal gusset plates, or wooden gusset plates.

Line-type detector A type of detection device that uses a tube or wires running in different directions as the sensing element to provide coverage over a wide area.

Live load The weight of temporary loads that needs to be factored into the weight-carrying capacity of the structure, such as furniture, furnishings, equipment, machinery, snow, and rainwater.

Manual fire alarm box A type of initiating device that requires a person to pull a handle on the device to send an alarm signal.

Manual initiating device A type of initiating device that requires a person to make physical contact with the device to operate the device.

Manufactured housing A construction technique whereby the structure is built in one or more sections; these sections are then transported to, and assembled, at the building site.

Mill construction An early form of heavy timber construction influenced and developed largely by insurance companies that recognized a need to reduce large fire losses in factories.

Modular home A dwelling constructed in a factory and placed on a site-built foundation, all or in part, in accordance with a standard adopted, administered, and enforced by the regulatory agency, or under reciprocal agreement with the regulatory agency, for conventional site-built dwellings.

Noncoded signal A constant visual or audible signal that remains activated until reset.

Noncombustible construction A construction type used primarily in commercial and industrial storage and in high-rise construction. The major structural components (e.g., brick, stone, steel, masonry block, cast iron, or non-reinforced concrete) are noncombustible; the structure itself will not add fuel to the fire.

Non-water-based fire suppression A type of fire suppression system that uses gas- or chemical-based agents.

Notification appliances Constant visual or audible signals that remain activated until reset.

Ordinary construction A construction type in which exterior walls are masonry or other noncombustible material, and the roof, floor, and wall assemblies are wood. (NFPA 921)

Paddle-type water flow device A type of initiating device installed on a wet-pipe sprinkler system; a paddle inserted into the pipe moves to initiate an alarm signal when there is sustained water flow through the system piping.

Passive fire protection system Any portion of a building or structure that provides protection from fire or smoke without any type of system activation or movement. (NFPA 921)

Photoelectric detector A smoke detector that uses a light source and receiver within the sensing chamber to initiate an alarm when the light source reflects or is obscured by smoke particles.

Pipe schedule A list of pipe sizes and the number of fire sprinkler heads that the pipe can support based on the hazard classification.

Plank-and-beam construction A construction type in which a few large members replace the many small wood members used in typical wood framing; that is, large dimension beams, more widely spaced, replace the standard floor and/or roof framing of smaller dimensioned members.

Platform-frame construction The most common construction method currently used for residential and lightweight commercial construction. In this method of construction, separate platforms or floors are developed as the structure is built.

Post-and-frame construction Similar to plank-and-beam construction, but the structure uses larger elements. An example is a typical barn construction in which the posts provide a majority of the support, and the frame provides a place for the exterior finish to be applied.

Preaction sprinkler system A type of automatic fire sprinkler system equipped with automatic fire sprinkler heads that interface with fire detection equipment; it requires a fire detector to activate the system, and an automatic fire sprinkler head to flow water.

Pressure proportioner A proportioner that redirects some of the water supply into the foam concentrate tank to either exert pressure on a collapsible bladder or push the concentrate out of the tank to the proportioner for mixing.

Pressure switch A type of initiating device installed on dry, deluge, preaction, and older wet-pipe sprinkler systems; the device initiates an alarm signal once a water pressure threshold is met or a supervisory signal once an air pressure threshold is met.

Private mode notification An alarm signal that is sent to a location within a facility so that trained individuals can interpret and implement the appropriate response procedures.

Projected beam detector A fire detector that projects a light beam to a receiver over a hazard; once the beam is obscured or scattered, it activates an alarm.

Proprietary supervising station A group of fire alarm systems at one location or multiple locations that are under constant supervision and monitoring by the property owner's trained personnel.

Public mode notification An alarm signal that propagates throughout a building to audibly or visually alert the building occupants so they can take appropriate action.

Radiant energy–sensing detector A type of fire detector that is not dependent on smoke and heat plumes to operate; instead, the detector looks for specific portions of the visible and invisible light spectrum produced by flames, sparks, or embers.

Regular or ordinary dry chemicals Dry chemical agents rated only for Class B and C fires.

Remote area The minimum square footage of the most hydraulically remote pipe in a fire sprinkler system; the minimum design density must be available from all heads in that area.

Remote supervising station A supervising station to which alarm, supervisory, or trouble signals, or any combination of those signals, emanating from protected premises' fire alarm systems are received and where personnel are in attendance at all times to respond. (NFPA 72)

Residential fire sprinkler system A type of automatic fire sprinkler system equipped with fast response automatic sprinkler heads specifically made for low heat release and low water pressures.

Response time index (RTI) The amount of time a head takes to activate once exposed to temperatures above the predetermined activation temperature.

Smoke alarm A single- or multi-station smoke detector with an integrated power source, sensing device, and alarm device.

Smoke barrier A continuous membrane, either vertical or horizontal, designed and constructed to restrict the movement of smoke. (NFPA 92)

Smoke detector A fire detector that senses visible and invisible particles of combustion. (NFPA 72)

Soffits The horizontal undersides of the eaves or cornice.

Spark/ember detector A radiant energy detector that senses specific portions of the visible and invisible light spectrum.

Spot-type detector A type of detection device that provides coverage in a specific area where the sensing element is in a fixed location.

Steel-framed residential construction A construction type with characteristics similar to those of wood-frame construction, but that is noncombustible. Steel framing can lose its structural capacity during extreme exposure to heat.

Supervisory alarm signal A type of fire alarm system alert signal that indicates when the normal ready status of other fire protection systems or devices

connected to or integrated with the fire alarm panel has changed.

Tamper switch A device that detects changes in the normal position of a fire protection system valve and sends a supervisory signal when the valve is moved from this position.

Textual signal A messaging signal that provides constant and specific information via voice, pictorial, or alphanumeric means.

Trouble alarm signal A type of fire alarm system alert signal that indicates a problem with the system's integrity, such as a power or component failure, device removal, communication fault or failure, ground fault, or a break in the system wiring.

Venturi proportioner A proportioner that uses water moving over an open orifice to create a lower pressure at the opening, which draws the foam into the water stream.

Video detector A type of fire detector that uses high-definition video cameras to analyze digital images for changes in the pixels due to smoke and flame generation.

Water control valve A device used to control the flow of water.

Water flow detection device An attachment to a sprinkler system that detects water flow within the system piping and initiates an alarm signal; also called a water flow alarm device.

Water mist system A fixed fire protection system that uses specialized nozzles to discharge a very fine spray mist of water droplets that extinguish by cooling, displacing oxygen, or blocking radiant heat.

Wet chemical extinguishing agents Suppression agents that mix water with potassium acetate, potassium carbonate, potassium citrate, and, in some instances, a mixture of these agents and other additives; used primarily to suppress Class K fires.

Wet-pipe sprinkler system A type of automatic fire sprinkler system equipped with automatic fire sprinkler heads that has water in the pipes at all times; when a sprinkler head activates, water flow is immediate.

Wood-frame construction A construction type in which exterior walls and load-bearing components are wood. This type of construction is often associated with residential construction and contemporary lightweight commercial construction.

Wood I-beams Constructed with small-dimension or engineered lumber, as the top and bottom chord, with oriented strand board or plywood as the web of the beam.

REFERENCES

National Fire Protection Association. 2019. *NFPA 72: National Fire Alarm and Signaling Code,* 2019 ed. Quincy, MA: National Fire Protection Association.

National Fire Protection Association. 2019. *NFPA 85: Boiler and Combustion Systems Hazards Code,* 2019 ed. Quincy, MA: National Fire Protection Association.

National Fire Protection Association. 2020. *NFPA 92: Standard for Exhaust Systems for Air Conveying of Vapors, Gases, Mists, and Particulate Solids,* 2020 ed. Quincy, MA: National Fire Protection Association.

National Fire Protection Association. 2020. *NFPA 921: Guide for Fire and Explosion Investigations,* 2021 ed. Quincy, MA: National Fire Protection Association.

National Fire Protection Association. 2021. *NFPA 13: Standard for the Installation of Sprinkler Systems,* 2022 ed. Quincy, MA: National Fire Protection Association.

National Fire Protection Association. 2021. *NFPA 101: Life Safety Code,* 2021 ed. Quincy, MA: National Fire Protection Association.

National Fire Protection Association. 2021. *NFPA 221: Standard for Fire Walls and Fire Barrier Walls,* 2021 ed. Quincy, MA: National Fire Protection Association.

On Scene

1. List the types of building construction, and describe how you would envision a fire spreading through each type. Which aspects of each construction type could either enhance or impede the flow and spread of fire?

2. What reactions and responses might you see from the public in a building in which a fire alarm has activated? How might disability or impairment be a factor in an appropriate response to an alarm, and what could be done to minimize any problems you have identified?

3. What are some of the reasons and ways that a building occupant might change, modify, or eliminate components of an alarm or suppression system? What would the consequences potentially be from such behavior?

4. What buildings in your area might require unusual suppression systems simply due to their size or their contents?

CHAPTER 4

Fire Effects and Patterns

KNOWLEDGE OBJECTIVES

After studying this chapter, you should be able to:

- Identify and describe fire effects. (**NFPA 1033: 4.2.4**; **4.2.5**, pp. 96–105)
- Identify and describe fire patterns. (**NFPA 1033: 4.2.4**; **4.2.5**, pp. 105–108)
- Explain the significance of the fire pattern location. (**NFPA 1033: 4.2.4**; **4.2.5**, pp. 108–110)
- Describe the geometry of fire patterns. (**NFPA 1033: 4.2.4**; **4.2.5**, pp. 110–115)
- Describe the potential causes of, and misconceptions about, irregular fire patterns. (**NFPA 1033: 4.2.4**; **4.2.5**, pp. 114–115)
- Describe how to analyze fire patterns to produce a hypothesis. (**NFPA 1033: 4.2.4**; **4.2.5**, pp. 115–118)
- Describe the factors that influence fire pattern development. (**NFPA 1033: 4.2.4**; **4.2.5**, pp. 115–118)
- Describe how to map fire patterns among arced electrical conductors. (**NFPA 1033: 4.2.4**; **4.2.5**, pp. 117–118)

SKILLS OBJECTIVES

After studying this chapter, you should be able to:

- Interpret fire patterns to determine the point of origin. (**NFPA 1033: 4.2.4**; **4.2.5**, pp. 115–118)
- Analyze fire patterns to produce a hypothesis. (**NFPA 1033: 4.2.4**; **4.2.5**, pp. 115–118)

You are the Fire Investigator

You are investigating a residential structure fire. On the exterior of the residence, you observe heat and smoke damage to the outside wall above a basement bedroom window. Upon entering the residence, you observe heat and smoke damage throughout the dwelling. The basement bedroom sustained significant damage, with the heaviest damage seen near the door and on the wall with the window. The hall near the basement door sustained heat, smoke, and flame damage, and the door itself has partially burned away. The room across from the bedroom is relatively undamaged.

1. Why is the room across from the bedroom undamaged?

2. How will you determine the point of origin?

3. What information might the responding firefighters be able to provide to assist with your investigation?

Access Navigate for more practice activities.

Introduction

The collection of fire scene data requires the recognition and identification of fire effects and fire patterns. Understanding the nature of fire effects and patterns, and the difference between the two, is critical to the fire investigator's ability to correctly evaluate the origin, development, and spread of a fire. **Fire effects** are changes in or on a material as a result of the fire. When fire effects are found in the form of a shape, a fire pattern is discerned. NFPA 921, *Guide for Fire and Explosion Investigations*, defines a **fire pattern** as identifiable shape(s) or progression of fire effects or of a group of fire effects.

While a layperson can usually identify that a fire has occurred from visual cues such as burned material, the role of the fire investigator is much more complex. The fire investigator uses knowledge of fire dynamics, patterns, and effects to determine how the fire developed, and traces the development back to attempt to identify the fire's origin. In most instances, the origin of the fire must be determined before the cause can be correctly identified. The primary physical evidence available for these determinations is the materials that have been burned or that have been subject to combustion by-products such as heat, smoke, and soot.

The interpretation of fire patterns is one of the primary processes used in fire investigation. In the past, knowledge about the meaning of fire patterns was typically gained through experience and training. As the field developed, however, many traditional interpretations of fire patterns and fire effects, when subjected to testing under controlled conditions, were found to be incorrect. For example, it was once believed that crazed glass, spalled concrete, and wide-base V-patterns were a direct result of the application of an ignitible liquid.

These myths have now been soundly disproved. More research is being done on fire patterns to establish the fire dynamics that produced them. It is important for the fire investigator to maintain current knowledge of the fire science literature and new research as well as knowledge pertaining to fire patterns, including a thorough knowledge of the information provided in NFPA 921.

Fire Effects

The fire investigator should understand the various fire effects and the science underlying them. Not only do these effects form the basis of the fire patterns analyzed in almost every fire, but they also may be significant in themselves—for example, in the estimation of temperature or when like objects are compared to determine whether one has sustained greater melting or damage.

NFPA 921 classifies the various fire effects into four categories, although some effects may fit into more than one category **TABLE 4-1**. The categories are **discoloration**, including color changes, oxidation, rainbow effect, and staining; **deformation**, including melting, bending and buckling, thermal expansion, alloying, breaking of glass, collapsed furniture springs, and distorted lightbulbs; **deposition** of smoke and soot on surfaces and agglomeration of soot in smoke alarms; and **mass loss** due to materials being converted to gas by the application of heat, including calcination, loss of mass, char, and spalling. Two fire effects, clean burn and victim injuries, were not placed into categories in NFPA 921's 2021 edition.

The main factor in the damage to materials that manifests itself as fire effects is the cumulative heat exposure received by the item in question during the fire. The heat exposure is the result of heat transfer from flaming combustion, including the fire plume and

TABLE 4-1 Fire Effect Categories

Discoloration	Deformation	Deposition	Mass Loss
Color change	Alloying	Deposition of smoke on surfaces Smoke alarms—acoustic soot agglomeration	Calcination Char
Oxidation	Breaking of glass		Mass loss
Rainbow effect Staining of glass	Collapsed furniture springs Discolored lightbulbs Melting Thermal explosion		Spalling

Reprinted with permission from *NFPA 921-2021, Guide for Fire and Explosion Investigations,* Copyright © 2020, National Fire Protection Association. This reprinted material is not the complete and official position of the NFPA on the referenced subject, which is represented only by the standard in its entirety.

flame spread through a material, high-temperature combustion gases such as those in the upper layer, and radiant heat transfer. All three forms of heat transfer—radiation, convection, and conduction—can be involved in this exposure. Materials can also show damage from exposure to combustion products, such as with the deposition of smoke, which contains particulates along with liquid aerosols and gases, and/or soot onto an item.

The fire investigator learns to recognize when fire or other damage has occurred and learns to determine from observation or measurement the relative severity of the damage compared to other damage or to other items. The investigator must also recognize and document areas and lines of demarcation. **Areas of demarcation** are areas in which the fire damage or effects are similar in magnitude or characteristics. **Lines of demarcation**, when visible, are the borders between more- and less-damaged areas. When a sharp difference in damage levels is seen on either side of a line of demarcation, it indicates a significant change or difference in heat exposure.

In addition to noting the existence of a fire effect, the fire investigator should analyze and document the characteristics of the effect, such as its magnitude (depth), shape, size, location, and orientation. The investigator should also note the properties of the material—for example, the type and finish of wood, or type of plastic—as these often determine how a material reacts to heat and fire. If a fire effect in an object is being compared to the effect in another, like object, note any characteristics that might be different between the two objects. Likewise, note any context or circumstances in which the effect is found, as that might explain an unusual location or characteristic of the fire effect.

Sometimes, the fire investigator may be able to estimate the temperature a material reached in a fire based on the effect observed in the fire's aftermath, if the investigator knows the approximate temperature needed to produce that effect. This can assist in understanding the intensity or duration of heating, or the relative heat release rate from various fuels. For example, Table 6.3.10.2 of NFPA 921 provides the melting temperatures of many common materials. Keep in mind, however, that many materials have a broad range of responses to increased temperatures, and that one type of metal or plastic may show effects at a far different temperature than another type. Thus, temperature estimation may require an expert analysis to determine the nature of the metal or plastic.

Mass Loss

As a fuel item is consumed during the combustion process, the mass of that object begins to decrease. The observable loss of mass may indicate that the area of the material with the most mass loss suffered the greatest exposure to heat in terms of duration, intensity, or both. Many additional factors, however, must be considered. For example, if the more-damaged side of an object is in an area near a significant ventilation point, especially in a ventilation-controlled fire environment, then the damage may be due to fire activity around ventilation points; thus, it may not be a reliable indicator of the fire's origin.

Importantly, for fire investigators to be able to accurately draw conclusions about mass loss in an object, they must understand the characteristics of that object prior to the fire. For example, if a chair exhibits greater mass loss to one arm than the other, the investigator might conclude that there was greater heat on the side with the more damaged arm. However, if the less damaged arm was protected by fabric or another object or was in a different configuration than the more damaged arm, this conclusion may not be accurate. Conclusions related to mass loss are often easier with objects that are known to have been symmetrical before the fire.

Postfire analysis of the damaged areas can be conducted either by using exemplar items or by examining undamaged portions of the object. Prefire photographs or interviews can also be used to determine an object's prefire shape and orientation. Additionally, if the object is commercially available and the investigator knows its brand and model, an online search may reveal drawings or photographs of the item.

The rate of mass loss also changes during a fire due to various factors, including heat flux, fire growth rate, and the object's heat release rate (HRR). If a fire continues to intensify and grow, the rate of mass loss increases as well.

Char

Char is carbonaceous material that "has been burned or pyrolyzed [(decomposed by elevated temperatures)] and has a blackened appearance," as defined in NFPA 921, and convex segments separated by cracks on the surface. Wood is the most commonly found fuel item that chars and is present at most structure fires. As the wood is pyrolyzed, it begins to break down and lose mass. This breakdown results in the formation of hot gases, water vapor, and other products of combustion. The remaining layer of char or ash consists largely of carbon, which forms cracks and blisters on the surface of the fuel item. Other items that can char include wallpaper, the paper surface on gypsum wallboard, and vinyl or other plastic surfaces.

For many years, fire investigators misunderstood the meaning of the appearance of char. Specifically, it was incorrectly believed that large silver or shiny char blisters (called "alligatoring") were a positive indicator of the use of an ignitable liquid, or provided information about the fire's speed or intensity. Numerous test fires and experiments have failed to produce scientific evidence to support the claim that this type of charring reflects the presence of an ignitable liquid. Indeed, the appearance of char can vary widely, even between boards made of the same wood in close proximity FIGURE 4-1. Once the wood has dried, the age of the wood has no bearing on its char and burn rates.

Similarly, the depth of the char alone does not provide reliable information about how long the wood was exposed to heat. A correlation of 1 inch (2.5 cm) of char per 45 minutes of burning was used in the past but has since been determined to be unreliable. Laboratory tests have shown a wide variety of char rates, ranging from 0.4 inch (1 cm) per hour to 10 inches

FIGURE 4-1 Variability of char blisters on wood boards exposed to the same fire.
© Vladimir Zanadvorov/Shutterstock.

(25 cm) per hour. Many factors affect the rate at which wood may char, including the following:

- Rate and duration of heating
- Ventilation effects
- Surface-to-mass ratio
- Direction, orientation, and size of wood grain
- Species of wood (pine, maple, etc.)
- Wood density
- Moisture content of the wood
- Any surface coating on the wood
- Oxygen concentration within hot gases
- Velocity of hot gases
- Edge effects of material, gaps, cracks, and crevices

While no specific time of burning can be determined based on the depth of char alone, the depth of char across the same or similar wood can sometimes be used to assess fire movement and the relative rates of heat release to which those wood surfaces were exposed. The depth of char on like-wood surfaces may also be useful in identifying areas with a high concentration of fuel, such as one might find when a fuel gas jet is directed on the surface from a leakage point.

Char depth can be measured with thin, blunt probes such as dial calipers or tire-tread gauges. It should

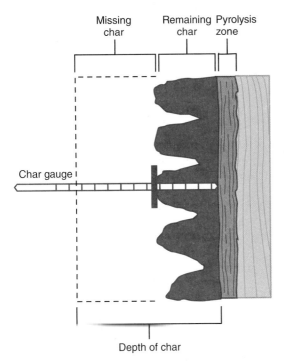

FIGURE 4-2 Depth of char is measured by adding the depth of the missing char, as determined from the original size of the wood item, to the charred material as measured by the probe.

Reproduced with permission from *NFPA 921-2021, Guide for Fire and Explosion Investigations*, Copyright © 2011, National Fire Protection Association. This reprinted material is not the complete and official position of the NFPA on the referenced subject, which is represented only by the standard in its entirety.

not be done with sharp objects such as knives, as these tend to give less reliable results. Press the tool in the center of the char blisters until resistance is felt, signaling the bottom of the charred material, and use the same tool and consistent pressure throughout the scene **FIGURE 4-2**. Measure the depth of the remaining char and add to it the depth of any missing wood that has completely burned away.

If the depth of char can be reliably measured throughout an area, the results may help the investigator evaluate the spread of fire or compare the heat exposure received at different points. With enough measurements, the investigator can even create a graph or diagram illustrating the differences.

Nevertheless, certain factors must be considered when drawing conclusions from the depth of char. The investigator can only compare like materials in like configurations; for example, wall studs should not be compared to wall paneling. Multiple heat sources can cause unexpected results, as can the presence of heavy fuel packages. The proximity of ventilation points can cause char depths unrelated to the fire's origin, and suppression and overhaul activities can destroy or modify char patterns. The investigator should consider these variables when interpreting the meaning of char within the fire scene. Does the char indicate the duration and intensity of the fire? Consider the type of

FIGURE 4-3 Spalling on the floor of a garage.
Courtesy of Gene Pietzak.

fuel, fuel load, fire conditions, and ventilation present during the fire.

Spalling

Spalling is the "[c]hipping or pitting of concrete or masonry surfaces," according to NFPA 921. It can also include cracking, breaking, and formation of craters, and can be found on concrete, masonry, rock, or brick. In a fire situation, temperature changes cause a loss of tensile strength in the material, which results in mechanical forces being generated within the material. These forces may cause the loss of surface material in the object, resulting in a chipped, pitted appearance **FIGURE 4-3**. In concrete, spalling may be more specifically due to moisture content, differential expansion between the concrete and reinforcing mesh, differential expansion between the concrete mix and aggregate, or differential expansion between the surface and the interior of the slab when heat is applied. Nonfire spalling can result from heat, freezing, chemicals, shock, force, or fatigue.

If spalling is seen in a fire, it may be related to greater loading or stress in a particular part of a concrete slab. Thus, this effect may have happened because of heat, but may not indicate where the heat was greatest.

The investigator should note and diagram any locations where spalling is seen and ask owners or witnesses whether the area being investigated already had spalling prior to the fire. Be cautious about attributing the spalling to the presence of an ignitable liquid. Such connections were often asserted in the past, but subsequent studies have shown that spalling may be caused by a number of factors that are unrelated to ignitable liquids. Indeed, practice burns have caused spalling in concrete when a pile of wood is placed directly on a concrete surface and allowed to burn fully without an ignitable liquid present.

Oxidation

Oxidation is "[t]he basic chemical process associated with combustion," as defined in NFPA 921. Oxidation of noncombustible materials, including metals, can produce changes in both their color and texture, often forming lines of demarcation and fire patterns. This reaction can continue even after the fire is extinguished. Large areas of oxidation can be seen in some fires—for example, in buildings constructed with metal sheathing. Oxidation can also be found in isolation, such as within a distinct area of a metal appliance, ductwork, or machinery panels. When found in isolation, its presence may signal elevated temperatures in a particular area in the equipment being examined, which in turn may provide a lead for further investigation.

A material may display more pronounced evidence of oxidation due to higher temperatures and longer periods of exposure. When the fire investigator is examining oxidation in a fire, it is more important to recognize the patterns and lines formed than it is to draw any conclusions based on the particular colors observed. Galvanized steel surfaces may appear dull white from oxidation. Often, the fire will affect the corrosion protection, so the surface may rust if exposed to moisture. Steel or uncoated iron may appear blue-gray in color. With elevated temperatures, the oxidation may appear black, and if the oxidation is thick, it may flake off. Moisture may also produce the expected rust-colored appearance. Stainless steel may appear dull gray with severe oxidation, but display a coloration with mild exposure. Steels that have been heavily oxidized display damage consistent with melting and may require a metallurgic exam to determine whether the object did, in fact, melt. Copper exposed to fire begins to oxidize, creating dark red or black oxide. As the copper object is subjected to higher temperatures and/or longer periods of exposure, the thickness of the oxide layer increases.

Objects such as rock and soil also change colors as a result of high temperature exposures, turning either yellow or red in color. Oxidation may also be caused by deposition of smoke aerosols that contain acidic components, or the application of fire suppression agents such as wet and dry chemicals.

Color Changes

Changes in color of materials, if caused by exposure to heat, may be a source of information at a fire scene. However, they can be subjective, and difficult to define and describe. All people do not see color from the same perspective. Factors such as lighting conditions, angle of view, and nature of color should all be considered when assessing color. In addition, coloration changes may occur from nonfire factors such as exposure to sunlight. Fabric dyes may change in appearance related to heat exposure. How a particular dye reacts to heat exposure may vary by product.

Although it is usually difficult to make specific conclusions about a temperature or other phenomenon based only on the color of a material, the fire investigator should note and document differences in color on like materials and any patterns formed by the differences in color.

Melting

Melting of a material is a physical change from solid to liquid caused by exposure to heat. Demarcation lines can be produced between the melted and unmelted portions of the material and may be useful fire patterns.

Each solid material has its own melting temperature. These temperatures can range from just above average room temperature to several thousand degrees. For example, steel has a melting temperature of 2660°F (1460°C), whereas thermoplastics melt at temperatures around 167°F to 750°F (75°C to 399°C). Note, however, that thermoplastics have a softening temperature as well as a melting temperature.

It is sometimes difficult, particularly in the case of steel, to distinguish visually between melting and oxidation. Heavily oxidized steel may have a bulbous appearance. The only way to demonstrate actual melting of steel is with a metallurgical examination. Additionally, small-diameter steel wire and springs may oxidize in a way that gives a false impression that they have melted.

The softening of glass items can be useful in determining temperatures sustained during a fire. However, these temperatures can vary, so the investigator should be cautious when drawing conclusions based on this factor.

Melting of materials may be useful to estimate the temperature at a particular location during a fire. Knowing the melting temperatures of common materials in a fire may help investigators better understand the minimum temperatures that were present in a given area of the fire, which can then provide insights helpful for determining the fire's duration and intensity. The melting temperatures of a variety of materials are listed in Table 6.3.10.2 of NFPA 921. A caveat applies here, however: Many materials have a broad range of responses to increased temperatures, and one type of metal, plastic, or glass may respond at a considerably different temperature than another type of metal, plastic, or glass. Still, even when temperatures cannot be estimated from the melting of a particular

FIGURE 4-4 Melted materials can provide insight into the temperatures that may have been produced during a fire.
© Stephen Roberts/Shutterstock.

FIGURE 4-5 Steel I-beams deformed by thermal effects.
Courtesy of Rodney J. Pevytoe.

material, the manner in which items melted may provide clues about the fire's direction or intensity. For example, if an object melted more on one side than the other, or if the degree of melting differs between two objects made of identical materials, the investigator may be able to form theories as to where the heat was greater during the fire **FIGURE 4-4**.

It is important to remember that the temperatures developed at a particular location do not necessarily indicate the type of fuel that burned there. For example, wood and gasoline burn at essentially the same flame temperature. The turbulent diffusion flame temperatures of all hydrocarbon fuels (plastics and ignitable liquids) and cellulosic fuels are approximately the same, although the fuels release heat at different rates. Under some circumstances, a fire involving an ignitable liquid may burn more rapidly or with a greater heat release rate than a fire without such liquid, but it does not burn at higher temperatures.

Alloying of Metals

Alloying, which can look very much like melting, must also be considered during postfire examination of damaged metal items. When two metals, one or both of which are in a liquefied state, come into contact with each other, they may begin to form a new

material—an alloy of the two. An example can be seen when zinc comes in contact with copper, creating a zinc–copper alloy. The melting temperature of zinc is lower than that of copper; however, the alloying of the two materials means that the copper will now melt at a lower temperature.

Alloying of metals with high melting temperatures does not necessarily indicate the presence of an accelerant or exceptionally high temperatures.

Thermal Expansion and Deformation

When exposed to heat, most materials begin to expand and change shape, either temporarily or permanently. In structures such as buildings, different materials are likely to have different expansion rates, which can compromise the integrity of the building or even cause structural failure.

When steel objects such as columns and beams reach temperatures in excess of approximately 900°F (500°C), they begin to buckle and bend. As steel is exposed to higher temperatures, its strength and load-carrying capability decrease until deformation of the material occurs. This deformation results from the object's inability to support the load placed on it (which may be the object itself). Steel and other materials may also expand and push on other building components or may bend due to thermal expansion—for example, when a steel beam is heated but restrained at the ends, as when used for the roof supports of a building.

Deformation should not be confused with melting—in deformation, the object has not liquefied but merely become distorted. Indeed, distortion or deformity of an object indicates that its melting temperature was never reached **FIGURE 4-5**.

Piping systems, specifically fittings, may deform or change in position when exposed to the heat of a fire. With the expansion and contraction caused by heating followed by cooling, a fitting may appear loose in a fire's aftermath, even though it was actually tight in the prefire condition. Sealants used in fitting connections may also be compromised by the fire.

Other materials that may become distorted because of thermal expansion include plastered wall and ceiling surfaces. When heated, plaster may begin to expand and detach from the lath support material. Likewise, drywall spackling and gypsum wallboard may separate from wall and ceiling surfaces as a result of thermal expansion.

Smoke Deposits

Smoke contains various products of combustion, including particulates, liquid aerosols, and gases. In a fire, smoke and soot collect on cooler surfaces such as walls and windows, producing **smoke deposits** or condensates. These materials may be wet and sticky and vary in thickness. The color and texture of smoke deposits are not reliable indicators of the material burning or the rate of heat release. A chemical analysis of the smoke deposits is the only method to determine the nature of the fuel that produced them. The smoke deposits will not necessarily indicate when the incident occurred; additional data may be needed to make that determination.

Once smoke settles out of the air, smoke deposition ceases. If a surface was covered by an object during the fire and is clear of smoke deposits afterward, it may indicate that the object has been moved since the fire occurred. This may be useful in determining the position of a victim during the fire. The configuration of smoke deposits can also aid in determining whether a switch was "on" or "off" during a fire.

Smoke Alarm Acoustic Soot Agglomeration

The nature of soot deposition on certain surfaces of smoke alarms can, in many cases, show whether the smoke alarm sounded during a fire. Enhanced soot deposition, also known as acoustic soot agglomeration, is a phenomenon in which soot particulates in smoke form identifiable patterns on internal and external surfaces of the smoke alarm cover near the edges of the horn and the surfaces of the horn disks themselves. When a smoke alarm displays this or other physical evidence of alarm activation, consider performing more detailed documentation, examination, and collection of the evidence.

Evidence of enhanced soot deposition can be delicate and easily disturbed or wiped away with careless handling or evidence packaging. Thus, investigators must take care not to disturb any suspected soot deposits. This evidence may be subtle and difficult to identify, and microscopic examination may be necessary to confirm its presence.

A smoke alarm should be preserved for examination when its performance may be an issue. This should be done in conformance with rules and laws pertaining to spoliation of evidence, and normally is performed by private-sector investigators trained in interested-party notification and joint exam protocols. Prior to any collection, the smoke alarm should be photographed in place and should not be altered by applying power, removing or inserting batteries, or pushing the test button. Alarms still on the wall or ceiling should be secured intact with their mounting hardware, electrical boxes, and wired connections.

Acoustic soot agglomeration will remain on a smoke alarm. Thus, while the presence of this fire effect indicates that the alarm sounded during a fire, it might not necessarily indicate that it sounded during the fire under investigation; additional data may be needed to settle that question.

Clean Burn

A **clean burn** is a distinct and visible fire effect generally apparent on noncombustible surfaces after combustible layers (such as paint, paper, or previously deposited soot) have been burned away from flame contact or intense heat. Clean burns are often white in areas surrounded by soot deposition, but may range from white to dark gray in color depending on the amount of heat exposure or may exhibit a speckled appearance. The lines of demarcation between clean-burned and darkened areas may be used to determine the direction of fire spread or differences in intensity or time of burning.

Although clean burn areas can be indicative of intense heating in an area, by themselves they do not necessarily indicate areas of origin, though such patterns should be carefully examined **FIGURE 4-6**. Clean burning can occur away from the origin if a vigorous heat or flame source is near the surface in question. Clean burning that results from ventilation will usually occur after the fire has become ventilation controlled. Such late-developing patterns may mislead investigators about the fire's origin.

Clean burning should be distinguished from spalling, which involves the loss of surface material. Also, water spray from fire suppression activities can result in clean surfaces, which can often be distinguished

FIGURE 4-6 Example of a clean burn.
Courtesy of Robert Schaal.

from clean burning based on shapes on the surface that have the appearance of droplets or rivulets of water.

Calcination

Calcination occurs in gypsum wall surfaces when the free and chemically bound water is driven out of the gypsum by the heat of the fire. Gypsum wallboard will react to fire exposure in a predictable manner: The fire will cause a difference in the color and composition of the material, where it may form a line of demarcation. Upon first exposure, the paper surface will darken, char, and then burn off. As the gypsum beneath the paper is exposed to the fire, the organics and other materials will change in color, often becoming whiter in appearance. This condition may, with sufficient exposure, go through the thickness of the material. With exposure of the total thickness of the material, the board becomes crumbly and often will drop off wall and ceiling surfaces, especially when exposed to mechanical forces and water during fire suppression. Fire-rated gypsum board contains additives that strengthen the board even after fire exposure.

The rate of calcination is not a reliable indicator for estimating the time of burning. Likewise, the depth of calcination is not a reliable method for determining burn times, although comparing the relative depth of calcinations can often provide support to a hypothesis about the direction of fire spread. Calcination can be measured in a like manner as depth of char, by inserting a probe survey with a blunt tip to reach the place where calcined and virgin gypsum meet. A force gauge can be attached to ensure even pressure throughout the measurement process.

NFPA 921 contains a suggested system to rate levels of drywall damage, reducing the potential for subjectivity in the analysis of this material. Drywall in a fire scene is rated from "DOFD (degree of fire damage) 0" (no visible damage) to "DOFD 5 (clean burn)," with "DOFD 6" representing complete destruction.

Other effects that can be seen in gypsum wallboard (and are discussed elsewhere in this section) include clean burning, charred paper, consumed paper, color changes, and smoke and soot deposition.

Glass Effects

Glass fragments found to be free of any soot deposits of smoke condensates usually indicate early failure of the glass prior to the accumulation of smoke. This effect may result from rapid heating, damage prior to the fire, or direct flame impingement. Other factors include the location of the heat source relative to the glass and ventilation effects.

Windows coated with thick, oily deposits, including the presence of hydrocarbon products, have in the past been attributed to the use of ignitable liquids. This is not an appropriate conclusion, as this type of staining may also occur from incomplete combustion and/or the burning of plastics.

Fractured glass is found in most structure fires. Glass panes that are fractured by an impact object will fracture most often from the point of force, leading to the typical "cobweb" pattern consisting of numerous straight fracture lines. Heat may also fracture glass. Research has shown that a temperature difference of 126°F (52°C) between the center of a glass pane and the protected edge of the pane can cause fractures. The fractures start at the edge of the pane and appear as smooth, undulating lines that may spread and then join together. Glass may also fracture when one side is exposed to flames and the opposite side is cool, causing stress to the glass. Tempered glass, such as that found in commercial buildings and shower and patio doors, will fracture into many small cubes when broken by impact or heat.

Because of the expansion caused by heat, some windowpanes may pop from their frames. Pressures developed by structure fires are generally insufficient to cause windowpanes to be blown out, but when an overpressure occurs from a backdraft, deflagration, or detonation, glass fragments may be found many feet away from the building.

When hot glass is rapidly cooled, usually with water, the cooling can produce **crazing**, a complicated pattern of short cracks that can be either straight or crescent shaped and can extend through the entire thickness of the glass **FIGURE 4-7**. In the past, it was believed that crazing might also be created by rapid heating of the glass; however, research has revealed that this is not the case.

Incandescent lightbulbs may show a "pulling" effect that can be used to determine the direction of heat impingement and fire travel. When sides of lightbulbs 25 W or greater are heated and softened, the gases within the glass envelope push outward toward the heated side. This pulled portion of the bulb will "point" toward the direction of the heat source **FIGURE 4-8**. In contrast, bulbs of less than 25 W collapse inward on the heated side due to the vacuum placed on the glass envelope. Bulbs that have survived the fire and extinguishment activities may be useful in determining the direction of fire travel; however, you must also consider the possibility that the fixture or bulb itself was moved or displaced during or after the fire.

Furniture Springs

Examining the damage sustained by furniture springs may produce clues about the intensity of the fire and

FIGURE 4-7 Crazed window glass.
Courtesy of Rodney J. Pevytoe.

its duration and direction of travel **FIGURE 4-9**. This damage, however, does not provide an indication of the type of fire (such as smoldering ignition) or the presence of ignitable liquids, as was once believed. Annealed (collapsed) springs have been determined through laboratory testing to be caused by the application of heat, which causes a loss in tension of the spring. Laboratory tests have revealed that the intensity and time of exposure both play roles in the loss of tensile strength to the springs. Additionally, the presence of a load or weight exposed to heat may cause a failure of the springs.

By comparing the differences in springs from one area of a mattress or cushion to another, you can develop a hypothesis regarding the relative exposure to a heat source. If one end of a bed mattress displays significant loss of strength yet the other end is relatively intact, you might hypothesize that the heat source or fire was closer to the damaged end. However, you should also consider the effects of ventilation, direction of fire travel, and other evidence located in the area of origin, and whether bedding, pillows, or other objects either protected the springs from exposure or provided fuel items that intensified heat exposure at a particular location. In addition, consider whether the springs had begun to weaken because of age or extensive use prior to the fire.

Rainbow Effect

Oily substances that do not mix with water can often be seen floating on the surface of water, creating an interference pattern that produces a **rainbow effect**. This effect, which manifests as a rainbow or sheen appearance, is often present at fire scenes and should not be relied on as an indicator of the presence of an

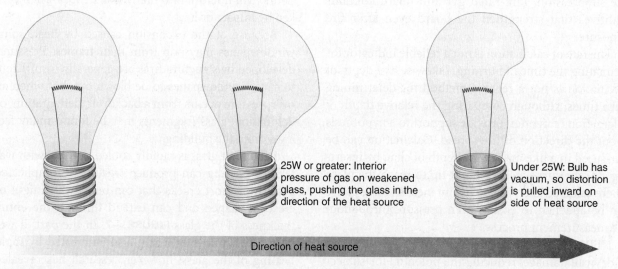

25W or greater: Interior pressure of gas on weakened glass, pushing the glass in the direction of the heat source

Under 25W: Bulb has vacuum, so distortion is pulled inward on side of heat source

Direction of heat source

FIGURE 4-8 Damage to lightbulbs indicating the direction of the heat source.

FIGURE 4-9 Relative damage to furniture springs can indicate the intensity, duration, and direction of fire.
Courtesy of Robert Schaal.

ignitable liquid. Many materials, such as asphalt, plastic, and wood products, can produce rainbow effects as a result of pyrolysis.

Fire Victims' Injuries

Although there may be a natural instinct to remove a deceased victim quickly from a fire scene, the body should not be moved until it has been analyzed and documented. As with any other object in the fire, the body could have effects and display fire patterns that can assist in the evaluation of heat exposure and fire propagation. (See the discussion of fire deaths and injuries in Chapter 12, *Complex Investigations.*) When considering effects and patterns on a body or victim, consider whether the person moved or changed location during the incident.

Heat will have numerous effects on the body, although these effects can vary based on the victim's age, weight, and overall health. Skin will redden, darken, blister, split, and char. Although blistering of skin is thought to be an action of living tissue, it may happen to a limited degree in postmortem exposure. The muscles of the body will dehydrate and begin to contract and shrink. This contraction in the limbs of the fire victim will often lead to the development of the pugilistic pose (boxer's stance). Bone that is exposed will change in color and, with enough exposure, in mass as it is consumed.

Calcination of the bone can occur as the organic components of the bone are consumed. The skull can fracture for various reasons, including heat and trauma. In general, the body is a poor fuel; body fat can melt and burn but typically requires contact with a porous, wick-like material (such as fabric, wood, or carpet) to do so. Bodies located or situated above a fire, such as on a chair or bed, are more susceptible to consumption as a result of their positioning.

Bodies found in fire scenes should be treated as evidence and interpreted with respect to the fire patterns present. When possible, attend autopsies to document and examine the effects of the fire. In fires where individuals survive but receive injuries, you should document those injuries as well.

Fire Patterns

Introduction

As noted earlier, fire patterns are identifiable shapes or progressions of fire effects. Every fire is different and produces unique fire patterns that fire investigators must interpret. Factors such as ventilation, ignition factors, fuel loads, and airflow all influence fire patterns.

When examining a fire scene, the fire patterns that remain can be used to determine the sequence of events that occurred during the fire. As the fire continues to grow from the point of origin, other fuel packages can ignite, producing additional fire patterns. When fires increase in size or burn for an extended period of time, the fire patterns located at the point of origin can be more difficult to identify.

Each fire pattern requires its own analysis to generate, if possible, one or more hypotheses as to what produced it. Determining which heat source produced each pattern assists the fire investigator in tracing the fire's progress and, more specifically, tracing backward in an attempt to find its origin. This process requires the investigator to have a solid understanding of fire dynamics and development, the modes of heat transfer, and the mechanisms of flame and smoke movement.

Lines and Areas of Demarcation

Recognizing fire patterns includes the identification of lines and areas of demarcation, discussed earlier in this chapter. These patterns and areas may be produced by exposure to heat, deposition of products of combustion, or consumption of materials. The appearance of these lines and areas depends on many factors, including the temperature and HRR of the fuel source, the heat flux absorbed by the material, and exposure time. Longer exposure to lower heat may potentially result in the same or similar appearance as a shorter exposure of the same material to greater heat.

The analysis of fire patterns should occur in conjunction with other data and information gathered during the investigation. For example, interviews with the owners, occupants, and witnesses can be essential to understanding the factors involved, such as the movement and development of the fire, location of fuel packages, occupant activities, and building construction.

Without this contextual information, the investigator risks relying too heavily on fire patterns that may not be relevant to the origin and cause of a particular fire.

Understanding the process and physics that underlie the formation of each fire pattern is important to the accurate analysis of a fire scene. Although many patterns can be traced to fire plumes or the release of heat from fuels, they can also be influenced, or even caused, by phenomena found in the development of compartment fires, such as the hot gas layer, flashover, full-room involvement, and ventilation. In addition, fire patterns can be caused or influenced by fire suppression activities. Therefore, all of these issues must be considered by the fire investigator when examining a fire.

Moreover, the relative sizes and energy content of the various fuel packages may be a factor in pattern formation and interpretation. For example, if a small or smoldering fire of origin is able to ignite a fuel package that is capable of producing large amounts of heat, the damage may be more severe from this fuel package than from the fire at the point of origin.

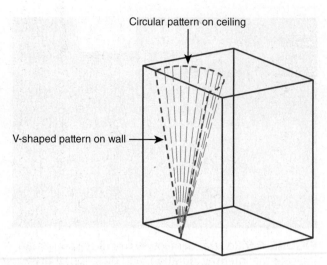

FIGURE 4-10 Conceptualization of the formation of a two-dimensional V-shaped pattern by a three-dimensional cone. Fire patterns such as this are often referred to as "truncated cone patterns."

Reproduced with permission from *NFPA 921-2021, 2020. Guide for Fire and Explosion Investigations,* Copyright © 2016. National Fire Protection Association. This reprinted material is not the complete and official position of the NFPA on the referenced subject, which is represented only by the standard in its entirety.

Plume-Generated Patterns

Although the patterns left on a single surface are essentially two dimensional, it is important to be able to visualize the fire in three dimensions. The pattern that remains on a surface, such as a wall, may be a V-shaped pattern, but the fire plume that created the pattern had a three-dimensional shape. When a given fuel package is ignited, a column of buoyant hot gases, smoke, and flame rises upward. If the fire is located in a compartment, the plume may reach ceiling height, causing the lateral extension of the fire and/or combustion products. As more fuel packages begin to be involved in a fire, new plumes are created.

Fire plumes are generally thought of as being cone-shaped, but when a fire plume intersects with a surface such as a wall or ceiling, it may leave a fire pattern in the shape of the plume being cut by the surface (a "truncated cone" pattern). Thus, the heat of the plume cone may leave a V shape as it interacts with the wall, and a circular shape when it intersects the ceiling **FIGURE 4-10**.

The creation of a fire pattern will be greatly affected by the HRR of the fuel burning. The fire pattern is most pronounced when the surface that displays the pattern was impacted by plume temperatures that were at or near the surface's pyrolysis temperatures. The shape of the pattern left on a surface may appear in various ways, including V-shaped, inverted cone, hourglass, U-shaped, circular, or pointer and arrow pattern; each of these patterns is discussed further in this section.

As the fire plume develops, the size and shape of the pattern will change. Early-stage (incipient) fires may produce an inverted cone shape pattern. As the fire progresses, a columnar pattern may evolve, followed by an hourglass pattern, a V-pattern, or a U-pattern. Eventually, a noticeable pattern may be lost with the full-room involvement of a flashover-stage fire. Early patterns may be partially or fully erased by new fire patterns caused during fire growth and spread or by heavy burning of large adjacent fuel packages. Further, dramatic patterns can be produced in post-flashover fire scenarios that are attributable to ventilation rather than to the presence of a heat source.

Ventilation-Generated Patterns

As pressure is created during the combustion process, hot gases and fire itself may escape through windows, doors, and other restricted openings with an increased velocity and flow over, under, and around combustible items. This phenomenon increases the rate of damage to materials at those locations. Well-ventilated fires burn with higher HRRs, which can also cause greater damage. When severe damage is noted at these locations, consider these factors as a possible explanation rather than an indication of the point of origin.

In the past, holes burned through floors have sometimes been mistakenly attributed to the application of an ignitable liquid. Causes of floor burn-through other than ignitable liquids can include ventilation, glowing combustion, and radiation, as well as building materials collapsing and smoldering in the debris. These holes allow the movement of air and increase the burning rate at these locations.

When fire and heat gases are able to escape the compartment, the fire intensifies, which in turn changes the shape and magnitude of fire patterns. Localized heavy damage is often found at open windows and doors with combustible items nearby heavily damaged or consumed. Although the damage might be the most severe in this area, the point of origin may not be at this location.

In a compartment fire that is not fully developed, the hot fire gases may escape in the top space between a door and the door frame, causing charring. Cool air enters the compartment under the closed door. In a full-involvement fire, the hot fire gases will extend to the floor level and may escape and char a door bottom. Ignitable liquids burning under a door edge will also char the surface **FIGURES 4-11** and **4-12**.

Anytime a compartment fire becomes ventilation-controlled, dramatic patterns may be formed near and around any ventilation openings. In this situation, earlier fire may diminish or disappear due to lack of oxygen, leaving only the areas near ventilation openings

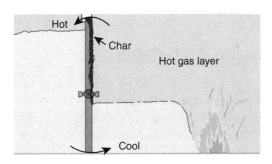

FIGURE 4-11 Airflow around a door.

Reproduced with permission from *NFPA 921-2021, 2020. Guide for Fire and Explosion Investigations,* Copyright © 2016. National Fire Protection Association. This reprinted material is not the complete and official position of the NFPA on the referenced subject, which is represented only by the standard in its entirety.

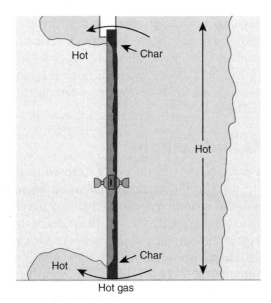

FIGURE 4-12 Hot gases under a door.

Reproduced with permission from *NFPA 921-2021, 2020. Guide for Fire and Explosion Investigations,* Copyright © 2016. National Fire Protection Association. This reprinted material is not the complete and official position of the NFPA on the referenced subject, which is represented only by the standard in its entirety.

with enough oxygen to continue the combustion process. If other areas exist where fuel-rich smoke can mix with fresh air, such as on a wall opposite a doorway opening, ventilation patterns may also appear there. The ventilation-controlled scenario is likely to happen after flashover and full-room involvement if the fire has exhausted the oxygen supply. Common patterns seen in this situation include large surface areas of damage near airflow or on surfaces across from the airflow opening, an increased magnitude of damage in these areas, lines of demarcation angled away from ventilation openings, large areas and increased magnitude of damage under a window, and increased damage around the seams of gypsum wallboard.

Identification of ventilation factors (e.g., door/window position, fire suppression strategy, fire stage, oxygen availability) is crucial in evaluating fire patterns. When performing sequential pattern analysis, the investigator must determine whether a pattern could be accounted for by ventilation effects. Full-room involvement does not obliterate all patterns. As a fire progresses toward sources of ventilation, some patterns, such as mass loss–type patterns on furniture in or around the area of origin, may be preserved.

Hot Gas Layer–Generated Patterns

Prior to a flashover event, the hot gas layer of a fire begins to descend, and radiant heat flux can damage both the upper surfaces of objects and the floor surfaces. If the gas layer descends to floor level, damage may also be present outside the room at floor level due to hot gases escaping under the door itself. The level of descent of the hot gas layer can be determined by examining the line of demarcation present on the walls and on the vertical surfaces of objects within the room. The damage begins high in the room and is usually level and uniform throughout the room; however, lines of demarcation may increase near vent openings. Drop-down materials can create isolated areas of damage. Objects such as dressers and televisions can protect both vertical surfaces and floor surfaces from exposure to the hot gas layer **FIGURE 4-13**.

Patterns Generated by Full-Room Involvement

Patterns created by full-room involvement can usually be found on all exposed surfaces throughout the room, all the way down to the floor level. Traditional fire patterns, such as V-patterns, may be more difficult to document and analyze if the fire continues to burn for a long time.

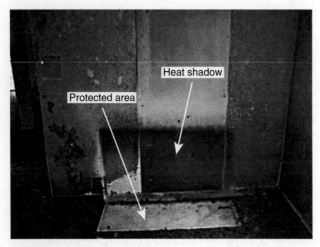

FIGURE 4-13 A portion of a plume pattern generated by the hot gases of the fire.
Courtesy of Rodney J. Pevytoe.

FIGURE 4-14 Patterns on the wall created by hose streams.
Reprinted with permission from *NFPA 921-2021, Guide for Fire and Explosion Investigations*, Copyright © 2020, National Fire Protection Association. This reprinted material is not the complete and official position of the NFPA on the referenced subject, which is represented only by the standard in its entirety.

TIP

Holes burned through floors are not necessarily the result of an ignitible liquid. It is possible that building materials may have collapsed and that smoldering in the debris created holes in the floor.

When full-room involvement occurs, the damage sustained to the room, including the floor, will become more extensive as radiated and convected heat ignites fuel packages within the room. As the hot gas layer continues to descend to the floor, damage to the objects within the room intensifies (including charring to the underside of objects, intensified damage to corners of the room, ignition of the carpet and flooring, and charring to the underside of doors), with greater damage occurring to surfaces that are directly exposed to the hot gas layer. Although full-room involvement can produce holes in the flooring, these patterns can also result from objects or debris burning on the floor or objects creating protected areas. Damage to the room increases as the fire continues to burn; however, major damage can occur within minutes in extreme conditions depending on the effects of ventilation and the specific fuel items available.

Suppression-Generated Patterns

Water and other agents used for fire suppression are capable of producing or altering fire patterns. Hose streams can leave distinct patterns on walls and ceilings that resemble water jets or droplets **FIGURE 4-14**. They can also alter the direction of the spread of the fire and create fire damage in places where the fire would not move in the absence of the hose stream. Likewise, fire department ventilation operations can influence the fire patterns present, although the patterns resulting from fire department entry or from the placement of positive-pressure ventilation are likely to resemble other ventilation-related patterns. Determine the actions of individuals at the scene and of the fire service to understand whether the actions taken by firefighters may have caused changes to the original patterns.

Pattern Location

Because fire is a three-dimensional event, fire patterns may be found on virtually any surface that has been exposed to the fire or to products of combustion. While investigators commonly look for patterns on walls, ceilings, and furniture in a building fire, patterns may also be evident on personal property, plants, clothing, toys, appliances, and more. It is necessary to examine all affected surfaces, including those surfaces that are not readily apparent. The scene examination includes looking for large-scale patterns such as can be seen with aerial views of the roof as well as small-scale patterns such as the heat effects on the insulation of wiring. Examination of room surfaces may show a pattern on the surface covering or on the construction material beneath.

Walls and Ceilings

When examining walls that were exposed to fire or heat, the investigator should keep in mind the conical shapes of the various fire plumes that may develop in a compartment fire, and the fact that lines of demarcation often show an intersection between such a cone and the surface being examined. Thus, wall patterns most often are in a V, hourglass, or U shape. Patterns on a ceiling (or other horizontal surface) may be circular, consistent with the wide area of the "cone" of the plume below, or they may take the form of a partial circle. If lines of demarcation are not distinct on the ceiling, patterns may still be discerned in some cases by looking for greater intensity of damage. Such patterns may be located above the area of fire origin, or they may be seen above a fuel package that burned more vigorously than surrounding fuels but is remote from the fire origin. Examine areas below these patterns to identify a possible heat source.

The transfer of heat by conduction may cause damage to components within walls and ceilings. Heat may be conducted through drywall and cause damage to wall studs behind it. If fire department overhaul has caused the removal of wall pieces, such as drywall, it may be possible to reconstruct them on a flat surface to determine whether they show patterns.

Floors and Other Horizontal Surfaces

Fire patterns may appear on floors and other horizontal surfaces and should be documented. When considering the source of a pattern on the floor, the fire investigator should determine whether the compartment in question reached flashover and/or full-room involvement, as these phenomena can add, change, and destroy fire patterns. At flashover, 20 kW/m² of radiant heat flux is common at floor level. Since flashover and full-room involvement typically involve the burning of horizontal surfaces, relatively uniform damage may appear on these surfaces shortly after flashover; later, as the burning progresses, uniformity may be lost. Patterns may develop on the floor from a variety of causes, including close-by fuel packages such as furniture or nearby ventilation sources. Conversely, the floor area may be protected or shielded from nearby fuels or ventilation, resulting in a different appearance.

Full-room involvement may cause not only burning on the floor but also burning around baseboards and door thresholds and sills. In addition, these components may show damage because they are located near ventilation openings or because gaps in construction provided sufficient air for combustion.

Where the overall fire damage is limited and small, or isolated, and irregular patterns are found, the investigator can consider the possibility of ignitable liquids. However, even the presence of irregular patterns is not in itself an indicator of the presence of an ignitable liquid (see the discussion of irregular patterns later in this chapter). Further, floor-level patterns and damage can result from a variety of sources, including nearby fuel packages, drop-down items, ventilation, glowing combustion, heat from below, and preexisting damage. Whenever the presence of an ignitable liquid is being considered, the investigator should take samples of the area as well as comparison samples for examination by a laboratory.

Since 1970, carpet sold in the United States has been required to be manufactured to specific fire resistance standards. Dropping a match or a cigarette onto these carpets is generally insufficient to cause horizontal flame spread. Generally, carpet requires the input of greater energy to burn, and even then may only burn to the point where the energy available drops below the minimum level needed to support flame spread. In flashover conditions, the radiant heat impact on carpet will cause ignition and burning. Vinyl floor tiles may curl at the edges for several reasons. The burning of ignitable liquids, radiant heat, or natural shrinkage may cause this kind of damage to tile.

Burning between the seams of floor boards is not necessarily proof of the presence of an ignitable liquid. This burning has been observed in numerous tests due to radiant heat alone where no ignitable liquid was involved. Indeed, the surface below a liquid remains cool, or at least below the boiling point of the liquid, until the liquid is consumed. In fully developed fires, gases may be forced downward, creating penetrations in the floor.

Holes discovered in floor areas could be the result of preexisting conditions, glowing combustion, isolated smoldering objects (such as nearby furniture, drop-down objects, or roof sections), radiant heat, ignitable liquids, or the effects of ventilation. Evidence other than the hole or its shape is necessary to confirm the cause of a given pattern.

To determine the direction of fire travel through a horizontal surface, examine the sides of the hole and the slope created by the fire. When the hole is wide and slopes downward from the top surface, the direction of fire travel would be from above. In the case of a fire advancing from below a surface, the sides will be wider on the bottom and slope upward. Be cautious when determining the direction of fire travel through these penetrations, because fire movement may have occurred in both directions, leaving only the last direction of travel visible in the hole **FIGURE 4-15**.

Remote Fire Patterns

The fire investigator should look for and document any fire patterns that are separate from the main body of the fire. Such patterns can be attributable to a number of factors, including fire extension, heat transfer igniting new fuels, fire brands, drop-down objects, separate areas of origin, or even previous fires. A knowledge of building construction is important when considering why a pattern might appear remotely from a fire.

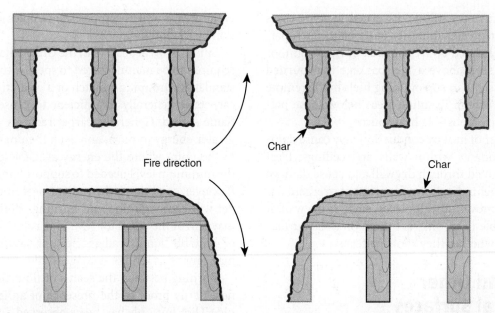

FIGURE 4-15 Fire penetration of a horizontal surface.

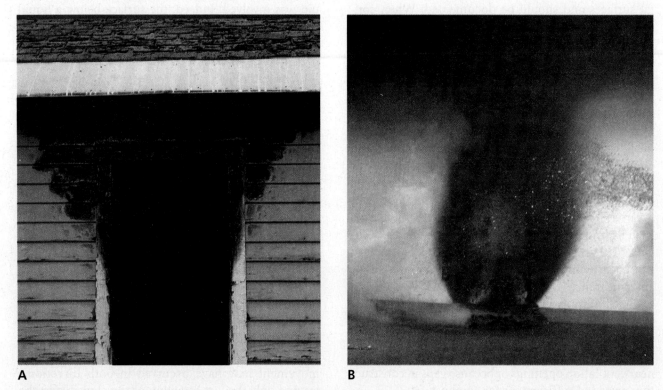

FIGURE 4-16 Common fire patterns. **A.** Partial V-pattern created by flames venting from the window opening. **B.** A pattern generated by a fuel package burning on the floor.

(**A**) Courtesy of Rodney J. Pevytoe. (**B**) Courtesy of Jamie Novak, Novak Investigations Inc. and the St. Paul Fire Department.

Pattern Geometry

Fire and smoke produce a variety of distinctive patterns, which are described based on their geometric shapes. The interpretation of these patterns is not based on scientific research, so alternative interpretations are always possible. With the diverse nature of fire patterns, many patterns are possible beyond those described in this text **FIGURE 4-16**.

V- and U-Shaped Patterns

V-shaped patterns are commonly found on vertical surfaces, such as walls, in structural fires. They are created by the heat, smoke, or hot gases of a fire plume. The presence of a V-pattern reflects thermal damage from nearby burning, but does not necessarily indicate the origin of the fire under investigation. Distinct fuel packages can ignite during the progression of a

FIGURE 4-17 U-shaped pattern on wallboard and studs.

Reproduced with permission from *NFPA 921-2017, Guide for Fire and Explosion Investigations,* Copyright © 2016. National Fire Protection Association. This reprinted material is not the complete and official position of the NFPA on the referenced subject, which is represented only by the standard in its entirety.

FIGURE 4-18 Inverted cone pattern.

Reproduced with permission from *NFPA 921-2017, Guide for Fire and Explosion Investigations,* Copyright © 2016. National Fire Protection Association. This reprinted material is not the complete and official position of the NFPA on the referenced subject, which is represented only by the standard in its entirety.

fire, and when they are undergoing combustion, these packages will have their own plumes and can create V-patterns remote from the fire origin.

The angle of the pattern is affected by several factors, including the HRR of the fuel package, the geometry of the fuel, the effects of ventilation, the combustibility of the surface where the pattern is observed, and horizontal obstructions. The angles of the lines of demarcation should not be interpreted as the sole indicator of the speed of the fire growth or the HRR of the fuel.

U-shaped patterns display a gently shaped line of demarcation, in contrast to the sharp lines seen with V-shaped patterns **FIGURE 4-17**. Although, as mentioned earlier, the angle formed by the lines of demarcation does not impart specific information about the fire's heat release or growth, at times U-patterns may be found farther from a heat source than V-patterns, and at a higher level, as might be expected when a cone is intersected farther away from its vertical axis.

Inverted Cone Patterns

Inverted cone patterns are often created by a vertical flame plume not reaching the ceiling level. They will appear as a two-dimensional triangle shape, with its base at the bottom. These patterns form from the heat or combustion products from a nearby fire; fires generating inverted cone patterns are often short lived or involve a fuel package with low heat release. They are not indicative of the presence of ignitable liquids, and in fact can be formed anytime a fuel produces a flame zone that is not intersected by a horizontal surface above.

Inverted cone patterns may also be created by a natural gas leak occurring below the floor level and

rising, then burning where the floor and wall intersect **FIGURE 4-18**.

Hourglass Patterns

Similar to inverted cone patterns, hourglass patterns can be created when a fuel package is burning next to or near a vertical surface such as a wall. The fuel package is at the base of the hourglass, and the vertical surface will show a pattern extending lower than the flame zone. In this type of pattern, the flame zone of the fire causes an inverted V or inverted cone pattern, and the plume above the fire creates a V-pattern connected to, and extending from, the top of the inverted cone. Typically, the V-pattern above is larger than the inverted V below.

Truncated Cone Patterns

Truncated cone patterns, also known as truncated plumes, result from the partial intersection of a cone pattern on both vertical and horizontal surfaces **FIGURE 4-19**. The truncation, or intersection, of a cone-shaped (or hourglass-shaped) plume by the horizontal and vertical surfaces (often walls and ceilings) mirrors the three-dimensional shape of the fire plume. Like the plume, the cone gets wider with increasing height, accounting for the shape of V-patterns. When the vertical surface is farther from the plume, the pattern may take the form of a U. When intersected by a ceiling or other horizontal surface, hot gases begin to move horizontally; thus, at the ceiling level, the fire investigator is likely to see damage beyond the circular line of demarcation that might be formed above the plume itself. An idealized cone pattern viewed with nearby horizontal and vertical surfaces can illustrate the various truncated cone patterns, along with typical V- and U-patterns, that might be found at these intersections **FIGURE 4-20**.

Pointer and Arrow Patterns

Beveling can often be seen in vertical wood wall studs that are consumed by heat and flame more on the side

FIGURE 4-19 Truncated cone pattern on perpendicular walls.

Reproduced with permission from *NFPA 921-2017, Guide for Fire and Explosion Investigations*, Copyright © 2016. National Fire Protection Association. This reprinted material is not the complete and official position of the NFPA on the referenced subject, which is represented only by the standard in its entirety.

where the heat is emanating from or from which the fire is traveling. A similar pattern may be seen in a line of similar exposed wall studs. "Pointer and arrow" refers to the pattern in which the studs closer to the heat source are shorter than ones farther away. Such patterns may be seen in an area where studs are exposed either because the area is unfinished or because the surface sheathing has been destroyed by the fire. Examination of the lines of demarcation on several of the wall studs may enable the investigator to interpret the direction of fire travel. The most severe charring to the studs should be expected on the side from which the heat is impacting the area.

Care should be taken in examining such patterns to ensure that the studs that are being compared to each other are similar in size. For example, a 2-inch (5-cm) by 4-inch (10-cm) wall stud would burn differently than a 2-inch (5-cm) by 6-inch (15-cm) stud, or a pair of 2-inch (5-cm) by 4-inch (10-cm) studs nailed together. Thus, the differences between the two might not be comparable. Further, the investigator must make sure that the ends of the studs being examined were not broken off mechanically during the fire or by firefighters.

The left portion of **FIGURE 4-21** shows studs that are beveled on top. Another type of beveling is seen when the heat source is directed toward the narrow

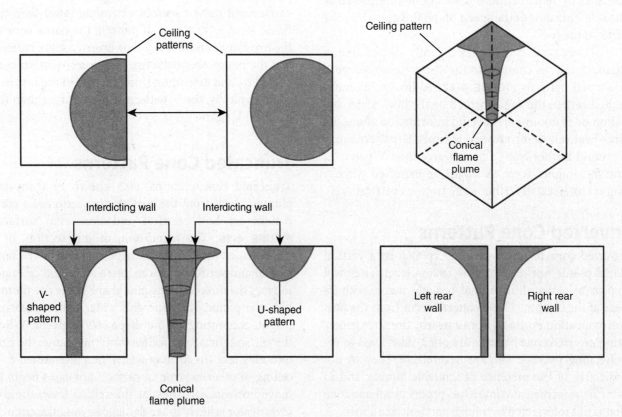

FIGURE 4-20 Idealized drawing of the formation of truncated cones.

Reproduced with permission from *NFPA 921-2021, Guide for Fire and Explosion Investigations*, Copyright © 2020, National Fire Protection Association. This reprinted material is not the complete and official position of the NFPA on the referenced subject, which is represented only by the standard in its entirety.

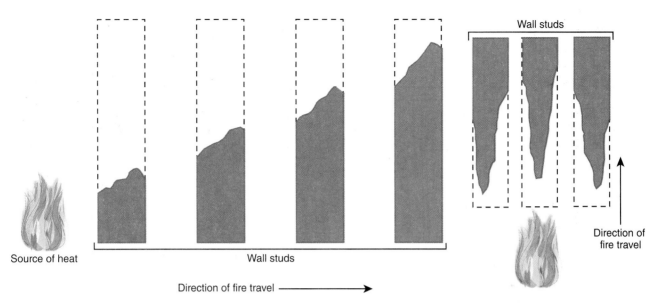

Source of heat

Wall studs

Direction of fire travel ⟶

Wall studs

Direction of
fire travel

FIGURE 4-21 Wood wall studs showing the direction of fire travel.

surface of the wall stud, as shown in the right portion of the figure. The combination of beveled studs on the left of the diagram, where the studs are lower to the left, is a pointer and arrow pattern indicating greater heat, or fire movement, from the left of the figure.

Circular Patterns

Circular-shaped patterns often are noted on the underside of horizontal surfaces. As the fire plume rises and reaches the surface above, it may leave a circular-shaped pattern, or part of a circle if the surface is not large enough to show the whole pattern. When the heat source is farther from a vertical surface, the pattern created will tend to be more circular in nature. The investigator should examine the area directly below the circle to determine which heat source caused the pattern.

Doughnut-Shaped Patterns

A doughnut-shaped pattern may be found when irregularly shaped areas of burning and damage surround an area of less-damaged material. This effect may be roughly circular or may take another form. It can be the result of an ignitable liquid and, therefore, should be closely examined for evidence that an ignitable liquid created the pattern. The "doughnut" pattern arises because the liquid itself cools the center of the area while the flames are present along the perimeter.

Saddle Burns

Saddle burns are curve-shaped, deeply charred patterns that are often localized on the top of floor joists.

These distinct U-shaped patterns may potentially result from the application of an ignitable liquid; however, they can also be created by burning materials on floor surfaces, objects that have melted and fallen on the floor, and floor openings that allow for downward ventilation.

Linear Patterns

Linear patterns are found on horizontal surfaces and may appear as lines of demarcation, extending and elongating across the surfaces. An elongated pattern of heat damage at floor level may be caused by the application of an ignitable liquid creating a **trailer**, which may connect one area to another. Trailers can consist of liquid or solid fuel; they are often used to connect multiple fires or "sets" or used to spread the fire to different floors. The fire investigator may be able to locate some unburned portions of a trailer. Other, non-incendiary causes of linear patterns include areas in flooring or carpet that have been heavily used, resulting in greater wear of the carpet or carpet pad, and clothing or other linens placed on the floor.

Furniture and other items may shield portions of a room's wall or floor surface, creating a "protected floor area" in which materials such as carpeting and flooring sustain little to no damage while the unprotected area is severely damaged. These linear patterns are sometimes mistaken for trailers; however, they could simply be the result of clothing, boxes, or any other material that may have been present. It is important to determine which materials or items were present that could have created these patterns. Also, when liquefied petroleum gas and propane burn, they may produce demarcation lines or linear patterns across noncombustible surfaces that direct back to the source.

Area Patterns

Sometimes the fire investigator may find patterns that appear to cover large areas or even rooms. This phenomenon can be due to the burning of fuel gases that had been dispersed within a room prior to their ignition. Such patterns can also be created by a rapidly moving fire, such as a flash fire. Flash fires may produce little to no surface damage within a space and may not create an explosion or any subsequent burning. Instead, the areas of greatest damage are likely to appear where secondary fuels are ignited and may be far from the origin. Such damage may produce fire movement patterns that are determined by the secondary fuel, room geometry, and HRRs. If a flashover occurs, the evidence of a flash fire may be altered or consumed. Flash fires may consume only the fuel gas, causing little damage to the surrounding combustible items. Pocketing of gas may also be present, creating the appearance of unusual movement patterns that make tracing the fire's spread more difficult. When attempting to establish the origin of a flash fire, investigators should obtain accurate witness statements and identify potential fuel and ignition sources.

Patterns in Flashover

As the fire transitions from growth to full development, a flashover event may occur and spread fire rapidly to all exposed combustible surfaces to floor level. Uniform depths of char and calcinations may be present; however, the effects of ventilation and the locations of objects such as furniture or fixtures may produce uneven damage to wall and ceiling surfaces.

Heat Shadowing

Heat shadowing is a discontinuous pattern that may mask lines of demarcation on that surface due to the interruption of that heat transfer. It can be caused by an object blocking the travel of radiated heat, convected heat, or direct flame to a surface.

Protected Areas

When an object is physically in contact with another material in such a way that the material shields the object from the effects of heat transfer, combustion, or deposition, the object is said to be in a protected area. Protected areas can be useful in reconstructing the fire scene. **FIGURE 4-22** shows the protected area beneath an ottoman.

FIGURE 4-22 Protected areas can be useful in reconstructing the fire scene.
© Jones & Bartlett Learning. Photographed by Glen E. Ellman.

Irregular Patterns

NFPA 921 offers numerous warnings about misinterpreting irregular patterns found in flooring material. While such patterns can be created by pooled ignitable liquids, they can also be created by myriad other conditions. In a fire that has progressed to flashover or collapse, or has required long extinguishment times, it is common to find irregular patterns as the result of hot gases, flaming and smoldering debris, and pooling of melted plastics, without the presence of an ignitable liquid. Long irregular patterns may also be caused by wear of carpet or carpet padding in high-traffic areas. Even without flashover or full-room involvement, irregular patterns may be found that are not from such liquids—they can be caused by fallen debris, radiant heating, clothing, plastic items, and other irregularly shaped materials. The investigator should always work to identify the fuel that gave rise to the pattern seen.

Investigators should not automatically conclude that an irregularly shaped or "pool"-type pattern in flooring materials is evidence of ignitable liquid based only on its visual appearance, and the use of the legacy term "pour pattern" is discouraged. If an ignitable liquid is suspected, submit samples for laboratory analysis. Further, investigators should collect and submit comparison samples from an area of like flooring where it is not believed an ignitable liquid was present, since pyrolysis products such as hydrocarbons may be detected in flooring and carpet even in the absence of an ignitable liquid.

The lines of demarcation of an irregularly shaped pattern may vary with the properties of the material and the intensity of the heat exposure. For example, a dense material such as hardwood flooring may show sharper lines as opposed to some carpets.

NFPA 921 contains photographs of test fires in which the postfire patterns appeared to be identical to the patterns produced by ignitable liquids but no ignitable liquid was actually used. For example, irregular patterns were produced by newspapers on vinyl tile, a cardboard box on an oak floor, and flooring materials in a room that was allowed a lengthy burn time.

Fire Pattern Analysis

Fire pattern analysis is the process of identifying and interpreting fire patterns to determine how those patterns were created and their significance. This analysis relies on an understanding of fire dynamics and fire development to ensure an investigator can properly recognize, identify, and analyze the patterns observed. Two basic types of fire patterns are distinguished: intensity patterns and movement patterns (although some patterns may display aspects defining both intensity and movement). Generally, no individual pattern is definitive in and of itself, and each pattern observed should be analyzed in relation to the other patterns at the fire scene. A comparison of the patterns helps determine the path of fire progression, leading back to the area of origin.

Analysis of all patterns and other data found at a scene must consider the context in which they appear. For example, clean burning from a small plastic object on the floor may suggest a fire originating from the object. However, if it is obvious that the object fell to the floor from a burning area above, the patterns associated with the object have a much different meaning.

Every fire pattern in a fully involved compartment should be analyzed to determine whether it could have resulted from ventilation. Patterns that can be accounted for by ventilation may provide little insight into the behavior of the fire in its early stages.

Because fire patterns are visual shapes, their recognition and interpretation can be subjective. Generally, it is considered a best practice for fire investigators to analyze fire patterns using the scientific method procedure already in place for the analysis of origin and cause. This procedure removes subjectivity and reminds investigators to evaluate not only the patterns themselves but also all of the context and circumstances in which they appear, and all of the other data from the scene **FIGURE 4-23**.

Heat (Intensity) Patterns

As a fuel item is exposed to heat and flames, patterns are created, producing lines of demarcation and mass loss that may be useful in determining the direction of fire travel as well as the characteristics and amounts of fuel materials present. For example, the levels of overall damage and smoke deposition will often assist an investigator in making an initial determination of the area of greater fire. Further, when one area or side of an object has lost more mass than other areas, this can often indicate a greater heat intensity coming from the area of greater loss. Often, fire investigators compare like materials (such as wall framing studs if they were configured identically) to gain information about where the heat was greater. V- and U-patterns show the presence of heat, although they may simply be indicating a source of heavy fuel rather than the origin of the fire. These lines of demarcation may suggest an area was exposed to more heat or exposed for a longer duration, resulting in greater consumption of an item in relation to a similar item or greater consumption of one area of an item in relation to another area.

Fire Spread (Movement) Patterns

As a fire grows and spreads, products of combustion (including smoke and hot gases) flow away from the initial heat source. Fire often leaves distinctive patterns as it travels through interior openings and doorways. Pointer and arrow patterns can be analyzed for information about fire movement within an area. As combustion products move throughout a compartment, they impact boundary surfaces and contents to varying degrees. This impact and the resulting patterns can be analyzed in relation to each other, allowing investigators to trace the patterns back to the heat source that produced them. This information is helpful in analyzing fire growth and development, identifying the area of origin, and determining fire causation.

Effects of Fuel Packages in Pattern Interpretation

The fire investigator must always take into account the impact that the size and location of fuel packages have on fire patterns and their interpretation. As mentioned elsewhere, large or high-energy fuels can be ignited during the progression of a fire and leave major fire patterns that are not related to the fire's origin. These major fire patterns can at times obscure or overwrite smaller, earlier patterns. Thus, the size and energy of a fuel must be considered during fire pattern analysis, and all of the patterns must be analyzed in context and in relation to each other. The locations of fuels and fuel packages can affect their ability to ignite and the time required for ignition. Indeed, if such fuels are located near walls or corners, they will have different

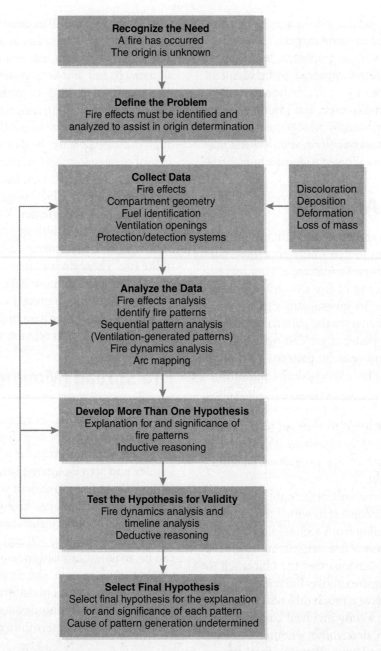

FIGURE 4-23 The steps of the scientific method as applied to the analysis of fire patterns.

Reproduced with permission from *NFPA 921-2021, Guide for Fire and Explosion Investigations*, Copyright © 2020, National Fire Protection Association. This reprinted material is not the complete and official position of the NFPA on the referenced subject, which is represented only by the standard in its entirety.

flame heights and potentially leave different patterns than if they were located elsewhere.

Arc Mapping

Arc mapping is the identification and documentation of a fire pattern derived from the identification of arc sites, which is used to aid in determination of fire origin and spread. It can be performed by the fire investigator in both large- and small-scale events. Arc mapping requires a careful examination of the electrical circuits to identify arcing that occurred within the circuit, taking care to avoid damage of the wiring

being examined. Mapping the locations of possible electrical activity can help identify an area that needs additional examination. The investigator should not conclude that arcs are more likely to be prevalent in the area of the fire's origin, compared to other areas.

Arc surveys can identify areas where the fire damaged energized electrical conductors, and areas where the fire occurred before electrical energy was cut off. When arc damage is found on a conductor, it is evidence that the conductor was energized and that the fire was intense enough to damage the insulation and create an unintentional path between conductors. However, while the presence of an arc demonstrates that the circuit was energized, the absence of an arc does not demonstrate that the circuit was not energized. Moreover, even when the site includes electrical energy and sufficient fire, there is no guarantee that arcing will occur.

If a **sever arc** is discovered on a conductor and other arcs are discovered downstream (away from the source) on the same conductor, the fire investigator can conclude that the downstream arcs occurred no later than the sever arc, as the sever arc would have prevented energy from reaching the downstream areas once it occurred. In this case, it may be possible to reach a conclusion about the fire movement along the circuit, but the investigator cannot say with certainty that a downstream arc is at or near the area of origin just by virtue of it being downstream. Additionally, this technique is not valid if the circuit is energized downstream due to backfeeding of power or contact with other energized conductors, or when "ring"-type wiring is used, such as the wiring found in some countries outside of North America.

Holes can be melted in metal conduits or in the sides of panelboards because of an arcing event. These holes usually result from fire impingement as the heat from the fire degrades the wiring insulation of the conductors inside. Blowouts are a result of the arcing between the conductors inside and the enclosure. Note, however, that a lack of blowouts or blow holes does not indicate that arcing did not occur. To confirm whether arcing did or did not occur, the conductors would need to be removed from the conduit. Because this process can damage the conductors, it can lead to spoliation of evidence. Thus, the investigator should consider leaving these conductors intact until the building owner's insurer and investigators can arrange for examination with any interested parties.

Arc mapping requires the investigator to examine the exposed electrical conductors closely. This should be done with extreme care so that the wiring is not damaged and is available for inspection by other investigators or engineers. Examination may involve carefully removing debris from the conductors, using a soft tool such as a brush or wooden implement so as to not damage the wiring. Inspect the length of conductors visually and by hand to identify any damage or other anomalies, and determine whether any anomalies found represent arcs, melting, or eutectic phenomena (see Chapter 7, *Building Electrical and Fuel Gas Systems*, for discussion of the differences between arc melting and melting as the result of fire). Arc damage often involves loss of metal from the conductor or the deposition of metal onto a conductor from an adjoining one. Note that differentiating between melting and arc sites is often difficult and may

require expert examination. Be sure to inspect the full length of any available conductors.

Document all damage with notes and photos, including photos of adjacent conductors. If damage (melting) by fire is discovered on conductors, document it as well. Note where the circuit was fed from, where it ran, and which loads and receptacles were on the circuit. Check the overcurrent protection device (fuse or circuit breaker) to see whether it opened.

Ties or flags can be placed at arc sites to assist in photographic documentation. For an arc survey, the investigator can divide the area of interest into zones, documenting the condition of any conductors passing through as well as the condition of all devices and boxes. Document the physical characteristics of any anomaly found (e.g., severed, faulted to another conductor in the same cable).

Documentation in the form of an **arc survey diagram** can be useful, and precision in preparing this diagram will minimize the risk of errors in the subsequent analysis of the data. Both plan and elevation views are recommended, where appropriate. It is also important to document the type and status of the overcurrent protection device for each involved circuit (e.g., fuse, breaker, ground fault, or arc fault circuit interruption) **FIGURE 4-24**.

If it is necessary to collect conductors from an arc survey site, ensure that the conductors are properly identified, labeled, and photographed before their removal. These conductors can be very brittle and broken easily. Associated devices can fall off the conductors, potentially eliminating the possibility of future circuit tracing.

While arc mapping is a useful tool in fire investigation, it has some limitations. Prolonged fire exposure resulting in thermal melting of conductors or structural collapse may prohibit the analysis of conductors. It is also possible that de-energized conductors (such as when a circuit breaker is in the tripped or off position) might have come in contact with other energized conductors, leading to an arcing event. This event may produce misleading evidence, so investigators should take special care to determine whether it occurred.

The visibility of arc damage is related to the duration of the arcing event and the timing of the arc. In alternating current (AC) circuits, the potential difference between two conductors depends on the point in the AC cycle at which the arc occurs. During the typical 120-volt AC cycle, the potential between the "hot" conductor and the neutral or ground conductor ranges between +170 V and −170 V, 60 times per second. As the potential approaches 0 volts, physical damage from the arcing event may be less severe and therefore more difficult to see.

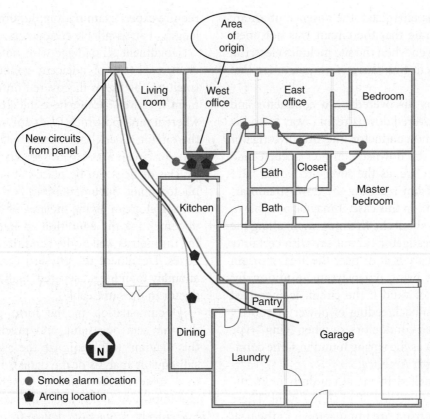

FIGURE 4-24 Electrical arc survey diagram.

Once all data have been collected, they need to be analyzed. Fundamentals of fire dynamics can be applied to the analysis of the collected data and to aid in the development of origin hypotheses. Analysis of the sequence of events during a fire can be useful in determining the fire's origin. Analyses can develop information such as potential first fuels ignited, the sequence of subsequent fuel involvement, recognition of lacking data to be collected, and identification of potential competent ignition sources.

After-Action REVIEW

IN SUMMARY

- Understanding the nature of fire effects and patterns, and the difference between the two, is critical to the fire investigator's ability to correctly evaluate the origin, development, and spread of a fire.
- In most instances, the origin of the fire must be determined before its cause can be correctly identified.
- Old myths suggesting that crazed glass, spalled concrete, and wide-base V-patterns are always a direct result of the application of an ignitable liquid have been disproved.
- NFPA 921 classifies the various fire effects into four categories: discoloration, deformation, deposition, and mass loss.
- When a sharp difference in damage levels is seen on either side of a line of demarcation, it indicates a significant change or difference in heat exposure.
- Sometimes, the fire investigator may be able to estimate the temperature that a material reached in a fire based on the effect observed afterward, if the investigator knows the approximate temperature needed to produce that effect.
- The observable loss of mass may be used as an indicator that the area of the material with the most mass loss suffered the greatest exposure to heat, either in duration, intensity, or both.

- Prefire photographs, interviews, and Internet searches can be effective ways to determine an object's prefire shape and orientation.

- In the past, large silver or shiny char blisters ("alligatoring") were considered a positive indicator of the use of an ignitable liquid, but this relationship has been disproved in numerous tests and experiments.

- Many factors influence the depth of char, so by itself this measurement does not provide reliable information about how long the wood was exposed to heat.

- Studies have shown that spalling may be caused by a number of factors that are unrelated to ignitable liquids.

- When the investigator is examining oxidation in a fire, it is more important to recognize the patterns and lines formed than to draw any conclusions from the particular colors present.

- Melting of materials may be useful to estimate the temperature at a particular location during a fire.

- Distortion or deformity of an object indicates that its melting temperature was never reached.

- Soot deposits occurring on or around the horn of a smoke alarm (acoustic soot agglomeration) may indicate that a smoke alarm sounded during a fire.

- Clean burn areas can be indicative of intense heating in an area, but they do not by themselves indicate areas of origin.

- Calcination occurs in gypsum wall surfaces when the free and chemically bound water is driven out of the gypsum by the heat of the fire.

- Incandescent lightbulbs may show a "pulling" effect that can be used to determine the direction of heat impingement and fire travel, pushing toward the heated side in bulbs of 25 W or greater, and pulling in when the bulb is less than 25 W.

- Annealed (collapsed) furniture springs result from the application of heat, which causes a loss in tension of the spring.

- A human body can display effects and patterns that assist the investigator in the evaluation of heat exposure and fire propagation.

- When examining a fire scene, the fire patterns that remain can be used to determine the sequence of events that occurred during the fire.

- When fires increase in size or burn for an extended period of time, the fire patterns located at the point of origin may be hidden or destroyed.

- The three-dimensional plume of a fire will produce two-dimensional shapes on nearby surfaces, such as V-patterns, circles, and other patterns.

- When a fire plume intersects with a surface such as a wall or ceiling, it may leave a fire pattern in the shape of the plume being cut by the surface (a "truncated cone" pattern).

- Causes of floor burn-through other than ignitable liquids can include ventilation, glowing combustion, and radiation, as well as building materials collapsing and smoldering in the debris.

- In a ventilation-controlled compartment fire, dramatic patterns may form near and around any ventilation openings.

- Water and other agents used for fire suppression are capable of producing or altering patterns and can leave distinct patterns on walls and ceilings that resemble water jets or droplets.

- Because fire is a three-dimensional event, fire patterns may be found on virtually any surface that has been exposed to the fire or to products of combustion.

- Full-room involvement may cause burning on the floor, baseboards, and door thresholds and sills that is unrelated to the application of an ignitable liquid.

- V-shaped patterns are commonly found on vertical surfaces, such as walls, in structural fires. They are created by the heat, smoke, or hot gases of a fire plume.

- Inverted cone patterns are often created by a vertical flame plume not reaching the ceiling level. They appear as a two-dimensional triangle shape, with its base at the bottom.

- A doughnut-shaped pattern may be the result of an ignitable liquid and should be closely examined for evidence that an ignitable liquid created the pattern.

- Trailers can be used to connect multiple fires or "sets" or used to spread the fire to different floors.
- When an object is physically in contact with another material in such a way that the material shields the object from the effects of heat transfer, combustion, or deposition, the object is said to be in a protected area.
- While pooled ignitable liquids may create irregular patterns, such patterns can also result from myriad other conditions.
- To reduce subjectivity, fire patterns can be analyzed using the scientific method, using a similar process to that provided in NFPA 921 for the analysis of origin and cause.
- Arc surveys can identify areas where the fire damaged energized electrical conductors and identify areas where the fire occurred before electrical energy was cut off.
- When arc damage is found on a conductor, it is evidence that the conductor was energized and that the fire was intense enough to damage the insulation and create an unintentional path between conductors.
- While the presence of an arc demonstrates that the circuit was energized, the absence of an arc does not necessarily mean that the circuit was not energized.

KEY TERMS

Alloying Mixing of two or more metals when one or more are in a liquefied state, forming an alloy.

Arc mapping Identification and documentation of a fire pattern derived from the identification of arc sites; aids in the determination of fire origin and spread. (NFPA 921)

Arc survey diagram A diagram of the affected area of the structure that identifies and plots the locations of electrical arcing. The spatial relationship of the arc sites can create a pattern and help establish the sequence of damage. This analysis can be used on building circuits and electrical devices within a compartment to help with origin analysis.

Areas of demarcation Areas in which the fire damage or effects are similar in magnitude or characteristics.

Beveling A fire pattern that indicates fire direction on wood wall studs. The bevel (angled side) points toward the general source of the heat.

Calcination A fire effect realized in gypsum products, including wallboard, when exposure to heat drives off free and chemically bound water. (NFPA 921)

Char Carbonaceous material that has been burned or pyrolyzed and has a blackened appearance. (NFPA 921)

Clean burn A fire effect that appears on noncombustible surfaces after any combustible layers (such as soot, paint, and paper) have been burned away. The effect may also appear where soot was not deposited due to high surface temperatures.

Crazing A complicated pattern of short cracks in glass that can be either straight or crescent shaped and can extend through the entire thickness of the glass.

Deformation Change in the shape characteristics of an object.

Deposition Collection and adherence of smoke particulates and liquid aerosols on surfaces during a fire.

Discoloration Change of color on the surface of a material affected by fire or heat.

Fire effects The observable or measurable changes in or on a material as a result of the fire. (NFPA 921)

Fire pattern The visible or measurable physical changes or identifiable shapes formed by a fire effect or group of fire effects. (NFPA 921)

Fire pattern analysis The process of identifying and interpreting fire patterns to determine how the patterns were created and their significance.

Heat shadowing A pattern that results from an object blocking the travel of radiant heat from its source to a target material on which the pattern is produced.

Lines of demarcation The borders defining the differences in fire effects on materials between the affected area and adjacent, less-affected areas. (NFPA 921)

Mass loss Loss of mass of a material due to consumption by fire or heat.

Melting A physical change caused by exposure to heat. (NFPA 921)

Rainbow effect A diffraction pattern formed when hydrocarbons float on a surface.

Sever arc An arc site where one or more of the circuit conductors were physically severed by the arcing event at that location. (NFPA 921)

Smoke deposits Hot products of combustion that may adhere upon collision with a surface.

Spalling The chipping or pitting of concrete or masonry surfaces. (NFPA 921)

Trailer Solid or liquid fuel used to intentionally spread or accelerate the spread of a fire from one area to another. (NFPA 921)

REFERENCE

National Fire Protection Association. 2020. *NFPA 921: Guide for Fire and Explosion Investigations*, 2021 ed. Quincy, MA: National Fire Protection Association.

On Scene

1. In your own words, explain the difference between fire effects and fire patterns.

2. Given the current knowledge reflected in NFPA 921, how can the fire investigator use depth-of-char or depth-of-calcination measurements?

3. Explain some of the reasons that a fire investigator might find irregular patterns on the floor of a burned structure, other than the application of an ignitable liquid.

4. List the types of fire patterns that may be created on surfaces by heat transfer from the typical "cone" shape of a fire plume.

JONES & BARTLETT LEARNING
NAVIGATE™

Chapter Opener: © Jones & Bartlett Learning. Photographed by Glen E. Ellman; On Scene siren: © Bildgigant/Shutterstock.

CHAPTER 5

Legal Considerations for Fire Investigators

KNOWLEDGE OBJECTIVES

After studying this chapter, you should be able to:

- Identify the basis of authority for conducting a fire investigation. (**NFPA 1033: 4.1.5**, pp. 124–125)
- Explain how a fire investigator may gain a right of entry to a fire scene. (**NFPA 1033: 4.1.5**, pp. 125–127)
- Identify the legal considerations when interviewing witnesses and suspects. (**NFPA 1033: 4.1.5**, pp. 128–129)
- Identify the legal considerations when detaining or arresting a subject. (**NFPA 1033: 4.1.5**, pp. 129–130)
- Define spoliation of evidence. (**NFPA 1033: 4.1.5**, pp. 130–132)
- Explain how to prevent spoliation of evidence at a fire scene. (**NFPA 1033: 4.1.5**, pp. 130–131)
- Explain the importance of cooperation between public and private investigators. (**NFPA 1033: 4.1.4; 4.1.5**, p. 132)
- Identify and describe the types of evidence in an investigation. (**NFPA 1033: 4.1.5; 4.2.1; 4.7.2; 4.7.3**, pp. 132–135)
- Categorize the types of witnesses in a legal proceeding. (**NFPA 1033: 4.1.5; 4.7.3**, pp. 135–139)
- Discuss arson, its forms, and its legal repercussions. (**NFPA 1033: 4.1.5; 4.7.2; 4.7.3**, pp. 140–141)
- Identify fire-related criminal acts. (**NFPA 1033: 4.1.5; 4.7.2; 4.7.3**, pp. 139–140)
- Characterize the burden of proof in criminal and civil cases. (**NFPA 1033: 4.1.5; 4.7.3**, pp. 140–141)
- Describe civil litigation that may arise around fire investigation cases. (**NFPA 1033: 4.1.5; 4.7.2; 4.7.3**, pp. 141–144)
- Describe the purpose of a deposition. (**NFPA 1033: 4.1.5; 4.7.2; 4.7.3**, p. 145)
- Describe the role of an investigator as a trial witness. (**NFPA 1033: 4.1.4; 4.1.5; 4.7.3**, p. 145)
- Describe potential uses of investigators' reports in a legal proceeding. (**NFPA 1033: 4.1.5; 4.4.3; 4.7.2; 4.7.3**, pp. 144–145)
- Explain the considerations when testifying as a fire investigator. (**NFPA 1033: 4.1.4; 4.1.5; 4.7.3**, p. 145)

SKILLS OBJECTIVES

After studying this chapter, you should be able to:

- Prevent the spoliation of evidence. (**NFPA 1033: 4.2.1, 4.2.2, 4.4.2**, pp. 130–131)
- Deliver accurate and effective verbal reports of investigative findings. (**NFPA 1033: 4.7.2**, pp. 144–145)
- Demonstrate effective techniques for delivering testimony. (**NFPA 1033: 4.7.3**, p. 145)

You are the Fire Investigator

As a public agency fire investigator, you are called to investigate a house fire that has resulted in extensive damage to the structure. While the firefighters work to extinguish the fire, you interview witnesses and document the fire and the building from the exterior. When the extinguishment and overhaul are complete, the incident commander (IC) informs you that the firefighters are leaving the scene, and that you will soon be the only public agency representative left on site. However, the building's owner, who is also its sole occupant, is on vacation and has not been reached yet.

1. What statutes, fire codes, or other documents provide you with the authority to perform a fire investigation?

2. Given the circumstances just described, do you have the authority to remain on scene to investigate the fire?

3. Are there steps you can take to prolong your right to remain on scene? Does this answer change if the fire is believed to be the result of a criminal act?

 *Access Navigate for more practice activities.*

Introduction

Fires are destructive by nature, and they often result in litigation. Fire-related legal proceedings may involve criminal prosecution for arson or other crimes, or they may involve civil lawsuits for negligence or other torts, product liability, or breach of contract. Sometimes a fire case can result in both criminal and civil litigation.

Fire investigators are likely to become involved as witnesses in legal proceedings during the course of their career. During these proceedings, they may find that all aspects of the investigation potentially come under scrutiny, including their training and qualifications, procedures used during the investigation, and their final opinions and determinations. Adhering to NFPA 1033, *Standard for Professional Qualifications for Fire Investigator*, and NFPA 921, *Guide for Fire and Explosion Investigations*, is critical to ensure that the investigator's testimony is reliable and admissible in court.

This chapter provides an overview of the legal landscape facing fire investigators. Note, however, that laws are in a constant state of flux and can vary by jurisdiction. Fire investigators are advised to become familiar with the laws that apply in their own locality or seek guidance from local counsel.

Role of NFPA 921 and 1033

Most qualified fire investigators receive training based on NFPA 1033 and NFPA 921. Nevertheless, many are surprised to learn about the potential impact of these documents during litigation. NFPA 921 and NFPA 1033 are closely related, and it is useful to read them together. NFPA 1033 specifically cross-references NFPA 921 as a source of the information for the many topics that form the foundation of an investigator's knowledge base. Further, NFPA 1033 recommends that investigators consult NFPA 921 for basic methodology and procedural matters, which are relevant to the mandatory fire investigator skills listed in NFPA 1033. Although NFPA 921 is defined as a "guide" and NFPA 1033 is designated as a "standard," it is NFPA 921 that is most frequently cited in court cases as a source used to evaluate the qualifications of fire investigators as expert witnesses and to assess the reliability of their opinions. NFPA 921 also contains a "Legal Considerations" chapter that touches on most of the legal topics listed in NFPA 1033.

Authority to Conduct the Investigation

The legal authority to conduct a fire scene investigation can be granted by either law or contract. For law enforcement and public-sector investigations, the authority is conferred by statute, ordinance, relevant case law, or agency rule. The scope and limitations on the power of the fire investigator are laid out in those laws and the court decisions that construe those laws. Not all public-sector investigators have the same powers. For example, some carry the status of peace officer or police officer by virtue of their position or by virtue of the applicable law. Other investigators do not have such status and may lack the power to arrest or swear to a search warrant. The investigator's employer retains further rights to define or restrict the investigator's authority, within the bounds of its own legal authority.

In the private sector, the investigator's authority is a contractual right that is derived from the person or organization authorizing the investigation. For example, a property owner, or an attorney acting as agent for the owner, may hire an investigator and define the scope and purpose of the investigation.

Sometimes the property owner is not the party directly responsible for instigating an investigation. In cases in which the owner has property or liability insurance on the property where the fire occurred, the owner's insurance company is usually the party that retains and instructs the investigators. In this situation, the property owner is said to be a third party to the investigation, and additional legal considerations may apply. Authority to investigate insurance claims is granted under the terms of the insurance policy, which generally requires that a claimant allow the insurance company to conduct a fire scene investigation before the claimant is eligible to receive payment for the loss (in the case of property insurance) or as a defense to liability claims (in the case of liability insurance).

Licensing Requirements

In many states, non–public-sector fire investigators must be licensed as private investigators or private fire investigators to conduct investigations. Obtaining such a license may require successful completion of a written exam and payment of a fee. Some statutes provide for limited exemptions from these requirements—for example, for employees of insurance companies or for insurance adjusters investigating insurance claims. Other statutes provide reciprocity between some states. It is best to check the legal requirements of each state where the investigation will be conducted to ensure compliance.

Also, be aware that states regulate the licensing of professional engineers. Similarly, check the relevant statutes in each state where the intended work will be done to see how the state defines the respective roles of investigator and engineer and which licensing requirements must be met.

Right of Entry

Even if a fire investigator has the legal authority to conduct a fire or explosion investigation, a right of entry onto property must also exist before that investigation can be undertaken at the scene. This right can be granted through the consent of a person with authority or by other means, as discussed later in this section. In the case of public-sector investigations, investigators must be legally authorized to enter the property that is the scene of the fire to examine it or to gather evidence. In other words, although a statute may require investigation of any fire of a certain size or type, this does not give the public investigator who is authorized by this statute the automatic right to enter the scene itself. Constitutional protections might require a search warrant to be issued before an investigator has a legal right of entry. Understanding state and local laws and statutes and consulting with the appropriate prosecuting authority will ensure the fire investigator correctly understands the right of entry and their ability to conduct an investigation and collect pertinent evidence.

In the private sector, the authority to enter a scene is derived from the authority of the person who retains the investigator. For example, a property owner who has full control of a scene after a fire can grant the investigator authority to enter their private property. In contrast, if the investigator is retained by a manufacturer of an appliance found within the scene, the investigator may have to request the consent of the owner or other entity in control before entering the property. It is always good practice for investigators to make inquiries about their authority and that of the entity retaining them, to ensure that proper authority exists and that no misunderstandings result.

During both the public-sector (governmental) investigation and the insurance investigation, the security of the scene and the preservation of fire patterns and potential evidence are of great importance. Government investigators should also try to communicate with the party who takes custody of the fire scene after the fire suppression efforts are terminated. The responsible party should be advised of the need to preserve the scene pending further investigation by other interested parties. Fire scenes with significant damage are often investigated by both public and private investigators.

Public-Sector Investigators' Right of Entry

The authority to investigate a fire does not necessarily grant the right of entry at the fire scene.

TIP

Public-sector investigators should get to know the prosecuting attorney in their jurisdiction who handles fire cases and should create a protocol for when the prosecutor should be notified of a criminal fire investigation, the circumstances in which a search warrant should be obtained, and the procedures for obtaining approval and signature of a warrant.

Public-sector/law enforcement investigators are bound by constitutional limitations under search and seizure law. Entry onto the property of another by a government representative is considered a **search and seizure**, and this definition certainly includes a fire scene investigation or any other crime scene investigation.

Under the Fourth Amendment of the U.S. Constitution and parallel provisions under state constitutional law, every entry onto a fire scene must be justified. There are four general circumstances that justify a search and seizure to conduct a fire scene investigation:

- Consent
- Exigent circumstances
- An administrative search warrant
- A criminal search warrant

Consent

Consent, defined as permission from the owner or occupant to enter the property, is perhaps the most convenient method of entry to ensure compliance with the Fourth Amendment. Even when investigators believe that exigent circumstances justify entry onto a fire scene (discussed later), they should consider asking the appropriate parties for their consent to be on the premises. Consent to search, when validly obtained, provides strong independent grounds for scene entry and alleviates any concerns that may be raised later about the timing and duration of the investigation (for example, allegations that there was an unreasonable delay in starting the scene investigation).

Strict requirements apply when obtaining lawful consent to search. First and foremost, consent must be obtained from the proper party. Although the owner of the property is often the proper person to give consent, this is not always the case. More specifically, the test to be applied is "common authority and control" over the premises. Property that has been leased or rented to an individual usually requires consent from the tenant or occupant rather than from the owner, even though the owner or landlord may have the right to enter and inspect the premises under the terms of a lease agreement.

The person who grants consent can authorize the search of only those areas under their common authority and control (such as common areas, like a kitchen, and the person's own private room). In the case of roommates, one roommate cannot authorize a search of the other roommate's private areas, such as a separate bedroom or personal closet. In all consent searches, only those areas within the common authority and control of the person giving consent may be searched.

While the courts consider a properly executed warrant to be the safest method of entry, and even though one of the other methods of entry might be appropriate to the situation, consent should always be requested whenever possible. Bear in mind that when a fire scene investigation is conducted pursuant to consent, the person who gives the consent may limit the scope of the search and may even revoke consent at any time. Consent to search a fire scene should always be obtained in writing and be witnessed.

Exigent Circumstance

Public-sector/law enforcement investigators have the authority to conduct a search and seizure whenever an exigent circumstance exists. **Exigent circumstance** allows investigators immediate access in the interest of public safety. The U.S. Supreme Court decisions in *Michigan v. Tyler* and *Michigan v. Clifford* outline the scope of a fire investigator's right of entry under this rule. Generally, a fire is considered a public safety emergency that requires a response by the fire service. When a fire is reported, the exigent circumstance allows entry onto the property to suppress the fire. Similarly, determining a fire's cause is part of the government response and is supported by strong public safety justifications. Thus, the investigation of a fire scene is authorized as part of the exigent circumstance surrounding every fire.

There is no open-ended authority to enter a fire scene and conduct an investigation, however. Instead, the investigation must take place within a "reasonable time" following suppression and extinguishment of the fire. Whether a fire scene investigation has been conducted within a reasonable time depends on a variety of facts and circumstances. If the investigator's actions are later challenged in court, whether the investigation was conducted within a "reasonable time" will be determined by the judge.

Generally, the investigation must commence at the first available opportunity following suppression and extinguishment of the fire. Any delay in beginning the investigation must be justified for the investigation to be considered part of the exigent circumstances surrounding the fire. A delay to wait for daylight is generally considered acceptable; however, waiting until later in the day when the investigation could easily have begun earlier may result in a finding of an illegal search and seizure. A prompt response by investigators is important in these situations.

Additionally, the investigation may continue only for a "reasonable time." Investigations must progress at an appropriate pace and not become unnecessarily drawn out. The investigator relying on the exigent circumstances doctrine must be able to document and prove why an investigation reasonably lasted the time that it did. In appropriate circumstances, these factors could include a large or complex fire scene, inclement weather, safety issues present at the scene, the requirements of a complete investigation (as outlined by NFPA 921 and NFPA 1033), and other applicable information.

Importantly, investigators performing a fire investigation under the exigent circumstances doctrine must ensure that government representatives, such as members of the police or the fire department, remain present on the scene from extinguishment until the end of the investigation. If the government representatives leave the scene, the exigent circumstance expires, and investigators will need consent or a search warrant to reenter the scene.

When in doubt as to what is "reasonable time," fire investigators should seek legal advice. Also, rather than relying exclusively on the existence of an exigent circumstance, it is advisable to seek consent whenever possible.

Administrative Search Warrant

An administrative search warrant is a third method of entry for the public-sector/law enforcement investigator, expressly authorized by the U.S. Supreme Court in the *Michigan v. Tyler* case. It is seldom used by fire investigators or the courts. An administrative search warrant can be requested in situations in which consent to search is refused or is otherwise not available, and there is concern that the "reasonable time" permitted by the exigent circumstances doctrine may expire. It requires a sworn affidavit presented to a judge in a court of competent jurisdiction for the limited purpose of gaining entry to a fire scene to determine the fire's origin and cause. The applicant must show in the affidavit that four conditions have been met:

1. They have the administrative authority to conduct fire scene investigations in the jurisdiction, such as the authority granted by an applicable fire code.
2. A fire has occurred, triggering the need for the entry onto premises for such investigation.
3. The investigator has not been legally able to enter the premises because consent was refused or because the appropriate person(s) is not available to provide consent.
4. The court's authorization is needed to fulfill the administrative responsibilities of investigating the fire to determine its origin and cause.

An administrative search warrant is not applicable once there is evidence of arson or some other crime, as outlined in *Michigan v. Clifford*. When a fire scene investigation pursuant to an administrative search warrant uncovers evidence of a crime, any further search of the premises requires a criminal search warrant. In this circumstance, the investigator should halt the investigation and proceed immediately to request a criminal warrant. Government representatives must remain on scene until that warrant is granted.

Criminal Search Warrant

A criminal search warrant is the fourth alternative for gaining entry to a fire scene for the purpose of conducting an investigation. This method is a judicial authorization based on the filing of a sworn affidavit by a government agent, who must be legally authorized to apply for the warrant.

Unlike an administrative search warrant, a criminal search warrant is issued only upon a showing of probable cause, establishing that a crime may have occurred, and that evidence of the possible crime is believed to be located in the premises to be searched. The affidavit must include reliable information and specific facts known to the applicant who is seeking the warrant. The premises and the specific areas within those premises where the search will be conducted must be precisely identified.

It is strongly recommended that a prosecutor or legal advisor assist in the preparation of the affidavit and application for warrant. Certain prerequisites to a search warrant must be fulfilled. For example, the address and description of the premises to be searched must be exact, and the facts shown in the affidavit must not be "stale" (meaning that the evidence sought will likely still be located in the premises to be searched when the warrant is executed).

Private Investigators' Right of Entry

In the private sector, the investigator's right of entry flows from the authority of the person retaining the fire investigator. This authority can vary depending on the role occupied by such a person. In the case of an investigator retained by an insurance company insuring the loss at issue, property insurance policies typically require the insured party to cooperate with a fire investigation as a condition of being able to recover payment for the loss. A tenant in a burned structure might hire an investigator to conduct a scene investigation of the building; however, that investigator might need consent of the public authorities who have control of the scene, or of the owner of the property, before entering to investigate.

Witness Interviews and Right to Counsel

Public-sector investigators should be aware of their obligations concerning the constitutional rights of persons suspected or charged with crimes. Both the Fifth and Sixth Amendments to the U.S. Constitution provide a right to counsel **FIGURE 5-1**. There are important differences between these rights, explained in the following sections.

The Fifth Amendment and Miranda Warnings

The **Fifth Amendment** provides a right to counsel and protection against self-incrimination. Under this amendment, a person cannot be compelled to testify against themselves or to provide an incriminating statement that the government could use against that person in a criminal prosecution. The U.S. Supreme Court recognized these rights in *Miranda v. Arizona*, which sets forth interrogation guidelines known as the Miranda rule.

The Miranda rule requires that a person must be given a "Miranda warning" before a government agent interrogates a person who is in custody. Specifically, before such questioning, a person must be told they have the following rights:

1. They have the right to remain silent.
2. Anything they say can be used against them in a court of law.
3. They have a right to the presence of an attorney.
4. If they cannot afford an attorney, one will be appointed for them prior to any questioning if they desire.

FIGURE 5-1 Fire investigators should be familiar with the Fifth and Sixth Amendment requirements when questioning a person suspected of a crime, including the right to have an attorney present.
© Jones & Bartlett Learning. Photographed by Glen E. Ellman.

The recommended practice is to read a statement of these rights directly from a preprinted document, such as a card provided by the prosecutor's office. This ensures that the investigator has phrased the rights correctly. Further, if a card is used in the Miranda warning, the investigator may be allowed to bring the card onto the witness stand when testifying, permitting the investigator to accurately recount the warning that was given.

The necessity of giving a Miranda warning is triggered when a government agent—someone representing the government, such as a police officer, a fire department investigator, or a person who is acting at the request of a government investigator—is about to begin a custodial interrogation of a suspect. The test for whether someone is "in custody" is whether a reasonable person in similar circumstances would believe that they were not free to leave. An arrest situation will generally lead a court to find a person "in custody," but a person can also be in custody under the Miranda analysis without being under arrest. The law on this topic is complex and situation dependent.

If a criminal defendant challenges the admissibility of a statement on the grounds that it was taken in a custodial setting without a Miranda warning being given, a court will consider various factors, including the following:

- Apparent authority of the interviewer to detain or arrest the witness. A private citizen or a public official who lacks arrest powers will generally not be required to comply with the Miranda rule.
- Location of the interview. The interview is more likely to be considered custodial if the witness does not have complete freedom of movement, including the ability to walk away from the interview. This could occur if the interview takes place in a locked room or a locked facility, if the exits are blocked by government representatives, or if the interviewee is handcuffed.
- Length and context of the interview itself. The longer the interview lasts and the more intense or accusatory it becomes, the more likely that it will be considered a custodial interrogation—even if it began as a simple witness interview.
- Participants in the interview. A large number of interviewers or other officials, especially law enforcement officials, can be considered a circumstance intimidating enough to transform a mere witness interview into a custodial interrogation.

To avoid creating the impression that an interviewee is in custody, the fire investigator can structure an interview to minimize the likelihood that the interview will be found to be custodial. For example, the investigator can inform the interviewee that they are not in custody and are free to leave; conduct the interview in a setting where the participants are not locked in a room; minimize the appearance of uniforms, weapons, and other shows of authority; and limit the number of interviewers. Additionally, if the interviewee is released after the interview, this factor helps to argue that the person was not in custody.

Not only must an interviewee be in custody for the Miranda rule to be triggered, but there must also be "interrogation," in which the questions asked or other conduct might reasonably cause a person to provide incriminating statements. Asking someone in custody, "Would you like a cup of coffee?" would not trigger a Miranda warning, whereas asking, "Did you set the fire?" most likely would. If a Miranda warning is not given to a person when required, any information given by that person during the interrogation is not admissible in court.

When any doubt arises about whether the interviewee might be considered to be in custody, the best practice is to advise the individual of all Miranda rights. When a person has been given a Miranda warning but waives those rights and agrees to make a statement without an attorney present, a written waiver of Miranda rights should be signed and witnessed. The interview itself should be tape recorded or videotaped whenever possible.

The Sixth Amendment

Like the Fifth Amendment, the **Sixth Amendment** affords protection of a person's right to counsel. As discussed earlier, a person's Miranda rights are triggered when a person in custody is about to be interrogated. This often happens before a person is arrested or charged with a crime. By contrast, the Sixth Amendment right to counsel is triggered only after criminal proceedings have begun against that person. Criminal proceedings can commence in a number of ways, such as when charges are filed or when a person is indicted. Once this happens, the Sixth Amendment entitles the accused individual to have an attorney present during interrogations or other proceedings related to that offense. Note that the Sixth Amendment right to counsel is "offense specific," meaning that while the person is entitled to have an attorney present during interrogations or court proceedings related to that specific offense, the police may question that person about unrelated matters without their attorney present.

The information provided here is merely a brief overview of the rights provided by the Fifth and Sixth Amendments. Public-sector investigators who may need to interrogate suspects in custody or question individuals regarding offenses for which criminal proceedings have begun should seek further training in procedures designed to protect the suspects' Fifth and Sixth Amendment rights.

Arrests and Detentions

The Fourth Amendment prohibits unreasonable searches and seizures, including seizures of persons. Thus, when a government official such as a law enforcement officer arrests or otherwise detains a person, a seizure has occurred. Whether a person is arrested or merely detained is determined by the specific circumstances involved.

Over time, the Supreme Court has developed a body of rules to govern detentions and arrests. What follows is a summary of those rules that are most applicable to fire investigators exercising law enforcement powers. Remember that constitutional rules apply to all government representatives (not just police officers), so any person acting on behalf of a governmental entity must be aware of these rules and how they may impact their interactions with persons in the field.

Arrests

A law enforcement officer making an arrest must have probable cause to make the arrest, meaning that they are in possession of acceptably trustworthy information that would lead a reasonable person to believe that the suspect has committed (or is committing) a crime. The information supporting probable cause can be based on the officer's personal knowledge or on another source, such as communications from other law enforcement officers, posted bulletins, witness statements, or information from an informant who appears to be reliable.

If a person's constitutional rights are violated and the arrest is deemed unlawful, a number of remedies are possible. For example, a judge might dismiss the charges against the person, or evidence flowing from the arrest might be excluded from trial.

Detentions

An officer may briefly detain a person if the officer has a reasonable suspicion that this person has committed (or is committing) a crime. A reasonable suspicion is something less than probable cause, but more than a vague notion. Whether the officer has information that gives them a reasonable suspicion will depend on the entirety of the circumstances. The detention must be brief and last only as long as is necessary for the officer to conduct a limited investigation to confirm or disprove the suspicion that caused the officer to stop the person. A detention can happen anytime a government representative (including a member of the fire department) appears to be physically or verbally stopping a person, even briefly; accordingly, all government fire investigators must be careful not to create the impression that they are detaining a person unless they are explicitly doing so for law enforcement purposes.

Spoliation of Evidence

Spoliation of evidence is the "loss, destruction, or material alteration of an object or document that is evidence or potential evidence in a legal proceeding by [a person] who has the responsibility for its preservation," according to NFPA 921. Spoliation is an issue that arises regularly in fire scene investigations, given the fragile nature of a fire scene and the fact that many parties potentially have a legal interest or liability in the outcome of an investigation. In general, spoliation is a term that applies in civil litigation, though it can also appear in criminal matters.

Alteration of evidence is an inevitable by-product of any fire scene investigation **FIGURE 5-2**. In a very real sense, an entire fire scene is critical evidence in a case. When an investigator has conducted a scene examination by moving fire debris and other evidence, the scene itself is necessarily altered. When a thorough fire scene examination has been conducted with the movement of debris using a methodical layering process, the result can be a scene in which most of the evidence is no longer in its original position.

FIGURE 5-2 Spoliation of evidence can be an issue at almost every fire scene. This dryer has been dismantled by fire crews.

When evidence is collected and removed from the scene, further changes to its state may occur. The examination of equipment and appliances found at a fire scene may involve destructive testing, which constitutes another form of change or alteration. In these situations, an investigator can face claims of spoliation when evidence is destroyed or altered because it denies other potential parties to a litigation the opportunity to independently evaluate the evidence, since that evidence is no longer available in its original postfire condition.

Court Sanctions for Spoliation

When spoliation has been found to occur, a court presiding over a fire-related lawsuit may take any number of actions against the responsible party as sanctions for the conduct that caused the spoliation—including sanctions for the conduct of an entity that reports to the responsible party, such as a fire investigator. Those sanctions may include the following:

- Prohibiting any testimony about the spoliated evidence by the responsible party
- Prohibiting the responsible party from presenting any testimony at all about the fire scene investigation
- Instructing the members of a jury that they may infer that the spoliated evidence would have been damaging to the responsible party's case
- Striking the responsible party's evidence and entering a judgment in favor of the party victimized by the spoliation

Moreover, courts in some states are now recognizing the tort of spoliation—a legal cause of action allowing a spoliator to be sued for their actions in an independent court proceeding. The concept of spoliation is a matter of fundamental fairness. In cases in which one party has lost, destroyed, or altered physical evidence in a way that disfavors another party, the courts will use the doctrine of spoliation to balance the scales. The sanctions that a court may impose are designed to correct the prejudice that the injured party has suffered by being unable to examine and test independently the evidence so as to defend itself against claims relating to the evidence or to use the evidence to assert its own claims against other parties.

Avoiding Claims of Spoliation

For all fire investigators, claims of spoliation can best be avoided by using a commonsense approach. Evidence should never be needlessly discarded, destroyed, or lost. Whenever evidence is uncovered that could be important to the interests of another party (such as in product liability cases), that evidence should be immediately preserved, and notification should be sent to all parties that have a potential interest in the evidence to allow them the opportunity to examine it. During this portion of the scene examination, the evidence should be secured in place, and the entire fire scene should be clearly marked and protected to prevent unauthorized movement of the evidence. This security should remain in place until the evidence is removed from the scene or released from the custody of investigators.

Collection and handling of evidence should be done carefully to minimize its alteration or degradation. Public-sector investigators should avoid taking appliances and items apart, especially in noncriminal cases, and should leave such items on scene where possible for the examination of other investigators.

Destructive testing of physical evidence should be preceded by notice to all known interested parties, permitting them an opportunity to have experts present and to observe or participate in the testing. NFPA 921 defines an **interested party** as "any person, entity, or organization, including their representatives, with statutory obligations or whose legal rights or interests may be affected by the investigation of a specific incident" (Section 3.3.123, p. 17). (Notice to interested parties does not necessarily apply to evidence collected during a criminal investigation.) Private-sector investigators should recommend to their clients that they notify interested parties and allow them the opportunity to participate in, or to witness and record, any testing that might damage the original evidence.

The inevitable alteration of evidence that occurs when fire debris is examined and moved during a fire scene examination generally does not constitute actionable spoliation, especially if the scene has been properly documented and photographed before, during, and after movement of the evidence. Thus, proper fire scene documentation—which should be employed in *every* investigation—is the first line of defense against claims of spoliation. When specific evidence (such as equipment or an appliance) is uncovered that is believed to have been a factor in the fire, those items should always be fully documented through photos and diagrams and carefully preserved, and notification should be immediately sent to all interested parties. When investigators engage in this practice routinely and view it as an established investigative protocol, the threat of spoliation is substantially reduced.

The nature and content of the notice that should be given to all known interested parties at the outset of a scene investigation, to allow them the opportunity to retain experts and attend the investigation if desired, are as follows:

- Notification to interested parties (via telephone, letter, or e-mail; oral notification should be confirmed in writing)
- Nature of notice to interested parties
- Content of notice
  - Date of the incident
  - Nature of the incident
  - Incident location
  - Nature and extent of loss (damage, death, or injury)
  - The interested party's potential connection to the incident
  - The next action date
  - Circumstances affecting the scene (such as pending demolition orders or environmental conditions)
  - A request to reply by a certain date
  - Contact information as to whom the notified person is to reply
  - The identity of the individual or entity controlling the scene
  - A roster of all parties to whom notice has been provided
  - Whether evidence has been removed from the scene and if so, its current location

NFPA 921 recognizes that public-sector investigators may have different notification responsibilities from those in the private sector. Furthermore, the duty to provide such notice will vary based on the jurisdiction, scope, and procedures of the investigators, as well as the circumstances surrounding the fire.

In criminal investigations, public-sector investigators generally have the right to proceed independently to determine whether a crime has occurred and to collect evidence for prosecution. Further, local procedures may permit the public investigators to proceed even in a noncriminal case so as to fulfill their public duties to determine the causes of fires. However, when a public investigation determines that a crime has not occurred, strong consideration should be given to when the scene should be released to permit the private-sector investigation, especially a scene that will require engineering or other expert analysis.

Even if the public investigator has the right to proceed without the attendance of the private-sector investigators, the best practice is to carefully document all actions taken at the scene, identify and preserve or set aside evidence that might be of interest to the private-sector investigators, and perform layering and other scene activities in a manner that allows private investigation professionals to be able to see and analyze the evidence, too. These simple steps will help foster goodwill in the eyes of the public and the private entities and promote the common goal of establishing a complete understanding of what happened at the fire scene.

Indeed, if a public entity such as a fire department does not possess the expertise to perform, for example, an engineering analysis on an appliance in a fire scene, the interests of all parties are furthered by leaving the appliance at the scene if possible, protecting it if necessary, and allowing the private or insurance entities to arrange for the necessary examination. The results of such an examination may be able to be relied upon by the public agency in its determinations regarding the scene.

TIP

A fire investigator should not discuss potential findings with anyone not involved in the investigation. "Under investigation" is the proper response to any inquiries.

Demolition

Local code provisions often allow a municipality to demolish a fire-damaged building immediately in the interest of public safety. The private-sector investigator may resort to court-ordered relief to protect against the destruction of evidence and allow time for documentation of the scene and potential evidence. In the event that demolition is authorized, the fire investigator should strive to document the scene as thoroughly as possible under the circumstances.

Public–Private Sector Relations

NFPA 921 encourages communication and cooperation between investigators in the public and private sectors. As first responders, public investigators are in the best position to record and document the postfire scene and to identify, document, and preserve evidence. Private-sector investigators often do not have the opportunity to view a scene in its original condition, and in some circumstances, a private investigator or engineer may not have the chance to see the scene at all. As a result, the reports, interviews, and photographs obtained by public investigators can be critical to other investigators coming in afterward, and a thorough job documenting and photographing a scene permits all parties to be better informed. Public-sector departments should have procedures in place by which

they can share reports and photographs with other interested parties under the appropriate circumstances.

Immunity Reporting Acts

Every U.S. state and the District of Columbia have adopted some form of immunity reporting act that requires insurance companies to report to local law enforcement authorities any fires that may have been the result of criminal conduct. These acts are known as arson-reporting/immunity statutes. Most of the statutes are based on the model act created by the National Association of Insurance Commissioners. Such laws are intended to facilitate the exchange of information between law enforcement authorities and insurance companies in the investigation of arson fires. The arson-reporting/immunity statutes require insurers to report to specified law enforcement agencies any loss due to an incendiary fire. Some acts also permit insurers to report fires of a suspicious origin (note that NFPA 921 does not recognize a "suspicious" origin as an accurate or acceptable level of proof for fire cause determinations; however, this is the language used by some of these statutes).

A designated public agency can make a written request listing the types of information and documents it requires the insurance company to release. This can include not only information about the fire in question, but also information about the insurance policy and the insured's claim history. In most jurisdictions, the arson-reporting/immunity acts require that this information remain confidential until it is required for use in civil or criminal litigation. In exchange for providing this information, the insurance companies are granted statutory immunity from civil or criminal liability.

In addition to the requirement to provide information to public agencies, many of the arson-reporting/immunity acts contain reciprocal provisions authorizing the exchange of information from those agencies to the insurance companies to assist in their civil investigations. In addition, some authorize the exchange of information directly between insurance companies.

Evidence and Testimony

Evidence can be categorized in a number of different ways. Four generally recognized categories of evidence will form the foundation of any civil or criminal fire trial:

- Real or physical evidence
- Demonstrative evidence
- Documentary evidence
- Testimonial evidence

These categories are not exclusive, and some evidence may fall into more than one category.

Real or Physical Evidence

Real evidence refers to physical items that are preserved from the scene of a crime or a relevant location in a civil case and may be produced in court for inspection by the judge and jury. A large range of such evidence might be relevant in a fire case, from minuscule cloth fibers to a large section of a wall that is removed from the scene and preserved for later analysis **FIGURE 5-3**.

Real evidence is commonly examined by experts and may be subjected to testing. Chapter 17 of NFPA 921 addresses physical evidence that may be relevant to the origin, cause, spread, or responsibility for a fire, along with recommendations for recognizing, collecting, and preserving this type of evidence at a fire scene. In cases involving fire-related deaths or injuries, physical evidence may include the deceased's remains and the victim's injuries. This type of evidence is covered by NFPA 921, Chapter 24, and other sections of this text.

The actual method used to collect physical evidence must comply with the technical standards outlined in NFPA 921, as well as the applicable ASTM standards and local requirements. To be admissible at trial, evidence must be properly transported and stored, and must be authenticated by establishing a **chain of custody**, a record of who had the evidence in their custody, to whom it was transferred, and where it was held. Even if the chain of custody is proven, some real evidence may be considered too prejudicial to be allowed into evidence and shown to the jury, such as a burned corpse.

In criminal cases, the collection of physical evidence at the fire scene is governed by the constitutional limitations on search and seizure. The right to collection and possession of physical evidence requires compliance with those constitutional standards.

If physical evidence has been altered, destroyed, or lost, or if the chain of custody has been compromised, a party to litigation may file a motion with the court seeking exclusion of the evidence at trial, or requesting a sanction or remedy for spoliation of evidence.

Demonstrative Evidence

Demonstrative evidence includes any type of tangible evidence, such as a photograph, diagram, or chart, that is utilized to demonstrate an issue relevant to the case; thus, it is an important form of evidence used at trial. This term can also refer to items that were gathered from the fire scene (real or physical evidence), as well as other tangible evidence items (as opposed to witness testimony).

Demonstrative evidence commonly introduced in a trial includes diagrams, charts, photos, timelines, models, and other such evidence that an expert witness or attorney specially creates to illustrate or demonstrate a point to the jury. The basic legal requirements for such evidence are relevance to the

FIGURE 5-3 Collected physical evidence.
Courtesy of Mike Dalton.

TIP

Demonstrative evidence can be very powerful because it is interesting and, if properly prepared and presented, will capture the attention of the jury. If you are presenting demonstrative evidence such as a chart, computer model, simulation, or even a PowerPoint presentation that includes photographs and graphical summaries of evidence, here are some suggestions:

- Ensure that you show the evidence or demonstrate the simulation to the attorney who calls you as a witness well in advance of trial. The attorney may need to disclose this evidence to the court or opposing side or make a **motion** (a request for the court to take action) before the trial begins to confirm its admissibility.
- Be prepared to testify that the chart, graph, model, or simulation is accurate in every detail.
- Recognize that one way to challenge demonstrative evidence is to show that the underlying facts or data on which such evidence is based are false or incorrect. Therefore, be prepared to provide details of the underlying facts and where they are found in the trial evidence (i.e., from witness testimony, documents, or physical evidence).
- If you are running a simulation or conducting a demonstration, be sure that you have practiced it and that it will work as predicted.

issues at trial and usefulness to the jury in understanding those issues. More generally, the admissibility of such evidence will turn on the accuracy or reliability of the assumptions or facts underlying the creation of these demonstrative aids.

Although proving the chain of custody of such evidence is normally not required, photographs, videos, slides, or digital media must be properly authenticated by showing that they are a fair and accurate depiction of the subject. This authentication need not be done by the person taking the photographs (although it is usually preferable to do so). Such evidence is generally found to be admissible when it has been properly authenticated in this way.

Documentary Evidence

The third form of evidence used at trial is **documentary evidence**. This category includes any type of written record or document that is relevant to the case. Business records, incident reports, telephone records, bank records, insurance policies, claim documents, correspondence, written statements from witnesses, investigative notes, and transcripts of recorded interviews are all examples of documentary evidence **FIGURE 5-4**.

All documentary evidence must be properly authenticated at trial under the standards of relevance and reliability. In the case of business records, bank records, telephone records, and other such documentary evidence, authentication will usually require the testimony of a "records custodian" to confirm the authenticity and accuracy of the documents. Documentary evidence may be subject to a legal challenge as hearsay if the information was obtained from an out-of-court source; however, there are many exceptions to the rule against hearsay evidence that may allow the documentary evidence to be used once a proper legal foundation has been laid. To avoid hearsay challenges, you should work with the prosecutor or attorney handling the trial to ensure that documentary evidence is properly presented.

Testimonial Evidence

The fourth form of evidence at trial is **testimonial evidence**, which is verbal testimony of a witness given under oath or affirmation and subject to cross-examination by the opposing party **FIGURE 5-5**. Although relevance and reliability are the baseline requirements for admitting testimonial evidence at trial, it is subject to the full range of requirements under the rules of evidence and may be prohibited for a variety of reasons. For instance, a witness may not be competent to testify at all, a witness may not be competent to testify about a specific issue, the testimony may be irrelevant or immaterial to the case, the testimony may include improper hearsay statements from other persons, the testimony may be speculative or unfounded, or the testimony may include the opinion of a witness who is not qualified to offer opinion testimony. Usually only an expert witness is qualified to offer opinion testimony. In the case of expert testimony, especially the testimony of a fire origin and cause expert, there are several requirements for qualifying an expert and presenting expert testimony, as addressed in the "Expert Witnesses" section later in this chapter.

FIGURE 5-4 Documentary evidence.
© Jones & Bartlett Publishers. Photographed by Glen E. Ellman.

FIGURE 5-5 Testimony.
© Guy Cali/Corbis/Getty Images.

During the testimonial process, the fire investigator may be involved in procedures such as affidavits, interrogatories, depositions, and trial testimony. An **affidavit** is a voluntary written statement of fact or opinion, made under oath and signed by the author. **Interrogatories** are a set of questions served by one party involved in litigation to another involved party. The response, called "answers to interrogatories," is generated by the served party and signed under oath by that party.

A **deposition** is a type of oral testimony made under oath. The witness, or deponent, is notified of the deposition by subpoena. Although the proceedings are usually conducted in an office setting, they are nevertheless legal proceedings made under oath with the use of a stenographer. Legal counsel for numerous parties may be involved in a deposition, and each is given an opportunity to question the deponent. A transcript is generated that is later reviewed and corrected by the deponent. The deposition transcript may be used if the proceeding continues to trial.

If the civil or criminal matter is not resolved through plea or settlement, the matter proceeds to trial. The trial is presided over by a judge and frequently includes a jury. After the judgment is made, appeals can be made by parties not satisfied with the result.

Witnesses

Witnesses called to testify in a court hearing or an out-of-court proceeding such as a deposition fall into two categories: fact and expert witnesses. A fact witness testifies as to factual events from personal knowledge. An expert witness presents opinions regarding the subject matter of the litigation that are outside of the knowledge of the average person.

Fact Witnesses

A **lay witness**, or **fact witness**, is a person who testifies about matters within their own firsthand knowledge. This type of witness has personally experienced the fact(s) to which they are testifying—that is, something the witness has seen, heard, touched, smelled, or tasted. Fact witnesses are limited in their testimony as to the types of opinions that they can give.

Like all witnesses, fact witnesses must be competent to testify. In other words, they must be able to observe, remember, and communicate their testimony to the court. They must also be able to understand and appreciate the nature of an oath and the obligation to tell the truth. Witnesses are generally presumed to be competent unless their competency is successfully challenged or a federal or state rule provides otherwise. An attorney may challenge the competency of a witness who may not be able to meet the criteria in any of these areas, such as a very young child or a person in the advanced stages of dementia.

Even when a witness is deemed competent, an opposing attorney can challenge the testimony by cross-examining the witness in an effort to show that the testimony is not credible or reliable. Common grounds for challenging fact testimony include the following:

- The witness is biased or has a vested interest in the outcome of the case. For example, the witness may have a close personal relationship with the party on whose behalf they are testifying, or the witness may have a general prejudice against opposing witnesses, such as the police.
- The witness's powers of perception are not trustworthy. For example, in a case of eyewitness testimony, the witness may have been too far away to have heard what they allege to have heard, or the lighting and obstacles at the time make it unlikely that the witness saw what they testify that they saw.
- The witness's testimony at trial contradicts prior oral or written statements made by the witness concerning the same events.
- The witness's testimony contradicts other evidence, such as physical evidence, documentary evidence, or the testimony of other witnesses. For example, perhaps a witness testifies to having seen the outbreak of fire at 1:00 PM, yet the 911 call reporting the fire was registered at 12:00 PM, or a witness says all of their possessions were in the house when it burned, but fire investigators sifting the scene find that pictures were missing from walls and clothing was missing from closets and dresser drawers.

Fact witnesses are not limited to bystanders or eyewitnesses to an event. Indeed, the fire investigator will very likely become a fact witness in a criminal or civil trial at some point. An investigator may be called to testify as a fact witness, if they were present before the fire was extinguished or before the fire crews left, or to testify about what they observed during the fire suppression and overhaul activities. If investigators testify about having collected, handled, transported, or stored physical evidence, they do so as lay witnesses, because this aspect of their testimony involves factual happenings, not expert opinion.

Fact witnesses are restricted in their testimony by Federal Rules of Evidence (FRE) Rule 701 or a comparable rule in state court. Rule 701 effectively limits the types of opinions that fact witnesses can give. For example, they may give an opinion about things like the speed a vehicle was traveling when they saw it, why a person appeared to be behaving in a certain way, or that a building was "fully involved" in flames. However, a fact witness may not give an opinion about the origin, cause, spread, or responsibility for a fire unless that witness is qualified by the trial court as an expert witness.

Note that **evidence technicians** require special training to do their jobs properly—that is, training based on NFPA 921 and NFPA 1033. Nevertheless, they may not have sufficient scientific, technical, or other specialized knowledge to qualify them to give opinion evidence about the implications of the fire or explosion as would an expert.

Expert Witnesses

An **expert witness** assumes a special role in the courtroom. Expert witnesses may testify about facts in their own personal knowledge (as lay witnesses do), as well as facts made known to them before or at trial, facts that they are asked to assume, or facts that are generally agreed on by experts in the particular field. Most significantly, experts are permitted to render an opinion respecting the fire or explosion **FIGURE 5-6**. Because fires often result in litigation, many public and private fire investigators will find themselves testifying as expert witnesses during their careers.

The nature of the opinion that an expert may render will be limited to some extent by the rules in the jurisdiction where the expert is testifying. In some states, an expert may not give an opinion on the ultimate issue that the jury is there to decide. For example, in some states, an expert may give an opinion about the mechanics of how an incendiary fire was started, but is not allowed to say that it is an arson fire started by the **defendant** (the person sued in a civil proceeding or accused in a criminal proceeding). Investigators scheduled to testify as an expert should speak with the attorneys calling them as a witness about any limits on their testimony and how their opinions should be presented.

Another limit to expert opinions in all jurisdictions is that the opinion must be relevant to the issues in the case and fall within the field in which the court has recognized the expert as qualified. Therefore, if a fire origin and cause expert has no specialized experience, knowledge, or training related to photocopier machines, the court may not permit that expert to testify about how a design fault in the photocopier caused a fire. Similarly, a trial judge may not allow a person who is qualified to testify about only the mechanics of propane explosions to give an opinion about why people did not smell the propane or whether it was odorized properly.

Before an expert may testify in court, several steps must be completed.

Admissibility and Relevance of Expert Testimony

As with all other forms of evidence, testimonial evidence from an expert must be relevant to the issues in the case and must be reliable; however, the measure of reliability for an expert witness can involve a complex analysis. In federal court cases, and in many state court jurisdictions, expert testimony is governed by the Daubert rule (stemming from *Daubert v. Merrell Dow Pharmaceuticals*, a 1993 U.S. Supreme Court case). Under the Daubert standard, the trial judge takes a proactive role in screening the testimony of prospective expert witnesses. When an expert is challenged, the judge must assess the reliability of both the conclusions (opinions) to be presented and the techniques and methods used to reach those conclusions. This assessment may happen long before the fire investigator is scheduled to testify in trial. If the judge is not satisfied with the reliability of the expert's conclusions or methods, the expert's testimony may be limited or prohibited.

Federal Rule of Evidence (FRE) 702 was amended to encompass the factors articulated by the court in the *Daubert* case. FRE 702 provides that expert witnesses may provide opinion testimony when they are qualified by "knowledge, skill, experience, training, or education," provided that the specialized knowledge

FIGURE 5-6 An expert witness assumes a special role in the courtroom.
© Guy Cali/Corbis/Getty Images.

will assist the jury to understand the evidence or determine a fact at issue. Per FRE 702, the judge must address three factors before allowing the expert testimony:

- The testimony must be based on sufficient facts or data,
- The testimony must be the product of "reliable principles and methods,"
- The expert has reliably applied the principles and methods to the facts of the case.

Thus, in federal and many state courts, before allowing an expert to testify, the judge must find not only that the expert is qualified, but also that the expert's testimony has a basis both in the facts and in a reliable methodology. In addition, the judge must find that the expert reliably used the methodology in making their determination in the question at hand.

The judge may consider various factors in evaluating an expert's testimony. The Daubert rule (*Daubert v. Merrell Dow Pharm., Inc.*, 509 U.S. 579, 113 S. Ct. 2786 [1993]) requires the judge to consider the following factors in determining whether the investigation was based on reliable principles and methods:

- Whether the theory or technique has been or can be tested and whether the hypothesis underlying the theory or technique can be falsified
- Whether the theory or technique has been subjected to peer review or publication
- Whether the theory or technique has a known or potential rate of error
- The existence and maintenance of standards controlling the technique's operation
- Whether there has been general acceptance of the theory or technique in the relevant scientific community

The fire investigator defending an investigation in a Daubert challenge should be prepared to address these factors with regard to the system used in the investigation. NFPA 921 provides a comprehensive, widely accepted fire investigation system. Investigators who employ NFPA 921 should be able to demonstrate that their system was reliable.

In addition to those general criteria, a trial judge may ask any other questions that are appropriate to the particular facts of the case to ensure that the testimony is reliable enough to be heard by a jury. Specifically, as outlined in FRE 702, the judge or counsel will inquire whether the fire investigator, in addition to relying on a reliable methodology such as NFPA 921, properly and carefully applied it to the specific investigation. Moreover, the investigator may be asked questions designed to ascertain whether they had, and relied upon, reliable facts and data in arriving at their conclusions.

Some jurisdictions apply other standards to measure the reliability of expert evidence. Prior to the *Daubert* decision, to be admissible, expert testimony based on novel scientific theories or techniques had to pass a test that came from a 1923 federal appellate court decision in *Frye v. United States*, known as the Frye "general acceptance" test. Pursuant to the Frye test, evidence was necessary to show that the principle or technique on which the opinion was based "was sufficiently established to have gained general acceptance in the particular field in which it belongs" (*Frye v. United States*, 293 F. 1013 [D.C. Cir. 1923]).

Some states continue to apply the Frye test for novel scientific testimony. Others use a modified test, combining elements of the Daubert rule and the Frye test. Regardless of whether the Frye or Daubert test is the standard in a particular court, NFPA 921 can be utilized in every jurisdiction in the United States to measure whether a fire investigator's theory or technique has gained general acceptance in the relevant scientific community (i.e., the community of fire investigators).

In jurisdictions that rely on the Daubert rule (and many other jurisdictions), expert witnesses can face vigorous challenges to their testimony and findings. In responding to a Daubert challenge, the fire investigator must be able to demonstrate that both the methods used in the investigation and the conclusions reached can be proven to be reliable by reference to objective and documented sources. Thus, a significant amount of preparation is essential anytime a fire investigator is expected to testify as an expert in a criminal or civil case. Although a Daubert challenge may happen long before trial, an investigator must be prepared at that time to defend and support their investigation. Investigators should perform a full and comprehensive review of the case facts, all hypotheses of fire origin and cause considered in the investigation, supporting data, and the pertinent science of ignition and fire development. A negative finding by a court in a Daubert challenge of their work can have a negative effect on an investigator's professional credibility in the case at hand and, potentially, in the future.

Familiarity with the entire NFPA 921 document and the ability to convey an understanding of the appropriate sections addressing any issue in question will enable an investigator to respond effectively to a Daubert challenge. Because NFPA 1033 is the industry standard for fire investigators and because it references NFPA 921, NFPA 1033 lends additional weight to employing NFPA 921 as a measure of reliability.

Qualifications of Expert Witnesses

As we have seen, the first admissibility requirement in FRE Rule 702 is that to testify as an expert, a witness must be "qualified as an expert by knowledge, skill, experience, training, or education." The trial judge is the person who determines whether an expert is qualified. This decision is based on the exercise of the judge's discretion as guided by precedent cases. Sometimes counsel may stipulate to the qualifications of an opposing expert—in other words, admit that the expert is qualified, not challenging the admissibility of the expert's testimony on that ground. Alternatively, an attorney may try to prove either that a witness is not sufficiently qualified to give any expert opinion testimony or that a witness is qualified as an expert, but in a more limited field or on more limited issues than those for which the expert was originally proffered. If the qualifications of an expert are not challenged, the trial court will typically recognize the expert as qualified. If the qualifications are challenged, the court will hear evidence on the issue.

Challenges on the qualifications of experts in fire and explosion cases are becoming increasingly common. NFPA 1033 is the national professional qualification standard for anyone wanting to serve as a fire investigator. It sets forth the minimum job performance requirements to be a qualified fire investigator in either the public or private sector. As such, it serves as a benchmark, along with NFPA 921, for all fire investigators. It does this in four ways:

- By specifying numerous broad-ranging requisite knowledge topics about which investigators must remain current
- By identifying the main responsibilities of fire investigators
- By pinpointing the job performance requirements that encompass the job of a fire investigator
- By listing the knowledge and skills necessary to do each task involved in an investigator's job

In court cases, especially in civil litigation, investigators often see their qualifications and knowledge challenged using the framework of NFPA 1033. As a fire investigation career develops, an investigator should consult regularly with NFPA 1033 to ensure that they are able to perform the job performance requirements, and that they have ongoing training on the knowledge topics listed in this document.

The types of information attorneys and judges will consider when evaluating an expert's qualifications should appear on the fire investigator's curriculum vitae (CV). It is important to be scrupulous to ensure

TIP

Basic components of an expert's qualifications include the following items:

- Personal background information (name, age, current place of business or employer, duties in current position)
- Formal education (schools, colleges, or universities attended; dates attended; diplomas or degrees received; any specialization obtained)
- Skills and experience (emphasizing those that relate to your qualifications to testify as an expert in fire or explosion cases)
- Professional licenses or certifications (including the date of any license or certification, the licensing/certifying body, the dates of recertification, and the jurisdictions where the license/certification is recognized [i.e., state, national, international])
- Distinctions and awards (relating to your work in your field of expertise)
- Publications (using formal citation format and including the name[s] of any coauthors, title of the work, title of the journal or collection if your publication was part of a larger work, name and city of the publisher, publication date)
- Memberships and offices in professional associations, including committee memberships or chairs and other special appointments (Be careful about listing memberships that were in name only, such as committees to which you were appointed but in which you did not actively participate.)
- Professional presentations (speaking engagements at seminars, conferences, and the like)
- Academic or teaching appointments

You should also keep track of the following additional information, which might be the subject of questions raised by the attorney calling you as a witness or by the opposing attorney at your deposition or when you are cross-examined:

- Publications to which you subscribe or read regularly
- Books that you have read in your field or that were the basis of courses you have taken, including NFPA 921 and NFPA 1033, and other books such as *NFPA Fire Protection Handbook*
- Other cases in which you testified (including affidavits, depositions, testimony in preliminary motions, and trial testimony)

Disclose to the attorney who will call you as a witness any negative aspects of your background or qualifications so that the attorney can help you plan how to limit the potential damage from this type of information. Negative aspects might include courses that you failed, certifications that you allowed to lapse, successful challenges to the admissibility of your testimony in prior cases, or accusations of ethical improprieties.

that a CV is fair and accurate with respect to each item listed. When preparing to testify in a particular case, the investigator should review their CV, focusing on the specific aspects that relate to issues in the case. Although an investigator's qualifications should always be accurate, the investigator should present this information to a judge or jury in a manner that demonstrates the expert's qualifications and addresses the specific issues in the case at hand.

Remember that NFPA 1033 requires (and the legal system expects) experts to stay current in their fields. Fire investigators should devote adequate time to reading professional journals and other publications, attending seminars and association meetings, and obtaining further formal education to the extent necessary to maintain and improve their qualifications.

Examination of Witnesses

The questioning of witnesses customarily involves two or more parties, and follows a course of direct examination and then cross-examination. The direct examination is conducted first by the party who produced the witness. Questioning in direct examination is open-ended and allows for description and explanation. After direct examination, the opposing party is able to conduct a cross-examination. The purpose of the cross-examination is to extract more detail from the direct examination or to challenge the testimony. Cross-examination questions are typically leading and closed-ended. After cross-examination, the party calling the witness is permitted to conduct redirect examination (limited to the topics discussed in the cross-examination); in some jurisdictions, "recross" examination is also permitted.

Arson

Arson is one of the oldest criminal offenses under the law. It dates back to the early common law of England, when it was one of only three capital offenses (along with treason and murder). The common law of England originally defined the crime of arson as the malicious burning of another's home at night. For several centuries, this was the only form of intentional fire setting that was considered to be a crime. Criminal laws defining arson more broadly have since been codified and modified in the United States.

Today, the crime of **arson** can be committed in many forms. It can include causing fire or explosion to residences, buildings, vehicles, aircraft, watercraft, personal property, and other structures and materials. The elements of arson crimes vary among jurisdictions. Investigators need to be closely familiar with the arson statutes both in their state and in the federal jurisdiction.

Most arson offenses are felony crimes punishable by imprisonment and fines. Arson to an occupied structure is generally classified as the most serious arson offense; it is punishable by substantial prison time, ranging up to a life sentence in some jurisdictions. Less serious arson offenses are often punishable as felony offenses as well. Further, some jurisdictions have established arson offenses with special elements, such as arson for the purpose of defrauding another. Such offenses can be considered serious crimes and may require special proof of facts above the facts required to prove a standard arson charge. Finally, the intentional burning of other types of property that are not designated under the arson statutes may be considered vandalism or criminal mischief.

NFPA 921, Section 12.5.6.3, includes a list of factors for investigators to consider when determining whether the crime of arson has occurred. In addition, investigators must consult the arson statutes of their jurisdiction, which contain requirements specific to that jurisdiction.

Circumstantial Evidence

A prosecutor often has to prove arson entirely based on circumstantial evidence rather than by offering the testimony of an eyewitness to the start of the fire. **Circumstantial evidence** is evidence based on inference and not on personal observation. Arson crimes will usually require proof of three things:

- The fire was incendiary.
- The person who set the fire had the necessary criminal intent.
- The defendant set the fire or arranged for someone else to do so.

If there was no eyewitness who saw how the fire started, an investigator will rely on circumstantial evidence to determine the cause of the fire and who was at fault.

NFPA 921, Chapter 23, *Incendiary Fires*, identifies a number of conditions related to fire origin and spread that may provide evidence of an incendiary fire cause. A few examples are listed here:

- Multiple fires: Two or more separate, nonrelated, simultaneously burning fires
- Trailers: Deliberately introduced fuel or manipulation of existing fuels to aid in the spread of fire from one area to another
- Incendiary device: A mechanism used to initiate an incendiary fire, such as a lit cigarette placed in a book of paper matches and propped over a container of combustible materials or a Molotov cocktail

- The presence of ignitable liquids in the area of origin that cannot be explained other than as a device to start or spread the fire

Once a fire has been determined to be incendiary, the next step is to identify the fire-setter. This will likely require an investigation into who had both the opportunity and a motive to set the fire. **Motive** is defined as "an inner drive or impulse that is the cause, reason, or incentive that induces or prompts a specific behavior" (NFPA 921, Section 23.4.9.1.1). It is the reason behind an act. Motive is not the same as criminal intent, which is the state of mind to commit the criminal act (or sometimes omission) without any justification, excuse, or other defense. Thus, a person may have a motive to commit arson, yet have nothing to do with the fire. If an insured party had a financial motive to burn their home, but the home burned down due to an accident, there would be no criminal intent and no arson crime.

Although motive is generally not a required element of the crime of arson, a prosecutor will want evidence of motive to present to the jury. Without an explanation of why a person would commit such a serious crime as arson, jurors may be reluctant to convict the defendant. It is human nature to want to know why someone committed arson. NFPA 921, Chapter 23, contains a summary of the traditional classifications of motives for fire-setting.

In addition to seeking the underlying motive, an arson investigation often includes an effort to identify persons who had the opportunity to set the fire because, for example, they were known to have been present at the time the fire started and/or had access to the site. Witness reports about the whereabouts of suspects at the time of the fire and paper trails, such as gas purchases or other receipts and phone records, may shed light on this issue. For fires that occurred in locked premises with no evidence of a breaking and entering, the investigation should aim to determine which persons had keys or other access. A strong circumstantial case may result if the investigator identifies a person who had exclusive opportunity to set the fire.

Other Fire-Related Criminal Acts

In addition to arson, other criminal offenses may come to light through a fire scene investigation. The following are only some of the potential offenses that may be relevant in a fire investigation:

- Burning to defraud
- Insurance fraud
- Unauthorized burning
- Manufacture or possession of firebombs and incendiary devices
- Wildfire arson
- Domestic violence
- Child endangerment/child abuse
- Homicide and attempted homicide
- Endangerment/injury to firefighters
- Obstruction/interference with firefighters
- Disabling of fire suppression/fire alarm systems
- Reckless endangerment
- Failure to report a fire
- Vandalism or malicious mischief
- Burglary/trespassing

In addition to these offenses, virtually any other type of criminal activity may be uncovered in the course of the investigation. In particular, a fire may have been set to cover evidence of other crimes.

Burden of Proof

In a criminal matter, the prosecution must prove the defendant's guilt beyond a reasonable doubt. This is a heavier burden than the burden of proof in a civil case, such as a claim based on a tort or contract. In most civil cases, the plaintiff must prove their claim by a preponderance of the evidence, sometimes also termed "more likely than not" or proof on a "balance of probabilities."

However, in a civil claim where the defense alleges that the plaintiff was involved in wrongdoing comparable to a criminal act, the applicable law may require the defendant to establish the defense by "clear and convincing" evidence—a standard that is higher than the civil preponderance standard, but lower than the criminal "beyond a reasonable doubt" standard. For example, if an insured person sues an insurance company because it denied a claim, the insurance company might defend itself based on the allegation that the insured burned their own property or committed insurance fraud by inflating the claim made on the loss. In this situation, the defendant insurance company may be required to prove its defense by clear and convincing evidence, meaning it must show that it is highly probable this defense is true.

Prosecuting Attorney Relations and Brady Materials

As mentioned previously, the public fire investigator should make the effort to get to know the prosecutor(s) in their jurisdiction who are involved with arson cases, so that when issues arise, a working relationship is already in place. When a criminal case is filed, it is best practice for the fire investigator to contact the prosecutor involved and meet with them, if possible, to supply any information that might be needed and to answer questions.

When a criminal case is filed, discovery obligations are triggered, although the extent of the materials that the prosecution is required to turn over to the defense varies by jurisdiction. The prosecuting authority may be required to turn over to the defense all of the materials generated by the public-sector fire investigator, including reports, photos, notes, and the like. Therefore, it is important that the public-sector investigator make sure that all such documents are turned over to the prosecution. This can be done through the investigating police department or directly to the prosecutor's office, depending on the local protocol. When the investigator generates new reports or photos, these need to be turned over to the prosecution just as the original ones were. Failure to supply the prosecution with all relevant investigative documents can result in the inability to introduce important evidence at trial or even sanctions against the prosecuting authority at the discretion of the judge.

Even in jurisdictions where discovery obligations are less comprehensive, the U.S. Constitution always requires the prosecution to turn over materials identified by the *Brady v. Maryland* case. **Brady materials** comprise any document or other evidence in the government's file that tends to contradict the guilt of the defendant (e.g., that is "exculpatory") or would act to mitigate their punishment. Failure to turn over such items can amount to an error of constitutional proportions, obligating the court to consider a severe response against the prosecution such as exclusion of evidence or worse. The prosecution is responsible for all of the evidence in government hands, including materials under the control of the fire department. If any question or reluctance about turning over evidence for an arson case arises, the investigator should consult with the prosecuting authority.

Civil Law and Litigation

Civil litigation in fire investigation cases is far more prevalent than criminal prosecution. Investigators for a public agency who have conducted a fire scene examination may become involved in civil litigation related to that fire, regardless of the outcome of their own investigation and the results of any subsequent prosecution. For private-sector investigators, the prospect of becoming involved in civil litigation at some point is nearly a certainty.

Civil litigation occurs when a lawsuit is filed, asserting one or more causes of action. A cause of action is a theory under which a person may be entitled to obtain legal redress from another person. Examples of some causes of action that may be cited as a result of a fire or explosion are breach of contract, product liability, negligence, a civil action based on fire code violations, or a civil claim for an intentional wrong such as arson. Every fire scene investigator must have an understanding of civil litigation principles to respond effectively when called to testify in a civil proceeding.

Of course, the criminal and civil courts differ in some fundamental ways. First and most importantly, the potential consequences of the proceeding are far more severe in a criminal case. An individual's life and liberty are at stake in a criminal case; a criminal defendant who is convicted at trial faces the prospect of imprisonment. Civil cases and their outcomes are usually concerned only with the award of monetary damages. Nevertheless, although life and personal freedom may not be at stake in a civil case, the individual's financial freedom may be at risk. Fire litigation cases in civil court are usually high-stakes cases. For this reason, they are seriously and aggressively litigated by the parties and their counsel.

A fire litigation case in civil court can directly involve the actions of the public-sector investigator and may ultimately rest on their findings. Frequently, one of the parties seeks to use a public investigator as an expert witness for its side of the case to have a presumably impartial witness support the case. Private-sector investigators, by comparison, are always challenged based on the fact that they have been paid for their testimony and findings.

Discovery in Civil Cases

The procedures in civil litigation are similar in many respects to those in a criminal case, but the discovery phase of a civil case is usually much longer and much more involved than that in a criminal case. Dozens of witnesses may be deposed. Hundreds of records may be subpoenaed, and numerous procedural motions may be filed by the parties.

The **discovery** process consists primarily of requests to produce, interrogatories, depositions, and reports. The request to produce will specify particular documents or document categories that are to be produced and must be responded to within a specified time frame. If documents or other items are needed from a nonparty to the litigation, subpoenas can be served upon them calling for the production of relevant documents on a particular date and at a specified location. Discovery between parties to civil litigation also includes interrogatories, requests for production, and depositions, as described previously.

Written reports made by experts can be used during all forms of discovery. The written report should address activities conducted, materials reviewed, the opinions the expert will address, and the basis for arriving at those opinions. Publications by the expert, prior expert testimony, and documentation of compensation for work conducted can also be included.

In federal court cases, witnesses who will be testifying as experts must make certain disclosures at the early stages of the litigation under Rule 26 of the Federal Rules of Civil Procedure. At the very outset of the litigation, all expert witnesses must be identified to the other side in a written disclosure. The substance of the expert's anticipated testimony must be disclosed, along with a copy of the expert's reports and analyses. An expert who has not yet prepared a report must do so to comply with Rule 26. This disclosure must include, among other things, the facts or data considered by the witness in reaching their opinions and conclusions about the case. Additionally, the expert must provide a current CV or résumé with complete information about their educational background, employment history, professional qualifications, fees charged, articles or publications written, and prior testimony as an expert. The information about prior testimony must be provided in detail, including the style (name) of each case, the court in which the case was tried, the attorneys who litigated the case, and the party who hired the witness. This information is required for every case in which the expert has testified as an expert witness within the preceding four years.

Civil Litigation Involving Fire Investigators

With increasing frequency, fire investigators are finding themselves more than mere spectators in civil litigation: Lawsuits have been filed against public-sector and private-sector investigators alike, as well as their agencies and companies. Negligence claims and professional liability (malpractice) claims have been asserted against fire scene investigators for improper investigations, incorrect findings, erroneous conclusions, false arrest, and malicious prosecution in criminal cases, as well as for bad faith in insurance claims cases and spoliation of evidence in both criminal and civil cases.

Civil Causes of Action

As mentioned earlier, various causes of action, or claims, can be asserted by a plaintiff in a civil case.

These are similar to charges in a criminal case, and generally must be proven by the party asserting the claims. At times, a defendant brings counterclaims against the plaintiff for causes of action it believes entitle the defendant to monetary relief.

Negligence

Negligence is the failure to exercise reasonable care toward another to whom the actor has a duty of care. Its essence is carelessness but, as with all causes of action, specific elements must be proven for the plaintiff (the person or entity bringing the civil action) to receive compensation, including proximate cause and damages or injury.

A cause of action for negligence from a fire or explosion can develop in countless ways. Negligence can range from unplanned accidents caused by mere carelessness to gross disregard for the rights, property, or safety of others. A number of defenses to negligence are also possible. The conduct of the plaintiff is a primary consideration. If the plaintiff's own negligence contributes to their injury, the effect will depend on the law in the given state. In most states, the plaintiff's damage award will be reduced by the percentage of the fault attributed to the plaintiff. In other states, the plaintiff's claim may be completely barred.

Some examples of fires that could result in negligence claims are as follows:

- Careless disposition of smoking materials (such as cigarettes), causing a fire
- Cooking accidents, such as grease fires
- Fires caused by inadequate lightning protection as required by law
- Fires that spread from fireplaces, campground fire pits, or improperly disposed-of hot fireplace ashes
- Fires caused by improper use, maintenance, or repair of gas or electrical systems or appliances

As described in the next section, negligence can sometimes be inferred or proven when a person breaches fire safety laws, causing a fire. Breach of safety codes may also give rise to a negligence action even when the cause of the fire is not known. If a life-safety code is breached, causing someone to be injured or killed in a fire, or if the spread of a fire can be traced back to violations of safety laws, the persons responsible may be liable for negligence.

Product Liability

Manufacturers or sellers of defective products may be liable in case persons are injured by those products. Claims for product liability can arise from manufacturing defects, inadequate warnings, or design defects. This area of law, which is sometimes called manufacturers' liability, is based on both common law principles and legislation.

Fire incidents frequently result in product liability claims. Products giving rise to such claims are many and varied—for example, motor vehicles, appliances, and components of building systems such as electrical, HVAC (heating, ventilation, and air conditioning), or gas systems. A product liability claim can also result when propane or natural gas for use by consumers is not properly infused with a readily recognized odor, allowing a leak to go undetected and ultimately cause a fire or explosion.

When any sort of a product is involved in the cause or spread of a fire, it is important for the investigator to explore a number of avenues in the ensuing investigation. Here is a brief overview of some investigative tasks that may be required:

- Reviewing NFPA 921 sections regarding spoliation of evidence before dealing with products at the fire scene. Note the requirements for notification to interested parties and the need to document the product prior to altering it or conducting any destructive testing.

- Establishing, as much as possible, the chain of custody of the product as it passed from the manufacturer to the seller and ultimately to the person who owned it at the time of the incident. This means collecting any of the following documents and information that are still available: receipts for the product's purchase; packaging (boxes, warning labels); users' manuals; warranty cards; serial number(s); and information on the make, model, and date of manufacture. In some instances, it may even be necessary to trace how the product was transported and warehoused before reaching the consumer.

- Identifying component parts of the product that may have been defective, resulting in the fire or explosion, and obtaining as many details as possible to determine whether a different company was responsible for manufacturing separate components.

- Tracing the history of the product, including the use (or misuse) by anyone, including the owner, and the service and maintenance history and records.

- Retaining one or more experts who can examine the product and/or its components and provide opinions on any defects in the manufacture, warning labels, or design of the product and/ or its components. Experts should also be asked to determine whether repairs or modifications were made to the product that contributed to the loss.

- Obtaining an exemplar of the product so that an undamaged product of the same make, model, and date of manufacture is available for comparison to the fire-damaged product.

Maintaining the chain of custody and the integrity of the product after the fire is also critical. The product should be documented in place at the scene, using appropriate methods, including photographing and diagramming and perhaps video recording. If a product will be further damaged if it is moved, it might be necessary to arrange for a portable X-ray to be taken before the product is moved so that the status of its internal components can be recorded. An investigator may need to be creative when packaging and preserving a damaged product. For example, moving a coffee maker whose plastic parts have melted, affixing it to a countertop, may require that the entire segment of the countertop holding the appliance be cut away so that the appliance is not damaged further by removing it from the countertop.

Strict Liability

Courts apply the concept of strict liability in product liability cases in which a seller is held liable for any and all defective or hazardous products that unduly threaten a consumer's personal safety. This concept applies to all entities involved in the manufacturing and selling of the product. It is based on the theory that by presenting a product to the public for sale, the manufacturer is representing that the product is safe for its intended use. To prove a product liability case when strict liability is the standard, the plaintiff typically does not have to show negligence on the part of the manufacturer or seller.

Subrogation

Subrogation is a legal concept that allows an insurance company or other entity that compensates a damaged party for a loss to proceed against the person who caused the loss and recover the payments it has made. For example, an insurance company that pays a homeowner for the losses incurred in a fire could use this principle to gain the right to file a claim or lawsuit against a manufacturer of a defective product that caused the fire, and in turn to recoup the money paid to the insured homeowner.

Subrogation is a common issue in fire losses, and property and casualty insurers will often consider the possibility of subrogation when a third party causes a fire loss. Preserving the fire scene and avoiding spoliation of evidence assists the parties involved in resolving these types of civil issues.

Fire Investigator Presentation in Court

To be a successful witness at trial, whether an investigator is a fact witness (such as an evidence technician) or an expert witness, it is important to become comfortable with the courtroom and with civil and criminal trial proceedings. This learning process begins with knowing the procedural steps that lead up to trial and the types of oral and written testimony that may be given at or before trial.

Reports

Whether expert witnesses prepare a report in a given case will depend on many factors, including the following:

- Whether the criminal or civil rules of procedure require witnesses who will testify at trial as experts to prepare a report (as does Federal Rules of Civil Procedure [FRCP] Rule 26[a][2][B]).
- The trial strategies decided on by the expert and the attorney who will call the expert as a witness.
- The advantages and disadvantages of preparing a report, as summarized later in this section.
- Consideration of whether the report is admissible in evidence and whether disclosure of the report is required to be given to any opposing parties.
- The dictates of any internal (company or departmental) procedures or industry standards that you observe, including NFPA 1033, ASTM E 620, and ASTM E 678. For example, public investigators typically prepare reports for every significant investigation performed. If these reports contain opinions and conclusions, they may be considered expert reports in subsequent legal proceedings.

The content of expert reports will be determined by various factors, including procedural rules, usual practice guidelines, and the requirements of the attorney who has retained the investigator or who is calling the investigator as an expert witness. In civil cases, the federal courts and many state courts require reports to be prepared and disclosed to other parties by any

expert who may testify at trial. In the federal jurisdiction, this requirement is reflected in FRCP Rule 26(a)(2)(B).

In addition to complying with procedural rules, it is very important that every report that is prepared be accurate, complete, and easy to read and understand. It should detail and describe every step of the scientific method as it was applied during the investigation and analysis. Check and recheck facts from the records and investigation documentation. Carefully proofread the report. An investigator's writing style should be professional and strong, couched in proper grammar, with no spelling errors. The report

TIP

In situations in which you are not required by law or by the attorney calling you as an expert witness to prepare a report, a number of considerations can help you decide whether to prepare a report anyway:

- Quite some time may pass between your investigation and the trial. If your report was prepared shortly after your investigation was completed, reviewing it can help refresh your memory prior to testifying at trial.
- A well-organized report can be used by you and the attorney who calls your testimony to structure your direct examination at trial.
- The report provides a useful summary of the materials you reviewed, the investigations and tests you conducted, and the analysis leading up to each opinion. This summary will help your attorney prepare for trial and coordinate your testimony with that of other experts and fact witnesses.
- To the extent that your report goes beyond your field of expertise or is inaccurate or incomplete in any way, it can be used to challenge the admissibility of your opinion or become an effective tool against you in cross-examination.
- If you prepare a report with your final opinions before all investigations are complete, the weight or admissibility of your opinion may be subject to challenge on the basis that it was not based on sufficient facts or data pursuant to FRE Rule 702 or other standards.
- If the attorney who calls your evidence works with you in drafting or preparing your report, this information can be discovered by the opponent and may substantially weaken the weight and credibility of your expert evidence, giving the appearance of bias.

Whether an expert report is discoverable by the opponent will affect the decision of whether and when to prepare a report. The best course is to discuss the matter of preparing a report and the nature of its contents with the attorney who will call you to testify.

should permit a person who is unfamiliar with fire terminology to understand the facts and conclusions being reported. See Chapter 11, *Documenting the Fire Scene*, for more information on report writing.

If necessary, an investigator should obtain some training on use of proper grammar, through either reading or taking courses. Many good books on writing are available at the public library. Have another investigator in the department or office critically review the report for internal consistency, grammar, and spelling.

The investigation report should be a complete and accurate summary of an investigator's actions, observations, and conclusions. Depending on the scope of practice for the individual or organization, the report may include a narrative, photographs, video, interviews, diagrams, and other documents created as supporting information. With the report alone, an investigator should be able to describe and support the determination of fire origin and cause.

Deposition Testimony

As mentioned previously, a fire investigator may be called to testify in a deposition prior to trial. A deposition of a fire investigator will likely be scheduled by the opposing party. Depositions can have many purposes, including the following:

- To learn all of the information the witness knows about the situation
- To establish all of the opinions held by a witness who is an opposing expert
- To test out those opinions and their limits
- To define areas in which the expert witness does not have an opinion, lacks enough facts for an opinion, or is unqualified to render an opinion

In addition to its fact-finding purposes, the deposition preserves the testimony of the deponent under oath. Later, if the person testifies in trial differently than in the deposition, the transcript of the deposition can be used as impeachment to question the witness about the discrepancy.

Because of its importance to the litigation, an investigator should prepare for a deposition similarly to preparing for trial: Review all of the facts and information in the case, understand them thoroughly, understand the legal theories involved in the litigation, review any weaknesses and issues with the potential testimony and opinions, and practice and prepare as necessary. These steps will help a fire investigator understand the questions in the deposition, and aid in illustrating the facts and their opinion.

Always review the transcript of the deposition after it has been prepared, to ensure it is accurate and to supplement or make changes necessary. In most jurisdictions, the requested changes from the witness are appended to the transcript, so that the reader can also see the original answer.

Testifying in Court

Whether testifying as an expert or as a lay witness, a fire investigator should stay in close contact with the attorney calling them as a witness to make sure there is a full understanding of the subject matter of the testimony and any expectations of the investigator as a witness. Similarly, an investigator should ensure the attorney calling them understands the scene being described, the science and reasoning behind any conclusions, and any technical terminology being used. In addition, the investigator must understand which documentation should be provided and brought to court.

Professional dress and demeanor are critical when testifying. Wear business attire or a uniform, as appropriate, and maintain a professional, discreet demeanor at all times while in the courthouse. Witnesses at trial or other hearings are frequently sequestered, meaning that they are not allowed in court to hear the testimony of other witnesses. To avoid a possible violation of a court sequestration order, witnesses should arrange to stay outside the courtroom until called and refrain from discussing the case with anyone other than the attorney who called them (and their own attorney, if they are represented).

All witnesses are typically sworn to tell the truth before they sit to testify. Questioning of the witness then occurs including direct examination, cross-examination, and redirect examination, as discussed previously. In some jurisdictions, judges and even jurors may also ask questions of the witness.

The fire investigator's demeanor while testifying should always be professional and objective. Answer questions clearly and succinctly, and make eye contact with the judge and jury. Avoid unnecessary movement, as well as nervous gestures and expressions. It is allowable to ask for a question to be repeated or explained. Treat all attorneys with equal professionalism. When permitted, leave the witness stand to give demonstrations and explain exhibits.

At times, an attorney may interpose an objection to a question or answer during testimony. Always stop answering if an objection is raised, until ruled upon by the judge. Typically, if the judge sustains an objection to a question asked, the witness should not answer. If the objection is overruled, an answer is permitted.

After-Action REVIEW

IN SUMMARY

- Fire-related legal proceedings may involve criminal prosecution for arson or other crimes, or may involve civil lawsuits for negligence or other torts, product liability, or breach of contract.
- In the private sector, the fire investigator's authority to conduct an investigation is a contractual right that is derived from the person or organization authorizing the investigation.
- Even when an investigator has the legal authority to conduct a fire or explosion investigation, a right of entry onto property must also exist.
- Under the Fourth Amendment, a public investigator may be permitted to enter and investigate a fire scene under the doctrines of consent, exigent circumstances, an administrative search warrant, or a criminal search warrant.
- A public-sector investigator seeking consent to search a premises must do so from the person who has the "common authority and control" over the premises.
- When a fire is reported, the exigent circumstance allows entry onto property to suppress the fire and to investigate it for a reasonable time.
- An administrative search warrant allows entry to a premises when the investigator has administrative (code) authority to investigate, a fire has occurred, and consent of the proper person to enter the property is not available.
- A criminal search warrant is issued only upon a showing of probable cause, establishing that a crime has occurred, and that evidence of the crime is believed to be located in the premises to be searched.
- In the private sector, the investigator's right of entry flows from the authority of the person retaining the investigator, such as a property owner or insurance company with a contractual right to investigate.
- Under the Fifth Amendment, a person cannot be compelled to testify against themself or to provide an incriminating statement that the government could use against that person in a criminal prosecution.
- If a suspect is in custody, a Miranda warning must be given before a government agent is permitted to interrogate the suspect.
- A fire investigator should avoid loss, destruction, material alteration, or destructive testing of fire scene evidence, so as to avoid accusations of spoliation of evidence.
- Destructive testing of physical evidence should be preceded by notice to all known interested parties, permitting them an opportunity to have experts present and to observe or participate in the testing.
- The inevitable examination and moving of fire debris evidence is not considered spoliation of evidence.
- Best practice for public-sector investigators is to carefully document all actions taken within the scene, identify and preserve or set aside evidence that might be of interest to private-sector investigators, and perform layering and other scene activities in a manner that allows private investigation professionals to be able to see and analyze the evidence.
- To be admissible at trial, physical evidence must be properly transported and stored; it must also be authenticated by establishing the chain of custody for the evidence.
- Demonstrative evidence commonly used at trial includes diagrams, charts, photos, timelines, models, and other such evidence that an expert witness or attorney specially creates to illustrate or demonstrate a point to the jury.
- In court hearings, a fact witness testifies about factual events from personal knowledge, while an expert witness usually offers opinions regarding the subject matter of the litigation that are outside of the knowledge of the average person.
- In many courts, expert testimony is governed by the *Daubert* Supreme Court case; the Daubert standard requires the judge to analyze the reliability of expert qualification, opinions, and methodology used.
- The questioning of witnesses customarily involves open-ended direct examination; it is followed by cross-examination, and then usually redirect examination.

- The common law of England originally defined the crime of arson as the malicious burning of another's home at night. Today, the crime of arson is codified in statutes and can be committed in many forms.

- Arson charges often require the prosecutor to rely on circumstantial evidence to prove the case.

- Although motive is generally not a required element of the crime of arson, a prosecutor will attempt to collect evidence of motive to present to the jury, so that the jury will understand why the defendant committed the crime.

- The prosecuting authority in a criminal case may be required to turn over to the defense materials generated by the public-sector fire investigator, including reports, photos, and notes.

- In a criminal matter, the prosecution must prove the defendant's guilt beyond a reasonable doubt. In a civil case, the usual burden is a preponderance of the evidence, sometimes also termed "more probable than not."

- Civil cases often involve substantial amounts of discovery, including requests to produce, interrogatories, depositions, and reports.

- Negligence is the failure to exercise reasonable care toward another to whom the actor has a duty of care. A cause of action for negligence from a fire or explosion can develop in many ways.

- A claim for product liability can arise from manufacturing defects, inadequate warnings, or design defects.

- In product liability investigations, maintaining the chain of custody and the integrity of the product after the fire is critical.

- All reports created by the fire investigator should be accurate, complete, and easy to read and understand, and should describe how the investigator's work followed the scientific method.

- A deposition has considerable importance in civil litigation, and the investigator should thoroughly prepare for a deposition.

- When testifying in court, make sure to treat all attorneys with equal professionalism and avoid displays of anger or disdain.

KEY TERMS

Affidavit A voluntary written statement of fact or opinion made under oath and signed by the author.

Arson The crime of maliciously and intentionally, or recklessly, starting a fire or causing an explosion. Legal definitions of "arson" are provided by statutes and judicial decisions that vary among jurisdictions. (NFPA 921)

Brady materials Information in the possession of the government in a criminal case that is favorable to the accused person, and is either material to the person's guilt or to their punishment if they should be convicted of the crime charged.

Chain of custody The trail of accountability that documents the possession of evidence in an investigation.

Circumstantial evidence Proof of a fact indirectly based on logical inference rather than personal knowledge.

Consent Permission from the owner or occupant to examine property.

Defendant The entity against whom a claim or prosecution is brought in a court.

Demonstrative evidence Any type of tangible evidence relevant to a case—for example, diagrams and photographs.

Deposition One type of pretrial oral testimony made under oath.

Discovery Pretrial procedure in a civil or criminal legal case in which evidence and information about the case is provided to the opposing party(ies) according to the rules of procedure.

Documentary evidence Any type of written record or document that is relevant to the case.

Evidence technicians Individuals who are specially trained to document, collect, preserve, and transport evidence.

Exigent circumstance A doctrine in criminal law in which law enforcement or other government agents may be permitted to enter property without a search warrant to respond to an emergency, dangerous condition, or other extraordinary circumstance.

Expert witness Someone who is recognized by a court as qualified by specialized knowledge, skills, experience, or education on an issue.

Fact witness Someone who testifies about facts from their personal knowledge or experiences; also called a lay witness.

Fifth Amendment An amendment to the U.S. Constitution, part of the Bill of Rights, that protects people from being compelled to be witnesses against themselves in criminal proceedings. It also specifies other rights, including the right to due process and protection against double jeopardy (prosecution more than once for the same offense).

Interested party Any person, entity, or organization, including their representatives, with statutory obligations or whose legal rights or interests may be affected by the investigation of a specific incident. (NFPA 921)

Interrogatories A set of questions served by one party involved in litigation to another involved party.

Lay witness Someone who testifies about facts from his or her personal knowledge or experiences; also called a fact witness.

Motion A request for court action regarding facts, documents, and evidence that are identified during the discovery phase of a legal proceeding. Motions may argue the admissibility of the involved facts, documents, or evidence.

Motive An inner drive or impulse that is the cause, reason, or incentive that induces or prompts a specific behavior. (NFPA 921)

Negligence Failure to provide the same care that a reasonably prudent person would take under similar circumstances.

Search and seizure A legal term encompassing the government's search or examination of a person, their property, or their land; or interference with the person's right to the use of the same.

Sixth Amendment An amendment to the U.S. Constitution, part of the Bill of Rights, that provides that an accused person shall have the assistance of counsel in their defense, a speedy and public trial, the right to "confront" opposing witnesses, and other rights in criminal cases.

Spoliation of evidence Loss, destruction, or material alteration of an object or document that is evidence or potential evidence in a legal proceeding by the person who has the responsibility for its preservation. (NFPA 921)

Testimonial evidence Verbal testimony of a witness given under oath or affirmation and subject to cross-examination by the opposing party.

REFERENCE

National Fire Protection Association. 2020. *NFPA 2021: Guide for Fire and Explosion Investigations*, 2021 ed. Quincy, MA: National Fire Protection Association.

On Scene

1. If you are working as a fire investigator, review the last report you prepared in light of the discussion in this chapter. Are there areas where you think an opposing attorney could attack the opinions in your report? What can you do on scene or during the reporting procedure to make your report stronger?

2. What are some of the ways a public-sector fire investigator could ensure that their department is fulfilling its mission to determine fire causes, without damaging or compromising evidence such as appliances or products that need to be preserved for examination by interested parties?

3. As a public investigator, what is your authority, if any, to be on a fire scene while the fire department is still conducting firefighting operations? After the operations are complete and the fire department is leaving the scene, what is your authority? What steps will you take to ensure that your presence on the scene continues to be permissible as the investigation goes on?

4. What are the elements, or components, of a charge of arson in your jurisdiction? What are the elements of a civil cause of action for negligence or product liability?

Chapter Opener: © Jones & Bartlett Learning. Photographed by Glen E. Ellman; On Scene siren: © Bildgigant/Shutterstock.

CHAPTER 6

Safety

KNOWLEDGE OBJECTIVES

After studying this chapter, you should be able to:

- Identify the fire investigator's responsibility for safety at a fire scene. (**NFPA 1033: 4.1.3**; **4.2.1**, pp. 150–152)
- Explain the possibility of safety concerns related to the public. (**NFPA 1033: 4.2.1**; **4.2.2**, p. 151)
- Describe how to conduct a hazard and risk assessment. (**NFPA 1033: 4.1.3**; **4.2.1**; **4.2.2**; **4.2.3**, pp. 152–153)
- Identify and describe factors that influence scene safety. (**NFPA 1033: 4.2.2**; **4.2.3**, pp. 152–156)
- Describe the impact of criminal acts or acts of terrorism on fire investigator safety. (**NFPA 1033: 4.2.1**; **4.2.2**; **4.2.3**, p. 156)
- Describe safety clothing and equipment to be used at fire and explosion sites. (**NFPA 1033: 4.2.1**; **4.2.2**; **4.2.3**, pp. 156–159)
- Identify the role of personal health and safety for the fire investigator. (**NFPA 1033: 4.1.3; 4.2.2**; **4.2.3**, pp. 159–162)
- Describe how the investigator fits into the incident command system. (**NFPA 1033: 4.1.6**, pp. 162–164)
- Identify the role of communication in fire investigator safety. (**NFPA 1033: 4.2.1**; **4.2.2**; **4.2.3**, pp. 162–164)
- Explain the role of Occupational Safety and Health Administration (OSHA) standards in keeping fire investigators safe. (**NFPA 1033: 4.1.3**, pp. 164–165)

SKILLS OBJECTIVES

After studying this chapter you should be able to:

- Perform a site safety assessment so that hazards are identified and removed or mitigated prior to the initiation of the fire scene examination. (**NFPA 1033: 4.1.3**; **4.2.1**; **4.2.2**; **4.2.3**, pp. 152–153)
- Secure the fire ground and safeguard personnel and bystanders. (**NFPA 1033: 4.2.1**; **4.2.2**, pp. 151–152)
- Select clothing and equipment to use to safeguard the fire investigators. (**NFPA 1033: 4.2.2**; **4.2.3**, pp. 156–159)

You are the Fire Investigator

You are on the scene to investigate a fire in a heavily damaged two-story residence. You conduct a site safety assessment prior to starting the investigation, and notice that two exterior walls of the residence appear to be free-standing and unsupported by other structural elements. A brick chimney also stands straight in the remains of the living room, but is no longer supported by the building's wood framing. A moderate wind is predicted for later in the day. Firefighters are still on scene and you meet with the incident commander (IC) to discuss your observations.

1. What safety hazards may exist at the scene that you must consider?
2. How does the suppression of the fire impact your safety during the investigation?
3. What steps can you take to protect yourself and others during the investigation?

 *Access Navigate for more practice activities.*

Introduction

Fire investigator safety must be the primary concern at every fire scene. The fire investigator is exposed to myriad hazards at the scene, many of them invisible, that threaten not only immediate safety, but long-term health as well. Safety must be considered and discussed by investigation teams prior to entry into a fire environment and must be considered and evaluated repeatedly throughout an investigation.

Safety threats at a scene may include structural collapse, falling building sections and objects, holes, sharp objects, energized electrical components, vapors and particulates, heat and fire, water and ice, hostile animals, and human threats, among others. Although fire suppression activities pose many threats to the firefighter, the fire investigator entering the scene after fire suppression is complete can encounter a far longer exposure to some of the hazards found on scene.

In recent years, awareness of safety issues within the fire investigation community has increased significantly. Indeed, fire scenes have arguably become more dangerous than they were decades ago, due to the products that typically burn in modern structural fires. In 2020, the International Association of Arson Investigators (IAAI) published *Fire Investigator Health and Safety Best Practices,* second edition, compiling industry knowledge on safety; this document is accessible to all fire investigators through the IAAI website, www.firearson.com. More studies of safety issues are also under way, particularly with regard to respiratory and absorption hazards.

To ensure the safety of all parties involved in fire scene investigation, the following categories of actions should be taken:

- Develop strong safety policies and procedures for investigating personnel, based on regional and national safety standards.
- Maintain current training and knowledge about safety topics.
- Maintain, understand, and utilize adequate personal protective equipment (PPE).
- Conduct site safety assessments on each scene; and reevaluate safety periodically throughout each investigation.

The following sections detail some of the safety issues posed at fire scenes and provide information on how to maximize short- and long-term safety and health.

Responsibility for Safety at a Fire Scene

Fire investigators may be expected to operate in areas that contain toxic atmospheres, compromised structural systems, and sometimes hostile people or animals, so they must always exercise caution. They should take every effort to protect themselves and the other personnel operating and/or assisting at the scene from hazards that are present.

Even though the emergency phase of an incident has likely passed when the investigation begins, it is nevertheless important to maintain the same safety standards as were in place during the initial phases of the incident.

TIP

Before entering a room with compromised structural members, assess the structure's stability and safety. The room should be shored up and made safe before any work begins. Any items that may still pose a hazard should be identified and marked, such as protruding studs, pipes, and other high and low hazards.

Investigating the Scene Alone

Ideally, a fire investigator should not work alone during any fire scene examination. The presence of at least two fire investigators provides for another set of eyes and hands to assess the dangers and handle problems that might be encountered. If an investigator becomes injured or trapped, the second investigator can provide assistance or call for help.

Unfortunately, because of monetary or staffing restrictions, fire investigators are often required to operate alone at fire scenes, even when an immediate hazard may be present. When this is the situation, an investigator needs to notify a responsible off-site contact person of the location, the nature and size of the scene, and the expected time required to complete the investigation. The investigator should also carry a two-way radio, cellphone, or other means of communication at all times. If there is an unexpected delay, notify the off-site contact person of the delay and adjust the estimated completion time. To prevent complacency, communicate periodically with the off-site contact person, and arrange for them to make contact at the end of an expected time interval.

Safeguarding Technical Personnel, Assistants, and Trainees

Sometimes, a fire scene investigation calls for the assistance of experts qualified in technical or specialized issues. These experts may not be trained in how to operate safely at the fire scene. Similarly, fire department trainees or assistants may have a lower degree of training or experience in safety issues. Thus, it is the responsibility of the fire investigator to help ensure that all of the personnel who assist with the investigation are made aware of the hazards present, the precautions that have been taken to make the work area safe, and any general or specific precautions that they should follow during the investigation.

Safeguarding Bystanders

A fire scene—especially a large-loss fire scene—often attracts large numbers of people who come to observe the activities and the devastation. These individuals often attempt to get as close as they can to the incident and could be injured by any one of the myriad hazards or exposures present at the fire scene. All bystanders, including the occupants of the fire-damaged structure and nonessential responders, should be kept at a safe distance from the scene. Police officers or fire personnel should be asked to place marking devices such as

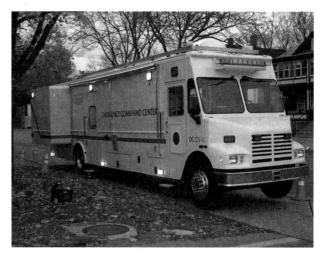

FIGURE 6-1 Identify a safe, secure place to conduct interviews, away from the distraction and hazards of the fire scene.
Courtesy of Lieutenant Bradley E. Olson, IAAI-CFI.

FIGURE 6-2 Example of a hazardous area.
Courtesy of Rodney J. Pevytoe.

rope, fire line tape, or barrier tape to encourage bystanders to maintain a safe distance. A fire investigator should also identify a safe, secure place to conduct interviews, away from the distraction and hazards of the fire scene **FIGURE 6-1**.

To the extent that owners and occupants are allowed on the scene, take special care to ensure that they are aware of the known and potential hazards there **FIGURE 6-2**. Occupants will typically have great familiarity with their buildings, but they must understand that the building is considerably different and less safe after a fire. Occupants who reenter a building to collect their belongings or assist fire investigators should be accompanied while they are inside.

Entry Control

While the scene is under government control, a fire investigator can request that the relevant law

TIP

The person in control at an explosion or fire investigation scene should have the following items and information available, at a minimum:

- First-aid kit
- Local emergency notification numbers
- Knowledge of the location of emergency medical care
- An emergency notification signal

enforcement agency assign an officer to remain stationed at the fire scene to prevent unauthorized entry by bystanders or others. To maintain proper security and to preserve the evidence, maintain (or delegate responsibility for) a written log of all authorized people who enter the scene. If any unauthorized people enter the scene, they should be required to leave immediately, and the person responsible for maintaining the scene should make note of their identities, the date and time of entry, and the area in which they entered.

Collapse Zones

Whenever collapse of a wall, chimney, or other part of a building is a risk, the fire investigator (or the incident commander, if applicable) should establish and enforce **collapse zones** in the affected area. The collapse zone must be identified by markers, barricades, vehicles, or specialized scene tape. Additionally, the incident commander and/or the investigator in charge should verbally inform all personnel of the collapse zones, because the presence of ordinary scene tape will not necessarily signal a collapse danger to police officers, firefighters, or contract personnel who are accustomed to crossing scene tape as part of their duties.

Hazard and Risk Assessment

One of the first tasks at any fire scene is to conduct an assessment of the hazards and risks. This is a three-step process:

1. Identify the hazards.
2. Determine the risk presented by the hazards.
3. Control the hazards.

Identify the Hazards

NFPA 1033 requires the fire investigator to perform a site safety assessment at all fire scenes. A **site safety assessment** is a full exterior and interior survey of a fire structure to identify the presence of physical, toxicological, and biological hazards that may pose a threat to the investigator. The hazards identified on scene may be classified as follows:

- Physical hazards: Are there many places that a person could slip, trip, and fall? Are there holes, sharp points, water, or ongoing combustion or smoldering?
- Structural hazards: Is the building structurally sound?
- Electrical and gas hazards: Is the electrical and/or gas system still connected/intact?
- Respiratory and absorption hazards: What is the condition of the atmosphere in the structure? Are materials still smoking? Is there natural or human-made ventilation?
- Chemical hazards: Are safety data sheets (SDSs) or other reference materials available to help manage the chemical hazards present?
- Biological hazards: Do bacteria, viruses, insects, plants, animals, or humans pose a risk?
- Mechanical hazards: Is the equipment still operational or functional? Is the machinery specialized, requiring a technical expert? Are all guards in place for all moving equipment?

Determine the Risk Presented by the Hazards

Although all fire scenes pose some form of hazard, the level of risk at each will vary. Fire investigators should assess the hazards present and determine the likelihood that contact may be made with the hazards and, if so, in what form. Based on this information, they can then institute measures to control the hazard.

Control the Hazards

After identifying a hazard, a fire investigator must determine the most effective manner to mitigate the risks by comparing the risk level to a suitable benchmark or acceptance criterion. Risks may be controlled through the following means:

- Engineering controls, such as shoring to reinforce damaged structural areas, demolition after documentation, or, in complex cases, obtaining the services of a structural engineer.
- Administrative controls, such as isolation of an area through the use of effective fencing, signage or barrier tape informing personnel of the hazards, or restricting entry to key personnel.

■ Proper selection and use of PPE, including boots, gloves, helmet, clothing, and eyewear, with a greater level of protection being used as deemed necessary by the scene assessment. Donning PPE is generally considered the least effective of the control measures, but it may be a suitable one under the conditions encountered by the investigator.

Factors Influencing Scene Safety

Many factors can influence the safety of a fire or explosion scene, including the status of the fire and fire suppression, the stability of the structure, and the status of the electrical system and other utilities. This section discusses these hazards in more detail.

Status of Suppression

Normally, a fire investigator should not enter a structure until the fire is completely extinguished. Occasionally, however, the primary goal of fire suppression can be done simultaneously with investigation if the circumstances require it **FIGURE 6-3**.

In the unusual event in which investigators' entry is necessary while fire suppression activities are

FIGURE 6-3 Fire suppression is required before the investigator can proceed with a full investigation.
© Glen E. Ellman.

ongoing, they should not do so unless they have the proper PPE and training for such entry and have received permission from the incident commander. It is also important to coordinate all ongoing investigation activities with the suppression crews and to keep the incident commander advised of any progress and of any additional areas investigators are entering.

When an investigation begins in what is believed to be an extinguished structure fire, be aware of the possibility of a "rekindle" occurring in the building. The rekindled fire could grow to the point that a fire investigator's safety is endangered and/or the escape route is cut off. Safety planning should include determining at least two means of exit, which should be at opposite ends of the work area, and hose lines should remain in place during the initial phase of the investigation. One means of escape could be a ladder being placed to a window or to the roof. If this type of escape might be necessary, the ladder must remain in place at all times when an investigator is occupying the area. The separate means of escape should be monitored to ensure they remain free of obstructions during the investigation.

TIP

The fire investigation community teaches firefighters not to destroy the fire scene. Classes on the firefighter's role in origin and cause investigations emphasize the need to keep overhaul to a minimum until the scene has been documented and examined by a fire investigator. Because of this training, overhaul by fire departments is often minimal, and firefighting operations are suspended when the forward progress of the fire is stopped.

Structural Stability

Before entering any fire scene, assess the stability of the structure and, if possible, eliminate physical hazards by either application of shoring materials or demolition. Depending on the nature of the structure, help from a qualified structural engineer may be needed. Do not take for granted that any portion of a building that remains standing after the fire is secure and safe; instead, continue to monitor the building throughout the investigation for changes or signs of a hazard. For example, a collapsing wall or other building component may fall without warning or noise. Also, be mindful of hidden holes or weaknesses in the floor caused by the fire itself; by water buildup from the suppression activities; or by weather-related weight factors such as ice, snow, wind, or rain.

Freestanding chimneys pose a substantial risk to both firefighters and fire investigators. These structures depend on the building structural members for their stability. When the building is compromised, so is the chimney. Investigators should consider implementing a collapse zone around any freestanding chimney. A freestanding wall, whether it is a parapet wall (the portion of a wall that extends above the level of a roof) or any wall that is no longer supported by other structural members, should not be trusted unless shoring has been implemented. The safest way to handle a freestanding wall is to knock it down in an appropriate manner and at a time that the fire investigator determines.

Portions of the building that are not supported by trusses or other support materials are also vulnerable to collapse. Most buildings are built with the minimum amount of materials required by engineering principles and local building codes. In some cases, a building's construction may have been deficient and not consistent with the applicable code, or a building that conformed to the applicable code when it was built would not be compliant with current codes. The removal of any portion of a building places unintended loads on the remainder of the building, which in turn puts the building in danger of collapse. All components of the building's construction are dependent on each other; the removal of any component alters the system that was created to form the building.

Building Load

A building has two different types of building loads: the dead load and the live load. (See Chapter 3, *Building Construction and Systems for Fire Investigators*, for more information on building loads.) When a structure is designed, the architect takes into account its intended use and the expected live load. Often, over time and because of changing occupancies, buildings are used in ways that the architect did not intend. Walls may be moved when occupancies change from residential to commercial to industrial. These changes may not take into account the design of the building, such that building components may become subject to unplanned stress. Poor maintenance, poor repairs, and aging of the building materials may induce further stress on the structure, increasing the potential for collapse when fire attacks the materials. When wood studs or joists are compromised or destroyed by fire, the surrounding studs or joists must carry the load. These components may be able to manage this feat for only a short period of time before failing.

Another consideration related to structural stability is the **impact load,** a sudden added load that can be caused by a firefighter jumping from a ladder onto a roof or a fire investigator jumping down onto a floor.

This sudden load may cause the collapse of already weakened or overburdened structural members. A partial building collapse may also provide the impact load that brings down the remainder of the building.

Controlled Demolition Approach

All threats to safety posed by the condition of the building should be mitigated prior to investigation, if possible. While many investigations can proceed in the face of some safety concerns, a building may potentially be too unsafe to permit an investigation to occur at all. Incident commanders and fire investigators should never hesitate to make the determination that the scenario presented does not permit investigation until hazards are mitigated, or even allow for any investigation at all.

At times, and subject to spoliation of evidence considerations, an investigation can proceed using a controlled demolition approach, in which investigator access and safety are supported by a piece-by-piece demolition of the unsafe areas of the structure. The demolition team should work under the direct supervision of a trained fire investigator who understands the scene and the goals of the investigation. This investigator directs the equipment operator to remove the portions of the structure that are unstable without destroying the areas of interest to the investigation team. Ideally, this can be done without destroying evidence **FIGURE 6-4**.

Utilities

Fire investigators often come into contact with the fuel and electrical systems in a fire scene and will need to examine these systems. They should not assume that

FIGURE 6-4 Technical personnel such as heavy-equipment operators can work in conjunction with investigators to remove hazardous debris or uncover evidence.
Courtesy of Mike Dalton.

these systems have been disconnected by fire crews. Instead, before entering a scene, investigators should always explicitly determine whether an electrical system is energized, the lines of a fuel system are charged, or the water mains are operating. Once inside the structure, they should verify the safety of utility systems there. Utilities to check may include the main electrical system, backup electrical and batteries, fuel oil, propane, and natural gas. Also, some solar equipment cannot be de-energized; the investigator should avoid contact with such equipment unless accompanied by an expert in such installations.

Electrical Hazards

Never assume anything is safe when it comes to electricity. The fire department may have disconnected the power, only for it to be reenergized later. In addition, some structures may have more than one electrical feed, allowing one system to be shut down and another energized within the same structure. Occasionally, surreptitious connections and wiring have been added to a building that bypass the meter and circuit protections. Temporary wiring may not be installed, grounded, or insulated correctly. Many electrical hazard scenarios could pose a risk to investigators, so remember the following when working at a fire scene:

- Assume that all wires are energized, even when the meter has been removed or disconnected.
- Be alert to fallen electrical wires.
- Look out for antennas that have fallen on power lines, for metal siding that has become energized, and for underground wiring.
- Operate ladders and other equipment cautiously when near overhead electric lines.
- Do not depend on rubber footwear as an insulator.
- Do not enter a flooded basement if the electrical system is energized.
- Shut off electric power remotely if the scene contains an explosive atmosphere.
- Communicate and cooperate with the utility company and make use of their expertise.
- Locate underground electric supply cables before digging or excavating.
- Be alert for additional electrical services that may not be disconnected, extension cords from neighboring buildings, emergency lighting units, backup generators, some solar equipment, and similar installations. These electrical units may remain energized after the building's electrical power has been shut off.

- Use a meter, such as a noncontact AC voltage tester, to determine whether the electricity is off. Always test the meter on a known live circuit before testing at the fire scene to ensure the meter is functioning properly.

Although fire investigators should not disconnect electrical power, they should ensure that the proper utility company does. Fire investigators must be knowledgeable about the safe use of testing equipment, such as a voltmeter, multimeter, or noncontact tester. This will allow them to test electrical systems for the presence of energy, if permitted by local protocols. Fire investigators can avoid electrocution by ensuring that the testing device is rated for the voltage supplied to the structure. Once electrical power has been disconnected or interrupted, the lockout/tagout process should be employed.

TIP

Put a small AC voltage detector in your pocket at the beginning of every structural investigation. It is less cumbersome than a multimeter and will come in handy because it can be used to test for power in circuit breaker panels, outlets, and electrical runs. This device can also detect voltage through the insulation of undamaged conductors.

TIP

Be alert for puddles of water near energized electrical wires, appliances, or other equipment.

Standing Water

Firefighting operations themselves can add to the stress on a building's structural members. Water weighs 8.35 lb/gal (1.05 kg/L). A 2½-inch (64-mm) or even a 1½-inch (44-mm) hose line putting out a flow up to 250 gal (946.4 L) per minute will place an additional live load of 2087.5 lb (946.9 kg) into the building every minute that the hose line is in operation. This additional load must be drained before the investigation can begin. During cold-weather investigations, the removal of this water can be difficult or even impossible. The water may be of unknown depth and may contain unseen hazards. Similarly, standing suppression foam can hide holes and hazards from the fire investigator's view.

Firefighters often cut holes in floors to remove water from the structure. Fire investigators must be aware of this possibility and identify the locations of any such holes. Some firefighters may use doors from the building to cover these holes and any holes caused by the fire itself. These doors must not be relied on to hold a person's weight.

Criminal Acts or Acts of Terrorism

Fires and explosions are powerful weapons that may be used in both criminal acts and acts of terrorism. In the event that an incendiary or explosive device initiated the fire, fire investigators should consider the potential for additional devices to be present in the area. Some devices may have failed to function or been intentionally left to harm personnel. The hazardous materials used in these devices' construction can also leave a residue, creating yet another exposure hazard.

Secondary Devices

A tactic commonly used by perpetrators of terrorist actions is to leave secondary devices that are meant to explode or deploy after the initial incident, with the aim of causing further damage, injuring first responders, and creating even more panic. Personnel on the scene should be wary of any unusual packages or containers. Officials may need the assistance of hazardous materials or explosive ordnance experts to address identified or potential risks.

Residue Chemicals

Fires and explosions may be initiated or enhanced with the use of various chemicals. Always use appropriate protective clothing and respiratory measures when handling any artifacts of a device or equipment to avoid exposure to these kinds of harmful substances.

Biological and Radiological Terrorism

Adding to the damage caused by a fire or explosion, perpetrators of terrorist acts may include biological or radiological components in their devices. Any scene that has been determined to pose a substantial risk of these types of components will require special safety and training measures for all personnel. This can include steps to render the scene safe prior to entry by a fire investigator. If the scene cannot be rendered safe, then only those investigators trained and properly equipped to work in such an environment should be allowed to enter the scene.

Safety Clothing and Equipment

Although there is rarely a good reason for fire investigators to enter a burning building before the fire has been brought under control, they must protect themselves from physical, biological, and chemical hazards while conducting an origin and cause investigation. Toxic gases, hazardous chemicals, asbestos, flammable gases, and oxygen-deficient atmospheres are a few of the dangers present at almost every fire scene. Although it is usually not necessary to enter the scene before the fire is extinguished, if such entry must be made, the fire investigator must be equipped with structural firefighting gear and a self-contained breathing apparatus (SCBA) and must be trained in its proper use **FIGURE 6-5**.

Protective clothing and equipment must meet or exceed the requirements established by NFPA standards. Just as police officers are saved by their bulletproof vests and firefighters are protected by their turnout gear and SCBA, fire investigators must have protective clothing and equipment available to them that should be worn and used at all fire scenes. It is critical to be trained in the correct usage, limitations, and decontamination (or disposal) of the professional clothing and equipment used at the scene **FIGURE 6-6**.

Protective clothing and equipment should be properly decontaminated prior to leaving the scene. If this is not possible, the equipment and clothing can be removed at the scene, transported in such a way that contamination cannot spread, and later washed according to the manufacturer's instructions.

FIGURE 6-5 The complete structural firefighting ensemble consists of a helmet, coat, trousers, protective hood, gloves, boots, self-contained breathing apparatus (SCBA), and personal alert safety system device.
© Jones & Bartlett Learning. Photographed by Glen E. Ellman.

FIGURE 6-6 Investigators wearing appropriate personal protective equipment (PPE) for the working conditions during a scene analysis.

Reproduced with permission from *NFPA 921-2021, Guide for Fire and Explosion Investigations*, Copyright © 2020. National Fire Protection Association. This reprinted material is not the complete and official position of the NFPA on the referenced subject, which is represented only by the standard in its entirety.

TIP

Fire investigators often wear firefighter PPE when working at a scene. The PPE for firefighters must be manufactured according to exacting standards. Each item must have a permanent label verifying that it meets the requirements of the standard. Usage limitations as well as cleaning and maintenance instructions should be provided as well.

The requirements for firefighting PPE are outlined in two standards:

- NFPA 1971, *Standard on Protective Ensembles for Structural Fire Fighting and Proximity Fire Fighting*
- NFPA 1977, *Standard on Protective Clothing and Equipment for Wildland Fire Fighting*

The requirements for SCBA and PASS devices are outlined in two standards:

- NFPA 1981, *Standard on Open-Circuit Self-Contained Breathing Apparatus (SCBA) for Emergency Services*
- NFPA 1982, *Standard on Personal Alert Safety Systems (PASS)*

Other requirements for training and maintenance are set forth in two additional standards:

- NFPA 1404, *Standard for Fire Service Respiratory Protection Training*
- NFPA 1852, *Standard on Selection, Care, and Maintenance of Open-Circuit Self-Contained Breathing Apparatus (SCBA)*

The Helmet

A helmet should be worn during all fire investigations. This helmet should have a suspension system to absorb impacts and should offer protection from electrical hazards. It should provide adequate levels of protection but also be comfortable enough to wear during extended operations. The helmet should have a chinstrap to hold it in place, and the chinstrap should be used whenever you wear the helmet in potentially hazardous areas. The helmet may also have some form of built-in eye protection that is easily deployed. This eye protection should be constructed of a material that resists scratches. An improperly constructed helmet or a helmet with unapproved modifications could cause injury.

Protective Clothing

Wearing protective clothing, such as firefighting turnout gear or coveralls, will prevent contamination of the investigator's street clothes. The fire investigator is exposed to toxins at every fire scene. These toxins can become embedded in clothing and in anything that the investigator may come in contact with after its exposure to the toxins—including vehicles, office furniture, and homes—if proper precautions are not taken. Such cross-contamination exposes investigators as well as coworkers and families to the toxins from the fire scene.

The coveralls or turnout gear should be washed after every fire scene examination in accordance with the manufacturer's instructions. This procedure often calls for machine washing with a mild soap and water, but not in a washing machine that is normally used to clean other clothing; instead, a separate washing machine or a company that specializes in cleaning firefighting equipment should be used. (Disposable coveralls are another option, which would eliminate the need for cleaning.) Certain toxins that are present at every fire scene (such as benzene and hydrogen cyanide) can be absorbed through the skin. The improper cleaning of protective clothing can cause repeated exposure to these carcinogens, as well as others.

Respiratory Protection

Respiratory protection must be considered at every fire scene. Even after fires are extinguished, the air at fire scenes is full of toxic smoke, chemicals, and other respiratory threats that can be extremely harmful to short- and long-term health. Further, some scenes also contain chemicals and other substances that were present before the fire and may not be known to the fire investigator.

Respiratory protection could take the form of SCBA, cartridge filter masks, or engineering controls such as mechanical ventilation. The ideal protection can be obtained with the use of engineering controls to remove the hazard from the work area. Positive-pressure ventilation will provide the work area with fresh air efficiently and effectively; however, it may not always be practical. During these investigations, a fire investigator can use a combination of positive-pressure ventilation and negative-pressure ventilation to help remove the toxic gases from the area or reduce their concentrations. All investigators should have a good grasp of the theory behind mechanical ventilation for any area where an investigation may take place.

Atmospheric monitoring for toxic gas, carbon monoxide, and oxygen levels must be performed by the fire department or by the fire investigator before anyone enters a fire-affected building without respiratory protection, and this monitoring should be repeated periodically during the investigation. Even if the atmosphere is found to be acceptable in a specific working area in a building, be aware that hazardous atmospheres could still exist in closed or hidden spaces, basements, and other below-grade areas.

When respiratory protection is used, it is necessary to have a written respiratory protection program in place. All individuals who will be using respiratory protection should be trained in the selection, use, and maintenance of the proper equipment for the hazards encountered. This training should also address emergency procedures in case of any equipment failure. In addition, medical surveillance is required to determine the physical abilities and limitations of the user. Periodic medical examinations may be necessary to determine whether any user has been exposed to a toxin that has the potential to cause a health issue.

Self-Contained Breathing Apparatus

On some occasions, ventilation alone cannot decrease the levels of the toxins to a safe level. During these investigations, respiratory protection must be employed. SCBA provides the highest level of respiratory protection.

When SCBA is used, the user is protected only for a limited time due to the limited amount of breathing air in the compressed air cylinder **FIGURE 6-7**. Usually, a compressed air cylinder is rated to provide 30 to 60 minutes of breathing air, although this time varies depending on the wearer's physical condition and the level of physical and mental strain the user is experiencing. The SCBA face piece also restricts the user's visual field, thereby limiting the investigator's ability to observe the scene. An alternative, the supplied air breathing apparatus (SABA), provides breathing air for longer periods, but requires a hose supplying air from a remote location limited to 300 feet from the user. When wearing any breathing apparatus, including SCBA and SABA, you should work in 30-minute periods and allow for rehydration, rest, and monitoring

FIGURE 6-7 Self-containing breathing apparatus (SCBA) carries its own air supply in a pressurized cylinder.

of physical condition caused by the stress and strain of wearing the apparatus.

Air-Purifying Respirator

The air-purifying respirator (APR) is useful at certain fire scenes; however, this type of respiratory protection has limitations. Air monitoring is a must when using an APR because the user is not protected from oxygen-deficient atmospheres (oxygen levels below 19.5 percent) or from areas where toxins exist at a level that is deemed immediately dangerous to life and health. The APR is toxin-specific, so the user must know all the hazards that are present at the scene to be able to select the most effective filter. Unfortunately, given all of the variables in heat, contents, and materials in a structure fire, it is virtually impossible to determine exactly what materials are created in a fire incident and what their level of toxicity might be. The APR can be selected based on general hazards, but the unknown is always present at any fire scene.

APRs are provided with and without eye protection. **FIGURE 6-8** shows a form of APR without built-in eye protection. If the APR does not have built-in eye protection, the user should wear vented goggles to protect the eyes. Some APRs also have a battery-powered fan to facilitate breathing efforts.

FIGURE 6-8 Air-purifying respirators can be used only when there is a sufficient amount of oxygen in the atmosphere.
© Alex459/Shutterstock.

Eye Protection

Eye protection that provides the maximum level of safety should always be worn. Snug-fitting goggles are more effective than the face shield on a helmet when fire debris becomes airborne and could contact the eyes. To protect the eyes fully, use safety goggles along with the eye protection provided by the helmet. Remember that beyond obvious eye injuries, any toxins that enter the eyes will quickly enter the circulatory and nervous systems. Eye protection should also shield from the sun's ultraviolet rays.

Foot Protection

Proper foot protection can prevent major and minor injuries. A safety shoe protects the foot from the hazards associated with stepping on nails or sharp objects, as well as from heavy objects falling on the toes. A properly waterproofed shoe protects against exposure to toxins that are present at fire scenes. Rubber boots provide a high level of protection, as do traditional thigh-high firefighting boots pulled up over the knees and used in association with hard kneepads. Boots worn at the investigation should conform to NFPA 1971.

Gloves

Gloves should be worn at every fire scene. Standard leather firefighting gloves provide limited protection from heat and physical injuries, but do not protect against chemical exposure. Therefore, consider double gloving during the investigation, wearing an inner glove that provides chemical protection. The inner glove should be made of a material chosen after assessing the hazards at the fire scene. For instance, latex gloves provide protection from biological agents but are not effective during extended periods of wear under wet conditions. For the average fire scene, nitrile gloves with some puncture resistance are commonly used and provide a higher level of protection. The inner glove should be worn from the beginning of the investigation. To prevent cross-contamination when collecting evidence, don a new pair of disposable gloves prior to collecting each individual evidence sample.

Safety Equipment

Each fire scene is different and, at times, additional equipment may be needed to provide a safe work environment. All work areas should be well lit to prevent trip-and-fall injuries and to allow for effective assessment of hazards throughout the investigation. Portable lighting is necessary at almost every fire scene, but at no time should a generator be brought into an enclosed work area; the buildup of carbon monoxide would present a serious health issue. Further, when an internal combustion engine is brought into the building, it can no longer be stated that the investigation team did not bring an ignitable liquid into the scene.

Some investigations require lifelines and fall protection. This equipment should be maintained and stored in accordance with the manufacturer's specifications. Only ladders brought by the fire investigator or the fire department should be relied on for use, as others may be unsafe, poorly maintained, or weakened by the fire. If a ladder is required, ensure that it is in good condition and of the proper height and design for the intended use.

Some safety equipment requires specialized training and experience for its proper use. For instance, shoring equipment or other urban search and rescue techniques may be required for protection at a fire scene. Only experienced operators of specialized equipment should be used to support an investigation scene.

Personal Health and Safety

Many hazards may exist at a fire scene, from chemical to biological and even, at times, radiological. Fire investigators may be exposed to these hazards in a number of ways, including inhalation, absorption, ingestion, and direct contact. Investigators must guard against complacency and make safety the first priority in all investigations.

In recent years, the fire service has become more keenly aware of the nature and extent of potential long-term harm from fire scenes, and the fire investigation industry is following suit. In 2016, the IAAI reestablished its Health and Safety Committee with the goals of studying the safety issues facing the fire investigation industry, compiling knowledge and best practices for safety, and training and educating the industry on hazards and best practices. In 2020, the Committee published *Fire Investigator Health and Safety Best Practices*, second edition, compiling industry knowledge on safety and making recommendations for safe practices; this document is accessible to all fire investigators through the IAAI website, www.firearson.com.

Respiratory and Absorption Hazards

As discussed previously, numerous toxic hazards are found at every fire scene that can pose a risk to the fire investigator; many of these fires might have a short duration, but exposure to their hazards can lead to serious long-term health consequences. These hazards remain on scene long after the fire is extinguished and can occur at a fire scene of any size. Dangerous exposure can occur during a single encounter with smoke or debris, or by means of long-term, repeated exposure to smaller amounts of toxins, which can eventually build up to harmful concentrations in the investigator's body. Fire investigators are at least as susceptible to dangerous exposure as are firefighters. In fact, investigators often spend more time in fire scenes than their first responder counterparts.

Smoke from a typical structural fire contains vapors, gases, and particulates that are hazardous, including carbon monoxide, hydrogen cyanide, formaldehyde, and approximately 100 known carcinogens. Fires in modern times involve more human-made substances than in years past, including plastics, petroleum products, and formaldehydes. When they burn, these substances give off toxic compounds—some of which are known, but many of which are unknown.

Importantly, the same toxic substances that are contained in smoke are likely present in the fire debris itself. Thus, the need for fire investigators to take precautions against these hazards does not end when the fire is extinguished.

One of the most common ways for toxic substances to enter the body of a fire investigator is through inhalation. The body has built-in defense mechanisms to protect against inhalation hazards, but these are not effective against some gases and particulates. For that reason, it is important to use breathing protection at every scene.

Absorption through the skin is another common route of entry of toxic substances into the body. Notably, over time in a fire investigation, the skin may become warmer, which increases absorption of toxic materials. Although the regular removal of soot and particulates from the skin is helpful in protecting from such entry, the best practice is to protect all skin from exposure during the investigation.

Although inhalation and absorption are common issues at fire scenes, fire investigators must also guard against ingestion of dangerous substances, which may happen when the investigator eats food or drinks liquid during a scene investigation without thoroughly removing PPE and washing hands and face. Likewise, investigators must protect against injection, in which toxins may be absorbed via puncture wounds—an injury that is all too common in a fire investigation. Knowledge of the effects and routes of exposure of chemicals can help those on scene protect themselves against toxins **TABLE 6-1**.

Carbon monoxide, a common by-product of fire, can cause chemical asphyxiation. It binds with hemoglobin 250 times more readily than oxygen does, thereby preventing the body's cells from receiving oxygen. This effect can be cumulative and will build after several exposures, potentially causing the investigator to collapse in less-than-toxic atmospheres. Hydrogen cyanide, another common by-product of fire, is considered one of the most dangerous gases at an investigation scene due to the level of exposure from smoldering synthetic materials. Consideration should be given to having a hydrogen cyanide detector present at every fire scene investigation.

TABLE 6-1 Chemical Exposure

Effects	Routes	Toxicity
■ Local (site of contact) ■ Systemic (affecting organs)	■ Inhalation (breathing) ■ Cutaneous (absorption through skin or eyes) ■ Ingestion (through the mouth) ■ Injection (penetration from contaminated object)	■ Acute (high exposure over a short time) ■ Chronic (repeated low exposures over a long time) ■ Cumulative (repeated exposures) ■ Latency (period between exposure and effects)

The fire investigation community is becoming increasingly aware of the amount of cancer-causing substances that are emitted even in an everyday structural fire. Fire investigators are encouraged to learn more about cancer prevention and the long-term ways in which they can potentially be harmed by the atmospheres at fire scenes. Resources on this topic are available through the Publications and Resources page of IAAI's website, www.firearson.com.

TIP

General safety meetings at an investigation scene should occur as often as necessary, but at least twice daily. Ideal times for safety meetings are the start of the day, at the end of the day, at organized breaks, and when beginning a new phase of the investigation.

Scene-Specific Hazards

In addition to the hazards encountered at virtually all fire scenes, certain types of fires can pose special hazards to the investigator. Fires in drug laboratories, for example, may expose the fire investigator to additional chemicals. A fire at a residence may expose the investigator to pesticides and other chemicals.

Asbestos, silica dust, heavy metals, and many other substances are present at many fire scenes. Some chemicals present at the fire scene may have been released because of the failure of their containers. Like other contaminants, these products may adhere to clothing and equipment, and create a risk of exposure if clothing and equipment are not properly decontaminated. Particles containing lead can be released into fire atmospheres and may adhere to the fire investigator's clothing, potentially posing a risk to family members if brought home.

Keep in mind that the atmosphere may change during an investigation. A safe atmosphere may turn toxic, and ongoing atmospheric monitoring should be provided. Also, because fire consumes oxygen, an oxygen-deficient atmosphere can continue to exist in confined areas even when other areas have been aired out.

Biological Hazards

Virtually every fatal fire scene contains biological hazards. It is highly unlikely that a fire will destroy all of the bodily fluids of a fire victim, and the trauma of fire can cause such fluids to be present not only in and on the victim, but elsewhere. The fire investigator should don appropriate protection whenever dealing with the body of a deceased fire victim.

Other materials on a fire scene can cause allergic reactions, infection, and diseases. Biological substances from plants, animals, insects, bacteria, and viruses can all pose potential hazards. Scenes may also contain other unexpected materials, including garbage, biological material, bedbugs, and human waste, among others. These hazards must all be anticipated and assessed by the investigator.

Radiological Hazards

Responding to a medical or construction facility can expose an investigator to radiological materials that are stored at these locations. After a fire at such a facility, attempt to determine whether radiological equipment was involved and whether a release of radioactive material has happened or may have happened. In some circumstances, the fire investigator may not be able to safely enter such a scene for investigation until the radiological hazards have been addressed.

Physical Hazards

Fire scenes can be cluttered and dangerous places, and the physical hazards that cause trip, fall, or crushing injuries are well known to fire investigators. Some of these hazards include holes in the floor, unsecured walls or chimneys, roof collapse, sharp objects, water, ice, piles of debris or belongings, wires, and energized electrical equipment. In addition to wearing proper PPE, investigators must exercise constant awareness of structural integrity, weather, footing, lighting, surface integrity, surface traction, and other issues.

After assessing the fire scene, choose the proper PPE to block the routes of exposure from the hazards present at each specific fire scene. A fire investigation is not a life-or-death situation; therefore, investigators should not risk their health and safety to determine the origin and cause of any fire. Use engineering controls available to protect the investigative team from harmful materials.

Asbestos and Hazardous Materials

Although asbestos is no longer used in building construction, it was commonly included in buildings before being banned from many applications in the 1970s and 1980s. As a result, many of today's buildings contain asbestos in various forms, including steam pipes, boilers, furnace ducts, tile, insulation, and wallboard texturing. These materials may become loosened during a fire incident or during the fire department's efforts to extinguish the fire and overhaul the scene, and their presence may not be immediately obvious. Once known to be present, asbestos hazards should be mitigated before the investigation begins.

Fire scenes can be hazardous materials scenes if there is the presence or release of a hazardous substance. Hazardous materials can take the form of chemicals, petroleum products, products used in drug labs, and many others. These materials are often identified while the fire department is fighting the fire, but occasionally they become evident during the fire investigation. A specialized hazardous materials team should be activated to respond to a hazardous materials scene if the protocol requires their participation, and the investigation will be required to wait for the remediation or mitigation of the threat.

Hazardous substances can also be used as a form of terrorism. Be alert and always consider the threat of terrorism or the possibility that a crime has been committed.

Discuss any possible hazards with the site manager and use safety data sheets (SDSs) and site safety plans before beginning the investigation **FIGURE 6-9**. Check with the relevant fire department to see if the location has a preincident plan that describes the occupancy and any hazardous materials or other safety issues that might have been identified. Keep in mind that several safety plans related to varying topics and of different complexities may be in use at a scene **TABLE 6-2**.

Human and Animal Threats

Fire investigators, both public and private, may face danger from hostile persons on fire scenes and during other investigative activities. Perpetrators of arson and other criminal acts may consider the fire investigator a threat to their freedom, and hostile persons often do not distinguish between uniformed fire personnel, uniformed private personnel, and the police. Moreover, some people unconnected with a fire may be willing to attack individuals they believe represent law enforcement or the establishment. As part of their overall safety strategy, fire investigators should consider requesting law enforcement to be present on the scene and should learn self-defense techniques appropriate to their jurisdiction and assignment.

Animals, including pets that are normally calm and sociable, can pose a potential threat to the fire investigator, especially when in distress. Animal control authorities may need to be called to help control animals on scene. Occasionally, a fire investigator might encounter a pet within a burned structure and be called upon to capture it or arrange for it to be returned to its owner.

Investigator Fatigue

The reconstruction and investigation of a fire scene are often long, tedious, and physically strenuous processes. Safety measures require the frequent use of heavy clothing and respiratory protection. All of these factors may combine to fatigue fire investigators. Fatigue may cause investigators to tire and lose strength and judgment, which could place them in jeopardy. In the interest of safety, it is important to seek periodic rest, food, and fluids in a safe environment. Access to sanitation and wash stations, preferably on-site but outside the hot zone, is also critical. Proper washing will also help prevent ingestion of contaminants while eating and drinking.

Post-Scene Hazards

Safety and potential exposure issues do not end when an investigator leaves the scene. All items removed from the scene, including evidence and equipment, should be handled, labeled, and stored in a safe manner.

Exposure to Tools and Equipment

Any items worn or used in an investigation are subjected to the risk of contamination when used in hazardous environments. All such items should be handled and cleaned with this risk in mind. Equipment that cannot be decontaminated should be disposed of in an appropriate manner.

The Fire Investigator Within the Incident Command System

Fire investigators should understand how to operate at the fire scene in conjunction with firefighters and other first responders. The investigation must be conducted within the **incident command system (ICS)** headed by the incident commander.

The fire investigator has the responsibility to contact the incident commander upon arrival at the scene. The incident commander can then relay any pertinent information about the fire that could be valuable to the investigation, such as structural stability, chemicals, or any other hazards present at the scene. A safety plan should be formulated and address such topics as PPE, emergency action plans, hazardous materials, and physical hazards. Depending on the complexity of the scene, the safety plan may include a formal organizational structure.

Nitrogen, refrigerated liquid

Safety Data Sheet P-4630

Nasal
cavity

This SDS conforms to U. S. Code of Federal Regulations 29 CFR 1910.1200, Hazard Communication.
Date of issue: 01/01/1979   Revision date: 10/21/2016   Supersedes: 10/03/2014

SECTION: 1. Product and company identification

1.1.   Product identifier

Product form	: Substance
Name	: Nitrogen, refrigerated liquid
CAS No	: 7727-37-9
Formula	: N2
Other means of identification	: Nitrogen (cryogenic liquid), Nitrogen, Medipure Liquid Nitrogen

1.2.   Relevant identified uses of the substance or mixture and uses advised against

Use of the substance/mixture   : Medical applications
Industrial use
Food applications

1.3.   Details of the supplier of the safety data sheet

XYZ Chemical, Inc.
10 Main Street
Anytown, NY 01234-5678 -USA
T 1-800-555-4321
www.xyzchemical.com

1.4.   Emergency telephone number

Emergency number   : Onsite Emergency: 1-800-555-0011
CHEMTREC, 24hr/day 7days/week
— Within USA: 1-800-424-9300, Outside USA: 001-703-527-3887
(collect calls accepted, Contract 17729)

SECTION 2: Hazard identification

2.1.   Classification of the substance or mixture

GHS-US classification

Refrigerated liquefied gas H281

2.2.   Label elements

GHS-US labeling

Hazard pictograms (GHS-US)   :

GHS04

Signal word (GHS-US)   : WARNING

Hazard statements (GHS-US)   : H281 - CONTAINS REFRIGERATED GAS; MAY CAUSE CRYOGENIC BURNS OR INJURY
OSHA - H01-MAY DISPLACE OXYGEN AND CAUSE RAPID SUFFOCATION

Precautionary statements (GHS-US)   : P202 - Do not handle until all safety precautions have been read and understood
P271 P403 - Use and store only outdoors or in a well-ventilated place
P282 - Wear cold insulating gloves, face shield, eye protection
CGA-PG05 - Use a back flow preventive device in the piping
CGA-PG24 - DO NOT change or force fit connections
CGA-PG06 - Close valve after each use and when empty
CGA-PG23 - Always keep container in upright position

2.3.   Other hazards

Other hazards not contributing to the   : Asphyxiant in high concentrations
classification   Contact with liquid may cause cold burns/frostbite

EN (English US)   SDS ID: P-4630   1/9

FIGURE 6-9 Example of a safety data sheet (SDS).

© Jones & Bartlett Learning.

TABLE 6-2 Emergency Action Plans

Plan	Description
Emergency evacuation plan	▪ Used when the scene suddenly becomes unsafe ▪ Includes exit routes, gathering location, and confirmation of complete evacuation
Medical emergency plan	▪ Used when someone on scene requires medical care ▪ Includes locations of emergency facilities, phone numbers, and a first-aid kit
Severe weather plan	▪ Used when there is severe weather ▪ Includes notification method and meeting place for those on scene
Fire emergency plan	▪ Used when a fire or explosion occurs during an investigation ▪ Includes fire department location, phone number, and notification method; exit routes; meeting place(s); and confirmation of complete evacuation for those on scene
Additional emergency action plan	▪ Used when scene-specific issues arise ▪ Includes information specific to the scene, as determined by the person in control of the scene

The emergency action plan should address escape routes, medical and fire emergencies, accountability, severe weather, and any other scene-specific potential hazard. When confined spaces are involved, a site program should be developed and implemented by confined space–trained personnel.

An ICS is necessary in incidents involving large buildings or in areas where it is difficult to keep track of the investigation team, especially if fire suppression activities are still in progress. The investigation team should participate in the accountability system in place for the ICS, and must check in with the designated officer and advise that officer of the areas where the investigators will be working and the jobs in which they will be involved. When they are finished in that location, the investigators should inform the accountability officer of the completion of that portion of the investigation and then verify the new location where they will be working. This information allows the accountability officer and the incident commander to monitor all personnel operating at the fire scene.

When the incident commander at a scene ends the ICS, there should be a face-to-face transfer of command from the incident commander to the fire investigator in charge of the investigation, during which they identify any areas of concern that can affect the safety of any personnel remaining at the scene. The investigator in charge then assumes control and responsibility for scene safety. Arriving personnel must be briefed on all of the known hazards, PPE requirements, evacuation signals, and emergency medical procedures. The lead investigator must also ensure scene security to prevent destruction or removal of evidence and to prevent bystanders from entering the scene.

Occupational Safety and Health Administration

The Occupational Safety and Health Administration (OSHA) requires that private and public employees have a workplace that is safe, and it has identified five elements instrumental in keeping these workplaces safe:

- Management commitment and employee participation
- Hazard and risk assessment
- Hazard prevention and control
- Safety and health training and education
- Long-term commitment

More information regarding hazard identification, evaluation, and prevention can be found in NFPA 1500, *Standard on Fire Department Occupational Safety and Health Program*.

The OSHA standards discussed in this section most directly affect the fire investigator. Countries outside of the United States may not use some or any of these standards, although they may have similar regulatory agencies. Investigators should identify the regulations that apply to their departmental or company policy, as well as the safety regulations established by the local agencies, state, province, or country.

29 CFR 1910.120, HAZWOPER

OSHA regulations in 29 CFR 1910.120, also known as the **HAZWOPER (Hazardous Waste Operations and Emergency Response) standard**, deal with operations at a hazardous waste site and the emergency response to hazardous materials incidents. By their nature, all fire scenes are potential hazardous waste sites. The HAZWOPER standard, as referenced in 40 CFR, Protection of Environment, Part 311, states that the provisions in 29 CFR 1910.120 apply to state and local employees who are involved in hazardous waste operations. This standard protects all employees, even public employees operating in states that do not have a state OSHA plan. All fire investigators should be familiar with the following NFPA documents, which can clarify the HAZWOPER standard:

- NFPA 472, *Standard for Competence of Responders to Hazardous Materials/Weapons of Mass Destruction Incidents*
- NFPA 473, *Standard for Competencies for EMS Personnel Responding to Hazardous Materials Incidents/Weapons of Mass Destruction Incidents*

As stated in these documents, the investigator is expected to understand the states of matter and to recognize hazards by occupancy, shape of containers, markings, and odors reported to the responder. As part of the Emergency Response and Responder Safety Consolidation Plan, NFPA 472 and NFPA 473 are currently being combined into a consolidated NFPA 470.

29 CFR 1910.146, Permit-Required Confined Spaces

The **Permit-Required Confined Space standard** deals with the safety aspects of entering a confined space. Before this standard was enacted, 60 percent of all confined space incident deaths were incurred by would-be rescuers. The fire investigator should be able to identify a confined space, and at no time should a confined space be entered without the proper precautions in place. A Permit-Required Confined Space, as defined by OSHA, meets the following criteria:

- A space where an employee can bodily enter and perform assigned work
- Limited or restricted means of entrance or exit
- Not designed for continuous employee occupancy

Entry into a confined space requires special training and further requires the continuous monitoring of the atmosphere, fall protection, ventilating the space, and wearing a SABA. Additional trained support and backup personnel must also be present.

29 CFR 1910.147, Control of Hazardous Energy (Lockout/Tagout)

Whenever an investigative team works around equipment or wiring where an unexpected energization, start-up of machines, or release of stored energy could result in injury, the team must consider OSHA's **Control of Hazardous Energy (Lockout/Tagout) standard**. This standard provides for placement of a lockout/tagout device that disables the equipment. Every fire investigator who works in the area places a padlock on the device, and each investigator retains the only key to their own lock. Then, on completing the investigation, each investigator leaving a hazardous area must unlock their individual device before power can be restored or energy released. This way, energy cannot be restored to the equipment until every investigator has left the hazardous area and has been accounted for. This standard should be followed whenever the investigator is working in or around heavy equipment or energized equipment or where a valve can be operated remotely, releasing a material that might engulf the investigator.

After-Action REVIEW

IN SUMMARY

- Fire investigator safety must be the primary concern at every fire scene.
- Fire investigators are exposed to myriad hazards at fire scenes, many of them invisible, that threaten both their immediate safety and their long-term health.
- Even though the emergency phase of an incident has likely passed when the investigation begins, investigators should maintain safety standards equivalent to those that were in place during the initial phases of the incident.

- Whenever possible, investigators should not work alone during the fire scene examination.
- A fire investigator working alone should notify a responsible off-site contact person of the location, the nature and size of the scene, and the expected time required to complete the investigation.
- Fire investigators on scene should take steps to protect assistants, trainees, specialized personnel, occupants, and bystanders who may be injured by a fire scene.
- Anytime there is a potential for collapse of a wall, chimney, or other part of a building, investigators should establish collapse zones to protect personnel.
- When a hazard is identified, investigators should find the most effective manner to mitigate the risks by comparing the risk level to a suitable benchmark or acceptance criterion.
- Risks may be controlled with engineering controls, administrative controls, and personal protective equipment.
- Normally, fire investigators should not enter a structure until the fire is completely extinguished.
- Investigator safety planning should include determining at least two means of exit, which should be at opposite ends of the work area.
- Before entering any fire scene, assess the stability of the structure and, if possible, eliminate physical hazards by either the use of shoring materials or demolition.
- Freestanding chimneys pose a substantial risk to both firefighters and fire investigators.
- Fire investigators should not assume that building utilities have been disconnected by fire crews.
- An investigator can use a meter, such as a non-contact AC voltage tester, to determine whether the building electricity is off and to test the status of components on a branch circuit.
- Firefighting operations can add to the stress on a building's structural members, through the application of water, which weighs 8.35 lb/gal (1.05 kg/L).
- If it is necessary to enter a fire scene before extinguishment, investigators must wear structural firefighting gear and a self-contained breathing apparatus.
- Protective clothing and equipment should be properly decontaminated prior to leaving the scene.
- A helmet should be worn during all fire investigations.
- Fire scene toxins become embedded in clothing, vehicles, furniture, and other items at the investigator's home if proper precautions are not taken.
- Contamination of clothing and equipment exposes the investigator, family, and coworkers to fire scene toxins.
- Even after fires are extinguished, fire scenes are full of toxic smoke, chemicals, and other respiratory threats that can be extremely harmful to short- and long-term health.
- Investigators should wear eye protection that provides the maximum level of safety.
- Respiratory and absorption hazards remain on scene long after the fire is extinguished and occur at fire scenes of any size.
- One of the most common ways for toxic substances to enter the body of a fire investigator is through inhalation.
- General safety meetings at an investigation scene should occur as often as necessary but at least twice daily.
- Virtually every fatal fire scene contains biological hazards.
- Physical hazards at a fire scene may include holes in the floor, unsecured walls or chimneys, roof collapse, sharp objects, water, ice, piles of debris or belongings, wires, and energized electrical equipment.
- Many buildings in use today contain asbestos in various forms, including in steam pipes, boilers, furnace ducts, tile, insulation, and wallboard texturing.
- Fire investigators, both public and private, may face danger from hostile persons at fire scenes and during other investigative activities.
- Fatigue may cause investigators to tire and lose strength and judgment, which could place them in jeopardy.
- An incident command system (ICS) is necessary in large buildings or in areas where it is difficult to keep track of the investigation team, especially if suppression activities are still in progress.
- Per OSHA, the fire investigator is expected to understand the states of matter and to recognize hazards by occupancy, shape of containers, markings, and reported odors.

KEY TERMS

Collapse zones Distances around a structure that may be affected by a structural collapse. The area should be identified by markers or specialized scene tape that indicates there is a potential for structural collapse.

Control of Hazardous Energy (Lockout/Tagout) standard The federal OSHA regulation that governs work around equipment or wiring where an unexpected energization, start-up of machines, or release of stored energy could result in the injury of the investigators. This standard utilizes a lockout/tagout device that disables the electrical equipment. Specifics can be found in 29 CFR 1910.147.

HAZWOPER (Hazardous Waste Operations and Emergency Response) standard The federal OSHA regulation that governs hazardous materials waste site and response training. Specifics can be found in 29 CFR 1910.120.

Impact load A sudden added load to a structure.

Incident command system (ICS) The combination of facilities, equipment, personnel, procedures, and communications under a standard organizational structure to manage assigned resources effectively to accomplish stated objectives for an incident.

Permit-Required Confined Space standard The federal OSHA regulation that governs any space that an employee can bodily enter and perform assigned work, with a limited or restricted means of entrance or exit, and that is not designed for continuous employee occupancy. Specifics can be found in 29 CFR 1910.146.

Site safety assessment A full exterior and interior survey of a fire structure to identify the presence of physical, toxicological, and biological hazards that may pose a threat to the investigator.

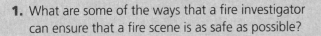

On Scene

1. What are some of the ways that a fire investigator can ensure that a fire scene is as safe as possible?

2. If you have fire investigation equipment in a vehicle or response kit, what personal safety equipment is included? Is it easy to access? Are there items that could be added or improved based on your review of this chapter?

3. List the hazards that you check for on arrival at a fire scene and describe the steps you take at the beginning of an investigation to ensure the safety of yourself and other persons. Do these steps represent an adequate site safety survey as described in this chapter?

4. If a safety issue should arise in your department's or team's operations, how would you bring it to the attention of others and/or the persons in charge of the scene?

Chapter Opener: © Jones & Bartlett Learning. Photographed by Glen E. Ellman; On Scene siren: © Bildgigant/Shutterstock.

CHAPTER 7

Building Electrical and Fuel Gas Systems

KNOWLEDGE OBJECTIVES

After studying this chapter, you should be able to:

- Explain basic electricity.
- Identify and describe the components of a building's electrical system. (**NFPA 1033: 4.2.5**; **4.2.6**; **4.2.8**, pp. 177–185)
- List the conditions that must exist for ignition from an electrical source. (**NFPA 1033: 4.2.6**; **4.2.8**, pp. 185–189)
- Describe how to interpret damage to electrical systems. (**NFPA 1033: 4.2.5**; **4.2.6**; **4.2.8**, pp. 189–197)
- Explain static electricity.
- Describe the elements of fuel gas systems and explain how they can contribute to fire and explosion investigations. (**NFPA 1033: 4.2.8**, pp. 199–204)
- Describe the characteristics of fuel gases. (**NFPA 1033: 4.2.8**, pp. 200–202)
- List and explain the components of natural gas systems. (**NFPA 1033: 4.2.8**, pp. 204–206)
- Describe the components of liquefied petroleum gas systems. (**NFPA 1033: 4.2.8**, pp. 200–204)
- Identify and describe components common to fuel gas systems. (**NFPA 1033: 4.2.8**, pp. 200–204)
- Describe common piping for fuel gas systems in buildings. (**NFPA 1033: 4.2.8**, pp. 206–207)
- Identify common appliance and equipment requirements. (**NFPA 1033: 4.2.8**, pp. 207–208)
- Identify fuel gas utilization equipment. (**NFPA 1033: 4.2.8**, pp. 208–209
- Describe how to investigate fuel gas systems. (**NFPA 1033: 4.2.8**, pp. 209–212)

SKILLS OBJECTIVES

After studying this chapter, you should be able to:

- Complete calculations based on Ohm's law. (**NFPA 1033: 4.2.6**; **4.2.8**, p. 173)
- Determine whether a circuit has proper overcurrent protection. (**NFPA 1033: 4.2.6**, pp. 179–181)
- Examine fire-damaged electrical conductors and determine whether the damage is the result of electrical activity or a result of the fire. (**NFPA 1033: 4.2.6**; **4.2.8**, pp. 182–184)
- Identify circuits with overcurrent conditions or that are overloaded based on a blown fuse or tripped circuit breaker in a panelboard. (**NFPA 1033: 4.2.6. 4.2.8**, p. 197)
- Conduct a fire and explosion investigation at a scene involving fuel gas systems. (**NFPA 1033: 4.2.8**, pp. 209–212)

You are the Fire Investigator

You are investigating a fire in a single-family residence that originated in a first-floor bedroom. Fire damage is present throughout the room, which contains a bed, nightstand, dresser, and desk. Based on your observations of fire patterns and your interview with the homeowner, you determine that the origin of the fire is located at the head of the bed. The homeowner states that she had been in the bedroom approximately 30 minutes prior to the fire working at the desk. A desk lamp, computer, and space heater were in use at the desk and were left on when the owner left the room. You have determined that these items were not ignition sources, as there was no evidence of a failure, and they were outside the suspected area of origin.

Examining the area around the bed, you find a duplex receptacle at the head of the bed. Nothing is plugged in to the receptacle. Charring can be seen to the wood stud on which the receptacle is installed, and you find evidence of melting on a screw connection of the receptacle. Further analysis of the wiring reveals that the items at the desk are on the same circuit as this receptacle, which is wired in series in the circuit.

1. Why is it important to understand electricity and electrical systems while investigating this fire?
2. How will the method of installation influence your ignition scenario?
3. What conditions must exist for this fire to have been caused by a failure of the outlet?

Access Navigate for more practice activities.

Introduction

This chapter provides an introduction to basic electricity and electrical systems. Electrical wiring and equipment are commonly found in building fires, and fire investigators must have a fundamental understanding of electricity and electrical systems to determine the involvement, if any, of electricity in a particular fire. The presence of electrical wiring or components in fire damage, or even in the origin of a fire, does not by itself indicate that an electrical event was the cause. When electrical damage is found, the investigator will need to examine and understand the system and the circuit(s) involved, and attempt to determine the answers to fundamental questions including whether electrical damage was the cause or an effect of the fire. An investigator finding an electrical cause to a fire event must always be able to demonstrate the specific failure and ignition mode that started the fire.

While fire investigators should understand fundamental electricity and systems, they must also exercise abundant caution in dealing with these systems. For example, always approach an electrical device or component initially as if it is energized, and always carry a noncontact AC voltage detector to check such systems before examining or handling them.

NFPA 70E, *Standard for Electrical Safety in the Workplace*, is a critical tool for protecting electrical workers from electric shock, arc flash, and arc blast hazards while doing electrical work; it should be used as a guide prior to approaching circuits of unknown energized status. NFPA 70E also provides several tables that can help electrical workers select the correct type of personal protective equipment (PPE) to wear, based on proximity to the electrical equipment, and the task that they are performing. This document outlines five different hazard/risk categories (HRCs): 0, 1, 2, 3, and 4.

Before analyzing an electrical circuit or equipment, assess the HRC of the installation and ensure that the system has been de-energized, and an electrically safe work condition has been created. For example, workers must wear PPE specified by the tables in NFPA 70E whenever they are within the flash protection boundary (for example, typically 48 inches [1.2 m] for 600-volt equipment), whether or not they are actually touching the live equipment. This includes when workers are performing tasks such as voltage testing to verify whether power has been turned off. Such exposure is considered "live work" that requires workers to wear PPE, including flame-resistant clothing. This PPE offers arc-flash and arc-blast protection, but does not protect against electric shock.

An investigator handling and examining electrical equipment should keep in mind the issues surrounding spoliation of evidence (see Chapter 5, *Legal Considerations for Fire Investigators*) and normally should not disassemble electrical equipment. If you are not qualified to analyze the electrical equipment, contact a qualified individual (usually an electrical engineer) to perform this task.

Basic Electricity

In an effort to explain the basics of electricity, a comparison to hydraulics is often used, although hydraulic and electrical systems are not entirely analogous. First, when making the comparison, it is important to envision the hydraulic system as a closed system that circulates water back to the source. A pool filtration system is a good analogy: Water flows from the pool (the source), through the filtration (the circuit), and then back to the pool (return/neutral). Comparisons to open systems that discharge water, such as a fire hose, are not good examples. NFPA 921, *Guide for Fire and Explosion Investigations*, contains a section on basic electricity, including graphics to help explain simple electrical circuits and electricity. **TABLE 7-1** compares a hydraulic system to an electrical system.

Ohm's Law

The units of electricity are voltage (volts [V]), current (amperes or amps [A]), resistance (ohms [Ω]), and power (watts [W]). The basic law of electricity

for a simple **resistive circuit** (one that does not contain inductance and capacitance) is **Ohm's law**, which defines the relationship among voltage, current, and resistance. If two of these three values are known, it is possible to determine the third. Knowledge of the values of these parameters and their relationship to one another is essential to understanding whether enough heat is present to ignite a fire or contribute to its ignition. The components of Ohm's law are shown in **TABLE 7-2**.

TIP

PPE is sometimes required when working on electrical systems that are not live.

Voltage is often represented in equations by either the letter *E* or *V*. These two letters are interchangeable in equations. When referring to the units of voltage, however, V is always used. Current always appears as *I* in equations, with units in amperes abbreviated

---

TABLE 7-1 Hydraulic System Comparison to Electrical System

Hydraulics	Electricity
A **pump** creates the force that moves the water.	A **generator/battery** creates the electromotive force that moves the electrons.
Pressure is measured in pounds per square inch (psi) and is measured with a pressure gauge.	**Voltage (*E*)** is measured in volts (V) and is measured with a voltmeter.
Water moves through the pipes and does the work.	**Electrons** move through the conductors and do the work.
Flow is measured in gallons per minute (gpm) and is measured with a flowmeter.	**Current (*I*)** is measured in amperes or amps (A) and is measured with an ammeter.
A **valve** controls the flow of water: ■ Open: water flowing ■ Closed: no water flowing	A **switch** controls the flow of electricity: ■ Off: open circuit; no electricity flowing ■ On: closed circuit; electricity flowing
Friction is the resistance of the pipe or hose to the water moving through it and is measured in psi.	**Resistance (*R*)** is the opposition of the conductors to the electrons moving through them and is measured in ohms (Ω).
Friction loss is the amount of pressure lost between two points in a pipe layout.	**Voltage drop** is the amount of voltage drop between two points in a circuit.
Pipe or hose size is measured in inches, based on the inside diameter: ■ Larger pipe = greater flow ■ Smaller pipe = lower flow	**Conductor size** is given in AWG or wire gauge size: ■ Larger wire = greater current ■ Smaller wire = lower current

TABLE 7-2 Ohm's Law Components

Value	Symbol	Units
Voltage	E	Volts
Current	I	Amperes (amps)
Resistance	R	Ohms
Power	P	Watts

Note the difference between symbols and units, as well as the relationship between the unit values when analyzing electrical circuits.

as A. The resistance is denoted by R in an equation, with units of ohms represented by the Greek letter omega, Ω.

Ohm's law is stated as follows:

$$\text{Voltage} = \text{current} \times \text{resistance} (E = I \times R)$$

$$\text{Volts} = \text{amperes (amps)} \times \text{ohms}$$

Simple algebra allows us to rearrange the values in this equation to determine current and resistance, as follows:

$$\text{Current} = \text{voltage/resistance} (I = E/R)$$
$$\text{Amps} = \text{volts/ohms}$$

$$\text{Resistance} = \text{voltage/current} (R = E/I)$$
$$\text{Ohms} = \text{volts/amps}$$

One of the more useful measurements when working with postfire circuits is the measurement of resistance. If the voltage is fixed and you know the resistance, you can estimate the power generated. You can then use the power (discussed in the next section) to find localized heating or higher respective heat fluxes that could increase the mass of a combustible to its corresponding ignition temperature. The electrical resistance of a simple heating appliance may be measured with an ohmmeter or multimeter, by taking a reading with the probes across the two spades of the male plug of an appliance power cord. The result is the resistance of that appliance's heating element in ohms. As stated earlier, you can then divide the voltage by the resistance (in ohms) to determine the current (in amps) used by that particular appliance.

You can determine the current of all appliances (if they are simple resistive devices) on a particular circuit by using this method. You can also determine what

the resistance should be for a particular appliance by obtaining the voltage and current and calculating the resistance. The voltage and current values are often printed on the appliance's nameplate, so their measurement may not be required. Apply Ohm's law by dividing the voltage by the current to determine what the resistance should be, as indicated in Table 7-2.

Power

Knowledge of voltage, current, and resistance leads to another relationship in defining the flow of energy—namely, calculation of power. **Electrical power** refers to the rate of doing work in an electrical circuit, such as in a hair dryer, electric motor, or lightbulb. Electrical power is measured in watts (or joules per second). Another unit for power with which most people are familiar is **horsepower (hp)**, which is used to express the rate of doing work for some mechanical objects. A third unit for power, used to describe heat sources such as furnaces, is Btu/hour, where Btu stands for British thermal unit. Each of these values is related in the following manner:

1 watt = 1 joule/second

1 watt = 3.4129 Btu/hour

1 horsepower = 745.7 watts

1 Btu = 1054.8 joules

The symbol for power is P. The formula for determining electrical power is as follows:

$$\text{Power} = \text{voltage} \times \text{current} (P = E \times I)$$

TIP

Often the resistance of a heating element cannot be measured directly from the male plug spades because of internal switches, relays, or fire-damaged conductors. Direct connections to the heating element inside the appliance are needed.

For a motor, both mechanical and electrical power are typically given on the nameplate of the device. The mechanical (output) power for motors is usually expressed in horsepower, while the electrical (input) power is expressed in kilowatts. If the mechanical power of the motor were to be converted to kilowatts, then a comparison would show that the mechanical output power of the motor is less than the electrical input power it received. This difference occurs because some of the energy is lost as heat during the operation of the motor.

Relationship Among Voltage, Current, Resistance, and Power

It is important to understand the relationship among voltage, current, and resistance and to know how to calculate power from these values. This knowledge will help you understand the potential of electricity to be the cause of a fire. For example, knowing two of these three values for an appliance in a circuit (voltage and resistance, or voltage and power, or power and resistance) will allow you to conclude how much current or energy was used by the appliance, whether the circuit was overloaded, or whether the overcurrent protection (e.g., fuse or circuit breaker) was properly matched to the power requirements for the circuit. The Ohm's law wheel is useful for calculating one parameter when you know the other two **FIGURE 7-1**.

Ideally, you should be able to calculate the total power requirements of a single circuit with multiple loads on it by inspecting each appliance or piece of equipment in the circuit, determining the power requirements of each, and then summing these values. Information specific to an appliance, such as amperage or wattage ratings, often appears on a nameplate or other label, or in the manual or packaging for the appliance. These values are typical maximum values and assume active operation of the equipment during peak power utilization. For example, a washing machine will draw its full rated current during the agitation or spin cycle, where the power required to overcome inertia is highest. At other times, it may not draw as much power.

In many appliances, therefore, power use is dynamic and can vary over time depending on the action of the appliance. As a result, even if appliances on a circuit are operating intermittently, as in the washing machine example, the multiple power draws can potentially result in overloads to the circuit.

If a fault occurs in a circuit (such as a short circuit or a ground fault), the appliance or other work-producing component(s) in the circuit can be bypassed. This results in the resistance in the circuit decreasing rapidly. Ohm's law confirms that if the circuit resistance decreases, the current increases to an abnormally high value. Under some circumstances, the abnormally high current will trip the overcurrent protection. However, if sufficient resistance remains in the electrical current path, the abnormally high fault current may not reach the level required to trip the overcurrent protection device. For example, if a fault develops across a carbon path between two conductors and there is 100 ohms of resistance, the current draw for that particular fault would be 1.2 amps (1.2 amps = 120 volts/100 ohms), and the circuit protection may not trip. However, if a fault occurs where there is very little resistance, such as 0.2 ohm, as much as 600 amps may flow through that fault, and the circuit protection should open immediately. In the former case, more than 100 watts is dissipated in the fault path, which can generate sufficient heat to ignite nearby combustible material.

FIGURE 7-2 illustrates the relationship among voltage (V), current (I), resistance (R), and power (P). Note that resistance must be measured with the voltage turned off (0 V).

> **TIP**
>
> Ignition by electrical energy requires an energized circuit, sufficient heat and temperature, proximity to a combustible material, and sufficient time to raise the fuel up to its ignition temperature.

Overload Situations

If the power needs in a circuit exceed the circuit's current carrying capacity, then an overload situation may occur. **Overload** is the persistent "operation of equipment in excess of [its] normal, full-load rating or of a conductor in excess of [its] rated ampacity, which, when it persists for a sufficient length of time, [may] cause damage or dangerous overheating," according to NFPA 921. A direct, short-duration fault, such as a short circuit or ground fault, is not an overload but rather an overcurrent situation.

An overload can cause the various conductors and/or components in a circuit to overheat. If the overloaded section of conductor or component

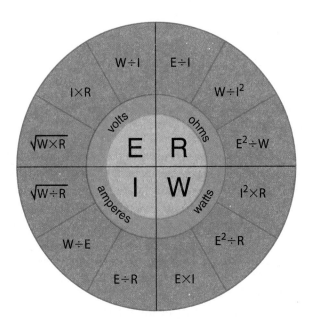

FIGURE 7-1 The Ohm's law wheel.

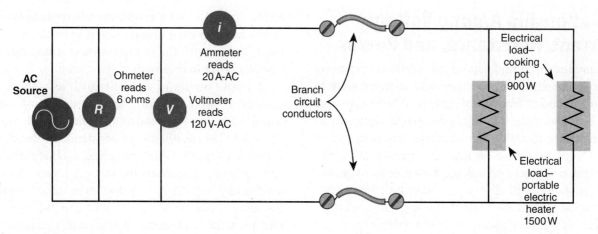

FIGURE 7-2 The relationship among voltage (*V*), current (*I*), resistance (*R*), and power (*P*).

has sufficient temperature, duration, and proximity to a fuel, ignition may occur. However, all three factors—temperature, duration, and proximity—must be considered when examining the circuit and appliances to determine whether an overloaded circuit started the fire. Overloading of the wiring in a house circuit by only a few amperes will not usually immediately open the overcurrent protection device or cause overheating because a safety factor is built into most electrical circuits. Significant overloading of a circuit, to the point of overheating, usually requires the circuit current to exceed the rating by a greater margin. More information about recognition and investigation of overloaded equipment appears later in the chapter.

Wire Gauge

The diameter of wire is commonly measured in sizes given by the American Wire Gauge (AWG). The smaller the AWG number, the larger the wire diameter is. Common household wiring is 14 and 12 AWG (copper). Some specific electrical circuits in the home draw larger currents and are served with larger wires. Appliances such as an electric range, dryer, or water heater are often supplied with 6, 8, or 10 AWG wire.

Wire gauge in an electrical system is similar to pipe diameter for a water system: Larger wire (smaller gauge number) equates to larger-diameter pipe. A blockage in a water pipe equates to a nick or cut in an electrical wire. If pressure (electrical voltage) is fixed, the amount of water flow (electrical current) is reduced.

NFPA 70, *The National Electrical Code* (NEC), requires that the size, type, and number of conductors be labeled on the insulation. For instance, residential cables designated as "NM-B 12/2 with ground" have a nonmetallic sheathed cable, 12 AWG, with two

current carrying conductors and a grounding conductor (i.e., a hot conductor, a neutral conductor, and a ground conductor).

Ampacity

Amperage is the amount of current flow measured in amperes (amps or A) and is similar to the flow of water in gallons per minute. **Ampacity** is the "current, in amperes, that a conductor can carry continuously under the conditions of use without exceeding its temperature rating," according to NFPA 70. The temperature rises with "current due to the power lost, in the form of heat, from resistance in the conductor. The ampacity of a conductor depends on several factors, including conductor size, the ambient temperature, the temperature rating of the insulation, the conductor's ability to dissipate heat to its surroundings, and the resistivity of the conductor material. Ampacity tables in NFPA 70, numbered Table 310.15(B)(16)(a) through Table 310.15(B)(21), indicate the conductors used in various types of insulation, ambient temperatures, and cable bundling/routing methods.

Alternating Current and Direct Current

Most electrical currents used in buildings, structures, and dwelling units take the form of **alternating current (AC)**, in which the voltage and current follow an alternating positive and negative cycle. The AC voltage and current waveforms are sinusoidal in nature. When a system is referred to as "AC," that designation refers to both the voltage and current: With a simple (resistive) load, both of them alternate in a similar fashion.

In the case of a residence, the service panelboard can be considered the local electrical source. With a simple load, electricity delivered to the branch circuits drives current from the source and then back again in a repeating cycle. One voltage cycle includes both a positive component and a negative component, going from peak positive voltage to peak negative voltage during the cycle **FIGURE 7-3**. The frequency is a measure of how many cycles occur in one second. A full cycle starts at zero, goes to the peak positive value, through zero to the peak negative, and then back to zero. The frequency is measured in cycles per second, with units of hertz (Hz). In the United States, the AC frequency is 60 Hz (60 cycles per second), while most other countries use 50 Hz.

The energized conductors, such as L1 and L2 in Figure 7-3, are sometimes called "hot" because they contain voltage with respect to the earth (ground). Typical residential installations utilize the "L1 to L2" configuration, which is rated at 240 V (340 V peak) between L1 and L2. However, most circuits in a typical residence are 120-V circuits, in which L1 or L2 is split and runs to neutral instead of to the other hot lead. In some situations, different voltages with various timing or phase relationships can be derived, but these are rarely used for residential service.

In the early days of electricity, when AC systems were becoming more prevalent, engineers devised a mathematical computation to equate the voltage level of an AC system to that of the more familiar direct current system. (Direct current is discussed in more detail following this section.) This computation is called **root mean square (RMS)**.

For a standard residential AC voltage source, the RMS value is equal to the peak voltage divided by the square root of 2 (or 0.707):

$$V\ RMS = V_{peak}/\sqrt{2}$$

The RMS value of an AC voltage is typically used and is assumed throughout this text unless otherwise specified. Thus, 120 V in residential use is actually 120 V RMS, with a peak voltage of 170 V.

In a 60-Hz wave, the AC voltage changes polarity twice in every cycle: once from negative to positive, and once from positive to negative. The AC current can follow the voltage and typically also looks like a **sine wave**, but can vary depending on the type of load connected. If the equipment or load is resistive (such as a heater), the current follows the voltage sine wave. In a complex device, such as a piece of electronic equipment, the current can be quite different because the internal circuits use different amounts of current at different times during the AC cycle. This can cause currents to flow that are significantly greater than the sine wave would predict, but they usually occur for very short durations of time and rarely stress the electrical system.

Modern electronic equipment often contains complex electronics with power, analog, and digital components. Often a microcontroller or application-specific integrated circuit (ASIC) component acts as the brain of the equipment. For the fire investigator, this means that simple calculations of power are not possible and that manuals from the manufacturer must be consulted to learn about power-up and power-down sequencing and about the various modes in which the equipment can run. A modern washing machine is an example of this kind of equipment. The power sequencing and mode switching are handled by the ASIC, and the power consumption is not steady, but intermittent. Overcurrent protection and conductor size are selected based on the typical maximum operating power levels.

A **direct current (DC)** system usually has current flow from the source to the load and back via the circuit return path with one polarity only. For example, the typical battery flows current from one pole to the other, and the current does not alternate. Besides being common in automobiles, DC-type systems may be found in fixed installations and devices requiring stable, controlled voltage levels, including some appliances and control systems such as those used in industrial settings. The voltage is held at a constant value, rather than varying with a sine wave. Mobile or portable equipment, such as electric vehicles and wheelchairs, also uses DC voltage.

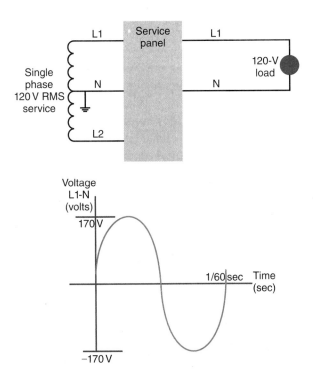

FIGURE 7-3 Single-phase AC sine wave: 120 V RMS residential use.

Some portable equipment can have two power sources—one AC and one DC—depending on how it is configured. An example is a wheelchair or an industrial floor cleaner that uses a battery when in normal operation but uses AC power when the batteries are being charged. In such a case, you must determine which electrical sources and loads are present to rule out possible ignition sources of a fire. For example, a battery charger found connected to batteries aboard a wheelchair may not require the chair motors to be energized, so they can be ruled out. The charger, however, is getting power from two sources—the AC line and the DC voltage, which is back-fed from the batteries. Both of these sources should be examined as potential ignition sources if the charger is found connected.

Single-Phase Service

A single-phase AC 120-V system, as it enters the building, requires three conductors: two ungrounded current-carrying (L1/L2) conductors (sometimes called **hot legs**) and a grounded (neutral) conductor, which is grounded near the source transformer located outside the building. A grounded distribution system has the neutral grounded at regular intervals if poles are used or at underground distribution cabinets. As mentioned earlier, the voltage between either of the hot conductors and the neutral or ground is 170 V peak or 120 V RMS. If you measure the voltage from the L1 conductor to neutral and from the L2 conductor to neutral, they appear to mirror each other. The voltage between the two hot conductors is 340 V peak or 240 V RMS. This kind of single-phase system is frequently found in residential buildings (single-family houses) and small commercial buildings.

Single-phase cables can be delivered to the structure either overhead or underground. Wiring coming in from an overhead pole is called a **service drop**. In a triplex service drop, the hot conductors may be wrapped around the neutral, which is typically not insulated **FIGURE 7-4**. The conductors may also be run separately overhead.

In wiring coming in underground, called a **service lateral**, the neutral is always insulated. The three conductors at the service entrance are typically multistranded aluminum and of a larger gauge than is typically found in branch circuits. The voltage between the hot conductors is a 240-V sine wave **FIGURE 7-5**.

Three-Phase System

In a three-phase system, electric power is provided by three current-carrying (hot) conductors and one

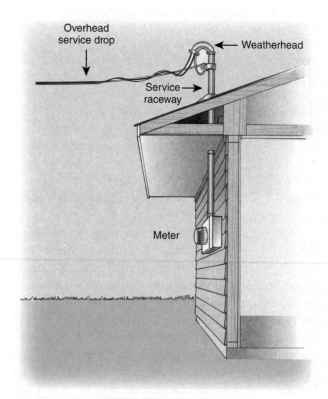

FIGURE 7-4 Triplex overhead service drop.

neutral. The sine wave of the three phases is out of phase, such that when one conductor has its highest voltage potential, the other two conductors are at a different position in the sine wave. This phase shift occurs in intervals of 120 electrical degrees. Thus, each successive phase's peak voltage is reached 120 degrees from the peak voltage of the previous phase **FIGURE 7-6**.

Three-phase systems are frequently found in industrial and large commercial buildings, as well as in large multifamily dwellings. The AC voltage between any of the hot (L1/L2/L3) conductors can be 480 V, 240 V, or 208 V. The voltage between one of the hot conductors and the neutral can be 277 V, 240 V, 208 V, or 120 V (these are RMS values; the values shown in Figure 7-6 are peak values). A common configuration in large commercial and industrial occupancies is 480/277 V (see Figure 7-6 for service connections and waveforms), where 277 V is used for lighting and 480 V is used for large equipment. Transformers are used to step down (or step up) voltages to meet the customer's needs.

In large buildings, there may be more than one service entrance for electrical power. In some buildings with very high electrical demands, the service entrance may have high voltage (e.g., 4000 V) with transformers inside the occupancy then reducing, or stepping down, the voltage.

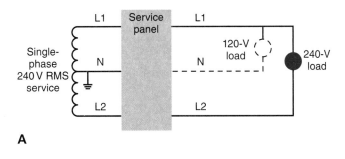

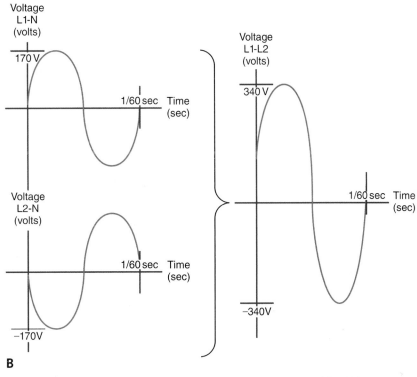

FIGURE 7-5 Single-phase AC sine wave: 240 V RMS, residential use.
A. Service entrance. **B.** Waveforms.

Building Electrical Systems

This section describes the components of a building's electrical system. The systems described here apply to residential and small commercial buildings. Larger buildings may have more complex systems.

Service Entrance (Meter and Base)

The **service entrance** is the point where the electrical service enters a building. It is often the transition point between the utility-provided power and the private owner's electrical distribution service. In most cases, there is no overcurrent protection between the utility transformer and the service entrance. Although a fuse can typically be found on the primary (high-voltage) side of the transformer, its purpose is to protect the utility company's equipment and wiring. Without overcurrent protection at the service entrance, any fault that

occurs can result in sustained high-energy faulting, possibly leading to significant damage or destruction of the service entrance equipment and cabling.

A service entrance consists of the following components:

- **Weatherhead:** The point where service entrance cables connect to the structure, which is designed to keep water out of the conduit that carries the conductors. An underground service does not have a weatherhead because the conductors enter from underground directly into the main panel.

- **Service mast:** The conduit between the weatherhead and meter base.

- **Meter base:** The enclosure that contains the connections for the utility's electrical meter and for the service entrance conductors.

- **Meter:** A watt-hour meter that plugs into the meter base to measure the flow of electricity.

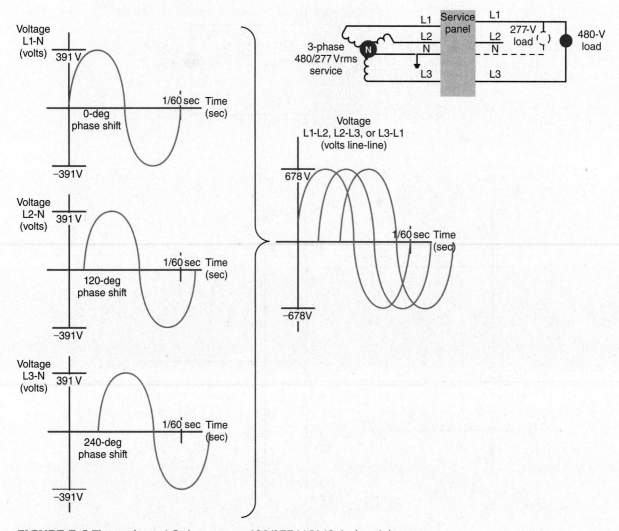

FIGURE 7-6 Three-phase AC sine waves: 480/277 V RMS, industrial use.

The **service equipment** is most often located close to where the service entrance cables enter the structure. The NEC does not specify a maximum distance from the location of the main disconnects to the point where the service cables enter the structure, but it is recommended to be as short as possible.

The service equipment includes the main service disconnect and overcurrent protection devices (fuses and circuit breakers), which are housed in electrical panels near the service entrance. The **main disconnect** provides the mechanism for shutting off the power and provides protection using overcurrent protection devices. There may be more than one main disconnect in some electrical services, or there may be none.

The main electrical distribution panelboard also distributes the power to different locations inside the structure via branch circuits **FIGURE 7-7**. This panel is sometimes referred to by different names, such as the circuit breaker panel, the service panel, the distribution center, or the load center. If the main disconnect is included, it is called the main panelboard. If a separate panelboard is used, it is called a subpanelboard. In larger structures, it is common to find the main panelboard feeding several smaller panelboards.

Grounding

Grounding is the mechanism for making an electrical connection between the system and the earth (ground). A solid ground connection provides a safe pathway for the electricity to follow if a fault occurs. Without a ground connection, the energy from a fault may flow through an undesirable path (such as a human body). When an energized conductor contacts a grounded component, such as the metal case of an appliance or water pipes, unimpeded electrical current flows to ground (fault current), and the overcurrent protection devices should open, causing the

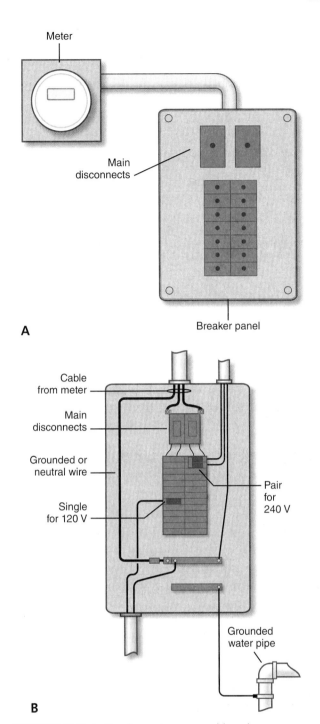

Meter

Main disconnects

Breaker panel

A

Cable from meter

Main disconnects

Grounded or neutral wire

Single for 120 V

Pair for 240 V

Grounded water pipe

B

FIGURE 7-7 A. Service entrance and breakers. **B.** Main electrical panel diagram.

when, for instance, someone touches a conduit that is not grounded. Additionally, the intended ground path could have moved from its original location because of the earth settling, a conductor having broken or come loose, or a poor ground connection, among other possibilities.

Current NEC requirements specify two alternative methods of creating a ground at the service entrance:

- Connecting the panelboard to a bare-metal cold-water pipe that extends at least 10 ft (3 m) into the soil
- Connecting the panelboard to a grounding electrode that may include a galvanized steel rod or pipe and/or a copper rod that is at least 8 ft (2 m) long that must be driven into the soil

In either case, there must be a secure bond between the panelboard and the electrode. This is often accomplished by connecting the grounding block in the service panelboard to the grounding rod or pipe with a copper or aluminum conductor and using proper connecting clamps to connect the conductors to the ground. Also, a minimum of two grounding electrodes must be connected to create an effective ground.

Because electricity flows from high potential (voltage) to ground and takes all available paths that allow the current flow, it is important to have a good grounding system to draw any fault current away from unintended paths. Although these currents may not be high enough to start a fire, the potential for personal injury still exists. In addition, all "containment" elements of the distributed electrical service, such as junction, switch, and outlet boxes, must be connected to the grounding conductor distributed with the current-carrying conductors or by means of a conduit. These boxes then act to shield combustible materials from possible arcing events at these connection points of connection. All connections and splices must be made within the box.

Overcurrent Protection

Overcurrent protection is provided by a device that stops the flow of electricity when an abnormally large amount of current is detected, thereby preventing damage to conductors and devices on the circuit involved. Overcurrent protection devices can be either resettable, such as a circuit breaker, or nonresettable, such as a fuse. Most modern overcurrent protection devices are resettable, meaning that they can be reused to protect the circuit after they have been tripped (activated).

Once a device has been tripped, the cause must be identified, whether it was a ground fault, overload, or

electricity to stop flowing through the circuit. By contrast, if the ground is not in place when the charged conductor faults to the metal case of an appliance or a water pipe, this metal component could become electrically energized.

If a person (electrically conductive and connected to the ground) comes into contact with an ungrounded circuit, they may become the path to carry the current to ground. This scenario may occur short circuit. Circuit breakers should not be reset until the cause for the activation is found and it is determined that no damage was done to the electrical system. Note that it is possible for some devices, such as ground-fault circuit interrupters (GFCIs), to trip due to age without the existence of a fault. GFCIs are discussed in more detail later in this section.

Overcurrent protective devices have two current ratings: a regular current rating and an interrupting current rating. The **regular current rating** is the maximum amount of current that the overcurrent protection device allows the circuit to carry under normal conditions. If the current exceeds this value, the device opens, stopping the flow of current. Common regular current ratings found in household use are 15 A, 20 A, 30 A, and 50 A. The time required for the device to open depends on the duration of the current, the magnitude of the current, and the ambient temperature. It is possible for an overcurrent protection device to carry a small amount of overcurrent for an extended time. It is also possible, in high ambient temperatures, for an overcurrent device to trip at a slightly lower current threshold (rating) over an extended period of time.

The **interrupting current rating** is the maximum amount of current the device is capable of interrupting. This rating can be so high as to seem impossible—for example, 3000 amps or more. It is essential, however, that the equipment be able to interrupt any possible excess current, as the result otherwise can be catastrophic.

Several types of circuit protection devices may be used, with the most common being fuses and circuit breakers. NFPA 921, Section 9.6.2.1, defines fuses as "nonmechanical devices with a fusible element in a small enclosure." **TABLE 7-3** provides examples of various fuses.

NFPA 921, Section 9.6.3.1, defines a circuit breaker as "a switch that opens either automatically with overcurrent or manually by pushing a handle." Circuit breakers open when exposed to heat or overcurrent that exceeds their rated trip current. To be properly sized, a circuit breaker should not exceed the ampacity of the conductors in the circuit **FIGURE 7-8**.

TABLE 7-3 Types of Nonresettable Circuit Protection Devices (Fuses)

Type	Application	Amps	Notes
Plug (also referred to as Edison)	■ Older application; not used in new installations	30 A or less	■ Possible to overfuse (e.g., put a 30-A fuse in a circuit that would require a 15-A fuse) because fuses with different ratings are all interchangeable ■ Possible to bypass the fuse by using a penny to complete the circuit
Type S	■ New installations		■ Designed to make bypassing more difficult ■ Can be used in older installations (Edison bases) with an adapter
Time-delay fuses	■ Allow short-duration overcurrent ■ May allow up to six times the normal current		
Cartridge fuses	■ Fast action or time delay ■ Single use or replaceable element ■ Can be used for high-current loads (e.g., water heaters, ranges) ■ Greater than 100-A fuses are found in commercial or industrial occupancies	Circuits greater than 30 A	

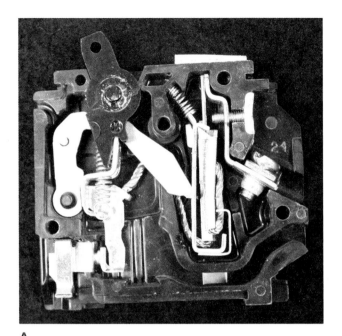

A

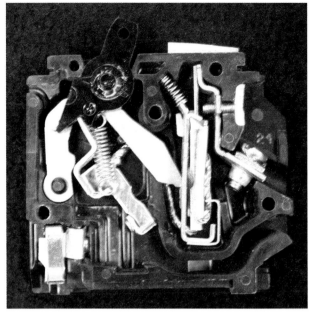

B

FIGURE 7-8 A. A 15-A residential-type circuit breaker in the closed (ON) position. **B.** A 15-A residential-type circuit breaker in the open (OFF) position.

© Jones & Bartlett Learning. Photographed by Jessica Elias.

Whereas it may take several minutes for the overcurrent protection device to activate for a small increase in current flow over its rating, it reacts more quickly to a large current flow. This phenomenon is described by the **time–current curve** (also known as the characteristic trip curve) for breakers and fuses, which defines the amount of time required for a device to interrupt at a specific level of current. The more current there is, the faster the device will trip. For example, the *American Electrician's Handbook* states that a typical 100-A breaker will take about 30 minutes to trip at 135 A and 10 seconds to trip at 500 A. Time–current curve information is available from the device's manufacturer.

In addition to standard circuit breakers, two other circuit protection devices are available: **ground-fault circuit interrupters (GFCIs)** and **arc-fault circuit interrupters (AFCIs)**. GFCI circuit breakers respond to overcurrent conditions like normal breakers, but also look for an imbalance of the current between the hot and neutral conductors. More specifically, if the electrical current leaving the GFCI is greater than the current returning to the GFCI, current must be flowing on some other path (a ground fault). Once this condition is detected, the GFCI breaker trips, interrupting the current flow. This helps to prevent electric shocks and electrocutions. Note that GFCI outlets and portable GFCIs (including those within

power plugs) provide only ground-fault protection, and not the overcurrent protection offered by GFCI circuit breakers.

AFCI circuit breakers monitor the circuit for abnormal conditions/waveforms that are indicative of arcing and aim to prevent electrical fires. Arcing at loads or in circuits can cause fires because the electrical arc can reach temperatures upward of 35,000°F (19,426°C). AFCIs are able to distinguish between normal circuit current flow and arcing current flow, and open when they detect an arcing condition. Arcing as a result of opening and closing switches will not cause an AFCI to operate.

You should consult the appropriate version of the NEC to determine the locations where AFCI protection was required in residential installations at the time of the building's construction or when electrical work was done. These requirements vary by NEC version and year. Depending on the applicable code, AFCI protection may have been required for such areas as kitchens, bedrooms, laundry areas, and other locations. **TABLE 7-4** lists types of resettable circuit protection devices.

Circuit Breaker Panels

Heat or electrical faulting can damage internal components of the circuit breaker panel. If they are made

TABLE 7-4 Types of Resettable Circuit Protection Devices

Type	Amps	Notes
Main breakers	100 A to 200 A (residential)	■ Interrupt the circuit at the main disconnect panel ■ Pair of breakers ■ Handles fastened together
Branch circuit breakers	15 A or 20 A (general lighting and receptacle circuits); 30 A, 40 A, or 50 A (e.g., ranges, water heaters)	■ Switch ■ Used on the branch circuits that distribute electricity throughout the structure ■ Rating generally imprinted on the handle ■ Cannot be manually tripped (can be placed in the "off" position manually); can be tripped by physical contact or external heat ■ Will trip even if the handle is locked in the "on" position ■ May trip to the "off" position ■ Have a body made of phenolic plastic ■ Body does not melt or sustain combustion ■ Can be destroyed by fire impingement ■ Do not rely on visual inspection to determine the status of the breaker

Ground-fault circuit interrupters (GFCIs)		■ Monitor the electrical flow for abnormal conditions
		■ Used in locations where a person might become grounded while using appliances:
		  a. Bathroom
		  b. Bedroom
		  c. Kitchen
		  d. Patio/garage

of standard plastic, insulation and molded circuit breaker components can melt and allow energized conductors or bus bars to contact each other or grounded components. The service entrance cables enter the panelboard and are connected to bus bars. Because these components are usually unprotected, significant arcing and damage, including blowouts in the panelboard wall, can occur if the separations and insulation are not maintained. Insulated conductors in a metal conduit, if overheated to the point of insulation breakdown, can also arc and cause a blowout in the conduit. Analysis of the circuits inside and outside an arc-damaged panel is needed to establish whether arcing inside the panel was the source of ignition or was the result of heat impingement on the panel by the fire.

Branch Circuits

The circuits that distribute electricity from the panelboard to the rest of the building are called **branch circuits**. Each branch circuit should have its own overcurrent protection device, either a fuse or a circuit breaker. These circuits get their name from the fact that they have a single connection point in the service panel, with conductors then fanning out to more than one load in the building, much like the branches of a tree. **TABLE 7-5** lists the conductors found in residential branch circuits.

Conductors

Many sizes of conductors are available. The AWG number or gauge refers to the wire diameter. The larger the gauge, the smaller the wire diameter is.

TABLE 7-5 Conductors in a Typical Two-Wire Plus Ground Branch Circuit

Type of Conductor	Function	Grounding Protection	Description	Notes
Ungrounded	Hot	Attached to the protective device	Carries the current to the load	All circuits have this conductor.
Grounded	Neutral	Attached to the grounding block	Returns the current to the source	All circuits have this conductor.
Grounding	Ground	Attached to the ground	Provides protection and allows the current to go to ground	This is not required to make a circuit function. Grounding may be provided through a metallic conduit instead of a separate conductor.

For instance, 15-A circuits should use no less than 14 AWG wire (12 AWG if aluminum), whereas 20-A circuits should use no less than 12 AWG wire (10 AWG if aluminum). Higher-power circuits use 8 and 6 AWG wire. The AWG describes the conductor's size, which determines its ampacity. The type of insulation, bundling of conductors, method of cable conveyance (e.g., proximity to insulating materials, embedded in walls), and ambient heat all affect the required conductor size.

Conductors may be larger than required but cannot be smaller than required. Undersized conductors have more resistance to current flow, so they can generate more heat when their ampacity is exceeded. Oversized circuit protection devices (sometimes called **overfusing**) create a dangerous condition when the circuit protection (fuse or circuit breaker) rating significantly exceeds the ampacity of the conductor, leading to a condition where increased heat can occur in the conductors.

The three most commonly used conductors are made of copper, aluminum, and copper-clad aluminum.

Copper Conductors

Pure copper is used in copper conductors. Its melting temperature is approximately 1980°F (1082°C). Surface melting of a copper conductor may occur below this temperature, however, and lead to the formation of copper oxide on the surface; the core may remain intact (unmelted) in such a case. When fire attacks a copper conductor, the conductor is heated over a larger area along its length (as opposed to an electrical arcing event, which is usually confined to a smaller, localized area on the conductor). When fire attacks a copper conductor from the exterior, you will observe surface melting and formation of copper oxide. Although it is not always easy to determine whether a conductor has melted from the heat of the fire or melted as a function of electrical arcing, there are numerous characteristics (discussed later) to look for.

Copper conductors can have a variety of appearances postfire. Some conductors lose their insulation and oxidize, leaving the surface blackened with cupric oxide. Other conductor surfaces may have no oxide or may be covered with reddish cuprous oxide. Other chemicals may also be present that can affect the coloration and condition of the conductors.

A copper conductor's melting temperature may be affected by "alloying," or **eutectic melting**, which occurs when other metals, such as aluminum or zinc, melt and come into contact with the copper and change its melting temperature, usually lowering it. This frequently happens during a fire when metals drip, splash onto, or come into contact with the copper conductors. Fire investigators can often, but not always, determine that this has occurred by noting a color change (brass or silver) on the melted copper. This change is not a cause of the fire, but rather a result of it.

Aluminum Conductors

Aluminum in its pure form is also used in conductors. Though primarily used as service entrance conductors on new installations, aluminum conductors may also be found in branch circuits in older homes and other buildings. In addition, circuits that supply power to larger appliances (e.g., water heaters, ranges, ovens, air conditioners) are often made of aluminum.

The conductivity of aluminum is lower than that of copper, so an aluminum conductor must be larger than a copper wire to carry the same current load (10 AWG aluminum = 12 AWG copper). The melting temperature of aluminum is approximately 1220°F (660°C), which is well below the temperatures found in a typical building fire. The heat of the fire often melts aluminum conductors entirely, rarely leaving evidence of any arc melting that may have preceded the conductor's destruction.

Copper-Clad Aluminum

A less common type of conductor is copper-clad aluminum. As the name implies, the aluminum conductor is encapsulated in copper. Its melting characteristics are similar to those of aluminum.

Insulation

Insulation on the conductors prevents faults or leakage via unwanted current paths. Insulation consists of any material that can be applied readily to conductors, does not conduct electricity, and retains its properties for an extended time at elevated temperatures. Air can be an insulator, even for high voltages, when the conductors are kept separate. An arc can occur when dust, pollution, or products of combustion contaminate air or the insulation. The conductor's applications and insulation ratings are marked on its insulation to identify its type and its temperature rating. Additional information of the conductor's construction and applications can be found in Table 310.104 (a) of the NEC.

Polyvinyl chloride (PVC), a commonly used insulator, can become brittle with age or heat and may give off hydrogen chloride in a fire. Rubber was frequently used as an insulating material until the 1950s. It can become brittle with age, at which point it can easily be broken off the conductor. Rubber chars and leaves an ash when exposed to fire. Other materials used as conductors include polyethylene and related polyolefins, which are often used on higher-current circuits in residential applications. Nylon jackets may be used to increase thermal stability.

The color of the conductor insulation typically indicates its function: Green indicates an equipment grounding conductor; white or light gray indicates a grounded conductor (neutral); and any color except green, white, or light gray indicates an ungrounded or hot conductor. 120-V circuits, the hot conductor is often black. In 240-V circuits with nonmetallic sheathed cable, the two hot conductors are usually black and red.

Investigators need to use caution when relying on colors of insulation as a guide. In certain applications,

FIGURE 7-9 Noncontact voltage detector.
© blitzstock/Shutterstock.

white or gray conductors could be hot conductors in three- or four-way switches and, therefore, could be energized. These conductors should be marked to designate this usage but may not be. Further, a building or circuit could have been wired improperly, such that the investigator might find a conductor of the incorrect color. It is wise to consider all conductors as being energized until proven otherwise. For this reason, it is recommended that fire investigators carry a **noncontact voltage tester FIGURE 7-9**. These devices should always be applied to a known live circuit before use to ensure that the tester is working properly, and the battery is good.

Outlets and Special Fixtures/Devices

Circuits will terminate at or connect to switches, receptacles, or appliances. Switches are used to control the flow of current to receptacles and/or appliances and will have either screw terminals, push-in terminals ("backstabbed" receptacles), or both. Permanent lighting fixtures are typically attached to electrical boxes in the wall or ceiling with a wall switch or dimmer to control the lighting circuit. Switches should be located on the hot side of lighting circuits, not on the neutral side.

Push-in terminal (backstabbed) receptacles are those in which the conductor is not attached to the screw terminal but is pushed in the back of the receptacle. Because of their small surface contact area, some push-in styles have experienced problems, such as resistance heating. Problems have also occurred in older backstabbed receptacles wired with 12 AWG conductors. Due to their increased stiffness compared to 14 AWG conductors, the 12 AWG conductors place increased strain on the connection points, which could lead to a poor connection. As previously discussed, terminals or connection points are the most likely areas for resistance-type heating to occur because of faulty connections.

Two types of 120-V receptacles are available: duplex receptacles on 15-A and 20-A circuits (two outlets per receptacle) and single receptacles on 30-A or greater circuits. Receptacles must now be polarized, so that the larger neutral blade of the plug can be inserted only into the larger neutral opening on the receptacle. Receptacles also must be grounded in new installations. Receptacles may have screw terminals, push-in terminals, or both. Hot conductors should be attached to the brass screw; neutral conductors should be attached to the colorless screw; and the green screw is for the grounding wire.

Special fixtures for lighting are usually connected to junction boxes in the wall or ceiling. Sometimes these fixtures contain thermal cutout (TCO) devices, which help prevent overheating of the device. Often these types of fixtures are insulation-contact (IC) rated; thus, they are known as IC fixtures.

GFCI outlets (receptacles) are used in bathrooms, kitchens, or other wet locations. These outlets monitor the amount of current going in and out and trip if the current differential exceeds 5 mA (0.005 A). Ideally, GFCI outlets are intended to interrupt current flow before the fault current reaches a point that could result in bodily injury. They do not provide the overcurrent (short circuit) protection that is provided by a circuit breaker or fuse. Explosion-proof outlets and fixtures should be used in all installations that are exposed to explosive atmospheres. Part III of Article 501 of the NEC discusses the equipment used for hazardous locations.

Portable GFCIs may sometimes be found where ground-fault protection is required or desired but not supplied within a building. Like GFCI outlets, these devices do not provide overcurrent (short circuit) protection.

Ignition by Electrical Energy and Electrical Failure Modes

Ignition from an electrical source can occur only when the following conditions are present:

- Wiring/equipment must be electrically energized (power must be on).
- Sufficient heat must be produced; that is, enough thermal energy is produced to ignite the first fuel (see Chapter 14, *Cause Determination*).
- The combustible material must be a material that can be ignited by the electrical energy produced by the electrical failure.
- The heat source and the combustible fuel must be close enough (sufficient thermal energy transfer) for a long-enough period to allow the combustible material to generate combustible vapors. Generally speaking, an arc does not produce enough energy to ignite most ordinary, higher-density combustibles.

Some conditions that may generate sufficient thermal energy include the following:

- Resistance heating: An electrical event that occurs when electrical current flows through something that provides high resistance to current flow, such as a heating element (intentional) or a resistive connection (unintentional).
- Ground fault: An electrical event in which equipment failure or poor grounding facilitates excessive current flow on an unintended path, as opposed to the intended circuit path.
- Parting arcs: Arcing that occurs when wires or switch contacts are separated, and current flow is interrupted. This will create a momentary arc of very short duration, with accompanying sparks.
- Excessive current flow: A condition in which abnormally high amounts of electrical current flow through an appliance or conductor, facilitating overheating (e.g., a 14 AWG wire carrying 50 A instead of 8 A).
- Lightbulb overlamping, close proximity of combustibles to heaters, or cooking equipment.
- Heating in a confined space, such as a heat-producing device that is normally exposed to ambient or cooling air but is somehow trapped under blankets or bunched up. Another example of confined heating is a heat-producing object such as a motor or ballast that is covered in dust, insulation, or other materials.

Electrical current flows from the source to the load, and back, via any path that allows this flow, even if it is not the intended path. When electricity flows through most materials, including wires and conductors, heat is produced because of the inherent resistance of the material it passes through. Under normal conditions, this heat is dissipated if the current is within the safety margins calculated for the size of the conductors (ampacity). Sometimes this heating is desirable, such as in the heating element on a stove. At other times it is not desirable, such as current flowing through an undersized conductor.

Heat-Producing Devices

Some devices are designed to generate heat. When this is the case, the design of the devices also allows for the safe dissipation of the heat. However, in cases where the heat is not allowed to dissipate safely and where a combustible fuel is located nearby, ignition may be possible. For example, if a cloth or piece of clothing is placed over a halogen lamp, the heat that is absorbed by the material may begin to produce combustible vapors, which may then be ignited if the heat is high enough (autoignition) or by arcing nearby (piloted ignition). An incandescent lamp, even one that has a small wattage, can reach temperatures that will ignite combustible materials. This heating is increased by immersing the lamp in cellulose or wrapping it in clothing. In some cases, a heat-producing device may deliberately be used to cause a fire, either by sabotaging its safety features or by placing it in proximity to a combustible fuel load.

If a heat-producing device is being considered as the cause of a fire, the fire investigator should determine the temperatures produced by the device to see if they are sufficient to cause the ignition of nearby materials. In other cases, the device may malfunction, allowing the heating elements to become hotter than they are designed to be, causing a fire. An example could be a deep-fat fryer whose controls fail, allowing the grease to heat up to its ignition temperature.

Poor Connections and Resistance Heating

Poor connections can allow heating and the formation of an oxide at the point of the poor connection. A poor connection can occur at the time of installation or can develop over time with movement of the wires or equipment. It may appear at screw or backstabbed connections at outlet receptacles, or at other connections. The oxide that forms at the connection increases the resistance of the connection, which contributes to the production of heat—a phenomenon called **resistance heating**. The circuit often remains functional even though resistance is increased. A glowing hot spot can develop, and if combustible fuels are in close proximity to this spot, they may potentially ignite.

Resistance heating can happen at a poor connection that is remote from the load that is drawing current through the circuit. For example, if outlets are connected in series such that current passes through the connections on each outlet as it moves through the circuit, then an appliance drawing current from a downstream outlet could result in resistance heating at a poor connection in an upstream outlet, even if the latter does not have any devices connected to it.

Aluminum wiring, which is often found in residences built in the 1960s and 1970s, expands at a different rate than copper. If connected to incorrectly selected equipment, such as receptacles designed for copper wiring, it can potentially come loose through expansion and contraction. This could create a high-resistance connection.

With currents of 15 to 20 A, for example, power of up to 40 W can be produced at a poor connection, with temperatures exceeding 1000°F (538°C). This may not seem significant, but the heat is very focused, rather than spread over an area.

Structure wiring connections are required to be made within an electrical box or appliance enclosure. This setup can protect a point of resistance heating from being in proximity with combustibles, reducing the chances of ignition.

Overcurrent

Overcurrent is current flow in excess of the rated capacity of equipment or of the ampacity of a conductor. Overcurrent happens as the result of a fault, such as a short circuit or ground fault, or an *overload*, which is the operation of equipment in excess of its normal rating or in excess of the ampacity of a conductor (e.g., an excessive number of appliances or other equipment drawing current through a circuit). Circuits can handle these types of overcurrents for short durations with little or no negative effects.

A **short circuit** is an "abnormal connection of low resistance between normal circuit conductors where the resistance . . . normally [would be] much greater," according to NFPA 921. For example, if a direct connection is made between an energized "hot" conductor in a nonmetallic cable and the grounded (neutral) conductor, a short circuit can result. The electricity bypasses the equipment or load on the circuit that would have otherwise provided resistance, so the resistance of the circuit decreases dramatically; this, in turn, causes the current in the circuit to increase to an abnormally high level. This high current would normally be expected to trip the overcurrent protection, so a short circuit should last for only a very short time. If a short circuit is being considered as a potential hypothesis for fire cause, the investigator should verify whether the overcurrent protection opened the circuit.

A short circuit can produce heat and arcing at the point of contact, which may cause the metal to melt and open the connection. If it persists, the excessive current causes the conductors to heat up.

A **ground fault** involves an inadvertent connection of current-carrying equipment or wiring to ground, causing the current to run to ground instead of returning through the neutral cable. When the current runs to ground, resistance decreases, which in turn allows for excessive current. As in the case of the short circuit, the overcurrent protection may then trip. A ground fault can also occur if a fault within an appliance or equipment causes the appliance's case to become electrically charged. In this situation, a ground fault could occur if a person touches the case, producing electric shock.

Factors that influence the potential for fire include the magnitude and duration of the overload. Large, persistent overcurrents can cause the conductor to melt the insulation and can heat the surrounding material.

Undersized cords can produce overheating if used far beyond their capacity. A 1200-W appliance requires a 12 AWG cable connecting it to a 120-V outlet. If an 18 AWG cable is used instead of 12 AWG, the normal 10-A current draw is much greater than the 5-A rating of the 18 AWG wire. Because the breaker is sized at 20 A, the circuit will continue to function, allowing the undersized cord to increase in temperature. If the cord is bunched, coiled, or covered, it can cause an overheating condition, leading to ignition **FIGURE 7-10**.

Overloads that cause fires are uncommon in circuits with properly sized and correctly operating overcurrent protection. However, an overload may happen when a severely undersized conductor is part of the circuit, as described earlier, or if the overcurrent protection is rated too high for the circuit it protects. For example, a 30-A circuit breaker is an improper choice to protect a circuit that has an ampacity of only 15 A, because it would permit current to flow in excess of the ampacity of the circuit without tripping the breaker.

The temperature of a heat-producing device, or the heat generated from an undersized conductor, can degrade the insulation between conductors. This situation is aggravated if the conductors are not cooled by free air—for example, if they are covered with blankets. If the overcurrent is allowed to continue, carbonization of insulation occurs, creating a path for current to flow. When heated to about 392°F (200°C), PVC insulation begins to pyrolyze, and combustible vapors escape. If an arc then occurs because of insulation breakdown in the high-heat environment, a fire can result if suitable combustible material is nearby.

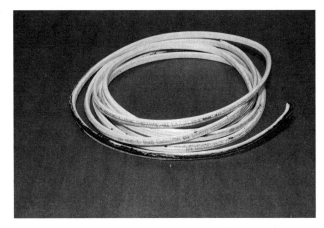

FIGURE 7-10 Overload damage to wiring.
Courtesy of Cameron Novak.

FIGURE 7-11 Overcurrent protection bypassed.
Courtesy of OnSite Engineering & Forensic Services, Inc.

In **FIGURE 7-11**, the overcurrent protection of an air-conditioner fuse disconnect has been bypassed by two copper pipes. This could allow a dangerous overload condition to develop in the event of a fault or locked rotor condition in the air conditioner's compressor motor.

Contaminated water or corrosion at connection points or near bare conductors can cause or contribute to a carbon tracking process. This weakness in the normal insulation properties of the device or conductors allows leakage current to flow. With time, heat, or vibration, this weakness can develop into a short circuit to a nearby ground connection.

Arcs

An **arc** is a "high-temperature discharge . . . across a gap or through a semiconductive medium such as charred insulation," according to NFPA 921. The temperature of an arc can be several thousand degrees, depending on the available fault current, the voltage drop, the length of the circuit, and the materials involved. For the spontaneous initiation of an arc to occur across an air gap, a relatively large voltage potential must be present. In 120/240 V systems, arcs do not occur spontaneously under normal conditions. However, a parting arc can occur when conductors separate while current is flowing, even in 120/240 V systems. This may happen under normal conditions (e.g., when a switch is activated or a plug is pulled from an outlet) or abnormal ones (e.g., when an energized conductor contacts a grounded conductor or other item with near-zero circuit resistance, causing a surge of current that melts the points of contact and creates a gap). Because of their typically short duration, arcs may not be competent ignition sources unless ignitable vapors or gases arc also present.

Solid fuels are much less likely to ignite from an arc due to its short duration. Arcing is brief and localized because of the cycling of AC voltage from the positive peak value to the negative peak value, going through zero in between these peaks. Adjacent combustible fuel that has a low surface-to-mass ratio and that cannot be heated sufficiently to sustain combustion (e.g., dimensional lumber) is unlikely to ignite. Fuels with a high surface-to-mass ratio, such as a gas or vapor, can sustain combustion, but must also have an acceptable ratio of air to gas to do so. This ratio must fall within a range whose endpoints are defined as the lower explosive limit (LEL) and the upper explosive limit (UEL). Fuels that may ignite from arcing include cotton batting, dust, tissue paper, volatile gases and vapors, and lint. Short-duration arcs often have insufficient energy available to ignite solid fuels under normal conditions. A conductor-to-conductor fault—for example, a metal knife or staple piercing an energized extension cord—will short the branch circuit and open the circuit protection device.

High-Voltage Arcs

High voltages exist at the transformer connection near the service entrance of the building and can accidentally be impressed on conductive elements that enter or are part of the building. Lightning can also cause high-voltage and high-current surges. These events can lead to voltage overloads on the building wiring, which is designed to accommodate a few hundred volts, and will cause insulation breakdown with arcing when the voltage of the surge exceeds the insulation's breakdown strength. Because voltages and high-current surges from lightning strikes are so high, arcs can jump at numerous places, resulting in various kinds of damage and igniting several kinds of combustibles.

Parting Arcs

Parting arcs are brief discharges created when the electrical path is opened while current is flowing. Examples of events that may produce parting arcs include opening a switch, pulling a plug, and running a motor that uses brushes, which can produce continuous arcing. Pulling a plug while current is flowing in the conductors often causes a visible arc. Once the conductor separates, the heating stops because the current flow has stopped. Contacts with surfaces that open and close are exposed to this kind of arc and often show pitting and roughness.

Parting arcs can also occur within temperature-control devices or thermostats. These devices are often placed in the housings of heat-producing appliances. They are designed to limit the temperature of a device by opening at a preset temperature, stopping the flow of current, and allowing the device to cool down to another preset temperature, at which point the thermostat resets and closes the circuit. Note that a thermostat does not control the actual temperature of the device, but rather limits the time during which the heat-producing element is on. Such temperature-control devices are designed to last for many thousands of cycles of operation, but can fail at the extremes of operation at elevated temperatures or if the protected component is confined by insulating dust. When this happens, arcing is accompanied by sparking as more material is removed from the contact surfaces. Additionally, the metal contacts can fatigue to the point of breakage. At this point, the cutoff action may not occur, or the device may simply heat up or increase the series resistance of a motor circuit, allowing it to stall or not start. Any of these failure modes has the potential to cause a fire.

Short circuits or ground faults may cause metals to melt at the initial point of contact, leading to the ejection of sparks. An arc occurs as the metals part, and the arc is quenched immediately. Some power supplies and motor drives with large capacitors will arc when plugged in because of the high in-rush current.

Arc Tracking

Arc tracking has been observed in high-voltage systems and has been demonstrated experimentally in 120/240 V systems. It may occur on surfaces of noncombustible materials and lead to the development of a path of electrical current that starts small and grows over time. This phenomenon begins when the surfaces become contaminated with salts, conductive dusts, and liquids; the contamination causes degradation of the base material. Arc tracking can also occur when a surface is contaminated with water-containing materials, such as dirt, dusts, salts, or mineral deposits. Tap water may contain these types of contaminants. Often the arcing will create sufficient heat to dry the wet path and stop the flow of current. If the water is replenished, thereby reestablishing the current, deposits of metals or corrosion can form. Arc tracking is more pronounced in DC systems.

In low-voltage systems with feedback temperature control, circuit board contamination can cause erroneous signals to be read in sensitive analog circuits. These circuits are sometimes coated with moisture-resistant coatings to prevent contamination from reaching printed circuit (PC) board surfaces. If no such coating is present, contamination can disrupt control sequences in heating appliances and lead to malfunction, even fires. An erroneous **thermistor** reading can drive a heating element to an over-temperature situation if contamination occurs in the sensitive portions of the measurement circuit. (Thermistors are sensors that measure temperature through changes in their resistance.) Even though the thermistor itself may be protected, the measurement circuit on the PC board to which it connects can be contaminated by moisture if not protected.

Sparks

Sparks are particles of solid material that emit radiant energy. This radiant energy may come from the particle's temperature or be produced through the process of combustion on its surface. When a high-current ground fault occurs, hundreds or thousands of amperes may be flowing. This energy is sufficient to melt the metal and throw out sparks. Protective devices generally open almost immediately, so the event should happen only once. If the metal involved is copper or steel, the particles will cool as they fly through the air. If the metal involved is aluminum, the particles may burn as they fly through the air and are more likely to ignite nearby combustibles.

In general, sparks are not competent ignition sources for ordinary combustibles, although they can ignite fine fuels. The size of the spark particle is important because it determines the total energy content. Arcing in service entrance conductors can produce more and larger sparks than branch circuits do.

High-Resistance Faults

In a **high-resistance fault**, the current flow through the fault is not sufficient to trip the protective devices, at least in the initial stage, so the current keeps flowing. These types of faults can develop more heat, depending on how long the fault persists, and may be capable of igniting combustibles. It is difficult to find evidence of a high-resistance fault after a fire because the ensuing fire and heat damage will likely mask the evidence of the original failure. Arcing through charred insulation of conductors is a type of high-resistance fault that may not open the circuit protection. Multiple events with sustained arcing can occur under these conditions.

Floating Neutral

Certain neutral connection problems can occur that create electrical hazards in a building. For example, in a 120/240 V system, if the neutral conductor is disconnected or has a poor connection, a floating or an open neutral condition can lead to unbalanced voltages at the loads. When this happens, each phase could have a current ranging from 0 to 240 V across it. There will still be 240 V between the two phases, but instead of the voltages of the two phases being fixed at 120 V to neutral each, they may vary (though the voltages will always add up to 240 V). This can cause damage to appliances and sensitive circuits in the building and pose a shock hazard to occupants. Signs of this problem include lamps burning too brightly or dimly or appliances that surge, overheat, or simply fail. A floating neutral condition does not result from improper grounding of the building's electrical service.

Interpreting Damage to Electrical Systems

Because of the high potential energy of electrical circuits, significant observable damage will occur from faulting or arcing elements and components in the circuits. The ability to differentiate this damage from that due to the heat of fire is a fundamental skill for fire investigators to develop; it can often make the difference between success and failure in determining whether electricity played a role in causing the fire.

Investigation Steps

The fire investigator should review the full electrical system in any building fire in which an electrical system is found and energized, and should document and sketch the relative locations of all circuit components and wiring that may be involved in a fire event. The status of the circuit breakers or fuses within a building is important to record. The investigator should take care not to engage in the spoliation of evidence (see Chapter 5, *Legal Considerations for Fire Investigators*), and should not manipulate electrical switches, knobs, or breakers. Ideally, electrical components should be left in place until appropriate coordination has been accomplished with all interested parties. If it is necessary to collect electrical components as evidence, the investigator should follow procedures set forth in Chapter 5 and in Chapter 9, *Identification, Collection, and Preservation of Physical Evidence*.

Short Circuit and Ground-Fault Parting Arcs

A short circuit occurs when current passes from an energized conductor to an object of a different potential, resulting in a surge of current. Melting occurs at the point of contact, creating a small void and a parting arc. Melting occurs only at the point of contact; adjacent or surrounding areas are left unmelted. When examining a point of contact, it is important to observe whether melting is localized to the point of contact or extends to adjacent areas. If adjacent areas are melted, you can conclude that the melting was caused by the fire. However, keep in mind that the point of contact may have experienced localized arc melting, which was then masked by subsequent fire.

It may be difficult to identify the area of the initial parting arc if subsequent melting occurred. Under microscopic examination, the surface of an arced contact point typically appears to be notched or beaded. A corresponding point of damage on the other conductor confirms that the fault occurred between the two points of arc damage. For an arc to occur on insulated conductors, the insulation had to fail in the first place. A significant fault may cause the wires to melt and sever. This **arc severed** condition can break the wires into segments, which may drop and become lost. Thus, the fire investigator must determine how the insulation failed or was removed and how the conductors came into contact with each other.

Damage to stranded cords is more difficult to read and interpret than damage to solid conductors. Cords used in lamps and appliances may have several strands notched or severed, or all strands may be severed or fused together. Due to their small size, stranded conductors melt easily in a fire, and caution should be used in their analysis.

There are many ways to interpret the damage to electrical conductors, including arcing through char, parting arcing, and so forth. Examples of such damage are shown in **TABLE 7-6**.

Service-entrance conductors have no overcurrent protection, and several feet of conductor can be partly melted or destroyed if they are subjected to electrical activity. Arcing can continue on service entrance conductors until the power is cut by severing of wires. The primary side fuse of the feed transformer may blow if the secondary side current is high enough, usually several hundreds of amps. A 2-A fuse at a 12,000-V primary on a 35-KVA transformer, for example, will allow a sustained 200-A secondary side current without blowing. Conductors used at the service entrance connection are typically made of aluminum, which has higher resistivity than copper, so higher-gauge wire is used. Faulting at that point often results in temperatures high enough to melt aluminum, and significant melting may occur because of the high current.

Arcing Through Char

Insulation exposed to fire may char. As it chars, it may become more conductive, which will allow sporadic arcing between energized conductors. This arcing may leave surface melting at spots and can melt through the conductor, depending on the duration and repetition of the arcing. If the arcing persists, the conductors may become severed. Because the heat caused by arcing wires is very intense but confined to a very small area around the point of contact, it is called localized heating. Melting caused by nonlocalized heating has a different, more gradual, or globular appearance from beads formed by arcing through char **FIGURE 7-12**. If the melted conductors are contained in a conduit, holes can be melted in the conduit.

Often, multiple points of arcing will be apparent. Several inches of conductor may be destroyed. Ends of conductors may be severed and have beads on the ends; the beads may weld two conductors together. In a conduit, holes may be melted in the conduit at one point or along several inches. **TABLE 7-7** provides some general indicators to help the fire investigator determine whether the damage to the conductor is from the fire, arcing, or overload. This damage, by

TABLE 7-6 Types of Conductor Damage

Mode of Damage	Effects	Result	Cause of Fire?
Arcing through char of fire		Direct fire heating	No, always a result
Parting arcing		Heating at about 400°F (250°C) but no direct fire	Usually not
Overcurrent		Short circuit or failure in a device plus failure of overcurrent protection	Yes, but also may be a result of fire
Fire		Cable exposed to existing fire	N/A
Heating connection		Connection not tight	Yes
Mechanical		Scraping or gouging by something	No
Alloying		Melted aluminum on the wire	No

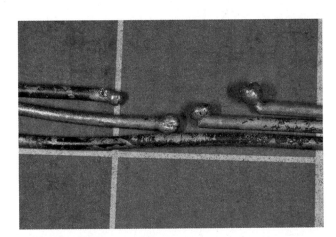

FIGURE 7-12 Beads formed at arc sites.

Courtesy of Cameron Novak.

TABLE 7-7 Potential Indicators on Conductors

Beads	Globules
Localized heating	Nonlocalized heating, such as overload or fire melting
Distinct line of demarcation	Extended area of damage without sharp lines of demarcation

itself, does not necessarily indicate whether the conductor was the cause of a fire.

Numerous studies have been conducted on the process of arcing through char. Studies conducted on energized, unloaded, nonmetallic-sheathed cables exposed to a radiant heat source yielded the following information:

- To form an arc bead on a cable within a period of investigative interest (i.e., minutes versus hours), the cables had to be exposed to heating conditions consistent with flashover, direct flame impingement, or exposure to the hot gas layer.

- The general appearance of arc beads is consistent over a wide range of heat fluxes. The characteristics of these arc beads are described in the "Melting by Electrical Arcing" section later in this chapter.

- A lack of oxygen, such as would be expected in flashover conditions, will neither prevent the formation of an arc bead nor influence the time required to create the bead. The physical appearance of arc beads formed in both normal and oxygen-deficient atmospheres is consistent with the characteristics listed in the "Melting by Electrical Arcing" section.

- Arcing through char can occur at voltages as low as 30 V (RMS). The formation of an arc bead within a timeline of investigative interest generally requires voltages greater than 60 V (RMS).

Overheating Connections

Poor connections are likely sites of overheating. The most probable causes include loose connections and the presence of oxides at the point of connection, creating resistance. Poor connections can often be verified by finding color changes at the point of the connection. At the point of the poor connection, portions of the metal connection may be deformed or destroyed. Poor contact connections can cause heating, oxidation, pitting at points of glowing, arcing, and even destruction of metals **FIGURE 7-13**. In duplex receptacles, a glowing connection can produce unique

evidence such as melted copper conductors around screw terminals, severed conductors at the screw head, or enlarged screw heads from severe corrosion **FIGURE 7-14**.

Overload

Overloads, as explained earlier, are overcurrents that are both large and persistent enough to cause damage and create danger of a fire. They occur when currents exceed the rated ampacity. The amount of damage depends on the degree and duration of the overcurrent. The most likely place for an overload to occur is on stranded cords, such as extension cords leading to large-draw appliances (e.g., air conditioners, space heaters, refrigerators). In contrast,

A

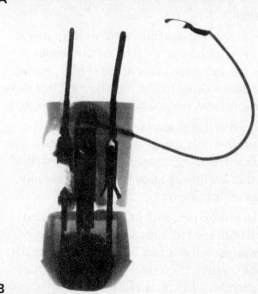

B

FIGURE 7-13 A. Photo of poor contact connections. **B.** X-ray of poor contact connections.
Courtesy of Cameron Novak.

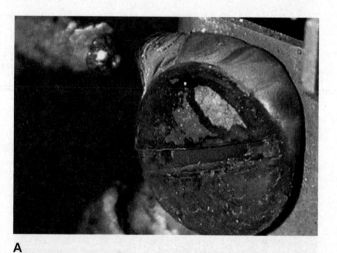

A

B

FIGURE 7-14 A. Severed conductors at the screw head. **B.** Enlarged screw head with severe corrosion.

Reproduced with permission from *NFPA 921-2021, Guide for Fire and Explosion Investigations*, Copyright © 2020. National Fire Protection Association. This reprinted material is not the complete and official position of the NFPA on the referenced subject, which is represented only by the standard in its entirety.

overloads are unlikely to happen on circuits with proper overcurrent protection.

An overload causes internal heating of the conductor, which occurs along the entire length of the conductor and may lead to sleeving. If the overloading is severe, the conductor can ignite fuels in the vicinity and melt the conductor. Once the conductor melts, the current stops flowing and heating stops. Overcurrent melting is not necessarily indicative of ignition, however, because the fire itself may cause an overcurrent on the circuit.

Overloads can bring a conductor to its melting temperature. The first point to sever stops both the flow of current and heating; other nearly melted spots then freeze as offsets. If the overload is caused by downstream faulting, evidence of this fault should be apparent.

Some effects on wires, however, are not caused by electricity. These effects, which include conductor surface colors (dark red to black oxidation, green, or blue), are of no value in determining cause but are always present in fires.

Melting by Electrical Arcing

Arcing produces very high temperatures, which will melt the conductors in a very small area. Because of fast overcurrent action, however, the duration of this effect is short. Thus, the damage caused by the arc is localized, and a sharp line of demarcation separates the point of damage and the surrounding conductor material. Gloves can be used that snag on the small notches on the conductor created by the arc. The fire investigator can run a gloved hand over the wires until it catches on the notch, at which point further investigation will identify whether the surface roughness encountered is caused by an arc or another event. Often, magnification is required to detect this damage.

Sometimes arcing produces sparks and dislodges material, which then collects on nearby surfaces. Further metallurgical analysis can be performed on wire segments containing suspicious notches to confirm melting patterns consistent with arcing.

In summary, the visual characteristics of arcing damage include:

- Sharp line of demarcation between the damaged and undamaged areas **FIGURE 7-15**
- Round, smooth shape **FIGURE 7-16**
- Localized point of contact **FIGURE 7-17**
- Identifiable corresponding area of damage on opposing conductor **FIGURE 7-18**
- Copper drawing lines visible outside the damaged area **FIGURE 7-19**

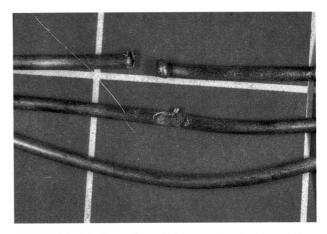

FIGURE 7-15 Sharp line of demarcation between the damaged and undamaged areas.
Courtesy of Cameron Novak.

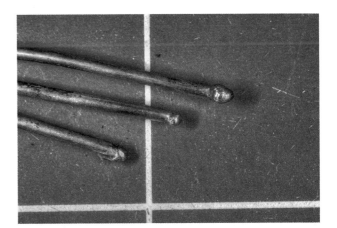

FIGURE 7-16 Round, smooth shape in area of arcing damage.
Courtesy of Cameron Novak.

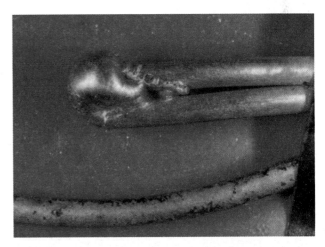

FIGURE 7-17 Localized point of contact.
Courtesy of Cameron Novak.

- Localized round depressions **FIGURE 7-20**
- Small beads and divots over a small area **FIGURE 7-21**
- High internal porosity when viewed in a cross section

FIGURE 7-18 Identifiable corresponding area of damage on opposing conductor.
Courtesy of Cameron Novak.

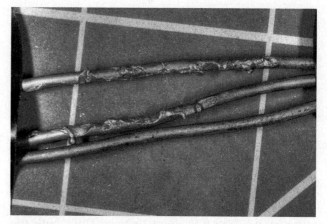

FIGURE 7-21 Small beads and divots over a small area.
Courtesy of Cameron Novak.

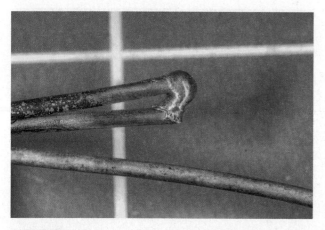

FIGURE 7-19 Copper drawing lines visible outside the damaged area.
Courtesy of Cameron Novak.

FIGURE 7-22 Extended area of damage without a sharp line of demarcation from undamaged material.
Courtesy of Cameron Novak.

FIGURE 7-20 Localized round depressions.
Courtesy of Cameron Novak.

Melting by Fire

Temperatures of sufficient intensity to melt copper can be reached in post-flashover fires. An important skill in fire investigation is the ability to differentiate melting caused by electrical activity from that caused by fire impingement. Characteristics of conductors melted by fire include:

- Extended area of damage without a sharp line of demarcation separating it from undamaged material **FIGURE 7-22**
- Visible effects of gravity on the artifact **FIGURE 7-23**
- Blisters on the surface **FIGURE 7-24**
- Gradual necking of the conductor
- Low internal porosity when viewed in a cross section

Stranded conductors melted by fire become stiff as they reach their melting temperature. Further heating then melts the individual strands together, and the surface becomes irregular, showing individual strands. Continued heating creates conditions similar to those found in solid conductors. In large-gauge stranded conductors, the individual strands may fuse, separate, and have bead-like globules.

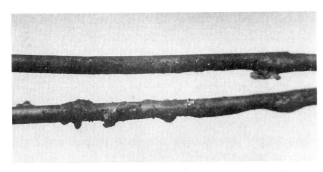

FIGURE 7-23 Visible effects of gravity in the artifact.
Courtesy of Cameron Novak.

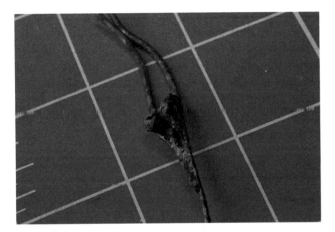

FIGURE 7-24 Blisters on the surface.
Courtesy of Cameron Novak.

As noted earlier, aluminum conductors have relatively low melting temperatures compared to the temperatures reached in the average structural fire. They are likely to melt in any fire and may solidify into irregular shapes. They are generally of little help in determining a fire's cause.

Alloying

Alloying is the mixing of metals with different physical properties. Aluminum–copper alloying can begin when aluminum drips onto the surface of a copper conductor and sticks lightly to the surface. If heating continues, the melted aluminum can penetrate the copper's surface and form an alloy. This alloy will have a melting point lower than that of pure copper or pure aluminum. You may observe an alloyed spot as either a dull gray or shiny silvery area. An aluminum–copper alloy is brittle and may break easily. If the alloyed material drips off during the fire, you may observe a pit lined with the alloy. Chemical analysis can verify the existence of an alloy.

Some other common alloys include brass (copper–zinc) and bronze (copper–tin). Silver is often used on the contact matting surfaces of electromechanical switches such as relays, thermostats, and contactors. Copper and silver on these surfaces can form an alloy at temperatures below their respective melting points.

Mechanical Gouges

Mechanical gouges can be differentiated from arcing marks by microscopic examination. They usually show scratch marks, dents in the insulation, and deformation of the conductors. They do not exhibit fused surfaces caused by electrical energy.

Considerations and Cautions

In recent years, some older beliefs about the role of electricity in fire have been disproved by laboratory studies based on knowledge of chemical, physical, and electrical sciences. Investigators need to be aware of the underlying research and studies supporting hypotheses that are used to describe electrical failures. This section highlights some cautionary notes when forming hypotheses about electrical fires.

Undersized Conductors

Conductor sizing is an important safety factor in the design of circuit ampacity; however, applying 20 A to a 15-A circuit does not always result in overheating. In fact, significant overcurrent may cause increased heating but not necessarily fire ignition. Carefully evaluate circuits that have more current flow than the NEC allows. You must also take into account the specific circumstances of the fire. Generally speaking, an undersized conductor in and of itself is not indicative of the fire cause. For example, a 20-A current flow in a 14 AWG conductor would not cause the wire to heat sufficiently to cause ignition if the wire were bare and uninsulated. However, fires have been observed in attics where an overcurrent of this magnitude exists (20 A on a 14 AWG wire) when the wiring is located underneath cellulose insulation.

Nicked or Stretched Conductors

Nicked or stretched conductors that reduce a conductor's diameter in a localized area usually cannot be said to cause heating at that location under normal load conditions. Nevertheless, the mechanical strength of the conductor is compromised in this situation, and flexing it near the damage could weaken it to the point where the conductor could fracture and arcing could occur. It is not easy to sufficiently stretch copper conductors in good condition to reduce the cross-sectional area without breaking the conductor.

Deteriorated or Damaged Insulation

Insulation deteriorates with both age and heating. Rubber deteriorates more quickly than thermoplastic material, and it may become brittle and will crack when bent. Cracks do not allow for conduction of electricity unless contamination should enter the cracks; however, the insulating strength of air (in a crack) is less than that of the insulation, leading to a greater potential for arcing.

Mechanical damage and vibration can also cause insulation to deteriorate. In **FIGURE 7-25**, the electrical installer had cut through the insulation, and a ground fault to the electrical panel occurred.

Overdriven or Misdriven Staple

Improperly driven staples will not damage the conductor to the point of faulting immediately. However, over time, if overdriven, a high-resistance contact can occur as the insulation gives way to pressure. This type of high-resistance contact can occur months or years after the original installation of the wire. Even if the fire investigator finds evidence of this kind of a fault, the question remains whether it is an ignition source

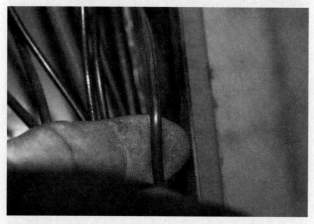

FIGURE 7-25 Ground fault from damaged (cut) insulation.
Courtesy of OnSite Engineering & Forensic Services, Inc.

or the effect of an advancing fire. Given that the steel staple will heat easily within an attic fire, the insulation on the wiring immediately under the staple can be affected early in the fire, allowing arcing immediately under the staple. For this reason, arcing is commonly observed in the immediate area of staples and commonly misdiagnosed as a fire cause.

Improperly positioned staples driven at an angle can cut across several conductors, creating a short circuit. This problem should be evident after the fire at the point where the staple was misdriven. Ideally, overcurrent protection should stop the flow of electricity and stop any heating. A parting arc from this event normally would not be sufficient to ignite the insulation or supporting wood. If a leg of the staple penetrates the insulation and contacts an energized conductor and a grounded conductor, however, a short circuit will occur. If the staple severs an energized conductor and a high-resistance fault develops, a heating connection may be formed.

Displaced Connections

Occasionally, during the investigation of a fire, twist-on wire connections (i.e., wire nuts) appear to be outside a junction box. In reality, the connection points likely ended up outside the junction box owing to the extent of the fire damage and/or collapse of the structure.

Short Circuit

Short circuits or conductor-to-conductor faults can generate sufficient heat to ignite combustibles, but activation of the overcurrent protection device should limit the duration of this heating. This prevents the fault from bringing the fuel up to its ignition temperature. If the overcurrent protection does not function correctly, then an overload may occur, which, along with arcing, may generate sufficient heat to ignite nearby material.

Beaded Conductors

A beaded conductor is not indicative, in and of itself, of the cause of a fire. Beading may be the result of the fire damaging the wire and then creating a failure. The only conclusion that can be definitively made after finding an arc-melted (beaded) conductor is that the conductors were energized at the time of the fire.

Arc Mapping

Arc mapping can provide valuable information for the fire investigator in both large- and small-scale events, as discussed in Chapter 4, *Fire Effects and Patterns*. Arc mapping requires a careful examination of the electrical circuits to identify arcing that occurred within the circuit. Mapping the locations of possible electrical activity can help the investigator identify an area that warrants additional examination.

If an arcing event causes a conductor to sever, then no current can flow downstream (electrically distal from the source) from the severance point of a conductor. Therefore, additional electrical damage cannot occur downstream. Unless damaged by heat, the conductors and insulation are likely to remain intact downstream from the severed conductors. Upstream from the severance point, the circuit may remain energized if the circuit protection does not open, and the circuit may continue to arc. If multiple arcs are found on a single circuit and one sever arc is located closer to the supply than the other arcs, the investigator can conclude that the downstream arcs occurred no later than the sever arc. Thus, the investigator should attempt to find as much of the conductor as possible to identify arcing damage.

Holes can be melted in metal conduits or in the sides of panelboards because of an arcing event. These holes usually result from fire impingement as the heat from the fire degrades the wiring insulation of the conductors inside. Blowouts result from arcing between the conductors inside and the enclosure. Note that a lack of blowouts or blow holes does not guarantee that arcing did not occur. The conductors must be removed from the conduit to confirm whether arcing did or did not occur.

Any observed damage should be documented through both notes and photographs. The investigator should also photograph the corresponding arcing damage on adjacent conductors. In any electrical examination, it is important to note the source that fed the circuit, the loads served by the circuit, the status of the circuit protection device, and the location of the arcing damage. In addition, document where arc melting was not found, as well as any melting caused by fire.

While arc mapping can be a useful tool in fire investigations, it has some limitations. Prolonged fire exposure resulting in thermal melting of conductors or structural collapse may prohibit the analysis of conductors. It is also possible for de-energized conductors (whether in the tripped or off position) to come into contact with other energized conductors, leading to an arcing event. This may be misleading, so investigators should take special care to determine whether such an event occurred. It may be inadvisable in some cases to attempt to remove conductors from conduits or other areas (see the discussion of spoliation of evidence in Chapter 5, *Legal Considerations for Fire Investigators*).

TIP

Electrical systems are usually complex, and new product advances in electronic controls increase the challenge in understanding them. Additional design information such as schematics and theory of operation may be necessary to unravel their workings. By knowing which areas to focus on and the characteristics of electrical energy flow, you can break the system down into workable chunks and, often as a part of a team of investigators, converge on a valid conclusion.

Static Electricity

Static electricity results from buildup of charge caused by the physical contact or movement of one object on the other. Examples of situations that can create static electricity include walking across a carpet, conveyor belts moving over rollers, and flowing liquids.

Static discharge can occur when two charged surfaces pass over each other and a sudden recombination of the separated positive and negative charges creates an electric arc. A static charge can build up over time. It is difficult to prevent static electricity entirely; however, to act as an ignition source, static discharge must occur in an ignitable atmosphere and have enough ignition energy for the fuels present.

Static charge can be generated by movement of liquids in relation to other objects and can occur in flow through pipes, mixing, pouring, pumping, spraying, filtering, and agitating. Static charge may accumulate in the liquid, particularly with liquid hydrocarbons. If sufficient charge builds up, an arc may occur; then, if a flammable vapor–air mixture is present, ignition results. Lower liquid conductivity implies a higher ability to create and hold a charge. The surface charge—that is, the charge on the surface of the liquid—is of the greatest concern.

Relaxation time is defined as the amount of time needed for a charge to dissipate. It can range from several seconds to several minutes and is dependent on the conductivity of the liquid, the rate at which the liquid is being introduced into a tank, and the manner in which the liquid is being introduced into the tank. If the electrical potential between the liquid and the metal tank shell reaches a high enough level, the air may ionize, and an arc may form between the liquid surface and the tank shell. In reality, an arc between the surface and the shell is less likely than an arc between the surface and a projection into the tank; projections are known as spark promoters. No bonding or grounding can remove these static charges. If the tank or container is ungrounded, an arc may occur between the tank and a nearby object.

Switch loading occurs when a liquid is introduced into a tank that previously held a liquid with different properties. Static discharge can ignite the vapors from the more flammable liquid.

Spraying operations can produce significant static charges on the surfaces being sprayed and on an ungrounded spraying nozzle or gun. If the material being sprayed is flammable, then ignition can occur. High-pressure spraying operations have a greater potential for generating static than low-pressure spraying operations do.

Static can build up when a flowing gas vapor mixes with metallic oxides, scale particles, dust, and liquid droplets or spray. When such a gas is directed against a conductive, ungrounded object, a static buildup may occur. If the static charge is sufficient and a flammable atmosphere is present, then an arc may occur to another nearby grounded object, causing the gases to ignite.

To be capable of causing ignition, the energy released in the discharge must be at least equal to the minimum ignition energy of the flammable mixture. Some of this energy is required to heat the conductive surfaces that will arc; thus, the minimum amount of charge required for static discharge ignition is generally 1500 V. Dusts and fibers require 10 to 100 times more energy to ignite than gases and vapors do. The minimum electrical charge required for ignition of a dust cloud is 10 to 100 millijoules (mJ). This range encompasses and can be less than the amount of energy in a typical static arc from the human body, which can reach 20 to 30 mJ. The human body can accumulate charges in atmospheres less than 50 percent relative humidity, with such charges being as high as several thousand volts. Charges can build up when layers of clothing are separated, moved away from the body, or removed entirely. This is common when the layers are of dissimilar fabric.

Controlling Accumulations of Static Electricity

Static charges can be removed or dissipated through humidification and by bonding and grounding. Humidification refers to the moisture present in a material. The more moisture that is found within a material, the more conductive it will be and the less likely it is to accumulate a static charge. The moisture content is related to the relative humidity of the surrounding air.

In atmospheres with high relative humidity (humidity of 50 percent or more), materials and air reach equilibrium, contain sufficient moisture to be adequately conductive, and do not have significant static electricity accumulation. In atmospheres with low relative humidity (humidity of 30 percent or less),

materials dry out and become good insulators, and static accumulations become more likely. The conductivity of the air itself is not changed by humidity.

Bonding refers to the electrical connection of two or more conductive objects in such a way as to reduce the electrical potential differences between them. Grounding is the process of electrically connecting an object to ground; it reduces the electrical potential differences between objects and the earth. Objects such as pipes and tanks that are embedded in the earth are naturally bonded. The grounding and bonding should be tested to determine their effectiveness.

Conditions Necessary for Static Arc Ignition

Five conditions must be present for static arc ignition:

1. A means of static charge generation
2. A means of accumulating and maintaining the charge
3. A static electric discharge arc with sufficient energy
4. A fuel source with the right air mixture and with a small enough ignition energy
5. Co-location of the arc and fuel source

According to NFPA 921, Section 9.12.5.2.4, if static arc ignition is suspected as an ignition source, an electrical inspector with proper qualifications (in compliance with NFPA 77, *Recommended Practice on Static Electricity*) should review the circuit grounding paths and related connections.

Investigating Static Electric Ignitions

Investigating static electric ignitions often requires the investigator to gather circumstantial evidence. The investigator must also determine whether five conditions listed previously were present. To determine how the static electricity was generated, identify the materials involved and analyze their conductivity, motion, contact, and separation. Additionally, the investigator must identify how the charge was able to accumulate through the presence or absence of grounding, bonding, or conductivity of the material. Determine the relative humidity, identify the potential location of the static arc, determine whether the arc contained sufficient energy to be a competent ignition source, and determine whether the arc and the fuel were located in close proximity. Eyewitness reports and circumstantial evidence should help establish, as precisely as possible, the location of the arc.

Also important are the actual discharge energy and the fuel present for ignition. Although arcing can occur at lower discharge voltages, 1500 V has been shown to be a lower bound for ignition to occur. Dusts and fibers typically require 10 to 100 times the energy of vapors and gases (in ideal air–fuel mixture) to ignite.

Lightning

A lightning bolt has the following characteristics:

- Core energy plasma of ½ to ¾ inch (12.7 to 19 mm) diameter
- Surrounded by a 4-inch-thick (102-mm) channel of superheated ionized air
- Average currents of 30,000 to 45,000 A, but can exceed 200,000 A
- Potentials that range up to 15,000,000 V

Lightning will strike any object that generates an upward-extending streamer that connects with a step-leader from the base of a cloud. According to NFPA 921, lightning can enter a structure in four ways: a direct strike to the structure or an attached item, a strike near a structure that couples or channels energy onto energized or nonenergized conductors, a direct strike to incoming conductors connected to the structure, or a strike near overhead conductors where current is conducted into the building via the conductors. Lightning strikes carry high electrical potentials (hundreds of thousands of volts) and large currents (thousands of amps). A strike may cause objects to be displaced or explode as a result of the energy content of the lightning.

The damage resulting from a lightning strike may take the form of damage to either the structure or the electrical system. Pay particular attention to any point where the building object may be grounded. Both line-voltage (120 V AC) and low-voltage (<50 V DC) systems in buildings are susceptible to lightning strike damage. Some equipment may contain surge protectors and surge monitors, though these devices are useful only if they are able to absorb or deflect the energy levels present on the equipment power conductors and signal interfaces.

Lightning detection networks and lightning strike reports may assist in determining the exact time and location of the strike; however, there are limitations to their use. It is important to discuss these limitations with the vendor.

Building Fuel Gas Systems

A building's **fuel gas** system can be used to control the indoor climate, to heat water, to cook, or to provide energy for manufacturing processes. The two fuel gases most commonly used within a structure are **natural gas** and **commercial propane**.

During every fire scene examination, investigators need to examine and document the presence and condition of the fuel gas systems. When a failure occurs within a fuel gas system, it can lead to dramatic events such as explosions and fires, which often destroy the structure and cause serious injuries to people within the building or nearby. All fire investigators should have knowledge of fuel gases and the systems that use them. Three documents are considered the primary resources for developing an understanding of building fuel gas systems:

1. NFPA 54, *National Fuel Gas Code*
2. 49 CFR Part 192, "Transportation of Natural and Other Gas by Pipeline: Minimum Federal Safety Standards"
3. NFPA 58, *Liquefied Petroleum Gas Code*

Fuel gas systems to either a building or an appliance can act as the initial fuel source, the initial ignition source, or both. The presence of gas can also cause the fire to spread rapidly through the structure. When the fuel delivery system fails and fugitive gas comes into contact with an ignition source, the fuel gases can become part of the original ignition sequence, acting as the first fuel **FIGURE 7-26**. Subsequently, these gases may influence the spread and growth of the fire. If fuel gases become involved in a fire, they are usually a factor in the fire's spread and growth.

Fuel Sources

Fuel involvement most frequently results from compromised fuel delivery, containment systems, or fuel-fired appliances. The fuel usually escapes from the containers that hold the bulk fuel supply, the

FIGURE 7-26 Natural gas explosion scene.

piping from the bulk supply to the appliance, or systems in the appliance itself. Fuel that has escaped in this way, referred to as **fugitive gas**, readily mixes with the air and is ignited easily, even by ignition sources that have very little energy. The ignition temperature of most fuel gases ranges from 723°F (384°C) to 1170°F (632°C). The minimum ignition energy for these gases can be as low as 0.2 mJ.

> **TIP**
>
> Because of the amount of damage that can occur at an explosion or fire incident, always examine the scene for potential hazards. Holes, voids, and even residual gas may be present, posing a significant risk to those personnel conducting the investigation.

Ignition Sources

Pilot lights and open flames from the appliances served by fugitive gases often function as ignition sources. Ignition sources can also take the form of static arcs from the appliances, electrical arcs, arcs from switches or contacts in appliances, or other electrical equipment. Often, it is difficult to isolate the specific source of ignition for fugitive fuel gases because there are so many possibilities.

Fugitive fuel gases may develop around gas appliances or equipment that is served by the fuel gas. Once these appliances become involved, the fire will ignite combustibles in their vicinity.

Gas Systems as Ignition and Fuel Sources

Fuel gas systems and apparatus, particularly those with burners and pilot lights, can serve as both the fuel source and the ignition source for the fire or explosion. This could occur when, for example, gas escapes from a damaged pipe or mechanism in a water heater, eventually reaching the pilot light, where it is ignited.

Additionally, fuel gases can serve as both the first fuel and the fuel for the ongoing fire. In such a circumstance, after providing the initial fuel for an ignition, the gas system continues to emit additional fuel that enables the fire to spread to other combustibles. The amount of energy provided by the fugitive fuel gases depends on the size of the leak or the amount of fuel that is escaping. Fugitive fuel gases can greatly accelerate the spread and growth of a fire, especially if the leak continues during the fire growth and combustibles in the area support the fire's spread. Fuel gases that are simply burning into the air with no additional combustibles nearby will not accelerate fire growth.

Additional Fire Spread

When driven by fugitive fuel gas, fire can spread to other areas of the building. Sometimes the scene may appear to include separate fires, which in reality are related to the release of the gas. This phenomenon often occurs when the gas is released for a period of time prior to ignition and is allowed to collect in other areas. It can also occur when the gas is released under pressure, spreading the fire to other areas of the building.

Characteristics of Fuel Gases

Be aware of the individual properties of the fuel gas that you suspect may have caused or contributed to a fire or explosion event. An understanding of each fuel gas's properties will allow you to interpret fire and damage patterns accurately. **TABLE 7-8** describes some of the characteristics of natural gas and propane (**liquefied petroleum [LP gas]**), the fuels most commonly encountered in residential use. (For more information, see NFPA 54 and NFPA 58.)

Natural Gas

Natural gas is a mixture of gases that is often recovered from underground pockets during drilling operations for crude petroleum. It is composed largely of methane (72 to 95 percent), though other gases such as nitrogen, ethane, propane, butane, hexane, carbon dioxide, and oxygen can be present in varying percentages.

Pure natural gas is lighter than air and has a specific gravity of 0.59 to 0.72. Its flammable range is between 3.9 and 15 percent, with an ignition temperature of 900°F to 1170°F (482°C to 632°C).

Commercial Propane

Propane, unlike natural gas, is recovered during the process of petroleum refining. Because of propane's ability to be liquefied at moderate pressures and normal temperature, liquefied propane (LP gas) can be more easily transported, stored, and used than natural gas, making it more suited to remote locations or portable consumer use. Commercial propane, at a minimum, is 95 percent propane and propylene, with the remaining 5 percent consisting of other gases.

Propane (undiluted) has a specific gravity greater than air (1.5 to 2.0), a flammable range of 2.15 to 9.6 percent, and an ignition temperature of 920°F to 1120°F (493°C to 604°C).

TABLE 7-8 Characteristics of Natural Gas and Propane

	Natural Gas	Propane
Composition	Hydrocarbon gas	95% propane and propylene
	Primarily methane	5% other gases
Density	Lighter than air	Heavier than air
	Vapor density of 0.59 to 0.72	Vapor density of 1.5 to 2.0
Lower explosive limit (LEL)	3.9% to 4.5%	2.15%
Upper explosive limit (UEL)	14.5% to 15%	9.6%
Ignition temperature	900°F to 1170°F (482°C to 632°C)	920°F to 1120°F (493°C to 604°C)
Btus (per cubic foot)	1030	2490
Delivery	Delivered via a distribution system directly to the customer	Stored in tanks on the customer's property and delivered via tank trucks

Other Fuel Gases

Other gases may also be of interest to fire investigators. Gasoline vapors in sufficient quantities and proper concentrations can be involved in combustion events. Butane, or combinations of butane with other gases, is found in marijuana concentrate extraction operations, including amateur home operations, and can also play a role in fires and explosions. In investigations at commercial and industrial facilities, fire investigators may encounter commercial butane, which is composed largely of butane and butylene (minimum 95 percent), with the remaining 5 percent consisting of other gases. Propane HD5 (95 percent propane) is used as a motor fuel and is subjected to more restrictive requirements than is commercial propane.

Specific manufacturing processes may also use or produce other gases, such as coke oven gas, hydrogen, and acetylene. Each gas has a unique specific gravity and flammable range with which the investigator should become familiar if that particular gas is under consideration in an investigation.

Odorization

In their natural state, LP gas and natural gas have no odor. Odorant is added during the distribution chain as a safety feature, however, so that people can be alerted by the odor to the presence of gas. Per NFPA 58, the odorant must be detectable when the fuel gas is at a concentration of not less than one-fifth of its LEL. Butyl mercaptan is most often used in natural gas for this purpose, whereas ethyl mercaptan or thiophane is most often used in LP gas. Typically, natural gas is not odorized while being transferred in a transmission pipeline system. Instead, odorization occurs when natural gas is introduced into a distribution system to be used by the end consumer. In the case of LP gas, odorant is added prior to being delivered to a bulk distribution facility.

If the investigation reveals that fuel gas may be involved, the investigator should verify the presence of an odorant through witness interviews or other means. Often the odorant can be detected during the investigation.

Sampling methods are available to verify the presence and concentration of odorant. Propane should be tested by obtaining a liquid sample for laboratory analysis, as gas samples will not provide an accurate measurement. Natural gas samples should be collected using Tedlar bags and tested at the scene with field detection equipment.

LP gas samples should be obtained using stain tubes and approved sampling techniques. Because there is no universal stain tube for all odorants, the investigator must identify the odorant and use the appropriate stain tube. It is recommended to seek assistance from someone who is experienced in this field to obtain the sample(s). Once collected, the sample(s) should be submitted for analysis as soon as possible to a laboratory that is able to analyze the sample for odorization. Methods for collecting and testing of the LP gas sample should comply with ASTM D1265 and ASTM D5305.

Gas Systems

The systems that transport various fuel gases are similar, in that they all pipe the gas directly into consumers' buildings and appliances. One notable difference, however, is that natural gas is supplied from a central location via underground supply service lines (a type of distribution line), whereas LP gas or other types of storage gases are stored in a bulk supply container, often at the consumer's location.

Natural Gas Systems

Natural gas is supplied through a transmission pipeline, then through main pipelines (mains), and then through service mains or service laterals, and finally connected to the meter at the consumer's location. **TABLE 7-9** provides a description of the natural gas supply system and the typical pressures found in its components.

LP Gas Systems

LP gas distribution systems are similar in operation to natural gas systems except that the LP gas storage supply is often located at the consumer's site. LP gas storage containers can also be housed in bulk storage locations and piped underground, similar to natural gas systems.

LP Gas Storage Containers

The differences among portable tanks, cargo tanks, cylinders, and containers are discussed in NFPA 58. This section provides an introduction to those systems.

Tanks. A tank is defined as a storage container constructed in accordance with the pressure vessel code **TABLE 7-10**. The design and construction of LP storage tanks are jointly governed by regulations of the American Petroleum Institute and the American Society of Mechanical Engineers **FIGURE 7-27**. The typical working pressure in the storage tanks varies depending on the temperature of the LP but can range as high as 200 to 250 psi (1400 to 1700 kPa). LP may not be transported with more than 5 percent liquid capacity.

Cargo tanks are containers that are permanently mounted on a chassis and are used for transporting LP gas. Portable tanks are also used for transporting LP gas, but are not mounted on a chassis. Their quantities exceed a 1000-lb (450-kg) water capacity.

TABLE 7-9 Natural Gas Supply System from Supplier to Consumer

Type of System	Typical Use	Pressure
Transmission pipeline	Used to transport the natural gas from the main facilities to the local utility	1200 psi (8275 kPa) gas from storage/production
Main pipelines	Used to distribute the gas in a centralized grid system (1000 kPa); typically 60 psi	Varies; seldom exceeds 150 psi (1000 kPa); typically 60 psi (400 kPa) or less
Service lines (service laterals)	Used to connect customers to the main pipelines	Ranges from 4- to 10-in. w.c. (1.0 to 2.5 kPa)
Metering	Measures the volume of gas passing through a pipe	Depends on the consumer's needs; ranges from 4-in. w.c. (2.74 kPa) to several psi

Abbreviation: w.c., water column.

TABLE 7-10 LP Gas Tank Chart

	150 Gallons (600 Liters)	250 Gallons (1000 Liters)	500 Gallons (1900 Liters)	1000 Gallons (3800 Liters)
Length	85 inches (215 cm)	92 inches (230 cm)	120 inches (300 cm)	190 inches (480 cm)
Diameter	25 inches (60 cm)	30 inches (75 cm)	37 inches (90 cm)	41 inches (100 cm)
Weight	320 lb (145 kg)	485 lb (220 kg)	950 lb (430 kg)	1750 lb (800 kg)

FIGURE 7-27 LP gas storage tank.

FIGURE 7-28 LP gas cylinder.

Cylinders. Cylinders are considered upright containers. They have a water capacity of 1000 lb (450 kg) or less. The design and construction of cylinders are governed by regulations of the U.S. Department of Transportation. The pressure in the cylinders can be the same as that in tanks or containers, ranging up to 200 to 250 psi (1400 to 1700 kPa). Cylinders are most frequently used in rural homes and businesses, mobile homes, and recreational vehicles and for outdoor barbecue grills and motor fuel **FIGURE 7-28**.

Container Appurtenances

The devices that are connected to the openings in tanks and other containers—such as pressure relief devices, liquid level gauges, and pressure gauges—are called **container appurtenances FIGURE 7-29**. These devices are briefly described here.

Pressure Relief Devices

Containers are governed by U.S. Department of Transportation regulations and are required to have pressure relief valves or fusible plugs. These pressure relief devices may be activated by pressure or temperature changes, and aim to prevent the pressure from exceeding a predetermined maximum and rupturing a container.

A **pressure relief valve** is designed to open at a specific pressure, usually around 250 psi (1700 kPa). This valve is generally placed in the container where it releases the vapor, although there are some exceptions. The pressure in a container is directly related to the temperature of the LP gas. As the temperature rises, the pressure rises. NFPA 58 provides methods to determine the pressure in a container if the temperature of the gas is known.

Fusible plugs are thermally activated devices that open and vent the contents of a container. Once activated, they cannot be reused. Aboveground storage

FIGURE 7-29 LP gas tank gauge.

tanks with less than 1200 lb (540 kg) of water capacity may have fusible plugs. The melting temperature of these plugs ranges between 208°F and 220°F (98°C and 104°C).

Connections for Flow Control

Some container appurtenances control the flow of gas from the container. The most common is a **shutoff valve**. In some installations, excess-flow check valves monitor the flow from the container; these valves are activated if the flow exceeds a set amount. In addition, **backflow valves** may prevent the gas from reentering the container or distribution system.

Liquid Level Gauging Devices

Often there are appurtenances on containers that indicate the liquid level of the propane. These types of devices include **fixed level gauges**, which are fixed at a maximum level, and **variable gauges**. Types of variable liquid level gauges include float, magnetic, rotary, and slip tube.

Pressure Gauges

Pressure gauges, another type of container appurtenance, indicate the internal pressure of the tank. These gauges are connected directly to the tank or sometimes through the valve. Pressure gauges do not indicate the quantity of liquid propane in the tank.

Pressure Regulators

Pressure regulators are attached through the gas supply system and aim to reduce the pressure so that the gas can be used by the appliance or utilization equipment. Some propane systems include multiple regulators that reduce the propane pressure in two stages. Often, when there are two stage regulators, one of the regulators will be located at the tank and the other will be placed near the appliances or where the fuel delivery system enters the structure. It is common to see the LP pressure reduced for use in the appliances to approximately 11-inch to 14-inch water column (w.c.; 2.7 to 3.5 kPa).

The propane pressure should be reduced from the tank pressure to a level at which it can be utilized by the equipment. The tank pressure will vary depending on the temperature of the liquid **TABLE 7-11**.

Vaporizers

Equipment is sometimes used to assist in vaporizing the liquid propane. **Vaporizers** are frequently employed when there is a demand for large quantities of propane, such as for industrial uses or in cold-weather environments **FIGURE 7-30**. Vaporizers heat the liquid propane, converting it to a gas.

Common Fuel Gas System Components

Whether the building is a single-family home, a commercial restaurant, or a large manufacturing facility, its fuel gas system will contain various distribution piping, valves, burners, and regulating devices. These various components are designed, installed, and used to deliver the fuel gas to the appliances safely. In a fire investigation, these components must be examined to determine whether they caused or contributed to the fire or explosion event.

Pressure Regulation (Reduction)

The most common types of pressure regulators are the diaphragm type, which contains a spring that is set to control the pressure, and the lever type. The amount of gas used may also be measured through a pressure regulator, commonly known as a gas meter **FIGURE 7-31**. The vents on the regulators must be clear

FIGURE 7-30 Vaporizer.

FIGURE 7-31 A large natural gas meter.

TABLE 7-11 Typical Pressure Inside Propane Tanks as It Relates to the Temperature of the Propane	
Temperature	**Pressure**
0°F (−18°C)	28 psi (190 kPa)
70°F (21°C)	127 psi (876 kPa)
130°F (54°C)	286 psi (1970 kPa)

for the device to operate properly. If the vent becomes plugged or obstructed, the pressure regulator might not function properly. In cold or flood-prone environments, it is important to place regulators where water or ice accumulations cannot obstruct the vent openings. Animals or other pests may build nests around the vent opening, causing an obstruction of the vent. **TABLE 7-12** identifies the general utilization pressures for appliances.

Service Piping Systems

Service piping provides for delivery of the fuel gas from the main lines to the user **FIGURE 7-32**.

The materials for these gas mains and services are typically constructed of wrought iron, copper, brass, aluminum alloy, or plastic. Service lines are often buried underground and are regulated by NFPA 58. Underground piping should be buried deep enough to prevent physical damage and should be protected from corrosion. If the piping runs under a building or areas that are used by the public, it should be encased in an approved conduit. Any plastic piping run underground must have a tracer wire buried with the pipe; however, the installer must ensure that it is not in direct contact with the pipe.

TABLE 7-12 Utilization Pressures for Appliances	
Use	**Pressure**
Nonindustrial natural gas applications	4- to 10-in. w.c. (1 to 2.5 kPa)
Nonindustrial propane applications	11- to 14-in. w.c. (2.7 to 3.5 kPa)

FIGURE 7-32 A 20-inch gas main pipeline.

Valves

Valves are installed on the piping system to control the flow of the gas. These valves are often installed where the gas piping exits the ground, prior to the meter or regulator, as well as prior to each appliance and in appliances. Most fuel gas systems include several types of valves. Some of the valves are automatic and are controlled by the appliance to be closed, open, or partially open, and some are configured to shut off automatically in an emergency condition, such as fire or excess flow. Valves typically found in appliances include individual burner valves, main burner control valves, and manual reset valves **TABLE 7-13**.

TABLE 7-13 Valve Types and Their Uses	
Valve Type	**Use**
Automatic valve	Includes a valve and an operating mechanism that controls the gas supply to a burner during operation of an appliance. The operating mechanism may be activated by gas pressure, electrical means, or mechanical means.
Automatic gas shutoff valve	Used in connection with an automatic gas shutoff device to shut off the gas supply to a fuel gas–burning appliance.
Individual burner valve	Controls the gas supply to an individual burner.
Main burner control valve	Controls the gas supply to a main burner manifold.
Manual reset valve	An automatic shutoff valve installed in the gas supply piping and set to shut off when unsafe conditions occur. It remains closed until manually reset.
Relief valve	A safety valve designed to prevent the rupture of a pressure vessel by relieving excess pressure.
Service shutoff valve	A valve usually installed by the utility gas supplier between the service meter or source of supply and the customer piping system; used to shut off gas to the entire piping system.
Shutoff valve	A valve located in the piping system and readily accessible and operable by the consumer; used to shut off individual appliances or equipment.

Gas Burners

Most gas utilization equipment uses some type of **gas burner**. The gas burners and/or orifices on appliances are not interchangeable between natural gas and propane. Ignition of the gas from a burner can occur by several methods, including manual ignition, pilot lights, and pilotless igniters. When a burner is ignited by manual ignition, a person provides the spark or flame. Gas burners may also be ignited from pilot lights—that is, small flames in the appliance found near the burners. The pilot light must be of sufficient size and close enough to ignite the gas as it escapes from the burner. In some designs, the pilot light burns constantly; in other designs, the burner is ignited electronically. In electronically lighted piloted appliances, electric arcs ignite the pilots, which in turn ignite the burners. Pilotless burners do not have pilots but are ignited by electronic arc or resistance heating elements. Many systems stop the flow of gas in the event that the burner does not ignite.

FIGURE 7-33 shows burners and the gas control valve for a typical furnace. **FIGURE 7-34** shows the pilot line and gas control valve on a water heater.

Common Piping in Buildings

This section discusses commonly used piping components in buildings and the utilization equipment for fuel gas piping systems. The size of the piping is determined by the maximum flow required by the equipment that is connected to the pipe. Piping in buildings is often similar to that described earlier in the section on service lines. Piping materials can include wrought iron (black pipe), copper, or brass.

Corrugated stainless steel tubing (CSST) is also used in commercial and residential applications and can be installed along floor joists, along ceiling joists, or inside wall cavities with more flexibility and fewer joints. Aluminum alloy may be used, but is not suitable for underground applications. Plastic may not be used indoors, although it may be used for exterior or underground applications.

TIP

Over the years, CSST has been found to be susceptible to leaks following lightning strikes or electric overcurrent events, which may result in a fuel-fed fire. Like all other gas distribution systems, CSST must be installed by qualified personnel, and the system must be appropriately bonded and grounded to reduce these risks.

TIP

Leaks created by physical damage or strain may develop at junctions far removed from the original location of physical impact. For example, if a gas meter is struck by a vehicle, the stress on the underground piping of the system may create a leak at a pipe union several feet away.

FIGURE 7-34 Water heater pilot line and gas control valve.

FIGURE 7-33 Furnace burners and gas control valve.

Joints and Fittings

Joints and fittings may be screwed, flanged, or welded. Screwed fittings are fittings with threads. Flanged fittings are connected with the use of a special device that flanges the end of the pipe and requires a fitting to complete the connection. Under some circumstances, compression fittings—which include a device that compresses the fitting to the pipe for a seal—may be used. Plastic piping, which may be used outdoors only, cannot be threaded.

Piping Installation

Piping installation is regulated by NFPA 54 and NFPA 58. Piping installations should not weaken the building structure. The piping must be supported with the proper devices, and not by other pipes or appliances. It should be equipped with drip legs where appropriate, usually at the lowest pipe point, and be equipped with sediment traps, usually near appliances, to capture foreign material. All unused outlets or openings in the piping should be capped—not simply controlled by a valve—to prevent gas from escaping.

Main Shutoff Valves

Main shutoff valves should be located on the exterior of the structure, although in older installations they may be located inside the building **FIGURE 7-35**. They should be placed upstream of each service regulator to allow for total shutdown if needed. If you suspect the main shutoff valve has leaked or failed, have a trained technician examine the appliance.

TIP

Locating the main shutoff valve may be difficult. It may be hidden in bushes, plants, or other concealing structures.

Prohibited Locations

As required by NFPA 54 and other recognized gas installation codes, natural gas piping cannot be run through air ducts, clothes chutes, chimneys, gas vents, ventilating ducts, dumbwaiters, or elevator shafts. Furthermore, gas piping should never be installed in areas that are subject to damage or exposed to corrosion.

Electrical Bonding and Grounding

Each aboveground portion of a piping system is required to be electrically bonded, and the system grounded according to the applicable code and installation guidelines.

FIGURE 7-35 The main shutoff valve controls the flow of gas throughout the entire building.

Common Appliance and Equipment Requirements

The appliance must be compatible with the type of gas it is using. When appliances are designed for natural gas but are used with propane, or when appliances are designed for propane but are used with natural gas, fire or explosion can result. The orifice sizes for natural gas connections differ from those for LP gas connections.

Appliances should be placed where they will not be subject to damage. In addition, they should not be placed where flammable vapors may accumulate because the appliance may ignite the gas vapors. In certain conditions, gas appliances can be installed in locations such as an automobile garage if they are located above floor level.

Appliance Installation

Where the system pressure is greater than the appliance is designed to handle, an additional regulator should be installed. (This situation was described earlier in this chapter in the discussion of multiple regulators.)

Although a gas appliance should be located where there is easy access for service or shutdown, this is not always the case. For example, commercial cooking appliances such as deep fryers and grill tops are usually supplied from the rear of the appliance, requiring the entire unit to be pulled away from the wall to shut down the supply of gas. Gas-fired commercial cooking appliances with casters must be secured with a chain or cable to prevent the appliance from being pulled too far out and damaging the gas line to the appliance.

The location should also have sufficient clearance between the appliance and any combustibles, including building construction, to ensure that the heat of the appliance does not ignite the combustible materials or items. The minimum required clearance is often displayed on the appliance itself. Appliances and accessories should also be approved as acceptable by the authority having jurisdiction and in accordance with NFPA 54. The electrical connections to the appliances should comply with NFPA 70, *National Electrical Code.*

Venting and Air Supply

Exhaust **venting** is required to prevent buildup of products of combustion inside the building. Codes require that fuel-burning appliances be vented. Some equipment, however, does not require venting—including ranges, ovens, and small space heaters. If used properly, these appliances will not produce significant products of combustion. By contrast, if they are used improperly, products of combustion can build up enough to cause harm.

Fresh air supply or combustion air supply venting from the exterior of the building is often necessary for proper operation of gas-fired appliances. If the volume of the room is sufficient, sometimes the installation of combustion air supply venting is not required. Many new appliances are designed so that the combustion air directly enters the appliance. When appliance burners begin to operate inefficiently because of an improper fuel–air mixture, carbon may begin to accumulate within the combustion chamber and flue **FIGURE 7-36**. With time, these deposits can block the flow of combustion products through the flue or possibly start a fire. The inefficient combustion may also produce carbon monoxide, a poisonous gas.

Appliance Controls

Fuel gas appliances have controls that are generally categorized as follows:

- Temperature controls
- Ignition and shutoff devices
- Gas appliance pressure regulators
- Gas flow control accessories

Fuel Gas Utilization Equipment

NFPA 921, *Guide for Fire and Explosion Investigations,* Sections 10.8.1 through 10.8.7, lists and describes commonly used fuel gas utilization equipment **TABLE 7-14**. You may encounter other types of gas utilization equipment, but all have similar features

FIGURE 7-36 Carbon deposits inside an improperly operating water heater burner.

TABLE 7-14 Fuel Gas Utilization Equipment	
Equipment	**Description**
Air heating	Boilers, forced-air furnaces, space heaters, floor furnaces, radiant heaters, duct furnaces, and clothing dryers
Water heating	Direct-flame burners for the heating of potable or industrial-process water
Cooking	Ranges, stove-top burners, broilers, and cooking ovens
Refrigeration and cooling	Absorption system refrigeration and cooling systems
Engines	Stationary and motor vehicle engines and auxiliary power units on service vehicles (e.g., pumps on tank trucks, electrical power generators)
Illumination	Gas lamps used for outdoor lighting; residential examples include driveways, patios, porches, and swimming pools; commercial examples include streets, shopping centers, and hotels
Incinerators, toilets, exhaust afterburners	Burn rubbish, refuse, garbage, animal solids, organic waste, and industrial-process waste

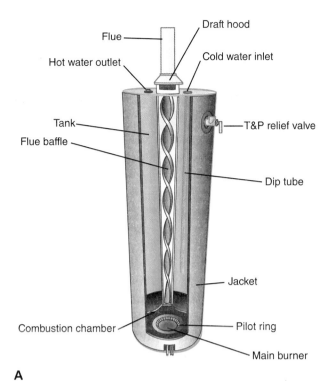

A

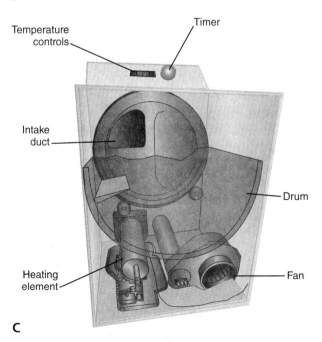

B

C

FIGURE 7-37 A. Water heater components.
B. Furnace components. **C.** Clothes dryer components.

designed to regulate the gas, control the flow of the gas, and burn the gas.

Several different types of appliances use fuel gas in their operation, including water heaters, furnaces, clothes dryers, and ranges **FIGURE 7-37**. Fire investigators should be familiar with the design features and operation of these four common fuel gas appliances.

Investigating Fuel Gas Systems

Analysis of the fuel gas system can provide valuable information if it was involved in the origin or cause of the explosion or fire. As with all parts of a fire investigation, the analysis of the fuel gas system and its components should be done in a systematic manner to ensure thoroughness. This systematic examination should include proper documentation—measuring and diagramming the system and identifying and measuring the piping and distribution system materials, including valves, couplings, or connectors. Ideally, the documentation should include detailed information on the system from the appliance back to, and including, the fuel gas source. Notations should be made of the positions of valves or other controls (open versus closed). Any faults in the system, such as breaks, cracks, or holes, should be noted and documented. The goal is to determine whether, and to what extent, the system operated or failed. Interview witnesses, property owners, and lessees to determine the condition of the fuel gas system and identify which activities might have occurred relative to the system, such as movement or repositioning of equipment, appliance service and repair, or installation of new appliances, components, and fixtures.

Compliance with Codes and Standards

Fire investigators should evaluate the design, manufacture, construction, and installation of the equipment and system to determine compliance with accepted codes and standards, calling on additional expertise or resources as necessary. Installation requirements for the various types of gas systems differ. The design and manufacture of the appliance itself are often regulated by an approval agency such as the American Gas Association. The installation manual for the appliance often provides information on its installation requirements, such as the proper clearances to combustibles.

Leakage Causes

The main causes of gas-fueled fires and explosions involve leakage in the fuel gas delivery system or in the

FIGURE 7-38 This water heater was damaged by a fire that involved combustibles stored next to the appliance.
Courtesy of Jeff Spaulding.

FIGURE 7-39 A natural gas line that ignited behind a commercial fryer that was not properly connected.
Courtesy of Jeff Spaulding.

appliance itself. The most common sources of leaks are pipe junctions, unlit pilot lights or burners, uncapped pipes, malfunctioning appliances and controls, areas of corrosion in pipes, and points of physical pipe damage **FIGURE 7-38**.

Pipe Junctions

Leaks can occur in pipe junctions if the threading is inadequate for a good connection, if the junction between the pipe and coupling is improperly joined or threaded, or if pipe joint compound was used improperly.

Unlit Pilot Lights or Burners

Most modern gas equipment stops the flow of the gas to the burners if the pilot light or the burner is not lit. Nevertheless, gas leakage may occur in these systems if the sensing system malfunctions or if the automatic shutoff valve does not operate correctly. The amount of gas that flows from an unlit pilot light is generally not sufficient to cause an explosion or fuel fire unless the gas is somehow confined to a space or leaks for an extended period of time. Burners that have been turned on but not lit can provide sufficient gas leakage to fuel an explosion or fire. Some equipment contains a safety device that monitors the flame from the burner and, if there is no flame, shuts off the flow of gas. This equipment

can stop the flow of gas to an unlit burner; however, if the sensing device or a gas valve fails, gas will continue to flow. It is common to see such burners without this sensing equipment, especially in cooking appliances.

Uncapped Pipes and Outlets

Appliances may have been disconnected from the fuel gas line without a cap being placed on the end of the gas pipe and outlets. If the shutoff valve is open, gas will escape from the uncapped pipe.

> **TIP**
>
> Misuse of the equipment or improper storage of combustible items near these pieces of equipment may also cause a fire.

Malfunctioning Appliances and Controls

Fugitive gas may leak around the controls, valves, fittings, and pipe junctions within the appliance control **FIGURE 7-39**. For example, accumulated dirt or debris could cause valves to allow gas to pass through even when they are set to a closed position. Additional expertise or resources may be needed to analyze the equipment if a malfunction is suspected.

Gas Pressure Regulators

A failure can occur within the internal diaphragm of a gas regulator, which then allows a leak from the regulator and potentially enables excess pressure to enter the system. Failures of the rubber-like seals that control the input of the gas can also occur, which would likewise allow a leak around the regulator and potentially excess pressure in the system. The vents to the regulator can become plugged or fail, allowing the system to become overpressurized.

Corrosion

Corrosion is believed to be the cause of many gas leaks. Corrosion can take place above or below ground. Generally, this type of failure takes a long time to develop. Corrosion is typically caused by rust, electrolysis, or microbiological organisms. Flexible brass appliance connectors may crack from corrosion and stress and are a factor in many fires and explosions.

Physical Damage to the System

Physical or mechanical damage to the gas system may also cause leaks. Such damage can occur from an auto collision with pipes or a meter, soil subsidence, the movement of appliances, and other causes. The point of the leak may be a long distance from the physical impact to the gas piping. Damage sufficient to allow gas leakage may appear at pipe junctions or unions, often at threaded portions of the pipe **FIGURE 7-40**. Hidden or underground pipes may be damaged by construction or other digging activities **FIGURE 7-41**. On occasion,

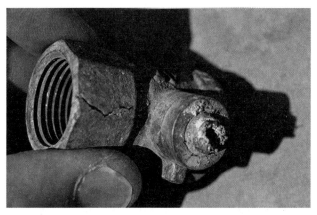

FIGURE 7-40 Damage is visible on the gas fitting.
Reproduced with permission from *NFPA 921-2021, Guide for Fire and Explosion Investigations*, Copyright © 2020. National Fire Protection Association. This reprinted material is not the complete and official position of the NFPA on the referenced subject, which is represented only by the standard in its entirety.

FIGURE 7-41 A gas line that has passed through a broken waste line. The auger of a "drain snake" that cut the plastic yellow gas line is visible behind the red tape.
Courtesy of Jeff Spaulding.

the area where the leak occurred may be remote from the point of ignition and the actual explosion or fire damage encountered by the investigator, as the gas may travel and spread out until it finds an ignition source.

Testing of Gas Systems

If fuel gas may be a factor in the fire's origin or spread, the investigator should test for leaks of the gas supply system prior to its disassembly. Expertise in conducting such evaluations is needed.

Several methods to test gas piping are available to examine for leaks. Typically the gas supply system must be pressurized during the test, with the pressure not exceeding the pressure ordinarily contained within that system. Damaged sections of the piping should be isolated so that undamaged sections can be tested. Take care not to unscrew or disassemble components before testing, as evidence can be inadvertently destroyed at the point where the leak occurred. Try to isolate the piping for testing in an area where the leak did not occur and where the proper fittings can be attached.

Gas Meter Testing

Testing can be done with a gas meter, but only if it is safe to reintroduce gas into the system. The meter can detect the gas escaping from the system **FIGURE 7-42**.

Pressure Drop Method

To check the gas piping, the system is pressurized with air or an inert gas. This type of test is appropriate for operating systems of 0.5 psi (3.4 kPa) or less. Once the system is pressurized, the air supply is capped, and the pressure is monitored by a pressure gauge. A drop in pressure indicates a potential leak within the system.

Locating Leaks

A number of methods can be used to locate leaks in fuel gas piping—for example, the soap bubble test or the use of a gas detector survey.

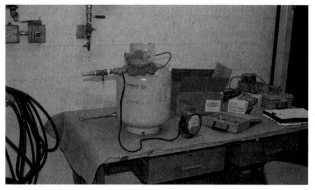

FIGURE 7-42 Gas regulator test.
Courtesy of Jeff Spaulding.

Soap Bubble Solution

The soap bubble solution is a simple method to locate leaks in a pressurized system. It entails the application of soap to the piping in question. If bubbling is observed, then there is a leak at that location **FIGURE 7-43**.

Gas Detector Surveys

Survey devices such as flammable gas indicators, combustible gas indicators, explosion meters, or "sniffers" may be used to detect the presence of gas, hydrocarbon gas, or vapors. A **combustible gas indicator** is used in gas detector surveys and may also detect the presence of other gases. When using this method on the exterior of the building, the fire investigator should test for fugitive gases at every possible ground opening, including all drains or pipes that exit the ground. In gas detector surveys performed on the interior, test junction boxes and unions of gas piping.

If an underground gas leak is suspected, an appropriate surveying procedure is to create **bar holes**. After a series of holes are drilled in the vicinity of the underground piping, readings are taken at each hole and the results are graphed or charted. This type of survey can potentially lead the investigator to the area of the leak. The utility company uses an electronic locator device to locate the underground gas lines.

Exterior surveys also include vegetation surveys. Vegetation situated in the vicinity of a long-term gas leak may die, turn brown, or be stunted.

Testing Flow Rates and Pressures

Anytime a fuel gas system is suspected to be involved in the ignition of a fire, components of the system that survive the fire should be tested to determine whether they were operating properly. Testing may be conducted either in the field or in a laboratory to analyze regulator operating pressures and safety devices and to locate leaks within a system. Testing should be performed only by those personnel who are trained and knowledgeable about these tests. If a flammable gas is being tested, all potential ignition sources should be removed.

Removal and Preservation of Gas Piping

It is important to maintain the evidentiary value of piping and its components when conducting an investigation. When possible, gas systems and piping should not be disturbed until all parties of interest can examine them.

Do not begin to test or disassemble fuel gas systems without proper training. Seek out a trained gas expert or investigator experienced with the investigation of a fuel gas system–related event to check the system for leaks. Prematurely removing system components or altering the system prior to proper analysis and testing may lead to a claim of spoliation that can adversely affect court proceedings.

If a determination is made that section(s) of gas piping should be collected, take care to preserve the evidentiary value of the components. If a section of interest must be removed from the system, it is preferable to cut the pipes outside the area to be isolated, taking care not to disturb the area of interest **FIGURE 7-44**. Screw junctions, unions, tees, or elbows should not be unscrewed, as this action may inadvertently damage or destroy evidence of a loose or improper connection. When removing the section of interest, preserve the orientation and configuration of the isolated area related to the overall system, and carefully note all changes that are made or that occur during the procedure.

FIGURE 7-43 Soap bubble solution test.
Courtesy of Jeff Spaulding.

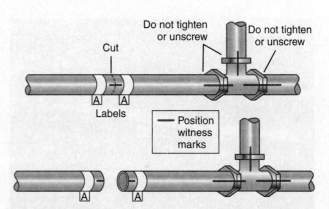

FIGURE 7-44 A nonintrusive, nondestructive method of marking and cutting of fuel gas pipes so that the relationships of the severed cut ends can be recorded.

Reproduced with permission from *NFPA 921-2021, Guide for Fire and Explosion Investigations*, Copyright © 2020. National Fire Protection Association. This reprinted material is not the complete and official position of the NFPA on the referenced subject, which is represented only by the standard in its entirety.

After-Action REVIEW

IN SUMMARY

- The presence of electrical wiring or components in fire damage, or even in the origin of a fire, does not by itself indicate that an electrical event was the cause. An investigator finding an electrical cause to a fire event must always be able to demonstrate the specific failure and ignition mode that started the fire.

- Always approach an electrical device or component initially as if it is energized, and always carry a noncontact AC voltage detector to check such systems before any examination or handling.

- Wear appropriate PPE when in proximity to electrical equipment.

- An investigator handling and examining electrical equipment should keep in mind the issues surrounding spoliation of evidence, and typically should not disassemble electrical equipment.

- The units of electricity are voltage (volts [V]), current (amperes or amps [A]), resistance (ohms [Ω]), and power (watts [W]).

- Ohm's law defines the relationship among voltage, current, and resistance. If two of these three values are known, it is possible to determine the third.

- In an ideal situation, the investigator should be able to calculate the total power requirements of a single circuit with multiple loads on it by summing up the power requirements of each appliance or piece of equipment in the circuit.

- Ignition by electrical energy requires an energized circuit, sufficient heat and temperature, proximity to combustible material, and sufficient time to raise the fuel up to its ignition temperature.

- Overload is the persistent operation of equipment in excess of its normal, full-load rating or of a conductor in excess of rated ampacity, which over time can cause damage or dangerous overheating.

- Most buildings, structures, and dwelling units use alternating current (AC) electricity.

- There is usually no overcurrent protection between the utility transformer and the service entrance to a building.

- Because electricity flows from high potential (voltage) to ground and takes all available paths that allow the current flow, a good grounding system is needed to draw any fault current away from unintended paths.

- Overcurrent protection is provided by a device that stops the flow of electricity when an abnormally large amount of current is detected, preventing damage to conductors and devices on the circuit involved.

- Overcurrent protection devices can be either resettable, such as a circuit breaker, or nonresettable, such as a fuse.

- Whereas it may take several minutes for the overcurrent protection device to activate for a small increase in current flow over its rating, the activation occurs more quickly in the face of a large current flow.

- Each branch circuit should have its own overcurrent protection device, either a fuse or a circuit breaker.

- Insulation on conductors prevents faults or leakage via an unwanted current path.

- In push-in terminal (backstabbed) receptacles, the conductor is not attached to the screw terminal but is pushed in the back of the receptacle.

- For ignition from an electrical source to occur, the following conditions must exist: Wiring must be electrically energized, sufficient heat must be produced by the electrical event, the combustible material must be one that can be ignited by the energy produced by the failure, and the material must be close enough for a sufficient time to allow ignition.

- When electricity flows through most materials, including wires and conductors, heat is produced because of the inherent resistance of the material the current passes through. The heat is dissipated under normal conditions if used within the safety margins calculated for the size of the conductors (ampacity).

- If a heat-producing device is being considered as the cause of a fire, the investigator should determine the temperatures produced by the device to see if they are sufficient to ignite nearby materials.

- A poor electrical connection can allow heating at the connection point, resulting in the formation of an oxide that increases resistance; this oxide contributes to heating and can, under some circumstances, cause a glowing hot spot that can ignite nearby combustibles.

- Overcurrent happens as the result of a fault, such as a short circuit or ground fault, or an overload.

- In a properly protected circuit, a short circuit event is expected to trip the overcurrent protection.

- Undersized cords can produce overheating if used far beyond their capacity.
- The temperature of a heat-producing device, or the heat generated from an undersized conductor, can degrade the insulation between conductors, especially if they are insulated by other materials and not in free air.
- Because of their typically short duration, arcs may not be competent ignition sources unless ignitable vapors or gases are present.
- Arc tracking may begin when the surfaces of noncombustible surfaces become contaminated with salts, conductive dusts, or liquids, or with water that contains material such as dirt, dusts, salts, or mineral deposits.
- If a sparking metal is copper or steel, the particles cool as they fly through the air. If the sparking metal is aluminum, the particles may burn as they fly through the air and are more likely to ignite nearby combustibles.
- A floating (open) neutral condition can lead to unbalanced voltages at the loads on 120 V circuits, leading to damage or failure of appliances and sensitive circuits.
- It may be difficult to identify the area of the initial parting arc if subsequent melting occurred.
- Damage to stranded cords is more difficult to read and interpret than damage to solid conductors.
- Damage to conductors may provide some evidence as to whether the damage comes from fire, arcing, or overload, but does not necessarily indicate that the event causing the damage was the cause of a fire.
- Studies of arcing through char have shown that to form an arc bead on a cable in a relevant period of time, the cable must be exposed to heating conditions consistent with flashover, direct flame impingement, or exposure to a hot gas layer.
- If overloading is severe, a conductor can ignite fuels in the vicinity and melt the conductor.
- Damage from electrical arcing is localized, and a sharp line of demarcation separates the point of damage and the surrounding conductor material.
- Nicked or stretched conductors that reduce a conductor's diameter in a localized area usually will not cause heating at that location under normal load conditions.
- The only conclusion that can be definitively made after finding an arc-melted (beaded) conductor is that the conductors were energized at the time of the fire.
- Examples of situations that can create static electricity include walking across a carpet, conveyor belts moving over rollers, and flowing liquids.
- Static electricity can be generated by movement of liquids in relation to other objects, and can occur in flow through pipes, mixing, pouring, pumping, spraying, filtering, and agitating.
- Static electricity can build up when a flowing gas vapor mixes with metallic oxides, scale particles, dust, and liquid droplets or spray. When such a gas is directed against a conductive, ungrounded object, static buildup may occur.
- For static arc ignition to occur, the following conditions must be present: a means of generating the static charge, a means of accumulating and maintaining the charge, a discharge arc with sufficient energy, a fuel source with the correct air mixture and a small ignition energy, and a co-location of the arc and fuel source.
- Lightning strikes carry high electrical potentials (hundreds of thousands of volts) and large currents (thousands of amps). Both line-voltage (120 V AC) and low-voltage (<50 V DC) systems in buildings are susceptible to damage from lightning strikes.
- The two fuel gases most commonly used in structures are natural gas and commercial propane.
- During every fire scene examination, the presence and condition of fuel gas systems need to be examined and documented, since fire and explosion incidents may result from failures within these systems.
- A fuel gas system or appliance can act as the initial fuel source, the initial ignition source, or both, or can cause fire to spread rapidly through a structure.
- Ignition sources can include static arcs from appliances, electrical arcs, arcs from switches or contacts in appliances, or other electrical equipment, or the pilot light or open flame from a gas appliance itself.
- Natural gas is composed largely of methane and has a specific gravity that is lighter than air.
- Propane is 95 percent propane and propylene and has a specific gravity that is heavier than air.
- Odorant is added to natural gas and propane during the distribution chain as a safety feature so that people can be alerted by the odor to the presence of gas. It should be detectable when the gas or propane is at not less than one-fifth of its lower explosive limit.

- Gas systems in homes and commercial buildings contain various distribution piping, valves, burners, and regulating devices, all designed and installed to deliver gas safely to appliances.
- Gas piping is not permitted to be run through air ducts, clothes chutes, chimneys, gas vents, ventilating ducts, dumbwaiters, or elevator shafts, and should never be installed in areas that are subject to damage or exposed to corrosion.
- The investigation and analysis of the fuel gas system and its components should be done in a systematic manner to ensure thoroughness.
- Investigators documenting a fuel gas system should measure and diagram the system; they should also identify and measure the piping and distribution system materials, including valves, couplings, or connectors, ideally from an involved appliance back to the gas source.
- Leaks can occur in pipe junctions if the threading is inadequate for a good connection, if the junction between the pipe and coupling is improperly joined or threaded, or if pipe joint compound was used improperly.
- Fugitive gas may leak around the controls, valves, fittings, and pipe junctions within the appliance control when an appliance malfunctions or has an accumulation of dirt or debris.
- Physical or mechanical damage to the gas system may cause fugitive gas leaks. The point of the leak may be a long distance from the physical impact on the gas piping.

KEY TERMS

Alternating current (AC) An electric current that changes its direction many times per second at regular intervals.

Ampacity The current, in amperes, that a conductor can carry continuously under the conditions of use without exceeding its temperature rating. (NFPA 70)

Amperage The strength of an electric current, measured in amperes.

Arc A high-temperature luminous electric discharge across a gap or through a semiconductive medium such as charred insulation. (NFPA 921)

Arc-fault circuit interrupter (AFCI) A device designed to protect against fires caused by arcing faults in home electrical wiring. The circuitry continuously monitors current flow for the electrical waveforms characteristic of arcing.

Arc severed The result of an arcing event in which the conductors are severed due to the arc.

Backflow valve A valve that prevents gas from reentering a container or distribution system.

Bar holes Holes driven into the surface of the ground or pavement with either weighted metal bars or drills. Gas detectors are then inserted into the holes to detect the location of a gas leak.

Branch circuits The individual circuits that feed lighting, receptacles, and various fixed appliances. (NFPA 921)

Combustible gas indicator An instrument that samples air and indicates whether combustible vapors are present. (NFPA 921)

Commercial propane Derived from the refining of petroleum, a liquefied gas comprising 95 percent propane and propylene and 5 percent other gases.

Container appurtenances The devices that are connected to the openings in tanks and other containers; they include pressure relief devices, liquid level gauges, and pressure gauges.

Direct current (DC) An electrical system that has current flow from the source to the load and back via the circuit return path with one polarity only.

Electrical power The time rate at which work is finished or energy emitted or transferred. The unit of power is the watt.

Eutectic melting Any combination of metals with a melting point lower than that of any of the individual metals of which it consists.

Fixed level gauges Devices that are primarily used to indicate when the filling of a tank or cylinder has reached its maximum allowable fill volume. They do not indicate liquid levels above or below their fixed lengths. (NFPA 921)

Fuel gas Includes natural gas, liquefied petroleum gas in the vapor phase only, liquefied petroleum gas–air mixtures, manufactured gases, and mixtures of these gases, plus gas–air mixtures within the flammable range, with the fuel gas or the flammable component of a mixture being a commercially distributed product. (NFPA 500)

Fugitive gas Fuel gases that escape from their piping, storage, or utilization systems and serve as easily ignited fuels for fires and explosions. (NFPA 921)

Fusible plugs Thermally activated devices that open and vent the contents of a container.

Gas burner A device that allows fuel gases and air to properly mix and produce a flame.

Ground fault An inadvertent connection of current-carrying equipment or wiring to ground, causing the current to run to ground instead of returning through the neutral cable.

Ground-fault circuit interrupter (GFCI) A device intended for the protection of personnel, which functions to de-energize a circuit or portion thereof within an established period of time when a current is detected to be flowing outside of the intended current path.

Grounding A connection, whether intentional or accidental, between an electrical circuit or equipment and earth or to some conducting body that serves in place of the earth. (NFPA 921)

High-resistance fault An unintended path for electricity that allows sufficient current for heating to occur, but not enough to activate the overcurrent protection device.

Horsepower (hp) A unit of power that is used to express mechanical energy use or production. One kilowatt is equal to 1.34 horsepower.

Hot legs The energized conductors in a circuit.

Interrupting current rating The maximum amount of current a device is capable of interrupting.

Liquefied petroleum (LP gas) Petroleum gases condensed to a liquid state with moderate pressure and normal temperatures to allow for more efficient distribution; also called propane.

Main disconnect A component that provides a means to shut off power to the entire electrical system. It may incorporate a circuit protection device such as a fuse or circuit breaker.

Meter A device that plugs into the meter base to measure the amount of electricity consumed at a site.

Meter base The enclosure that contains the connections for the utility's electrical meter and for the service entrance conductors.

Natural gas A naturally occurring, largely hydrocarbon gas product recovered by drilling wells into underground pockets, often found in association with crude petroleum.

Noncontact voltage tester A device that will emit a visual and/or audio signal in the presence of an electromagnetic field produced by an AC voltage.

Ohm's law A basic law of electricity that defines the relationship among voltage, current, and resistance. If two of these three values are known, it is possible to determine the third.

Overcurrent Any current in gross excess of the rated current of equipment or the ampacity of a conductor; it may result from an overload, short circuit, or ground fault. (NFPA 921)

Overfusing A dangerous condition that occurs when the circuit protection (fuse or circuit breaker) rating significantly exceeds the ampacity of the conductor, leading to increased heat in the conductors.

Overload Operation of equipment in excess of normal, full-load rating or of a conductor in excess of rated ampacity; when it persists for a sufficient length of time, it could cause damage or dangerous overheating. (NFPA 921)

Parting arc An arc that occurs when conductors separate while current is flowing, either in normal conditions (e.g., when a switch is activated or a plug is pulled from an outlet) or abnormal ones (e.g., when an energized conductor contacts a grounded conductor or other item with near-zero circuit resistance, causing a surge of current that melts the points of contact and creates a gap).

Polyvinyl chloride (PVC) High-strength plastic that is used in many applications, including pipes and wire insulation.

Pressure gauge A type of container appurtenance depicting the internal pressure of a tank. The gauge is connected directly to the tank or sometimes through the valve. Pressure gauges do not indicate the quantity of liquid propane in the tank.

Pressure relief valve A valve designed to open at a specific pressure, usually around 250 psi (1700 kPa). It is generally placed in the container, where it releases the vapor.

Regular current rating The level of current above which the protective device will open, such as 15 A, 20 A, or 50 A.

Relaxation time The amount of time for a charge to dissipate.

Resistance heating Heating that occurs when current flows through a path that provides resistance to current flow, such as a heating element (intentional) or a resistive connection (unintentional).

Resistive circuit A circuit that does not contain inductive or capacitive elements.

Root mean square (RMS) A mathematical computation that equates the voltage level of an alternating current (AC) system to that of a direct current (DC) system.

Service drop The overhead service conductors between the power pole and the structure, up to and including any splices, which connect to the service entrance conductors.

Service entrance The area where the electrical service enters a building.

Service equipment The equipment used to shut off, distribute, and protect the wiring to a facility. It includes the main disconnect, overcurrent protection devices, and the main distribution panelboard.

Service lateral The service entrance wiring entering the structure from underground.

Service mast The conduit between the weatherhead and the meter base.

Short circuit An abnormal connection of low resistance between normal circuit conductors where the resistance normally would be much greater. (NFPA 921)

Shutoff valve A valve located in the piping system and readily accessible and operable by the consumer; it is used to shut off individual appliances or equipment.

Sine wave The waveform that an alternating current follows.

Static electricity The electrical charging of materials through physical contact and separation, and the various effects that result from the electrical charges formed by this process.

Switch loading Loading of a product into a tank or compartment that previously held a product with a different vapor pressure and flash point.

Thermistor A sensor whose resistance changes in response to temperature; used in temperature measurement.

Time–current curve A graph, provided by the manufacturer, used to determine the activation time of an overcurrent protection device when it is exposed to various currents.

Vaporizer A heater used to heat and vaporize propane where larger quantities of propane are required, such as for industrial applications.

Variable gauge A gauge that gives readings of the liquid contents of containers, primarily tanks or large cylinders. It gives readings at virtually any level of liquid volume.

Venting The escape of smoke and heat through openings in a building. (NFPA 921)

Weatherhead A conduit body at the service entrance used to prevent water from entering the service mast and meter base.

REFERENCES

National Fire Protection Association. 2020. *NFPA 921: Guide for Fire and Explosion Investigations*, 2021 ed. Quincy, MA: National Fire Protection Association.

National Fire Protection Association. 2019. *NFPA 70: National Electrical Code*, 2020 ed. Quincy, MA: National Fire Protection Association.

On Scene

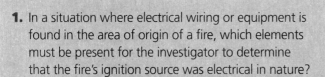

1. In a situation where electrical wiring or equipment is found in the area of origin of a fire, which elements must be present for the investigator to determine that the fire's ignition source was electrical in nature?

2. Untrained people often point to short circuits as a fire cause. In a typical residential electrical installation, is a short circuit a likely or unlikely fire cause, and why?

3. Under which circumstances could a parting arc potentially provide sufficient energy to be the ignition source for a fire or explosion?

4. How might the behavior of fugitive gas leaking from a building gas system differ depending on whether the gas was natural gas or propane?

5. If, during the investigation of an explosion within a building, you believe fugitive natural gas was involved, which points in the gas system should you inspect, and which potential causes might exist for a gas leak to have occurred?

Chapter Opener: © Jones & Bartlett Learning. Photographed by Glen E. Ellman; On Scene siren: © Bildgigant/Shutterstock.

CHAPTER 8

Examining the Fire Scene

KNOWLEDGE OBJECTIVES

After studying this chapter, you should be able to:

- Identify the basic information that should be gathered about an incident for use in a fire investigation.
- Describe how to organize the investigative functions. (**NFPA 1033: 4.1.6**, pp. 219–223)
- Describe the role of the pre-investigation team meeting. (**NFPA 1033: 4.1.4; 4.1.6**, p. 222)
- Identify the roles of specialized personnel and technical consultants. (**NFPA 1033: 4.1.4; 4.1.6; 4.6.3**, pp. 222–223)
- Explain the need for case management in fire investigations. (**NFPA 1033: 4.1.6**, pp. 223–224)
- Describe the process and considerations of an initial scene assessment. (**NFPA 1033: 4.2.1**, pp. 224–227)
- Describe methods of securing the fire scene. (**NFPA 1033: 4.2.1**, p. 226)
- Describe the process of and considerations for an exterior survey. (**NFPA 1033: 4.2.2**, p. 227)
- Describe the process of and considerations for an interior survey. (**NFPA 1033: 4.2.3**, pp. 227–228)
- Explain postfire evaluations of building systems. (**NFPA 1033: 4.2.8**, p. 228)
- Explain considerations related to identification and protection of evidence. (**NFPA 1033: 4.4.2**, pp. 227–228)
- Describe the debris removal process. (**NFPA 1033: 4.2.6**, pp. 232–235)
- Explain reconstruction of a fire scene. (**NFPA 1033: 4.2.7**, pp. 233–234)

SKILLS OBJECTIVES

After studying this chapter, you should be able to:

- Develop a preplan including resources to be prepared to conduct a fire investigation when the "team approach" is required. (**NFPA 1033: 4.1.4; 4.1.6; 4.6.3**, pp. 220–223)
- Describe the measures taken to physically secure a scene. (**NFPA 1033: 4.2.1**, p. 226)
- Describe how to perform an exterior assessment. (**NFPA 1033: 4.2.2**, p. 227)
- Describe how to perform an interior assessment. (**NFPA 1033: 4.2.3**, pp. 227–228
- Describe how to search through layered debris while preserving evidence. (**NFPA 1033: 4.2.6, 4.2.7**, pp. 232–235)

You are the Fire Investigator

Your public-sector investigation team has been called to the scene of a fire in a two-story, eight-unit apartment building, which has experienced significant damage throughout the building. You observe that the roof of the building has burned away in spots. Occupants have been displaced, and the police department has identified more than a dozen people with information who are awaiting interviews. The fire department is in the overhaul phase of its operations, and some of the building's contents have been moved to the front courtyard. You are the lead investigator and will be meeting with the incident commander (IC) to discuss the status of the fire and any safety issues encountered by firefighters.

1. What information would be useful while conducting the pre-investigation meeting?

2. Which investigation activities will likely need to be assigned for this investigation? What will your investigative priorities be?

3. Which types of equipment or supplies may be needed to complete the investigation?

Access Navigate for more practice activities.

Introduction

A successful scene examination requires the fire investigator to be well prepared; to have proper safety gear, tools, and equipment; and to follow a careful, thorough methodology to obtain as much data as possible for a correct determination of fire origin and cause. Like all investigation processes, the steps taken in the scene examination may depend on the jurisdiction and assignment of the investigator.

This chapter discusses the steps taken in scene examination. The evaluation and analysis of the data gathered in the examination are discussed in Chapter 13, *Determining the Origin*, and Chapter 14, *Cause Determination*. Other topics with which the fire investigator should be familiar when performing the scene investigation include entry and spoliation of evidence, safety, and documentation; all of which are addressed in Chapter 5, *Legal Considerations for Fire Investigators*; Chapter 6, *Safety*; and Chapter 11, *Documenting the Fire Scene*.

NFPA 921 specifies that the scientific method should be the overall approach used in any fire investigation (see the methodology discussion in Chapter 1, *Introduction to Fire Investigation*). It also identifies a basic investigative method and a recommended process, whose use will help ensure that data collection is done thoroughly and correctly. Investigators are encouraged to apply a systematic approach to each investigation to ensure that all issues and areas are covered thoroughly in an investigation, and that relevant avenues are not inadvertently ignored.

Planning and Preparation

Proper planning and preparation for the scene examination can make a big difference in the overall success of the investigation. Planning includes being prepared in advance for the investigation with the proper training, knowledge, tools, and safety equipment. It also includes a variety of prearrival tasks and information gathering that occur before responding to a particular scene, while responding, and when first on scene. These tasks will help the fire investigator be ready to address both expected and unexpected challenges found during the examination.

The purpose of the investigation may vary by position within a department or organization or by assignment. An assignment may require an investigator to define the origin only, to determine both origin and cause, to analyze fire spread issues, or to assess the function of particular equipment or safety devices. Public-sector investigators are also often charged with determining whether a crime has been committed. Each type of assignment may bring up unique issues requiring different resources. For example, if the assignment requires understanding not only the fire's origin and cause but also responsibility for its cause, post-scene investigation resources may be required.

Basic Incident Information

Although all fires are unique, certain information must be gathered about all fire scenes during the initial assignment and response to the investigation. While quick action may at times be necessary, the fire investigator should err on the side of accuracy when possible when gathering information, as this can improve safety, improve the quality of the information received, and reduce the potential for acting on bias instead of on reliable information. The accuracy of the information gathered at this point is paramount to determining many of the procedures and needs that may follow.

During the initial assignment and response, basic incident information obtained should include the following items:

- The location of the incident
- The date and time of the incident
- The weather conditions at the time of the incident
- The size and complexity of the incident
- The type and use of the structure involved in the incident
- The nature and extent of the damage
- Whether injuries to civilians or firefighters are reported
- The security of the scene
- The purpose of the investigation

This information enables the investigator to determine which resources will be needed for the initial response. For example, if the stability of the structure is compromised, structural evaluations or placement of shoring may be needed before the scene investigation can proceed; however, if severe structural compromise is present, a structural engineer or public building official may condemn (red tag) a structure, eliminating the possibility of interior evidence collection. If the fire created a large loss with the potential to involve multiple interested parties, the response may require heavy equipment, laborers, and environmental monitoring. Complete data are instrumental in providing the resources necessary to complete the task.

Weather may also play an important role in the type of equipment and resources needed. Where conditions are expected to reach freezing temperatures, there may be a need to seal the structure and to provide heat until the compartment is suitable to conduct an investigation without destroying data during the debris removal process. At the opposite end of the spectrum, hot weather conditions may require an early morning investigation with frequent breaks and shelter for investigators.

Safety Planning and Preparation

Planning for safety issues is critical. Fire scenes present a variety of safety hazards including, for example, overhead hazards, foot hazards, slips and falls, and electrical issues. The fire investigator must be prepared to address such issues at each scene and to use the appropriate personal protective equipment (PPE), including:

- Eye protection
- Flashlight
- Gloves
- Helmet or hard hat
- Respiratory protection (the type will depend on the exposure)
- Safety boots or shoes
- Turnout gear or coveralls
- Cell phone or two-way radio

Additional and unusual hazards may be present at some scenes, and the fire investigator should attempt to learn as much information about those hazards and issues as possible before entering. These hazards can include structural integrity compromise, hazardous chemicals and other materials on scene, hazards that have developed during the fire, and hidden or unusual hazards specific to the scene. Such dangers may require the use of special PPE. All known hazards should be reported to incoming team members, and the hazards should be monitored on an ongoing basis during the examination (see Chapter 6, *Safety*).

Organizing the Investigation Functions

The person who is responsible for planning the investigation should identify the functions and objectives of the specific investigation. As the fire investigator, this person can preplan for the investigation and identify potential resources and team members according to their skills and the investigation's needs **FIGURE 8-1**.

FIGURE 8-1 Team members assemble prior to an investigation.
© Jones & Bartlett Learning. Photographed by Glen E. Ellman.

Six basic functions are commonly performed in each investigation, and may need to be assigned to other team members:

1. Leadership and coordination
2. Safety assessment
3. Photography, note taking, mapping, and diagramming
4. Interviewing witnesses
5. Examining the scene and debris
6. Evidence collection and preservation

It may also be necessary to fill specialized roles to complete the investigation. For example, specialized personnel may be needed to evaluate the fire detection and control; to examine electrical, heating, or air-conditioning systems; or to evaluate fire spread or ignition dynamics.

Resources may also need to be planned for the follow-up investigation at the scene that is typically required in an assignment, and a fire investigator should be familiar with the data and resources that may be available to complete the postfire investigation. If the scene examination suggests that a criminal act occurred, an investigator should understand how to obtain data away from the scene, such as financial records, financial history, building records, and business statements. Similarly, if responsibility to determine spread issues is required, the investigator needs to be able to obtain information on building systems, fire spread experts, flame spread characteristics of products, and codes.

Pre-Investigation Team Meeting

The **pre-investigation team meeting** is the ideal forum for the team leader or lead investigator to explain the goals and objectives of the team and to introduce team members to one another. The team leader should address issues such as jurisdictional boundaries, legal authority to conduct the investigation, assignment of specific responsibilities, and collection of basic information on teammates for accountability and reporting purposes **FIGURE 8-2**.

The team leader should also ensure that each member of the investigation team is aware of the conditions at the scene and any safety considerations. The leader should address needed clothing and PPE, as well as other supporting equipment that may be necessary.

Once the investigation begins, there should be further briefings for all members to relay pertinent information that is gathered during the day. At these meetings, team members can get an update on the overall investigation and will have the opportunity to

FIGURE 8-2 The pre-investigation team meeting allows for direct communication among team members.
© Jones & Bartlett Learning. Photographed by Glen E. Ellman.

request additional information from other members, the assistance of specialized personnel, and so forth.

Tools and Equipment

Fire investigation also requires specific tools and pieces of equipment, as listed in **TABLE 8-1**.

Specialized Personnel and Technical Consultants

Planning the investigation should include consideration of which specialized personnel should be brought into the investigation, either at the outset or on an as-needed basis. Fire scene investigators must recognize the limitations of their expertise and plan for obtaining the services of personnel who can provide assistance in areas in which their own knowledge is lacking. Ideally, if specialized help is needed, a fire investigator can include persons with such abilities in the initial team. Otherwise, the investigator should know where to obtain them if the need arises.

Ideally, the fire investigator will have identified expert resources in advance of the incident, being mindful that there are different types of experts or technical consultants, and that a title or degree does not guarantee the consultant is the exact type of resource required. Even among experts, there are specialists in particular fields; for instance, simply because a person is an electrical engineer or mechanical engineer, that does not necessarily mean the individual can evaluate a building wiring system issue or a commercial oven fire.

When the data being analyzed include reports, documents, or financial data, the post-scene investigation

TABLE 8-1 Recommended Tools and Equipment

Absorbent material	Air blaster (found in camera stores)
Ax	Batteries
Broom	Calcination gauge
Camera	Char gauge/digital caliper
Claw hammer	Directional compass
Evidence-collection container	Evidence labels with adhesive
Hand towels	Hatchet
Hydrocarbon detector	Ladder
Lighting	Magnet
Marking pens	Metal detector/probe
Paintbrushes	Paper towels/wiping cloths
Pen knife	Pliers/wire cutters
Pry bar	Rake
Rope	Ruler/straight edge
Saw	Screwdrivers (various types)
Shovel	Sieve/sifting screens
Soap and hand cleaner or moist towelettes	Styrofoam cups
Tape measure	Tape recorder
Tongs	Tweezers
Twine	Voltage detector
Voltmeter/ohmmeter	Water
Writing/drawing equipment	

may require experts with competencies to analyze and identify or confirm theories or conclusions. For example, evaluation of building history, construction records, codes, and flame spread characteristics of materials may provide data to support theories of fire spread and affix responsibility to a building feature or material for the fire spreading and causing damage. If no one on the team is qualified to perform these functions, seek an outside expert.

TABLE 8-2 lists some examples of specialized personnel who might be needed in a fire investigation.

Management of Documents and Information

Information will be gathered and generated throughout the investigation, and an investigator will need to have a system in place to retain and save all of the information learned.

NFPA 921, *Guide for Fire and Explosion Investigations*, suggests that notes compiled during the investigation are evidence and need to be retained with the file, even after reports are generated from the information contained in them. Notes from other team members should likewise be saved. If electronic means are used for record retention, the notes can be scanned and saved. This helps ensure that all of the empirical data compiled by the team are recorded and retained, including notes and drawings made by all team members.

The complexity of the investigation will dictate the detail of sketches, drawings, or data collection. A simple, small fire may require only a sketch, whereas a large, complex investigation may require several drawings—fire flow drawings, photo position drawings—or other visual aids to support the investigation and the analysis.

The complexity of the report is often dictated by the responsibility of the investigator or mandated by jurisdictional requirements, but in all cases, the report should contain the empirical data relied on for the investigator's conclusions and opinions. Depending on the jurisdictional requirements, the fire scene investigation report defining the origin and cause may be separate from the postfire investigation, which may involve the analysis of responsibility (see Chapter 11, *Documenting the Fire Scene*, for more information).

Effective case management of the entire investigation from the fire scene to postfire investigation will ensure that all of the empirical data needed to meet the scope of the investigation are obtained. With proper management and data collection, a fire investigator can then publish the information as required—in reports or in litigation presentations **FIGURE 8-3**.

TABLE 8-2 Specialized Personnel	
Position	**Description**
Materials engineer or scientist	A person with specialized knowledge of how materials react to different conditions, including heat and fire (e.g., someone with a metallurgical background for metals or a polymer scientist or chemist for plastics)
Mechanical engineer	A person qualified to analyze complex mechanical systems or equipment (e.g., heating, ventilation, and air-conditioning systems) and who may also be able to perform strength-of-material tests
Electrical engineer	A person who can provide information about building fire alarm systems, energy systems, power supplies, or other electrical systems or components
Chemical engineer/chemist	A person qualified to help identify and analyze possible failure modes where chemicals are concerned
Fire science and engineering specialists	A variety of experts who can provide advice and assistance in understanding the dynamics of fire spread from the origin; the energy needed for ignition; issues related to causation, spread, and fire dynamics; and the reaction of fire protection and detection systems
Fire protection engineer	A person who assesses the relationship of ignition sources to materials to determine what may have started the fire; knows how fire affects materials and structures; is able to assist in the analysis of how a fire detection or suppression system may have failed; and knows crucial information about building and fire codes, fire test methods, fire performance of materials, computer modeling of fires, and failure analysis
Fire engineering technologist	A person with a bachelor of science degree in fire engineering technology, fire and safety engineering technology, or a similar discipline, who has studied various topics related to fire science, investigation, suppression, extinguishment, and prevention; hazardous materials; fire-related human behavior; safety and loss management; fire and safety codes and standards; and fire science research
Fire engineering technician	A person with an associate of science–level degree in fire and safety engineering technology or similar discipline, who has studied various topics related to fire science, investigation, suppression, and prevention; hazardous materials; fire-related human behavior; safety and loss management; fire and safety codes and standards; and fire science research
Industry expert	A person who is an expert in a field related to a specialized industry, piece of equipment, or processing system involved in an investigation
Attorney	A person who can provide legal assistance regarding rules of evidence, search and seizure laws, gaining access to a fire scene, obtaining court orders, and dealing with spoliation concerns
Insurance agent/adjuster	A person who can offer information regarding the building and its contents prior to the fire, fire protection systems in the building, and the condition of those systems
Canine-handler teams	Teams of ignitable liquid detection (IGL) canines with handlers, which are used to assist in collecting samples for laboratory analysis to identify the presence of ignitable liquids

Initial Assessment and Identifying the Scope of the Fire Scene

Initial Scene Assessment

Upon arrival, the investigator should perform the initial scene assessment. The **initial scene assessment** provides an overall look at the structure or occupancy, and surrounding area. As the scene assessment takes place, the collection of data is also occurring—indeed, it should have been taking place since the investigator received the initial call to respond (data collection is step 3 of the scientific method, as described in Chapter 1, *Introduction to Fire Investigation*). The main purpose of this assessment is to determine the scope of the investigation—for example, equipment, manpower, safety, and security requirements.

FIGURE 8-3 Organization is a major role of the fire investigation position.
© Jones & Bartlett Learning. Photographed by Glen E. Ellman.

Furthermore, it helps to identify areas that require detailed inspection.

Site Safety Assessment

Prior to entering the scene, the first step is to perform a **site safety assessment** that takes into account all of the hazards and factors discussed in NFPA 921 and Chapter 6, *Safety*. The overall goal is to determine whether it is safe to enter the scene, which steps are required to make the scene safe to enter, and which PPE will be required when entry is made. Only after appropriate safety measures have been implemented should the initial scene assessment continue.

Identifying the Scope

During the initial assessment, the overall scene in which the investigator(s) will be working should be defined, and the fire area should be determined. Specifically, the **fire area** is the boundary between damaged and undamaged areas, as demonstrated by the various fire effects observed at the scene. Examine the full site and surrounding areas, looking for fire patterns located away from the core of the scene, and evidence that might be on the edge of the scene or outside of it. Do not hesitate to expand the scene as necessary to encompass newly discovered evidence or damage. Important witnesses can be identified, including the reporting party and neighbors. It is important to document findings during this phase.

Walk around the entire exterior to evaluate the possibility of extension from an outside fire, to examine the building construction details and materials, to determine the occupancy or use of the building, and to identify areas that may require further study. Additionally, walk through the interior of the structure to observe the conditions of occupancy, such as

housekeeping and furnishings, and to compare areas of damage to the exterior. The extent of damage in each area should be noted as severe, minor, or none. During this and all subsequent phases of the investigation, a fire investigator should constantly be evaluating the safety of the building and identifying any conditions that need to be corrected. Such conditions may include securing utilities, shoring unstable areas, removing water, cordoning off unsafe areas, and noting inhalation dangers. If a new or more severe threat to safety is discovered, respond to the threat with an appropriate increase in PPE levels or retreat until the hazardous conditions can be remedied.

In an explosion investigation, the outer perimeter of the incident scene (or blast zone) should be established at one and one-half times the distance of the farthest piece of debris found. Blast debris may be propelled for great distances, and if debris is found outside the initial scene perimeter, the scene perimeter should be widened accordingly.

Building Identifiers

Beginning with the initial scene assessment, it may be beneficial to establish building identifiers and to use them throughout the investigation. First verify that the site is the correct property and document the identifiers in your report. These identifiers can provide consistency in various aspects of an investigation—such as in the description of photographs, diagram preparation, report writing, and testimony. Building identifiers or a building marking system can include any of the following:

- Points on a compass (such as the north side of the building)
- Front, back, left side, and right side
- A method similar to those used in the incident command system (Side A, Side B, etc.) or for urban search and rescue

TIP

Do not be hasty during the initial scene assessment. Time spent during this phase may save significant time and effort later. The steps performed during this assessment include determining the safety of the scene, identifying staffing and equipment requirements, and determining the areas that will require a detailed examination.

Weather Considerations

The weather at the time of the fire is an important component of the initial scene assessment and data collection. Wind can play a role in the development of the fire and the patterns left behind. At times, wind can also present a safety issue for investigators. Weather information can be obtained from a number of sources, including the local airport, local media outlets, the National Weather Service, and Internet sources. Another source would be the weather observations of witnesses and firefighters obtained through interviews.

It is important to consider the distance between the point at which the weather readings were taken and the fire scene. Weather conditions can vary dramatically from one location to another. Corroborate official reports with observations from local witnesses.

Securing the Scene

Securing the fire scene is an early responsibility of the fire investigator. This step enhances the safety of investigators and others; permits the scene and evidence in the scene to remain undisturbed, reducing the potential of a spoliation of evidence claim; and allows investigators to work unimpeded. In the case of public-sector investigation, scene securement and monitoring may be essential to preserving the investigator's right to remain on the scene and continue the investigation (see Chapter 5, *Legal Considerations for Fire Investigators*, for more information).

Steps to secure the scene may have already been taken by fire department or other personnel who arrived before the fire investigator. If not, the investigator should immediately seek to determine the scope of the scene and begin security measures. In addition, the investigator may determine that the area to secure is different or larger than the scene originally established by first responding personnel.

Overhaul and Fire Department Operations

At times, the fire investigator may become involved in a scene before fire department operations have ended. The investigator should establish communications with the fire incident command (IC) and determine which operations are under way and in which parts of the scene. Not only does this provide important information for the investigation, but it also allows the investigator to influence the fire operations by keeping them from damaging or destroying potential evidence in the investigation. Fire department personnel should be requested to limit **overhaul** to the minimum necessary and take care with the deployment and movement of hoses. Firefighters should report any item of potential evidence to the investigator or IC. They should also be admonished not to move any breakers, knobs, or switches on electrical or gas equipment unless immediately necessary for safety or suppression and, if any such movement occurs, to document it for the investigation.

Entry by Non-Investigators

Unauthorized persons should be kept out of the scene, and a record should be maintained of any person who enters during the investigation. In public-sector investigations, even members of the fire department or law enforcement agency without direct responsibility in the investigation should be asked not to enter the scene, due to the potential for cross-contamination or inadvertent destruction of evidence. At times, homeowners or tenants may have the right to be on property during the investigation. In these circumstances, they should be encouraged to permit the investigation to proceed and to remain clear of the area of origin and any appliances or equipment that may be involved in the fire.

Securement Methods

Scene securement can be accomplished in a number of ways. At a minimum, warning tape should be placed around the full scene, including all affected areas and debris as described earlier. In some circumstances, portable fencing can be erected to provide a more secure barrier. Such fencing, which is often provided by a property owner or their insurer, has numerous advantages: It reduces the risk of entry during the scene examination, helps preserve the scene for incoming investigators whose arrival may be delayed (e.g., insurance or private investigators retained after the public-sector investigation is complete), and prevents curious members of the public from altering the scene or sustaining injury.

In some cases, local law enforcement may be requested to remain on scene to maintain security and handle onlookers. Scene security is enhanced if personnel are designated specifically to security duties instead of attempting to maintain security and focus on investigative tasks at the same time. Even if personnel are dedicated to the security task, the placement of scene tape is still important.

TIP

Safety should be kept constantly in mind at every fire scene. Safety considerations include use of proper equipment, stability of the structure (including what you are walking on), weather conditions, status of utilities, standing water, air quality, biological and chemical exposure risks, and, if necessary, looking for additional hidden devices in an incendiary fire or explosion. All found hazards should be clearly marked to warn others.

Exterior Survey

Following the initial scene assessment, an **exterior survey** is performed for every structural fire scene. Although the exterior survey is geared toward gathering data and information about the fire and preserving evidence, the fire investigator should also continue to scan for hazards that could injure team members. In addition, the investigator should look for access points to the interior and note all points of building ingress and egress.

A detailed exterior surface examination is critical because even if the fire originated inside the building, some issues may be brought up later in the investigation that may be important and related to the exterior. For example, the location and condition of a window (open or closed) may have affected the ventilation and the subsequent fire growth and spread inside the structure. If this condition is not documented, then its value in determining the fire spread is lost.

Another critical reason to conduct an in-depth exterior examination is to look for additional fires that may have occurred at the scene. Investigators may mistakenly zero in on the first fire-damaged area they find or the area with the greatest damage, not realizing that there could be other areas where fires occurred.

Importantly, even when a fire appears to be concentrated inside a building, its origin could be on the building's exterior. For example, if a fire began on a deck or in vegetation outside the building and subsequently entered the building through roof components, windows, or wall penetrations, there could be substantial damage within the building that could incorrectly lead investigators to focus on the interior as the potential area of origin. The exterior survey may provide important evidence to lead to a hypothesis that the fire's origin was actually outside the building. In such a case, it is particularly important to note and document the patterns supporting the spread from exterior to interior (and all interior damage and patterns should be documented in any event).

As with the initial scene assessment, the fire investigator should adhere to a systematic procedure during the exterior survey. Even if the fire clearly originated in the interior, this process should be used to document various building features, including, but not limited to, the following:

- Construction type, construction materials, any mixed construction, and any evidence of remodeling
- The prefire state of repair of the building and its components, the condition of structural elements, and the condition of the structure's fire suppression and detection systems
- Utilities: Type, location, and meter readings
- Doors, windows, and other openings: Location, condition (open, closed, or broken and for how long prior to the fire), and security mechanisms
- Explosions: The presence or absence of evidence that explosions may have affected the exterior components
- Fire damage: Overall damage, damage to natural openings (windows, doors), damage resulting in unnatural openings (holes made by the fire or explosion, holes made during suppression efforts), smoke and flame venting

The exterior evaluation of fire damage should also account for the fire suppression efforts. These efforts could have a significant impact on the spread of the fire and fire patterns. It is important to interview firefighters to determine which actions were taken on the fire ground, such as forcible entry, ventilation, overhaul, and other related activities.

Interior Survey

In an **interior survey**, all interior rooms and other relevant areas are inspected systematically to identify and narrow down areas requiring more detailed inspection and to gather data for origin hypotheses. As with the exterior survey, this survey also includes identification of additional safety hazards, and structural stability and soundness should be evaluated on an ongoing basis.

Inspect all interior areas of the structure if possible, and at least all rooms and areas that are relevant and/or affected by fire or smoke and rooms adjacent to those areas. Make observations on damage levels between rooms and areas, identifying which rooms may have become fully involved in fire, and comparing interior damage with damage to the exterior.

Among the things to examine are the following:

- Prefire conditions of the structure
- The contents of the structure
- Storage of contents
- Housekeeping
- Maintenance
- Evidence of explosion damage, if any
- Areas of fire damage
- Relative extent of damage (severe, moderate, minor, none) in each area
- Building systems (heating, ventilation, and air conditioning [HVAC]; electrical; fuel gas; fire protection; alarms; and security)

- The composition of and the surface coverings of walls and floors
- Positions of windows, doors, and other openings (ventilation aspects)
- Indicators of smoke and heat movement (i.e., fire patterns) on various surfaces

Although this goal is sometimes difficult to achieve, every effort should be made to identify the prefire conditions of the structure. Various conditions, including construction details and defects; holes in floors, walls, or the roof; and missing doors or windows may have a significant effect on fire development and spread.

Evaluation of Building Systems in Interior Survey

It may also be necessary, when considering fire development and spread, to identify the prefire conditions of various building systems. These systems can include the electrical, fuel gas, HVAC, and fire protection systems, among others.

HVAC systems can enhance the development and spread of fire, particularly if they were in operation during the fire. Inspecting filters for soot accumulation or heat damage can help the investigator determine whether the system was operating during the fire. The location of HVAC vents, both supply and return, in any areas of interest should be documented. Documentation should include the location and dimensions of vent openings and ducts/plenums, as well as the settings of the thermostat and fan(s), and the power supply.

Some HVAC systems, particularly in commercial structures, are fitted with detectors and manual or automatic dampers designed to change or stop operation of the system in the event of fire. The condition, location, and position of these devices should be documented, and monitoring records (if any) should be sought.

The fuel gas service should be inspected and documented to include or exclude the possibility that it might have contributed to the fire. This system should be documented in detail when an ignition or spread scenario includes involvement of fuel gas. In such cases, inspection should include documentation of the position of valves and testing to determine the supply pressure (if possible) and to locate any leaks. Note that fires frequently compromise the integrity of gas piping, threaded joints, and other connections.

In some structures, there may be oil-fired heaters or other appliances requiring a liquid fuel. These systems may be permanent (large boiler) or portable (kerosene heater). The location of such devices, fuel reservoirs, and the quantity of fuel remaining should be documented. If the ignition hypothesis includes this system, a sample of the liquid fuel can be taken in case the type or use of fuel becomes relevant in the case.

The electrical system of the structure should be documented, beginning at the service entrance. Important information to gather includes the main panel amperage and voltage input, the conditions of circuit breakers or fuses (type, rating, position), and areas of flame or heat damage.

Any appliances plugged into receptacles in the area of origin should be identified. Note the location of the receptacles, which branch or circuit they are on, which appliances are plugged into them, and the positions of any knobs and switches. A more detailed inspection and documentation of the electrical system may be required where an electrical ignition scenario is hypothesized or when an arc map survey is warranted. It may become necessary to trace circuits back to the panel, noting the condition of the circuit protection and loads along the circuit.

Fire protection systems and the data recorded by such systems can be of tremendous value in reconstructing the sequence of events in a fire. Detector activation and sprinkler response can be utilized to help identify an area of origin and/or spread scenario. Off-site monitoring services can often provide these data. Some systems retain such data on site in the central panel, which can be downloaded and preserved. The data in such systems must be retrieved quickly because the electrical supply to the structure has likely been terminated and the backup battery life is limited. Special equipment and a qualified technician are generally required to retrieve these data, as they are easily compromised.

The location and condition of devices associated with the fire protection system should be documented. Pay particular attention to the height of wall-mounted devices or, on ceilings, the distance from walls. Additional insights can be obtained from the location of activated sprinklers.

Postfire Alterations

Note any apparent postfire alterations, which may include debris removal or movement, content removal or movement, alterations to electrical service, gas meter removal, and indications of prior investigations. Identify those persons who were responsible for the alterations, and interview them to determine, as closely as possible, the conditions before the alterations occurred.

Identification and Protection of Evidence

Fire scene evidence preservation should be initiated by first responders during the fire attack and overhaul. Ideally, the fire scene—and particularly the room or area of fire origin—should be left as intact and undisturbed as possible. Physical evidence can take many forms and can exist in a variety of locations in a fire or explosion scene examination. Although the most important physical evidence in a fire investigation is generally found within the area of origin, important evidence can also be found elsewhere within the fire scene and frequently outside the fire scene as well **FIGURE 8-4**.

The value of evidence preserved at the scene is often not known until the investigation ends. For this reason, it is best practice while going through fire debris (see the "Debris Removal (Layering) and Fire Scene Reconstruction" section later in this chapter) to methodically photograph the debris that is being examined and to contain it or otherwise place it in a known and marked location where it can be viewed again, and/or viewed by fire investigators who arrive later. Then, if a small piece of equipment is missing (or if its significance was not understood on first review), for example, it might later be found in a labeled container or location containing debris from that specific area.

Errors in identification, documentation, recovery, or preservation may preclude later analysis and use of evidence at trial. In addition, inadvertent or intentional spoliation (alteration or destruction) of physical evidence could potentially expose the fire investigator

to legal sanctions. See Chapter 5, *Legal Considerations for Fire Investigators*, for further discussion of spoliation of evidence.

Protecting Evidence

After a fire has been brought under control, interior crews may begin to discover potentially important physical evidence. There are a number of effective ways to protect physical evidence on a fire ground. The best way is to post a firefighter at the entrance to the area of fire origin or near a critical piece of evidence to restrict access, create a fire watch, and provide additional suppression as required. Trained fire investigators should be called to the fire scene as early as possible in this process.

Potential physical evidence, such as a liquid container, a gun, a tool that may have been used in a burglary, an appliance that shows evidence of electrical faulting, or similar item, can often be "taped off" or identified by an evidence cone. The incident commander (IC) should always be notified when potential evidence is discovered.

A common technique used by crime scene technicians for years has been to "flag, bag, and tag." In this process, the investigator seeks to identify items that may have evidentiary value and uses small flags or flagging tape (similar to that used by surveyors) to identify and protect items during the initial examination of the scene. The location of the items is documented, and then the technician returns and collects the items in the appropriate containers. The technician identifies the evidence with tags that note the contents of the container and its location, thus creating the chain of custody. To avoid evidence cross-contamination, a new pair of disposable gloves always should be worn for each piece of evidence collected and processed at the scene.

Roles and Responsibilities of Fire Suppression Personnel

Preservation of potential physical evidence on a fire ground begins with the firefighting or suppression operation. Although firefighters are not responsible for determining the origin or cause of fire, they play an integral part in the investigation. Firefighters should be trained in the fire academy and beyond to be cognizant of the evidentiary value of the fire scene, and to take steps, where possible, to preserve it for the fire investigators. Public-sector investigators should establish rapport with suppression personnel and commanders before fires occur and should respond as early as possible to fire scenes to guide and advise fire personnel

FIGURE 8-4 Physical evidence is generally found within the area of origin.
Courtesy of Captain David Jackson, Saginaw Township Fire Department.

on investigative needs and issues to avoid destruction and compromise of evidence. In addition, where necessary, investigators should interview firefighters and other witnesses while they are still on scene.

Because the specific cause of a fire may not be known until after a full scene examination is conducted, a professional approach would be to consider the entire fire scene as physical evidence to be protected to the extent possible by controlled firefighting and conservative overhaul operations, especially in the area of fire origin. Once life and safety issues are controlled, the IC should ensure that the fire scene is protected from any further destruction.

Suppression Activities

Well-managed fire-ground operations avoid actions that destroy evidence, such as directing high-pressure, straight streams of water into potential areas of fire origin, overhauling all room contents out a window, and pulling down all walls and ceilings. These tactics can create confusing fire damage patterns and can destroy physical evidence; thus, they should be used only in conditions in which there is no alternative. Putting salvage covers on top of the contents within the area of fire origin can potentially introduce a source of cross-contamination and should be avoided. Every effort must be made to leave items and debris on the fire ground, and especially within the area of fire origin, in their prefire positions.

Early suppression of fire within the area of origin should always be a priority to help preserve evidence there. In cases in which the area of fire origin is known and the fire has extended into other areas of a structure, it is desirable, whenever sufficient assets are present, to keep the area of origin "darkened down" by periodically directing suppression there. Fire attacks by indirect ceiling spray or fog streams are always best whenever possible because these methods are apt to have the least impact on potential evidence. Similarly, suppression personnel should take every precaution to preserve any deceased victims and avoid activities that could damage or move the body. In general, physical evidence should be left where it is found until trained fire investigators can assume responsibility for the scene and guide documentation and recovery.

Overhaul Operations

Management of overhaul is crucial, especially within the area of fire origin. It is at this stage that remaining evidence that was not damaged by fire is most susceptible to damage or displacement. Walls and ceilings that reveal fire movement and intensity pattern evidence as well as damage patterns to furniture, appliances, and

other equipment within the area of origin are critical evidence that should be documented and analyzed, if possible, before overhaul takes place. Highly effective fire knockdown and overhaul techniques have been developed that can both stop fire and minimize inadvertent destruction of evidence. Fire crews must be trained to leave as much as they can intact within potential areas of origin, including debris, and to permit documentation of debris and damage by investigators before moving items, wherever possible.

Protection of Valves, Knobs, Switches, and Breakers

Fire suppression personnel must avoid adjusting or trying valves or knobs and switches on appliances, because doing so loses valuable information for the fire investigator. The original position of a switch or valve is often of crucial importance. When an appliance is losing fuel or otherwise presenting a hazard and must be cut off, firefighting personnel should consider whether they can mitigate the hazard without losing evidence; in a worst-case scenario, they should document the position of valves, knobs, or switches before moving them. Individual gas appliances always have a shutoff valve behind the appliance, which should be used to cut the fuel gas flow **FIGURE 8-5**.

If electrical and gas service can be cut off at the street (or, in the case of electricity, with the utility company removing the meter), devices downstream may not need to be adjusted or compromised. The practice of throwing all breakers in a breaker panel should be avoided whenever possible, and usually the need to de-energize a building can be accomplished instead by simply throwing the main breaker to "off."

FIGURE 8-5 Note the positions of the knobs on this range.
Courtesy of the NJ State Fire Marshal's Office, Arson/K-9 Unit.

Additional Considerations with Fire Service Personnel

A standard law enforcement practice at crime scenes is to post personnel at potential entry points and to restrict entry to the scene to only those personnel whose job it is to examine, document, and collect evidence. This could be accomplished on a fire ground by posting firefighters at each entry point to the fire area.

Firefighters are often trained to look for indicators of incendiarism, including multiple fires, incendiary devices, or ignitable liquids, and should be interviewed by fire investigators for this information. Additionally, firefighters should be asked about existing conditions encountered on arrival at the scene, including audible alarms, open or unlocked doors or windows, odors, and objects that do not seem to belong where they are found. Suppression personnel should be trained and reminded to fuel any power tools that they may be using outside of the fire scene to avoid spilling fuel into the area under investigation.

The fire service IC may have formed a hypothesis of the area of fire origin during suppression activities and/or through experience and training, and the fire investigator should inquire about this possibility. The IC may also be able to provide information obtained from eyewitnesses who were present at the fire development or from the first public safety official to arrive at the scene.

Roles and Responsibilities of the Fire Investigator

Before any evidence is removed from the area of fire origin, it needs to be analyzed, photographically documented, and fixed in a diagram that indicates its location and position before its collection. Importantly, the fire investigator should avoid the temptation to pick up and examine a newly discovered item whose position and orientation may turn out to be important in the analysis of the fire. Before examination, the object must be photographed in place, and its position and location should be noted on the scene diagram.

Field notes should document the condition of evidence when it is discovered and list the other people who are present at its discovery. In **FIGURE 8-6**, the position of a Molotov cocktail is documented in a diagram. In **FIGURE 8-7**, a photograph is used to document the location and position of the evidence.

Skilled fire investigators, prosecutors, and medical examiners caution against moving physical evidence, such as containers of a suspected ignitable liquid or dead bodies, from a fire scene before they can be carefully examined and their location documented.

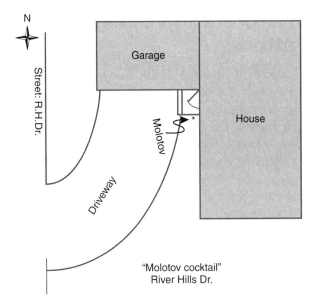

FIGURE 8-6 The position of evidence (a Molotov cocktail) is documented in this diagram.

FIGURE 8-7 The position of evidence is documented in this photograph.
© Jones & Bartlett Learning. Photographed by Glen E. Ellman.

Likewise, in cases in which there is evidence of forcible entry, pieces of an explosive device, or some other potential item of physical evidence, the general rule is that the evidence should be left undisturbed, and firefighting operations, whenever possible, should be directed to a secondary position away from potential evidence sites.

Interpretation of Fire Patterns

As discussed in more detail elsewhere in this text, a principal task of the fire investigator is to follow fire patterns in an effort to trace the growth and development of the fire back to its area of origin and, where possible, to its point of origin. The investigator will likely have noted and documented fire patterns during the earlier stages of the investigation, particularly during the comprehensive exterior and interior surveys. At this point, all of the relevant patterns must be noted and documented, as well as other data and information, and the investigator will begin to form hypotheses regarding the fire area and point of origin. For more information on this process, refer to Chapter 1, *Introduction to Fire Investigation*; Chapter 4, *Fire Effects and Patterns*; and Chapter 13, *Determining the Origin*. Once the area of origin is determined, debris removal (layering) will likely be necessary to develop further data and information and to determine a point of origin. However, after a careful analysis of area of origin, the scene will ideally have been narrowed down sufficiently that the fire investigator is not obliged to remove debris from a large area.

Debris Removal (Layering) and Fire Scene Reconstruction

Once an area of origin has been determined, **layering**— that is, the comprehensive removal of debris within the area to obtain further data and information, and to help in determining the fire's point of origin—is generally completed. Debris should be removed systematically and, when possible, as fully as possible to allow the fire investigator to discover items beneath the debris and to review patterns on walls and horizontal surfaces.

The fire investigator should keep in mind during this process that while the layering process often happens at floor level, fires do not necessarily start on the ground or floor. Indeed, patterns and other evidence might lead the investigator to an elevated point of origin.

Following debris removal, the fire investigator can reconstruct the fire scene by replacing furniture and items that were removed previously back into the scene, to attempt to recreate the prefire conditions without the interference of fire debris. This enables the investigator to observe patterns and provides an enhanced visual picture of the spread and movement of the fire **FIGURE 8-8**.

FIGURE 8-8 Fire scene reconstruction recreates the conditions that existed prior to the fire.
Courtesy of Rodney J. Pevytoe.

Debris Removal and Examination

Debris removal should be done in a careful and systematic manner to ensure that all items contained within an area have been reviewed, and all nearby patterns have been noted. A systematic removal also ensures that patterns are not obscured or changed, and that important evidence is not discarded. The debris should be moved only once, with the team digging and exposing materials from the top down, similar to an archaeological dig. Items should be removed in a systematic fashion, and the process should be fully documented through measurements, sketches, and photography to include the location, condition, and orientation of any contents that are uncovered. The investigation team should work together, deciding what is important and what is not before beginning the process. Hand tools, rather than heavy equipment, should be used to remove debris, when possible.

In many building fires, a debris pile may contain components of the building's roof, attic, and ceiling along with other fallen items. All of this material may need to be removed before evidence of origin is located below.

As mentioned previously, when items and debris are removed during the layering process, it is advisable to place them in an identifiable location, such as on a clean tarp or in a marked container, so that if the need arises to review the debris again, it will still be available. This location should be one that has already been examined and determined not to require further examination or documentation. Moving debris to a defined location is useful if an item is initially discarded but determined later to have evidentiary value. Just as importantly, while the investigator may have the right to perform debris removal and layering, other fire investigators may later be tasked to review the same evidence and area. Best practice and the need to avoid spoliation of evidence dictate that the fire debris be preserved for future scene investigators.

Safety remains important during the debris removal phase for many reasons. Removal of debris may cause other debris to fall and can make a structure unstable. Hazardous materials or toxic vapors may become apparent, or energized electrical wires may become exposed. Hands can be cut by sharp objects in debris. (See Chapter 6, *Safety*, for more information.)

TIP

Debris removal is a tedious and laborious task. Plan ahead so the debris has to be moved only once.

TIP

Debris removal can weaken a fire-damaged structure and cause collapse.

Heavy Equipment Use

Some fire scenes may require the use of heavy equipment to facilitate debris removal or to mitigate safety hazards. The fire scene should be fully documented prior to the introduction of heavy equipment and frequently during its use.

Fire scene inspections progress at a much slower rate than routine construction demolition. A briefing should be held with the equipment operator and crew prior to the commencement of work. The equipment operator should clearly understand the investigative goals, the specific area to be worked in, and evidence or items of interest to be recovered. In addition, heavy equipment operation is noisy, can damage evidence or contaminate the scene, and can create safety hazards. The equipment operator should be under the fire investigator's supervision at all times while the equipment is in use, either via radio or through use of hand signals.

If questions arise about the presence of ignitable liquids in the scene, sampling should be completed before heavy equipment is brought in. The equipment should be inspected for leaks and documented prior to use. Sampling of the equipment fluids can be performed to help address contamination concerns. Refueling or lubrication of the equipment should occur only in a designated location well away from any area(s) of interest.

It is preferable that heavy equipment be positioned outside the structure during use. If large amounts of debris or structure need to be removed, this process should be done in a way that does not change or add debris to the underlying area. Shoring of structural elements should be considered as an alternative to demolition where possible.

Where heavy equipment must be operated within the scene, the equipment should initially be positioned outside the area of interest. Once an area has been inspected, documented, and cleared and samples taken if indicated, the equipment can move progressively into the scene. Loading and unloading of the debris should be monitored to identify relevant items of evidence.

Large or heavy items can be lifted out of the scene by utilizing rigging to minimize potential damage. These items can be further inspected, documented, and even stored on scene (assuming proper security exists) for the duration of the investigation. Sometimes, however, circumstances may require removal of evidentiary items from the scene. Any such removal should be well documented and can complicate ongoing inspection of the scene.

Avoiding Contamination and Spoliation

Potential ignition sources, initial fuels, or other important evidence may be identified during excavation of the scene. Care should be taken to limit damage to and prevent contamination of any such items. Simple decontamination and processing techniques can prevent these problems.

Both public- and private-sector investigators should recognize that civil litigation may result from fire incidents. Parties that may be interested in such litigation should be identified, placed on notice, and invited to participate in the ongoing inspection. It is important that potentially interested parties be provided the opportunity to view the scene and evidentiary items in place where possible. Different requirements exist for criminal investigations.

Fire Scene Reconstruction

The **fire scene reconstruction** phase is important because it can allow for a strong visualization of how the fire developed and spread (sequential pattern analysis), aiding in the determination of a point of origin. During the previous phases of the investigation (initial scene assessment, in-depth exterior examination, and in-depth interior examination), the fire investigator likely has been able to generally narrow down an area of origin. The reconstruction efforts can therefore be limited to the probable area of origin, as opposed to the entire area of damage. Reconstruction can allow an investigator to better correlate observed fire patterns with each other, as well as with the building and contents. Fire scene reconstruction should happen only after the layering process is complete, the debris has been fully examined and photographed, and any evidentiary samples have been taken. In particular, the investigator must photograph both the pre- and post-reconstruction scenes.

Prior to replacing contents, the fire investigator should decide whether to "wash" the floor of the scene to further remove fire debris and allow any floor patterns to become more evident **FIGURE 8-9**. The decision to wash floors depends in part on whether there is an avenue for the water to flow back out of the area, and whether flowing water into the

FIGURE 8-9 Washing floors after debris removal may help assist in revealing fire patterns.

Reproduced with permission from *NFPA 921-2021, Guide for Fire and Explosion Investigations*, Copyright © 2020. National Fire Protection Association. This reprinted material is not the complete and official position of the NFPA on the referenced subject, which is represented only by the standard in its entirety.

space will cause further damage or inconvenience to the building owner. To avoid damaging evidence, the investigator should avoid applying high-pressure streams of water.

Reconstruction activities involve determining, as well as possible, the prefire positioning and orientation of the larger items on a scene, with the view toward examining the patterns on them. If, after layering, large items remain on scene that were moved during the firefighting efforts, the investigator should use all surrounding clues and patterns to carefully reorient these items. For example, protected areas or patterns on a floor or wall may indicate where an item sat prior to receiving fire damage. If there are several possible orientations for an item based on the patterns seen (e.g., spots in the floor where table legs sat), the investigator should try all possible combinations. If the true orientation of an item cannot be determined, it should not be included in the reconstruction, as doing so could create unreliable data.

In a similar manner, the fire investigator can attempt to replace other large items that have been taken out of the area of origin, if clues and patterns enable the investigator to confidently place the items back on the scene. With all items, witness statements can be useful in determining where the large items sat before the fire; these can include statements of property owners and tenants as to the location of furniture as well as statements of firefighters who moved the items. Prefire photos and videos are also useful in this determination, although they may be less reliable if they were taken in the distant past.

Once all large items are replaced at the scene, the fire investigator can examine the patterns that remain on each one to determine whether they provide further data and information about the movement and spread of the fire. For example, suppose the area of origin is a living room and the investigator is able to reposition or replace several large furniture items. The investigator may then be able to see directional patterns in context and draw further conclusions about the fire's origin—and to do so much more easily than if the furniture were out of position or off scene and the investigator was forced to imagine all of the items in place.

If appropriate, reconstruction can also include efforts to replace structural elements, such as wall sections, joists, studs, and doors, into their prefire positions to better understand and record the patterns that were left on these elements prior to them collapsing or being removed during fire suppression efforts.

If the origin area is not available, reconstruction of a room can be attempted in an open area outside of the scene, by placing furniture and other fuel packages in the orientation and at the distances provided by witnesses or observations from the scene. While it will not permit direct comparison with patterns on the walls and ceiling, such a reconstruction can sometimes provide valuable insights into the patterns and damage on the contents of the room, and their relationship to each other.

All reconstruction efforts should be reported on, photographed, and otherwise fully documented to the same extent as all other investigative steps.

After-Action REVIEW

IN SUMMARY

- Scene examination is one of the principal ways in which an investigator collects data and information for analysis.
- A successful scene examination requires the fire investigator to be well prepared; to use proper safety gear, tools, and equipment; and to follow a careful, thorough methodology so as to gain as much data as possible for a correct determination of origin and cause.
- Proper planning and preparation for the scene examination can make a big difference in the overall success of the investigation.
- During the initial assignment and response, the investigator can obtain basic incident information, including location, date and time, weather conditions, size and complexity, structure type and use, nature and extent of damage, and whether injuries occurred.
- Scene planning includes addressing common safety hazards, including overhead hazards, foot hazards, slips and falls, and electrical issues.
- The pre-investigation team meeting is the ideal forum for the team leader or lead investigator to explain the goals and objectives of the team and to introduce team members to one another.
- Planning the investigation should include consideration of which specialized personnel should be brought into an investigation to address areas of expertise that the fire investigators may lack.
- Upon arrival at the scene, the fire investigator should perform the initial scene assessment, taking an overall look at the structure or occupancy and surrounding area.
- Prior to entering a scene, the first step is to perform a site safety assessment that takes into account all of the hazards and factors discussed in NFPA 921.
- Securing the fire scene enhances safety, protects evidence, and allows investigators to work unimpeded.
- On arrival, the fire investigator should establish communications with the fire incident command and determine which operations are under way and in which parts of the scene.
- Unauthorized persons should be kept out of the scene, including members of the fire investigator's agency who do not have investigative responsibilities.
- Scene securement should include, at a minimum, placing scene warning tape around the full scene, including all affected areas and debris.
- Following the initial scene assessment, an exterior survey is performed for every structural fire scene.
- Following the exterior survey, all interior rooms and other relevant areas should be inspected systematically to identify and narrow down areas requiring more detailed inspection and to gather data for origin hypotheses.
- When examining debris, best practice is to methodically photograph it and place it in a known and marked location for later reexamination or examination by other investigators.
- Firefighters should be trained to be cognizant of the evidentiary value of the fire scene, and to take steps, where possible, to preserve it for the investigators.
- Fire crews performing overhaul should leave intact as much as they can within potential areas of origin, including debris, and permit documentation of debris and damage by investigators before moving items.
- Fire suppression personnel must avoid adjusting or trying valves or knobs and switches on appliances, because doing so loses valuable information for the fire investigator.

- Once an area of origin has been determined, it is normally necessary to perform comprehensive debris removal within the area to obtain further data and information, and to help in determining point of origin.
- Debris removal should be done systematically and fully documented.
- During debris removal, the investigator should avoid the temptation to pick up or move a newly discovered item of interest until its position and orientation are fully documented.
- Heavy equipment should be used sparingly in a fire investigation, and only after full documentation of the scene.
- The fire investigator must closely supervise the use of heavy equipment to remove debris or mitigate hazards.
- Prior to replacing contents, the investigator may decide to "wash" the floor of the scene to further remove fire debris and allow any floor patterns to become more evident.
- Following debris removal, the investigator can reconstruct the fire scene by replacing furniture and items that were removed previously back into the scene to attempt to recreate the prefire conditions.
- If, after layering, large items remain on scene that were moved during the firefighting efforts, the investigator can use the surrounding clues and patterns to carefully reorient these items and can bring displaced large items back on scene for reconstruction.
- Reconstruction efforts should be reported on, photographed, and otherwise fully documented to the same extent as all other investigative steps.

KEY TERMS

Exterior survey A detailed examination of the exterior and exterior surfaces of a structure involved in fire, which is performed to collect data and information, scan for hazards, and gather information for comparison with interior patterns.

Fire area The boundary of fire effects in the scene, characterized by the border between undamaged areas and areas that are damaged by flame, heat, and smoke.

Fire scene reconstruction The process of removing debris and replacing contents or structural elements in their prefire positions. Reconstruction can also include recreating the physical scene at a different location in an effort to more clearly assess damage and patterns. (NFPA 921)

Initial scene assessment A preliminary phase of the fire investigation that seeks to ensure preservation of the scene, identify safety hazards and personnel and equipment needs, and determine areas that require further inspection.

Interior survey A systematic inspection of building interior areas to collect data, narrow down inspection needs, and identify hazard and stability issues.

Layering Systematic removal of debris within an area of origin from the top down, to identify evidence and patterns.

Overhaul The final stages of extinguishment by the fire department after knockdown of the main body of fire, during which any pockets of fire are found and extinguished, often through cutting or pulling down wall and ceiling segments. (NFPA 921)

Pre-investigation team meeting A meeting that takes place prior to the on-scene investigation. The team leader or investigator addresses questions of jurisdictional boundaries and assigns specific responsibilities to the team members. Personnel are advised of the condition of the scene and the safety precautions required.

Site safety assessment A thorough assessment of all hazards and safety factors performed by the fire investigator prior to entering a fire scene to determine whether it is safe to enter, and which personal protective equipment will be required.

REFERENCE

National Fire Protection Association. 2020. *NFPA 921: Guide for Fire and Explosion Investigations*, 2021 ed. Quincy, MA: National Fire Protection Association.

On Scene

1. When performing the initial site safety assessment before beginning an investigation, which questions should you ask of the incident commander, the firefighters, and the occupants of a burned building that would help you establish the safety status of a fire scene? What would you advise other investigators and team members to watch for during the investigation to ensure their safety?

2. If you were asked to give a presentation to firefighters or police officers about steps they can take to assist fire investigators and ensure the best outcome for fire investigations, which topics would you cover?

3. Where can you or your team locate the specialized personnel who may be needed at a scene to help examine the electrical system, gas utilities, and mechanical systems?

4. If you are a public investigator, is your agency or jurisdiction aware of the need to preserve building contents and items of interest so that they may be viewed by private and insurance investigators who may need to investigate the scene after you are finished? What steps can be taken to improve this awareness?

JONES & BARTLETT LEARNING
NAVIGATE™

Chapter Opener: © Jones & Bartlett Learning. Photographed by Glen E. Ellman; On Scene siren: © Bildgigant/Shutterstock.

CHAPTER 9

Identification, Collection, and Preservation of Physical Evidence

KNOWLEDGE OBJECTIVES

After studying this chapter, you should be able to:

- Identify the types of physical evidence found at a fire investigation scene. (**NFPA 1033: 4.4.2**, pp. 240–241)
- Describe how to preserve the fire scene and physical evidence. (**NFPA 1033: 4.2.1; 4.2.2; 4.2.6; 4.4.2**, pp. 241–243)
- Describe how to avoid contaminating physical evidence. (**NFPA 1033: 4.4.2**, pp. 241–243)
- Describe how to collect physical evidence. (**NFPA 1033: 4.4.2; 4.4.4**, pp. 243–248)
- Describe documenting how evidence is collected. (**NFPA 1033: 4.4.2; 4.4.4**, pp. 243–244, 248)
- Describe how to identify collected physical evidence. (**NFPA 1033: 4.4.2; 4.4.4**, p. 248)
- Describe how to transport and store physical evidence. (**NFPA 1033: 4.4.2; 4.4.4**, pp. 248–249)
- Describe the types of examinations and testing processes that may be available for physical evidence. (**NFPA 1033: 4.4.3**, pp. 249–250)
- Describe how and when physical evidence should be disposed of. (**NFPA 1033: 4.4.5**, pp. 250–251)

SKILLS OBJECTIVES

After studying this chapter, you should be able to:

- Process physical evidence using best practices and accepted methods. (**NFPA 1033: 4.4.2; 4.4.3; 4.4.4; 4.4.5**, pp. 243–250)

You are the Fire Investigator

You are investigating a one-room house fire and locate the point of origin in the living room near some electronic components. A dog trained in the detection of ignitable liquids is called into the scene to check the area and indicates the potential existence of an ignitable liquid at the point of origin. Inspecting further, you find the remnants of a paper match and a cigarette lighter lying on the floor in the nearby dining room. A firefighter reports seeing a gasoline container in the attached garage.

1. How do you determine what physical evidence to collect?

2. How might physical evidence at a fire scene become relevant in a court case, and how should it be handled so as to minimize damage and alteration?

3. How does the approach to physical evidence differ between public-sector (police and fire department) fire investigators and investigators who are retained by the private sector, such as insurance companies?

4. With what types of evidence would you typically attempt to collect control or comparison samples along with the evidence itself?

JONES & BARTLETT LEARNING
NAVIGATE™   *Access Navigate for more practice activities.*

Introduction

Physical evidence is "any physical or tangible item that tends to prove or disprove a particular fact or issue," according to NFPA 921. At a fire scene, the entire scene, including the fire patterns, sources of ignition, security and fire detection equipment, and items associated with the cause of the fire, can initially be considered physical evidence. Physical evidence can be produced in court or other proceedings if it is properly identified, documented, collected, preserved, and analyzed.

In most cases, the fire investigator is responsible for locating, photographing, properly identifying, collecting, documenting, examining, storing, and arranging for testing of physical evidence from a fire or explosion scene. Preservation of potential physical evidence on a fire ground begins with the firefighting operation itself. Fire investigators in a jurisdiction can improve their investigations by offering training to fire crews and fire officers in how to recognize evidence, the importance of evidence preservation, and basic ways to avoid inadvertent destruction of physical evidence.

A common challenge for the fire investigator is locating important physical evidence once the fire has been extinguished and overhaul has been completed. The investigator should know basic techniques for preserving and collecting physical evidence and must be prepared to recognize situations in which the investigator will need to call on specialized professionals to correctly collect certain types of evidence **FIGURE 9-1**.

FIGURE 9-1 Physical evidence is usually found in the area of origin.
Courtesy of Captain David Jackson, Saginaw Township Fire Department.

Identification of Evidence

The range of physical evidence that can be found at a fire scene is far greater than the evidence typically associated with a crime scene. Physical evidence at a fire scene may include an area of flooring to be tested for ignitable liquids, a tool mark that is made at a point of forcible entry or that indicates adjustment of a critical valve, a faulted electrical appliance or circuit, fingerprints, blood, or other physical items or marks that help the investigator establish facts related to the incident.

Physical evidence can be obvious, such as a melted liquid container, or it can be latent (hidden), such as a fingerprint on a tool that might not even be visible until it is developed by a lab technician. **Artifact evidence** at a fire scene may include objects or items on which fire patterns are present, the remains of the first fuel ignited, the competent source of ignition that ignited the first fuel, and materials that influenced the fire's growth and development.

Which items of evidence are documented and collected will differ depending on the context of the scene and the investigation. The investigator obviously decides which evidence to collect, but this decision may also depend on the scope of the investigation, legal requirements, or limitations on jurisdiction. For example, the public-sector investigator tends to limit evidence collection to items of incendiary or criminal evidence. It is considered unnecessary to store evidence that is civil in nature. If the public investigation results in a finding of an accidental fire or a fire caused by a potentially defective product, the fire or police department generally will not advance any type of court case based on these findings. Thus, a scene with evidence in support of these theories can simply be secured and left for private interested parties to address.

The private-sector investigator often focuses on the failure of products or other types of civil responsibility, so the types of evidence collected in this case could be broader, and often will focus on appliances and electrical components. Anytime an appliance or component is believed to be the cause of a fire, the investigator's policies and procedures will likely require the article to be secured and collected (subject to notification of interested parties and other spoliation of evidence considerations). Other appliances and components in the area of origin might also need to be collected to rule them out as a cause through expert analysis.

Finally, sometimes certain types of evidence should be collected. If an ignitable liquid is believed to have been applied to an area, the investigator would be expected to collect debris and other samples from the area for laboratory testing for such a liquid. Containers with ignitable liquids that are believed to have played a role in fire cause should be sampled. Suspected open-flame ignition sources, such as matches and lighters, should be collected using established protocols. Even when the use of an ignitable liquid is uncertain, it may be appropriate for the investigator to take these types of samples to rule out the application of an ignitable liquid, or to hold them in storage in case further investigation establishes the need to test the samples.

The investigator should have a mindset of inclusiveness when it comes to physical evidence, giving consideration to any type or form of evidence that may either support the investigator's conclusions or rule them out in favor of an alternative theory. Once an investigation is complete, it is difficult to return to a scene to collect more evidence because the scene may have been cleaned up, and the investigator may no longer have the authority to be on the property.

Physical or other limitations may dictate what the investigator can collect from a practical standpoint. For example, proving failure of electrical components does not necessarily require collection of a full electrical system. Proving the contents of a container can be done through sampling rather than by attempting to store and process the full container. Proving a fire pattern can be done with photographs rather than by collecting a wall with fire patterns on it.

To effectively collect evidence, fire investigators should obtain training in physical evidence recognition and collection, and should be mindful of when collection of certain types of evidence would be better left to other professionals. For example, fire investigators are skilled at recognizing and collecting fire debris evidence. However, other types of evidence, such as tool marks, footwear or tire track impressions, and fingerprints, may be better left to crime scene technicians. Public-sector fire department investigators should establish a close working relationship with law enforcement crime scene evidence technicians who are already skilled in the recognition, collection, analysis, and preservation of trace evidence and who possess all of the equipment needed to properly perform these steps.

TIP

Always be aware of the hazards associated with fire investigative work, especially within fire scenes. Careful assessment of the structural integrity of a building and all utilities is constantly required. Any occupancy may contain dangerous substances such as pesticides, chemicals, or other hazards. Wear personal protective equipment such as respiration masks, gloves, helmets, and other gear and take all necessary steps to maintain your health and safety.

Contamination of Physical Evidence

Avoiding contamination of physical evidence is of concern in any scene investigation. Opportunities to contaminate evidence may occur during firefighting, overhaul and salvage operations, evidence handling, storage, and transportation.

Contamination During Collection

Most contamination of physical evidence occurs during its collection through the use of contaminated tools and protective equipment. This kind of **cross-contamination** involves the unintentional transfer of a substance from a location contaminated with a residue to an evidence collection site. Typical potential sources of cross-contamination at a fire scene include tools, protective equipment such as gloves, evidence containers, and emergency equipment. The following procedures will help prevent cross-contamination.

TIP

To limit contamination of evidence, consider using the evidence container itself to collect the evidence.

Contamination of Evidence Containers

New containers should be used to collect each piece of evidence, and the investigator should always don new disposable gloves for each new collection. Fire debris and other material to be tested for ignitable liquids should be collected in new, unused, lined paint-style evidence cans. The investigator can keep a variety of sizes of these containers, including one-pint, one-quart, and one-gallon cans, to store residue and comparison samples. These cans protect the evidence inside from drying out, which would result in the inadequate formation of any headspace vapors that could be tested by the laboratory.

Immediately on receipt from the supplier, and before putting them into a response vehicle, these evidence cans should be tapped shut to the point where they will not open spontaneously in the vehicle or other place they are stored. This minimizes the possibility that the interior of the cans could become contaminated during storage or transport by the vehicle, gear, or other equipment. The cans are not opened again until it is time to collect evidence.

Tool Contamination

Fire investigators should be equipped with a special tool kit to process fire scenes. These tools should be kept separate from other fire department equipment and must not be coated with rust-preventive material. When collecting evidence for ignitable liquid testing, the fire investigator should not use tools from fire apparatus or tools found at the scene. When fire investigation tools are first put into service, and after every fire scene examination, the tools must be cleaned according to protocols and stored in an uncontaminated location.

Steel blade tools such as shovels, chisels, and utility knives, and tools with hard rubber blades may be used for excavation and ignitable-liquid evidence sampling and then can be cleaned according to protocol for the next use. Other tools and equipment, such as bristle brooms and firefighter safety gloves, cannot be adequately cleaned once they have been contaminated and should not be used in evidence collection procedures.

Each excavation tool must be cleaned between fire scenes, and each tool used for evidence collection must be properly cleaned before each piece of evidence is collected. The July 2017 edition of the International Association of Arson Investigators' (IAAI's) *Fire and Arson Investigator* magazine includes an article reporting on research on tool and boot cleaners; the authors concluded that while the industry relied on simple dishwashing detergents for tool and boot cleaning in the past, detergents appeared to have become less effective for tool decontamination over time. In these authors' research, the best results were obtained with a 3:1 solution of water with Simple Green Pro HD brand (purple-colored) cleaner. The researchers cautioned that this cleaner in undiluted form could cause damage to boots and should not be used with dogs. Investigators should research and stay current with cleaning methods and solutions to ensure they are collecting uncontaminated samples.

When cleaning the fire investigator's tools and boots, make sure to use the recommended concentrations of cleaner to water, and to thoroughly and carefully scrub all areas of the item being cleaned, including recessed spaces. Repeat the scrubbing process, and rinse the item thoroughly with water. Once the item is clean, do not place it on a surface that may potentially contain petroleum products or their residues.

Once the tools have been cleaned, a properly trained accelerant detection canine can be used to check them to confirm that the cleaning was successful. A sensitive hydrocarbon detector may help with this confirmation as well.

Protective Equipment

It is important to clean boots before entering the area where samples are to be taken and during the scene examination process. Likewise, investigators should avoid walking through contaminated areas en route to the collection site. Boots are cleaned in a similar fashion to tools as described above.

Ignitable liquid residue samples should not be collected while wearing fire gloves or work gloves. Instead, a new clean pair of disposable gloves, such as latex or nitrile exam gloves, must be donned for the collection of each piece of evidence. Each investigation kit should contain an ample supply of these kinds of gloves to prevent any temptation to reuse gloves.

Emergency Equipment

Firefighters and investigators should not bring tools or generators powered by gasoline or other petroleum products into an area of fire origin. If such a tool must be brought by firefighters into an area of investigative interest, the incident commander should first notify and consult with the fire investigator. If an area of the fire scene has contained gasoline-powered tools brought in by firefighters, fire investigators should avoid walking in the area until it can be screened through a canine or hydrocarbon detector for the presence of fuel.

If a gasoline-powered tool was used during an earlier part of the investigation (such as a generator for lighting), investigators must decontaminate their gloves and boots, and any other affected equipment, before entering the area of origin.

Collection of Physical Evidence

The collection of physical evidence is governed by the type, form, size, and condition of the evidence. Take care to minimize the alteration or destruction of evidence during its collection and packaging. Further, always consider whether the packaging itself will inadvertently degrade a fragile form of evidence such as latent fingerprints or a tool mark. If any doubt arises about how to handle and preserve a piece of evidence properly, always obtain the assistance of a trained evidence technician.

Collection methods depend on the following factors:

- Physical state: Is it solid, liquid, or gas?
- Physical characteristics: What is the size, shape, and weight?
- Fragility: Will the evidence disintegrate or break?
- Volatility: Will the evidence evaporate?

In many jurisdictions, public safety fire investigators are principally concerned with collecting evidence related to fires of incendiary origin. In contrast, civil fire investigators focus on evidence of arson as well as evidence that establishes liability or negligence.

As a general rule, public fire investigators should not remove ordinary appliances, wiring, or electrical components from a fire scene when the fire is not determined to be incendiary, even when the electrical equipment may have been involved in the fire cause through malfunction, misuse, or other reasons. When a noncriminal fire cause is established, public safety investigators should determine the identity of the insurance carrier from the property owner and make appropriate notification. In most cases, the property owner can provide the name of a local insurance agent or broker who can quickly make the notification. The insurance carrier and other interested parties to a potential civil case will have a strong interest in examining the electrical components, which are normally best preserved in place and without being manipulated and/or moved.

In some cases, the fire scene is structurally unstable or there is a possibility that someone may intentionally destroy or steal evidence. In that situation, it may be appropriate to document the scene thoroughly and then, if the building cannot itself be secured for later examination by other interested parties, to remove the evidence to a secure storage facility. If this is done, interested parties should be notified in advance where practicable.

Documenting the Collection of Physical Evidence

Field notes, reports, sketches, and diagrams should be used to document the collection of physical evidence. This should include photographs and diagrams notated with measurements. Make a list that contains each piece of evidence removed from the scene, and document the names of the person(s) who collected the evidence (and any persons who handled the evidence prior to its storage). This documentation should begin before the physical evidence is moved. Doing this will help establish the following:

- The origin of the physical evidence, by providing its location, condition, and relationship to the investigation
- That the physical evidence has not been altered or contaminated

All fire scenes should be thoroughly photographed inside and outside before any debris excavation or evidence collection occurs. As pieces of potential evidence are recovered or samples are taken, photograph each in place and fix its location in a crime scene sketch.

When photographing physical evidence, it is considered good practice to include a permanent feature (e.g., a radiator, wall, valve, or door casing) in each evidence photograph.

When documenting collection of a sample in your diagram, measure the location of any movable item of evidence from two or more fixed, permanent objects whenever possible. Use a magnetic compass to orient the drawing to north.

After the physical evidence has been documented with field notes and photographs and fixed in a diagram, it is ready to be collected and preserved. Ensure that the documentation of evidence includes information on all parties who took custody of the evidence, even if for just a short time.

Collection of Traditional Forensic Physical Evidence

Traditional forensic physical evidence "includes, but is not limited to, finger and palm prints, bodily fluids such as blood and saliva, hair and fibers, footwear impressions, tool marks, soils and sand, woods and sawdust, glass, paint, metals, handwriting, questioned documents, and general types of trace evidence," per NFPA 921.

Specific procedures for recovery of traditional physical evidence are entirely dependent on the type and condition of the evidence. Fire investigators who are trained and equipped to make recoveries of specific types of evidence can do so. When the collection process exceeds their training or equipment capabilities, however, they should bring in trained evidence technicians from local, state, and federal law enforcement agencies. Follow ASTM International standards as well as policies and procedures of the laboratory that will look at or test the evidence.

On its website, the IAAI publishes guidelines for the collection of various types of evidence. Investigators should take care not to exceed their training when deciding whether to collect a certain piece of evidence themselves or ask for assistance from other trained individuals.

Accelerants

According to NFPA 921, an **accelerant** is "any fuel or oxidizer, often an ignitable liquid, used to initiate a fire or to increase the rate of growth or speed of the spread of a fire." Accelerants can be solids, liquids, or gases. ASTM E1387, *Standard Test Method for Ignitable Liquid Residues in Extracts from Fire Debris Samples by Gas Chromatography*, or ASTM E1618, *Standard Test Method for Ignitable Liquid Residues in Extracts from Fire Debris by Gas Chromatography—Mass Spectrometry*, should be followed for this type of collection or testing.

Ignitable Liquids

The fire investigator commonly needs to collect evidence for potential laboratory testing to determine whether an ignitable liquid was present in an area of origin. Although the collection of many types of evidence is within the purview of the trained crime scene technician, ignitable liquid debris collection is best left to the fire investigator, and all fire investigators must be trained in such collection.

Characteristics of Ignitable Liquids

The most commonly encountered arson accelerants are ignitable liquids such as gasoline or kerosene. These liquids share important physical properties that govern their behavior in a fire environment and as evidence that can be recovered by a skilled investigator using the proper tools. Knowing the properties of these liquids will help the investigator effectively collect samples for testing.

Most importantly for debris collection purposes, ignitable liquids are readily absorbed by structural components, furnishings, carpets, carpet pads, and other debris, and they run downhill, forming puddles in low areas. When absorbed by debris, they may remain in the area even after a fire in sufficient concentrations to be sampled and tested. Most of these liquids are lighter than water and may display "rainbow" coloration (i.e., a sheen floating on water). Other ignitable liquids (e.g., alcohol and acetone) are water soluble.

Almost all commonly used ignitable liquid accelerants tend to form flammable or explosive vapors at room temperature. Their vapors are usually heavier than air and tend to flow downward into stairwells, cellars, drains, pipe chases, elevator shafts, and so on. Ignition of such vapors requires that the vapor be within its flammable or explosive range at the point where it encounters an ignition source at or above its ignition temperature.

Most of the ignitable liquids used as fire accelerants do not ignite spontaneously. Some are powerful solvents that can dissolve or stain many floor surfaces, finishes, and adhesives. When they are poured onto a floor and ignited, many types of synthetic surfaces (such as vinyl) or surface treatments will mollify (soften) beneath the liquid. As the liquid pool boils off, its edge recedes. Floor surface charring (or melting and charring) follows the receding liquid edge. The floor area under the ignitable liquid is protected from the effects of burning until the liquid boils off that section. Some experiments have shown that maximum concentrations of ignitable liquid residues are found at the edges of the fire pattern and minimum concentrations toward the center; other research, however, has found just the opposite. Given this uncertainty, best practice is to take samples from both the edges and the center.

Collection of Fire Debris Evidence for Ignitable Liquid Testing

To collect fire debris for testing for ignitable liquid residue, the investigator must first choose a productive sampling area. A **productive sampling area** maximizes the potential for laboratory identification of any ignitable materials and has one or more of the following characteristics **TABLE 9-1**:

- It is in within an area containing indicators of a possible ignitable liquid, such as an irregular pattern suggestive of the use of an ignitable liquid or an area that has been identified by a canine properly trained in the detection of such liquids (see the discussion of canines later in this chapter).
- It contains absorbent material.
- It is in an area where ignitable liquids may collect, such as a low point on a floor.
- It is at the edge of a burn pattern or irregular pattern that suggests the use of an ignitable liquid.
- It contains non-smooth components such as edges, ends, cracks, knot holes, or other similar areas.

After photographing the debris in place and donning new clean gloves, collect the debris by hand or with a properly cleaned tool, placing it into a new lined, paint-type collection can. Ensure that the can is no more than two-thirds full. If a tool is not available, the debris may be collected by the gloved hand, if that can be safely done, or scraped into the can using the can's lid.

Debris that remains attached to other materials can be pried or cut clean with properly cleaned tools, such as a utility knife, wood chisel, or putty knife. A wood chisel is also useful for cutting baseboard and subfloor pieces as necessary. Debris can be broken up if that can be done safely. For carpet, cut both the carpet and the padding beneath it with a properly cleaned utility knife and chimney roll it into the can.

Do not add debris to the can simply to fill it higher; if the debris you are intending to collect does not seem to adequately fill the can, switch to a smaller can.

Clean debris off of the lid prior to placing the lid on the can, and then close the can tightly with a rubber mallet (use of a metal hammer should be avoided, as it has the potential to bend the can or lid and prevent the airtight seal). After the lid is affixed, mark and store the can properly. An airtight can is critical to preserve the material for later testing. The air remaining in the can after the collection of the materials is called **headspace** and is important—this space is used for vapor testing in most laboratory tests.

A lined can is recommended for evidence collection because it can retain debris evidence longer before it begins to rust. The liner is typically made of Teflon or other similar material. The can should be obtained from an evidence supply company, not a paint or hardware store. If a properly trained accelerant detection canine is available, the can should be cleared by the canine before being used for collection of evidence. Best practice states that an empty can of the type used for the evidence collection should be sealed shut and submitted to the laboratory for testing along with the evidence samples so that the laboratory can determine the specific chemical characteristics of the can.

If local protocol permits, sealable plastic bags can be used for fire debris collection, but the investigator must use only bags from an evidence company that are specifically designed to hold such debris. The laboratory may also recommend first placing the debris into a lined can, and then placing the can in the fire-specific sealable plastic bag.

Collection of Standing Liquid and Liquid Samples for Testing

Sometimes, the fire investigator may need to collect a sample of liquid either because it is encountered in a scene in a pool (e.g., it is believed to be mixed in with water from fire suppression activities) or because it is found in a container and may have been poured or applied as an accelerant. If the liquid is present at the

TABLE 9-1 Ignitable Liquid Collection Areas	
Most Desirable Collection Areas	**Least Desirable Collection Areas**
■ Lowest areas and insulated areas within the fire damage pour pattern ■ Samples taken from porous plastic or synthetic fibers ■ Cloth, paper, and cardboard in direct contact with the pattern ■ Inside seams, tears, and cracks ■ Floor drains and bases of load-bearing columns or walls	■ Deeply charred wood ■ Gray ash ■ Edge of a hole burned through a floor ■ Samples from absolutely nonporous surfaces ■ In general, areas that were exposed to the greatest heat or hose streams

scene and not found in a container, the investigator should choose to sample from a productive sampling area as mentioned earlier. The following characteristics suggest good choices for sampling areas:

- An area containing indicators of possible ignitable liquid, such as an irregular pattern suggestive of the use of an ignitable liquid
- An area where ignitable liquid may collect, such as a low point on a floor
- At the edge of a burn pattern or irregular pattern that evidence indicates might be from an ignitable liquid
- Near soaked debris

Whether the liquid is part of the scene or in a container, the fire investigator should choose either the eyedropper collection method or the absorption method to sample it. With the **eyedropper collection method**, after photographing the pool or container and after donning new clean gloves, use a new, sterile eyedropper or disposable pipette to collect ½ teaspoon (2 mL) of fluid into a new glass vial. A vial with a Teflon cap is preferred, but if such a cap is not available, place aluminum foil over the open top before screwing the cap down. Secure the vial into a metal paint-type collection can.

With the **absorption method**, after photographing the pool or container and after donning new clean gloves, use a new, sterile cotton gauze pad and skim the liquid to absorb it. Then, place the gauze into a suitable paint-type lined metal can and seal the can tightly with a rubber mallet. After the lid is affixed, mark and store the can properly. It is recommended to seal a new comparison sample of the gauze into a separate metal can and also submit it to the laboratory for testing.

TIP

A pure ignitable liquid sample should never be placed in a metal can. If the can were to become heated, the internal vapor pressure could pop off the lid.

Use of Canines in the Detection of Ignitable Liquids.

Canine–handler teams have demonstrated great effectiveness in identifying fire debris samples that have a higher probability of confirmation for ignitable liquids than samples that are not acquired with the use of such teams. A **canine–handler team** is a team composed of a professional handler and an ignitable liquid detection dog ("IGL canine") that has been properly trained and certified through a documented and recognized program and then recertified on an annual basis. (Use of the terms "accelerant detection dog" and "arson dog" is discouraged because these dogs are not trained to recognize the reasons why a liquid is located in an area or the intent of a person who might have placed the liquid in the area.)

A canine–handler team can be useful in locating the presence of ignitable liquids in fire debris, structures, open areas, vehicles, or personal effects. The areas to be reviewed by the team depend on the context of the case and should be discussed with the canine handler. Once an area is reviewed and the presence of an ignitable liquid is detected, the team can also assist in the collection of debris or other samples. Once the collection has occurred, the team can help verify whether the appropriate sample has been recovered. Moreover, such a team can be useful in checking tools, equipment, and footwear for contamination by ignitable liquids. Some investigators utilize a canine–handler team to check new lined evidence cans before such cans are tapped shut and stored in an office or response vehicle.

The use of a canine–handler team is within the discretion of the fire investigator and may depend on agency or employer protocols. This team will be subject to the same rules and limitations as the investigator in terms of permission to be on the fire-affected premises.

The investigator and the handler should always confer prior to the team's examination of the scene to discuss the needs of the investigation, the potential uses and benefits of using the team, the division of responsibility with regard to reporting and collection of evidence, access to the scene, and any safety issues that might impact the handler or canine.

While the IGL canine can detect some liquids at levels lower than those normally recognized in laboratory methods, its ability to distinguish such a liquid from background materials is at least as important as the dog's sensitivity. Ordinary synthetic materials produce many of the same compounds as those found in ignitable liquids, and the compounds that trigger a dog's alert are not specifically known.

The alert of an IGL canine is not sufficient evidence of the presence of an ignitable liquid, and any material collected from such an alert must be tested by a competent laboratory before any conclusions may be made about the presence or absence of such a liquid. Without laboratory confirmation, the dog's alert alone is not competent evidence of the presence of an ignitable liquid and may not be introduced in court as such. A confirmed canine alert is only part of the data and analysis that the investigator will rely on in deciding whether a fire is incendiary.

Collection of Solid Samples for Accelerant Testing

Solid accelerants may include commercial chemical compositions such as Thermite, Fire Starter paste, or composite fireplace logs available for consumer purchase. Chemical incendiary materials made from common household chemicals and cosmetics can also be used to set fires. These materials require special handling to prevent dangerous injuries as well as special packaging to preserve the sample. Call your regional or state crime laboratory for advice if these materials are encountered at the scene.

Comparison Samples and Exemplar Products

Comparison samples should be taken whenever samples of liquids or solid materials are believed to contain ignitable liquid residue. This is especially true when sampling from a surface believed to have a petrochemical base, as in the case of many common floor surfaces such as carpet or vinyl. Comparison samples can often be successfully obtained from the same carpet or surface, but away from the area where the ignitable liquid residue is believed to be present, by removing sample material from underneath large objects that sit on the floor, such as couches or cabinets. Taking a sample of uncontaminated carpet for laboratory testing helps to establish a baseline of chemical components to which the contaminated sample can be compared. Cut and package a comparison sample in the same manner and store it in the same type of container as the material to be tested for ignitable liquid.

Sometimes a comparison sample will not be available because of damage at the scene. At other times, a comparison sample is deemed unnecessary. It is up to the analyst to make the determination whether one is necessary. Because the investigator may find it difficult to return to the scene after the initial investigation to obtain the sample, however, a comparison sample should be obtained during the initial investigation if one is available.

If an appliance or control is believed to have caused a fire, investigators can often learn the model type and source by simply interviewing the occupants of the area of fire origin. In serious fire cases, obtaining an exemplar unit is often desirable. A good resource for product safety information is the U.S. Consumer Product Safety Commission website, which has information on product recalls by manufacturer and model. This information includes products that have been recalled for fire and electrocution hazards. Mechanical or electrical engineers are usually involved in the examination to determine the cause of such a failure.

Investigators are cautioned not to perform destructive examinations that exceed their training and experience. In these cases, consult with the insurance carrier and the person sustaining the loss about the need to perform any testing or plans to remove any evidence that could be destroyed or further damaged.

Collection of Gaseous Samples

Investigators should not enter any area where a potentially explosive atmosphere is present. Several methods can be used to collect samples of gaseous vapors, including commercially available sampling devices and evacuated air sampling devices. Contact state or federal laboratories for advice if confronted by this situation.

Collection of Electrical Equipment and System Components

Before beginning the examination of electrical equipment, the first priority is to make certain that there is no electrical current to the equipment. If there is any question about its status, it is advisable not to touch the equipment. A qualified electrician should be called to the scene.

Fire-damaged electrical equipment, such as electrical panels or switchgear, appliances, controls, and the like, needs to be examined, photographed, and fully documented on scene because movement will often degrade its evidentiary value. Whenever possible, the appliance or suspected defective circuit should be left in place and an expert brought back to examine it. If such a resource is not available and the building has fire insurance coverage, contact the insurance carrier. In most cases, if there are reasonable grounds to believe that an electrical malfunction caused a fire, the carrier will assign a qualified expert to assist fire investigators. Potential evidence of a nonincendiary fire cause should be left in place, pending notification of other interested parties. A group inspection of the scene and potential evidence can then be scheduled.

Any wire that is cut for evidence collection should be identifiable with reference to at least one of the following:

- The device or appliance from which the wire was taken
- The circuit breaker or fuse number or location from which the wire was taken
- The wire's path between the device and circuit protector

If an undamaged appliance or control can be found on the fire ground that is the same make and model as the one suspected of causing the fire, give strong consideration to collecting this exemplar for later comparison.

Identification of Collected Physical Evidence

The first stage in identifying collected physical evidence is photographing the evidence where it was found and placing the evidence into a scene drawing by measuring its exact position from fixed objects such as walls or door casings.

Photograph articles of evidence in a progressive fashion, starting with an overall photograph to show its relationship to other major objects in the room or vicinity, a second photo to show the object in its immediate surroundings, and then a close-up photo of the item. A fourth photo of the evidence should be taken with a measurement scale in the photo to establish size, and photos should also be taken of the collection process.

The collection of an item of evidence must include marking the evidence for positive identification **FIGURE 9-2**. Specifically, crime scene procedure calls for marking, tagging, or bagging of evidence. It is best to place the mark directly on the evidence itself if this will not be destructive. Otherwise, the item can be tagged or placed in an appropriate container. Be careful not to place a mark on a piece of evidence in a location where other evidence such as a latent fingerprint or tool mark may exist.

When marking an item of physical evidence, tag, or package for identification, the following data should be included:

- Date and time collected
- Case number
- Location
- Brief description of the evidence
- Where and at what time the item was discovered
- Name of the investigator(s) collecting the evidence

The materials used for marking the evidence should not be susceptible to removal, damage, or alteration.

Investigators should prepare and maintain complete documentation and a detailed list of all physical evidence collected. This list should include the following information:

- Identification of the evidence by particular case
- Individual evidence/item number or letter
- Evidence collection location
- A brief description of the evidence item

TIP

Fingerprints will often remain on surfaces even after a fire. You should always search any location, such as a window or door opening where there is evidence of forcible entry or drawers or other objects that have been moved, for latent fingerprints and tool marks. Investigators trained in retrieving latent fingerprints and casting tool marks and who have the equipment needed for such data collection should recover this evidence. Otherwise, contact a qualified crime scene technician for assistance.

Transportation and Storage of Physical Evidence

At some point, physical evidence may need to be transported for examination or laboratory testing. Maintaining the physical integrity of the evidence is a major concern during this process.

Fire investigators may use several methods to transport evidence. Personal delivery is the preferred method. When this is not possible, the evidence may be shipped via the post office or a common carrier. Specific requirements vary depending on the type of evidence.

The laboratory should be consulted on the safest way to ship a given category of evidence. Some materials, such as explosive devices or flammable liquids, or electrical evidence such as circuit breakers, relays, or thermostats, cannot be shipped under normal

FIGURE 9-2 Mark the evidence container.
© Jones & Bartlett Learning. Photographed by Glen E. Ellman.

circumstances. Consult with the postal service or the shipping company for regulations and limitations on shipping hazardous materials.

Evidence should be stored under conditions that will maintain it in the best possible condition, guarding it against excessive heat, humidity, or other sources of contamination. Dry, dark, and cool conditions are preferred for most evidence. Refrigeration or freezing may be recommended depending on the type of evidence.

A return receipt is necessary for all shipments of evidence, and a letter of transmittal should be prepared for the laboratory, listing the evidence submitted, the nature and scope of the testing requested, and, if requested or important to establish context, the circumstances surrounding the event in question. The name, address, and phone number of the investigator needs to be included so the examiner can contact them if necessary. If a fire investigator is mailing the package, ship it by registered mail or a commercial courier service, and request a return receipt and signature surveillance.

Chain of Custody

Forensic and legal requirements mandate that evidence be positively identified and maintained in a secure chain of custody from the point where the evidence is collected right to its appearance in the courtroom. The integrity of any piece of evidence may be challenged, so be sure to keep an accurate historical accounting of the evidence, and a record of every person who has handled it since its collection **FIGURE 9-3**. The inability to show the chain of custody was maintained for an evidentiary item can lead to challenges of the item's admissibility as evidence in court.

The fewer persons who handle and transfer an item of evidence, the stronger the chain of custody will be. Ideally, the same investigator who recovers the evidence should place it into storage, transport it to and from the laboratory, place it back into storage, and then bring it to court. If this is not possible, then the investigator must be certain to maintain a strict accountability for each person who has custody of the evidence in accordance with local requirements.

Crime Scene Search Evidence Report

Name of subject ____________________

Offense ____________________

Date of incident __________ Time __________ a.m. p.m.

Search officer ____________________

Evidence description ____________________

____

Location ____________________

____

Chain of Possession

Received from ____________________

By ____________________

Date __________ Time __________ a.m. p.m.

Received from ____________________

By ____________________

Date __________ Time __________ a.m. p.m.

Received from ____________________

By ____________________

Date __________ Time __________ a.m. p.m.

Received from ____________________

By ____________________

Date __________ Time __________ a.m. p.m.

FIGURE 9-3 A chain of custody form.

Reprinted with permission from *NFPA 921-2021, Guide for Fire and Explosion Investigations*, Copyright © 2020, National Fire Protection Association. This reprinted material is not the complete and official position of the NFPA on the referenced subject, which is represented only by the standard in its entirety.

Examination and Testing of Physical Evidence

After an item of evidence has been collected, it may be submitted for examination or testing. Testing is often performed to establish chemical composition or for failure analysis. Examination and testing often require that the evidence be altered somehow, but without impacting the item's value as evidence. For example, debris may be removed from the item or a case or cover may be opened or removed to allow full inspection of the item. The investigator is then able to conduct X-rays, depth of char, density, and other tests. If it is determined that testing may alter the evidence, interested parties should be notified prior to testing to afford them an opportunity to object or to have a representative present at the testing.

Laboratory Examination and Testing

Although many fire investigators use the laboratory strictly for analysis of suspected residue of ignitable liquids and materials, modern forensic laboratories can actually perform a wide variety of tests on many types of physical evidence. Investigators should learn the scope of these capabilities in the laboratories available to them. Some of the tests conducted on evidence are listed in **TABLE 9-2**.

TABLE 9-2 Methods of Testing Evidence

- Gas Chromatography (GC)
- Mass Spectrometry (MS)
- Infrared Spectrophotometer (IR)
- Atomic Absorption (AA)
- X-Ray Fluorescence
- Flash Point by Tag Closed Tester (ASTM D 56)
- Flash and Fire Points by Cleveland Open Cup (ASTM D 92)
- Flash Point by Pensky–Martens Closed Tester (ASTM D 93)
- Flash Point and Fire Point of Liquids by Tag Open-Cup Apparatus (ASTM D 1310)
- Flash Point by Setaflash Closed Tester (ASTM D 3828)
- Autoignition Temperature of Liquid Chemicals (ASTM E 659)
- Heat of Combustion of Hydrocarbon Fuels by Bomb Calorimeter (Precision Method) (ASTM D 4809)
- Flammability of Apparel Textiles (ASTM D 1230)
- Cigarette Ignition Resistance of Mock-up Upholstered Furniture Assemblies (NFPA 261)
- Cigarette Ignition Resistance of Components of Upholstered Furniture (NFPA 260)
- Flammability of Finished Textile Floor-Covering Materials (ASTM D 2859)
- Flammability of Aerosol Products (ASTM D 3065)
- Surface Burning Characteristics of Building Materials (ASTM E 84)
- Fire Tests of Roof Coverings (ASTM E 108)
- Critical Radiant Flux of Floor-Covering Systems Using a Radiant Heat Energy Source (ASTM E 648)
- Room Fire Experiments (ASTM E 603)
- Concentration Limits of Flammability of Chemicals (ASTM E 681)
- Measurement of Gases Present or Generated During Fires (ASTM E 800)
- Heat and Visible Smoke Release Rates for Materials and Products (ASTM E 906)
- Pressure and Rate of Pressure Rise for Combustible Dusts (ASTM E 1226)
- Heat and Visible Smoke Release Rates for Materials and Products Using an Oxygen Consumption Calorimeter (ASTM E 1354)
- Ignition Properties of Plastics (ASTM D 1929)
- Dielectric Withstand Voltage (Mil-Std–202F Method 301)
- Insulation Resistance (Mil-Std–202F Method 302)

Sufficiency of Samples

Become familiar with the purposes and standards established for each test. Each test requires a minimum amount of evidence to perform an accurate test. Investigators should confirm that the testing facility is conducting the appropriate test in accordance with the established standards before the testing begins.

Comparative Examination and Testing

Comparative examination is generally an engineering examination performed in a laboratory. In cases where a destructive examination of a complex appliance suspected of causing a fire is to be performed, the investigator will very often attempt to obtain an exemplar for comparison purposes.

Be aware that an exemplar may be present within the fire scene. Examples are lamps, extension cords, outlets, switches, or circuit breakers.

Disposition of Physical Evidence

Standard operating procedures should be established for how the agency or firm packages, labels, and stores evidence that it takes from a fire scene. After the evidence has been gathered, tested, and stored, a question may arise about how long the item needs to be maintained. Many factors influence the length of storage. Legally, some cases—such as a murder-related investigation—may have no statute of limitations, so evidence may be brought from storage many years after the fire occurred. Even after a conviction in a criminal case, the defendant is entitled to appeal, and the evidence should be maintained during the time allotted for an appeal. In other cases, the evidence may be returned to the owner. Return or disposal of evidence is best done in consultation with legal counsel.

Civil cases also have lengthy time frames during which evidence may be called into court. In general, an investigator should maintain all evidence until proper written authorization to dispose of the evidence has been received from all concerned parties involved in the investigation, and the investigator is certain that there is not other litigation filed or pending. Improper or premature disposal of evidence can give rise to spoliation concerns.

Once authorized, disposal of evidence should be done in a manner that is consistent with applicable laws and regulations, and appropriate for the type of evidence involved. Materials whose disposal may be subject to legal requirements or restrictions may include:

- Ignitible liquids
- Fuel
- Explosives
- Other hazardous materials
- Chemicals
- Electronics
- Biological materials
- Drugs

- Medical waste
- Firearms and ammunition
- Thermometers
- Certain types of batteries

The investigator should consult with local and state authorities to determine the proper method and location for disposal of the particular evidence involved. Depending on the material involved, the investigator may wish to involve a business that provides waste disposal services.

Evidence disposal should be carefully and fully documented so that the investigator can demonstrate compliance with applicable disposal regulations.

After-Action REVIEW

IN SUMMARY

- Many different types of physical evidence can be found at a fire scene.
- Physical evidence can be obvious, such as a melted liquid container, or it can be latent (hidden), such as a fingerprint on a tool that might not even be visible until it is developed.
- The investigator decides which evidence to collect, but that decision should take into account the scope of the investigation, legal requirements, and limitations on jurisdiction.
- The investigator should have a mindset of inclusiveness when it comes to physical evidence and give consideration to any type or form of evidence that may support the investigator's conclusions or rule them out in favor of an alternative theory.
- To effectively collect evidence, fire investigators should obtain training in physical evidence recognition and collection, and should be mindful of when collection of certain types of evidence would be better left to other professionals.
- Avoiding the contamination of physical evidence is a primary concern in any scene investigation.
- Most contamination of physical evidence occurs during its collection through the use of contaminated tools and protective equipment.
- New containers should be used to collect evidence, and the investigator should don new disposable gloves for each new collection.
- Tools used by investigators, especially for evidence collection, should be kept separate from other fire department equipment. The fire investigator should not use tools from fire apparatus or tools found at the scene.
- Each excavation tool must be cleaned between fire scenes, and each tool used for evidence collection must be properly cleaned before each piece of evidence is collected.
- It is important to clean boots before entering the area where samples are to be taken and during the scene examination process, and to avoid walking through contaminated areas en route to the collection site.
- Firefighters and investigators should not bring tools or generators powered by gasoline or other petroleum products into an area of fire origin.
- As a general rule, public fire investigators should not remove ordinary appliances, wiring, or electrical components from a fire scene when the fire is not determined to be incendiary, even when the electrical equipment may have been involved in the fire cause through malfunction, misuse, or other reasons.
- Field notes, reports, sketches, and diagrams should be used to document the collection of physical evidence, including photographs and diagrams with measurements.

- The most commonly encountered arson accelerants are ignitable liquids such as gasoline or kerosene.
- A productive sampling area in fire debris maximizes the potential for laboratory identification of any ignitable materials. Such an area may be found in conjunction with indicators of an ignitable liquid, contain absorbent material, be at a low point, be at the edge of an irregular pattern, and/or contain non-smooth components.
- To collect a sample of liquid that is present at the scene or in a container, the fire investigator should choose either the eyedropper collection method or the absorption method.
- A canine–handler team can be useful in locating the presence of ignitable liquids in fire debris, structures, open areas, vehicles, or personal effects.
- The canine–handler team will be subject to the same rules and limitations as the investigator in terms of permission to be on fire-affected premises.
- The alert of an accelerant detection canine is not sufficient evidence of the presence of an ignitable liquid, and any material collected from such an alert must be tested by a competent laboratory before any conclusions may be made about the presence or absence of such a liquid.
- Comparison samples should be taken whenever sampling liquid or solid materials are believed to contain ignitable liquid residue.
- Fire-damaged electrical equipment, such as electrical panels or switchgear, appliances, controls, and the like, needs to be examined, photographed, and fully documented on scene because movement will often degrade its evidentiary value.
- Whenever possible, an electrical appliance or suspected defective circuit should be left in place and an expert brought back to examine it.
- When marking an item of physical evidence, tag, or package for identification, the data should include date and time collected, case number, location, description, time of discovery, and investigator name.
- Fingerprints will often remain on surfaces even after a fire.
- Personal delivery is the preferred method to transfer physical evidence.
- Evidence should be stored under conditions that maintain it in the best possible condition, guarding it against excessive heat, humidity, or other sources of contamination.
- The fewer persons who handle and transfer an item of evidence, the stronger the chain of custody will be.
- Each laboratory test requires a minimum amount of evidence to perform an accurate test, which should be confirmed with the testing facility prior to collection of samples.
- In general, evidence should be kept secure until proper written authorization to dispose of the evidence has been received from all concerned parties involved in the investigation.
- Improper or premature disposal of evidence can give rise to spoliation concerns.
- Once authorized, disposal of evidence should be done in a manner that is consistent with applicable laws and regulations, and appropriate for the type of evidence involved.
- Evidence disposal should be carefully and fully documented.

KEY TERMS

Absorption method Collection method for liquid evidence in which the liquid is absorbed into an absorbent material prior to packaging.

Accelerant Any fuel or oxidizer, often an ignitable liquid, used to initiate a fire or increase the rate of growth or spread of fire. (NFPA 921)

Artifact evidence Objects or items on which fire patterns are present, the remains of the first fuel ignited, the competent source of ignition that ignited the first fuel, and materials that influenced the fire's growth and development.

Canine–handler team A team composed of a handler and an ignitable liquid detection canine that has been properly trained and certified through a documented and recognized program.

Cross-contamination The unintentional transfer of a substance contaminated with a residue from one fire scene or location to an evidence collection site.

Eyedropper collection method Collection method for liquid evidence in which the liquid is collected into a glass vial through the use of an eyedropper.

Headspace The zone inside a sealed evidence can between the top of fire debris and the bottom of the lid.

Physical evidence Any physical or tangible item that tends to prove or disprove a particular fact or issue. (NFPA 921)

Productive sampling area An area for fire debris or liquid sampling where the potential for laboratory identification of ignitable materials is maximized due to its location or contents.

Traditional forensic physical evidence Evidence that includes, but is not limited to, finger and palm prints, bodily fluids such as blood and saliva, hair and fibers, footwear impressions, tool marks, soils and sand, woods and sawdust, glass, paint, metals, handwriting, questioned documents, and general types of trace evidence. (NFPA 921)

REFERENCE

National Fire Protection Association. 2020. *NFPA 921: Guide for Fire and Explosion Investigations*, 2021 ed. Quincy, MA: National Fire Protection Association.

On Scene

1. Review your agency's or employer's evidence collection and storage protocols and facilities. What do they provide for the safe and proper collection and storage of evidence? Do they permit on-site examination of items that should not be moved due to their fragility or due to spoliation concerns?

2. Fire investigators often consider collecting comparison samples to submit for testing with evidence collected from scenes. List the various things that you might consider collecting comparison samples for, and the purpose for which you might collect such a sample.

3. In your company or jurisdiction, what personnel are available to assist with the collection of items that are traditionally collected by crime scene technicians, such as blood, hair and fiber evidence, or tool marks and impressions?

4. Explain the benefits of a careful documentation of the chain of custody of an item of evidence. What reasons do you see for the philosophy that as few people as possible should be within the chain?

CHAPTER 10

Data Collection Outside of the Fire Scene

KNOWLEDGE OBJECTIVES

After studying this chapter, you should be able to:

- Identify and describe different forms of information. (**NFPA 1033: 4.6.1**; **4.6.2**, p. 256)

- Discuss the reliability of information. (**NFPA 1033: 4.5.3**; **4.6.1**; **4.6.2**, p. 256)

- Describe how to prepare for and conduct interviews in an investigation. (**NFPA 1033: 4.5.1**; **4.5.2**; **4.5.3**, pp. 256–259)

- Explain how to gather, canvass, and survey evidence. (**NFPA 1033: 4.6.1**, p. 259)

- Describe considerations for collecting digital recordings and photographic evidence made by others. (**NFPA 1033: 4.6.1**, pp. 259–260)

- Describe systems and surveillance data that may be available. (**NFPA 1033: 4.6.1**, pp. 260–265)

- Identify legal considerations when finding sources of information. (**NFPA 1033: 4.5.2**; **4.6.1**; **4.6.2**, p. 265)

- Identify government sources of information. (**NFPA 1033: 4.6.1**; **4.6.2**, pp. 265–269)

- Identify private sources of information. (**NFPA 1033: 4.6.1**; **4.6.2**, p. 269)

- Describe data analytics. (**NFPA 1033: 4.6.2**, p. 271)

SKILLS OBJECTIVES

After studying this chapter, you should be able to:

- Expand a fire investigation beyond the immediate fire scene through a variety of sources of information. (**NFPA 1033: 4.6.1**, p. 256)

- Create a strategy and a plan for conducting investigation interviews. (**NFPA 1033: 4.5.1**, pp. 256–259)

- Conduct interviews to gather information pertinent to an investigation. (**NFPA 1033: 4.5.2**; **4.6.2**, pp. 256–259)

- Analyze the information gained in an interview. (**NFPA 1033: 4.5.3**, pp. 256–259)

You are the Fire Investigator

You are assigned to investigate a fire that occurred in the basement of a single-family home. The 20-year-old building has a finished basement, and the northeast corner of the basement appears to have the greatest fire damage. It contains a furnace, water heater, clothes dryer, and water conditioner, and is also used for storage of other personal belongings.

1. Where can you obtain basic information about the home?

2. Where can you obtain information about the appliances located in the affected compartment?

3. When interviewing the occupants of the home, what questions might you ask?

Access Navigate for more practice activities.

Introduction

While the examination and excavation of the fire scene are of great importance in most fire investigations, the collection of data goes far beyond scene examination, and the fire investigator who knows how to obtain the maximum amount of data and information relating to the circumstances of a fire will improve the chances of investigative success. The collection of data is one of the necessary elements of the scientific method, and is necessary to establish both origin and cause of a fire.

Data collection begins as soon as the investigator is tasked to investigate a fire, and may continue long after the scene has been examined. This chapter discusses some of the places and ways that the investigator can find information and data in support of an investigation. The investigator should never hesitate to ask more questions and look for more information on any question that comes up in the case being investigated, and should not hesitate to seek out or review technical or scientific data or contact experts in a field. Furthermore, the investigator should take whatever time and steps are necessary to ensure that all necessary data have been collected, and should not feel rushed to make premature conclusions.

Forms of Information

Forms of information that the investigator can gather include verbal information, usually from witnesses; written information, such as reports or documents and other reference materials; visual information, such as photos and videos; and electronic information, such as that gathered from computers, cell phones, and computer research. Upon receiving any information other than purely verbal information, the investigator should keep that information safe and maintain a chain of custody for original documents.

Certain legal limits will affect the investigator's ability to access some of the data referenced in this chapter. Thus, the investigator should be familiar with the applicable laws and limitations, including laws pertaining to confidentiality, privacy rights, and privileged communications.

Reliability of Information

Important information may be available through interviews, written records and electronic data, and visual and scientific documentation. It is important to evaluate the accuracy of any such information and the source. Common sense, knowledge, and experience will help with this evaluation. Also take into account the source's reputation and/or particular interest in the investigation.

Once information is received, the investigator will need to evaluate it for accuracy and reliability based on common sense, training and experience, consistency with other information and fire scene data, and consistency with the principles of fire dynamics.

Interviews

Information obtained during interviews may prove to be the most valuable information gathered during a fire investigation. Interviews of witnesses, firefighters, or even potential suspects can often provide critical information about the origin and cause of a fire **FIGURE 10-1**. Attempt to conduct interviews as soon as possible to ensure that witnesses are located, identified, and questioned in a timely manner. With eyewitnesses, early interviews can be the most reliable: At that point, the information is fresh in the witness's

FIGURE 10-1 Interview witnesses, firefighters, and potential suspects for information about the origin and cause of a fire.
© Glen E. Ellman.

mind, and has not yet been diluted or changed through conversations with others, or through bias or rationalization. Often, essential interviews are conducted at the scene or within a vehicle, with follow-up interviews performed later at the witness's home or work or within the investigator's office. Whether interviews are conducted by police or fire personnel may depend on the jurisdiction or on the nature of the investigation. Interviews regarding fire location, causation, circumstances, and development should, when possible, be conducted by persons trained in the determination of fire origin and cause.

The **credibility** of a witness should always be considered as part of the evaluation of the witness's statements. Evaluate each statement with caution and carefully compare it to the objective information available. The quality of the data gained from a witness can be negatively impacted if the witness has visual or emotional impairment, impairment from alcohol or drug use, prior knowledge of the event or of the statements of other witnesses, or a poor sight line to the incident being reported. Credibility issues may exist whenever witnesses have a motive to lie or distort facts in a manner favorable to themselves. This could include a business, financial, or personal interest in the outcome of the investigation. For example, a witness may hide their involvement in causing a fire out of fear of incurring civil or criminal liability.

TIP

It is your responsibility to evaluate the quality of the data obtained from each witness at the time of the interview. Often, witnesses provide false or inaccurate information either to mislead you or as a result of attempting to help you too much.

Interview Purpose and Preparation

Prior to beginning a witness interview, determine the purpose of the interview and the information that may be gained. Consider what the witness *should* know based on their involvement, and what the witness *might* know depending on their involvement with the scene or with other people. On-scene witnesses may have information about the circumstances resulting in the fire, the fire location and development, unusual occurrences and activities, and statements by other witnesses or persons. Owners and occupants can be asked about building layout, location of fuel packages and potential heat sources, building systems, utility troubles, performance of appliances, and insurance involvement. Neighbors may be asked about activities and the history surrounding a building. Interviews with the incident commander, police, and firefighters can provide information on the size, location, and extent of the fire on their arrival and unusual circumstances they may have encountered. Eyewitnesses should be asked about fire location and development, unusual sounds and odors, statements and behavior of occupants and witnesses, and any other topics that are suggested by their involvement in the incident.

The information learned on arrival and during the first steps of the investigation will help point the investigator to the witnesses who should be interviewed. The investigator should consider doing a preliminary walkthrough of the scene before interviewing an occupant, so as to better understand the layout of the building in question.

Make a list of witnesses, and as new names are found through investigation or other interviews, add those persons to the list. Names of witnesses to be contacted can be determined from dispatch calls, property listings, media reports, incident reports listing responding personnel, other witnesses, and cell phone records, among other sources.

Investigators must be prepared to conduct each interview and have a thorough understanding of the facts known about the incident to that point. Based on these facts, an investigator can create a plan for conducting the interview, including questions that are crafted to elicit information pertinent to the investigation. Prepare questions that are specific to what is believed to be the role of a particular witness or a checklist of information to be obtained.

Conducting the Interview

If possible, find a setting that is free from distractions and onlookers to conduct the interview. To ensure the most accurate responses, each witness should be interviewed, if possible, out of the hearing range of other witnesses. An investigator conducting an interview can be accompanied by a second investigator who takes notes so that the interviewer can concentrate on the discussion with the witness. However, it is not recommended to have more than two interviewers for a given witness, as this could create an intimidating atmosphere for the interview.

Investigators should begin by informing the witness who they are and provide appropriate credentials when requested. Then, establish the identity of all persons interviewed, as well as the information listed in **TABLE 10-1**. This information may prove useful at a later date if an investigator needs to conduct a follow-up interview or locate the witness for the purpose of testifying in court. It may be several months or years after the fire that an arrest or trial occurs. During that time, key witnesses may have changed jobs or relocated, or may have become reluctant to participate in the investigation and trial.

Public-sector investigators should consider whether any Fifth or Sixth Amendment issues may exist with the witness to be interviewed, and consult counsel as necessary. If a person being interviewed could potentially be considered a suspect in a crime, consider whether the need exists to give Miranda warnings or whether the setting of the interview would create the impression that the interviewee is not free to leave (see Chapter 5, *Legal Considerations for Fire Investigators*).

Scenes, settings, and people are all unique and require an investigator to adapt and be flexible during interviews. Even though a list of questions or a checklist of topics may have been prepared for the interview, the investigator should remain ready to pursue new avenues during the interview if the witness provides unexpected information.

Questions should be meaningful and designed to elicit information from the witness. Open-ended questions work best with most witnesses because they allow the witnesses to tell their answer like a story. Closed-ended questions generally will produce only one- and two-word answers that lack meaning and explanation. The difference between these types of questions is illustrated in a situation in which the investigator would like to ask what the witness saw. A closed-ended question would be "Did you see the fire?" whereas an open-ended question might be "What did you see?"

Both question types can be useful when appropriate. Open-ended questions work best when attempting to obtain information during initial interviews and in follow-ups when new information is requested. Closed-ended questions are useful for very distinct responses where "yes" or "no" will suffice, and can be helpful for clarification or establishing the precision of an answer.

In addition to the information provided in the interview, consider the wording the witness uses. Also observe nonverbal cues such as eye contact and body language, while being aware that these cues can vary depending on the witness's age, gender, and cultural background.

TIP

Be cautious of anyone you approach to interview. The person may mistake your intentions or may dislike people who wear uniforms. Always ensure that your personal safety remains the highest priority during an investigation, regardless of the location.

Documenting the Interview

Every interview should be documented in a manner that will allow the investigator to review the information provided at a later time. Often, investigators may speak to numerous witnesses or speak to a witness on more than one occasion during an investigation. Being able to review all of these statements allows them to analyze the information obtained and determine whether statements either align or conflict. This analysis may also reveal new facts or sources of information (leads) that can be further investigated.

Although there are numerous methods of documenting an interview, written notes and tape or digital recordings are the most commonly used. Written notes do not require any special equipment other than a writing utensil and paper; however, poor penmanship may

TABLE 10-1 Witness Information

- Full name
- Date of birth
- Driver's license number
- Physical description
- Home address and personal e-mail address
- Home and cellular telephone numbers
- Place of employment
- Business address and work e-mail address
- Business telephone number
- Other information that may be deemed pertinent to establish positive identification

make the analysis more difficult when attempting to decipher the notes at a later date. Tape recorders allow for a complete record of the interview; however, ambient noise, other people speaking, and equipment malfunctions may negatively impact the interview record.

Another method is to conduct video-recorded statements of interviews. The person speaking can be identified, unlike in a recorded statement. However, on-scene use of this technology is limited and not always practical.

State and local laws may prohibit the use of audio- or video-recorded statements without the consent of the person being interviewed. These laws may vary by jurisdiction.

Canvass and Survey Evidence

As every fire investigator knows, fire scenes are chaotic places and keeping track of the people involved can be difficult. A useful tactic that should be considered with a fire of any size is to conduct a canvass or survey of people found in the vicinity of the fire. For example, if a house fire occurs in a residential neighborhood, consider designating fellow investigators or police officers to speak to neighbors in surrounding houses and apartments. Often, such a survey will reveal the identities of people who were present during the fire and who left before the investigators had the opportunity to talk to them, or who observed the fire and surrounding activities from their property. These people may have photos or videos of the fire scene.

Even if they were not on the scene, neighbors can often provide important information about the occupants of the fire building, and may have observed activities around the fire building that might be of interest to investigators. They may also be able to direct the investigator to surveillance cameras or video recordings.

Remind anyone performing a canvass that the details of the fire and the fire building remain confidential, and that they should not share any factual information or impressions about the fire origin or cause with the persons being interviewed. All such persons should be advised simply that "the fire is under investigation."

Digital Recordings and Photographs

Prior to the popularity of digital photography, it was relatively rare for a fire investigator to obtain witness photographs or videos of a fire developing or burning

at a scene. Even when digital cameras proliferated, many people did not have them at the ready when fires broke out, and fire photos from witnesses continued to be rare. But today, due to technological advances, virtually anyone with a cell phone has the means to take high-quality photos and videos, and almost any fire scene accessible or visible to the public likely has been recorded, at least in part, by digital photos and/or video.

Accordingly, the fire investigator should ask any eyewitness to a fire, and any person contacted during the canvassing activities, whether they have photos or videos of the fire. Although it may be possible in a particular jurisdiction to obtain such photos or videos through search warrants, personal cell phones have a high expectation of privacy. Thus, the best approach in most cases may be to ask the witness to voluntarily provide any photos or videos, or voluntarily upload them to a sharing site if they are too large for transmission by e-mail or text.

Social media posts by witnesses and involved parties may also contain photos, videos, or descriptions of fire events. While these can be very useful, the fire investigator should consult with legal counsel about when it is permissible to access such media, especially if these media are protected by an individual's privacy settings. As a general rule, the fire investigator should not assume a false identity or take other surreptitious actions to persuade a person to allow access to any privacy-protected social media account or post **FIGURE 10-2**.

Fire scene video may also be available from the law enforcement agency with jurisdiction, as many agencies use dashboard cameras and body cameras that record emergency scenes such as fires. These resources are particularly useful because often the police arrive

FIGURE 10-2 Social media posts by witnesses and involved parties may contain photos, videos, or descriptions of fire events.
© Tero Vesalainen/Shutterstock.

very early in an incident, potentially recording relevant video evidence of fire development, occupant behavior, and fire suppression operations.

Press reports, such as television news and other recorded media appearing on a news web page, may also contain important investigative information, including video of the fire, video of the firefighting operation, and interviews with witnesses and occupants.

Alarm, Building Systems, and Surveillance Information

Modern fire alarm and burglar alarm systems are controlled by sophisticated electronic devices and can provide a large amount of data pertaining to the operation of the system, user settings and modifications, the times and nature of any alarm(s) or trouble events in the system, and occupant activities. If the electronic controls in a local alarm system have survived the fire, they may retain a substantial amount of data of interest to a fire investigation. However, fire and water damage, along with loss of line power, can mean that these systems will retain data for only a short time. If the investigator knows such data exist, they should take immediate steps to protect it; this may require contact with building management, the alarm company, or the manufacturer of the system.

Many alarm systems are controlled in full or in part from a remote location. The investigator should determine if this is the case, because data stored remotely may not be affected by the fire event in the building. However, the data stored remotely may not be as detailed as the local data, and due to storage limitations may be retained for only a limited amount of time.

Surveillance cameras have become increasingly common and can often provide valuable video of a scene. Once limited to only certain businesses, such cameras can now be found in many businesses and in residences as well. An occupant of the building where the fire occurred may know whether there was any video surveillance within the building. Check the vicinity of the fire for other businesses or commercial properties that might also have video surveillance, and ask people in surrounding areas during the canvassing operation. Remember that surveillance data are not stored indefinitely, so a request for data from surveillance systems should be made as early as possible.

Postfire Analysis of Fire Alarm Systems

If the investigation reveals that the building is equipped with any fire protection system, the fire investigator must thoroughly document and analyze the system to determine its role, if any, during the fire. The collection of data should include gathering information on building systems and contents, reviewing passive and active fire protections, obtaining data from systems, and interviews of occupants and first responders.

Installation Considerations

As part of a postfire analysis, you should attempt to gather all related fire alarm system documentation, such as manufacturers' data sheets; design drawings; installation, inspection, test, service, and maintenance records; and any dispatch and response records to the location, including any previous alarms. In addition, determine which building and fire codes and referenced design standards were in force at the time of installation. This information will allow you to compare the original system design with the installation found at the time of fire, to determine whether changes in the use and occupancy occurred, renovations resulted in changes to the building construction and building features, or any change affected the performance of the initiating devices and notification appliances.

Consider interviewing personnel from the authority responsible for review and approval of the system design and installation, including acceptance testing. In addition, interview any design professionals or contractors associated with the design and installation, whether they were responsible for the initial installation approval or periodic inspection, testing, service, and maintenance approval.

Operability

A critical part of any investigation is verifying the fire alarm system's status, operability, and functionality at the time of the fire. Establishing whether this system was in normal ready service or showing trouble or supervisory alarms could provide considerable insights into how the system responded to the fire event. A trouble or supervisory alarm could render some or all of the system incapable of performing properly. In addition, you should determine whether the system was powered by the primary power source or the secondary power source at the time of the incident. This distinction is important because when the system relies on a secondary power source such as batteries, it could experience inadequate operability due to limited power reserves. Fire alarm systems are required to be operational in normal ready condition for a certain amount of time after loss of primary power, typically at least 24 hours, but they can operate for only a limited amount of time when on standby power in full alarm mode, typically only 5 minutes—which translates to limited notification capabilities. Out-of-date or uncharged batteries may not allow a fire alarm system to function the way it was designed.

Another very important part of the investigation is determining whether the initiating devices functioned as required by activating in a timely manner and within the design parameters for the conditions to which they were exposed. Witnesses and occupants using the alarm system may have information about the system's response to the event or to other events.

Witnesses and the scene may also provide information as to whether all of the expected smoke alarms or detectors were in the places where they were expected. Occupants may modify or remove these devices for a variety of reasons, resulting in delays in occupant notification and response to a fire.

Whenever a fire alarm monitoring company is involved in the system's operation, obtain as much information from the company as possible on topics such as system history and status, time of alarm, history of the alarm, detector(s) activated and their locations, times of trouble or other alarms, and occupant response. This information may be available electronically or in printed form and helps establish useful timelines for the fire's start and development. Similarly, obtain any alarm information that might be available from the public-safety answering point (PSAP).

Analysis of Smoke Alarm Response

It is crucial to ascertain and understand the incident timeline, including how and when the fire was detected or discovered and how the fire department was notified. These considerations are especially important if the fire alarm system or smoke alarm activation was the initial source of the incident alarm, because this information can help determine the area of origin and whether notification appliances or smoke alarms sounded. Data on whether a particular initiation device or smoke alarm activated and sounded can be gathered from first-arriving firefighters, building occupants, the fire alarm control unit (FACU), PSAPs, and proprietary or central supervising stations.

You should also interview occupants and firefighters to document fire growth and spread, and attempt to correlate those observations with sounds and sights associated with the fire alarm system's notification devices. Sounds and sights can vary, from loud tones or ringing bells to voice directions and flashing strobe devices that may confirm system activation. Examine fire alarm system devices for signs of contamination by smoke or soot or thermal impact, which may indicate which device(s) operated earliest or if they failed to operate. Attempt to determine whether a smoke alarm

had a battery; even with heavy thermal damage, some evidence of the battery will usually be left behind.

Analysis of Smoke Deposition

Examine, analyze, and document smoke detector, smoke alarm, and other system components for soot accumulation, soot patterns, and thermal impact in much the same manner as you investigate other fire patterns found on various surfaces. In addition, account for localized or specific fuel packages and ventilation effects that may be in proximity to affected or exposed system devices. Analysis of these conditions may assist you in determining whether a smoke detector and smoke alarm device activated and operated during the event.

Alarm Response Time

It is important to understand and verify the response time of the fire alarm system's initiating devices. Investigators may sometimes be able to use the scene findings and data to develop a computer model of the fire scene and to calculate the fire alarm system device response. These models can also be helpful in confirming or disproving fire alarm system activation. A caveat applies, however: Exercise caution when using computer models because their correct use may require a significant amount of knowledge, understanding, and experience.

Estimation of Fire Size

Understanding the effects of fire size and propagation in relation to system response, especially regarding initiation devices such as smoke detectors, can help fire investigators determine fire size, origin, growth, and spread. It is important to account for fire alarm initiation device activation, but also equally important to account for nonactivation of initiation devices. By verifying these actions, you may be able to use the fire alarm system as a tool to determine the size of the fire at a particular point in the fire event. Numerous sources of information are available to help in this effort. Manufacturers' specifications and data sheets, design standards, and calculations can be employed to estimate system characteristics and responsiveness regarding fire size.

The activation point of a head can be estimated by the response time index of the head or nozzle. If a system can be shown to have activated and operated appropriately, the fire at the time of the alarm could be established as the minimum fire size. In contrast, if the system did not activate, assuming proper and correct installation, the fire could be established as the maximum fire size because fire growth did not reach a level sufficient to activate the system.

Development of the Fire Timeline

An active fire alarm system can provide multiple data points. For example, the fire alarm system control panel and annunciation panels are good sources of system response and activity information if not damaged or destroyed by fire. No matter the severity of the incident, the amount of damage to the fire alarm system and components, or whether the primary or secondary power has been disconnected or compromised, you should photograph the fire control panel and annunciation panel to capture the indications and the events that took place. If primary power is lost, some secondary power sources will sustain the system for only a limited amount of time. Thus, once the secondary source expires, system data could be lost—making timely recording of this information critical.

The amount of information available will vary based on the age and type of fire alarm system. Some systems and panels provide only general and basic information concerning the device that activated and the general location, whereas others indicate system status and the type, time, and specific location of an initiation device activation. The information may be presented as indicating lights associated with a device or zone, alphanumeric text, or a combination of both. Additionally, verification of system monitoring provided by remote, proprietary, and central supervising stations can supply useful information concerning when and how the fire alarm system responded.

Take care not to overstep your abilities regarding the use or operation of fire alarm control and annunciation panels. If you do not possess knowledge or competency on the use of such devices or systems, or if the system has been damaged, there is a risk of corrupting, altering, or losing data. To ensure that these data are safely accessed, only trained and competent individuals should determine the operational status and capabilities of the system. It is also important to interview occupants, witnesses, and fire responders to document their observations related to fire growth and spread, and to correlate those observations with the data gathered from the fire alarm system panels. This includes any information concerning sights and sounds associated with activation of the fire alarm system's occupant notification device(s), location of the interviewees, and whether the interviewees took any action to initiate an alarm, such as activating a manual fire alarm box. Both fire alarm system activation data and information obtained through interviews can be helpful in creating and verifying a timeline for the incident and, in turn, determining fire origin, growth, and spread.

Thermal Damage

All of the fire alarm system's components have a temperature range established by the manufacturer in which they can properly operate. Evaluate the entire system for signs of damage from smoke, heat, gases, and flames to determine whether they had any effect on the operation of the system. For example, thermal damage to fire alarm system notification appliances, including smoke alarms, could affect the device's ability to produce an audible alarm signal, or any alarm signal at all. Once a device is exposed to an environment that forces it to operate outside of its design parameters (including temperatures outside of the established operational range), that component is susceptible to decreased performance or failure.

> **TIP**
>
> During the course of any investigation, do not attempt to access or tamper with the internal components of the fire alarm system if you do not possess the necessary technical expertise. Work under the supervision of a trained individual to avoid injury from electrical shock and potential damage to the system or components, which could lead to the loss of valuable data. This is especially important when trouble or supervisory signals are present related to systems' power supply issues.

Postfire Analysis of Fire Supression Systems

Origin and Cause

You can apply information derived from the system examination to bolster and substantiate hypotheses about the fire event. Conversely, system data may also be used to exclude hypotheses. For example, if fire spreads in a structure protected with a fire suppression system, a detailed examination of the system to determine whether it failed to operate and suppress or control the fire, and how and why such a failure occurred, can help you develop a hypothesis or conclusion about the fire origin, cause, spread, and responsibility.

Timelines

To develop a hypothesis relating to fire origin, cause, and spread within a building protected with a fire suppression system, you must account for all relevant data associated with system operation. This includes any data gathered through electronic monitoring of system functions, which is typically available in the activity logs of modern fire alarm system control panels and from onsite proprietary monitoring and offsite monitoring companies. You must correlate all data sources and scene evidence to formulate and substantiate one or more hypotheses for the fire incident. In addition, these data can aid in the development of a timeline for the fire, which can then contribute to explaining fire ignition and spread. Beyond the physical examination of the system, ascertain all data related to electronic monitoring of system functions, if provided.

Most water-based and non-water-based systems are required by code to be electronically monitored at a communications center, proprietary facility, or third-party contractor. Monitoring is used to communicate changes in the normal ready state of a system. Changes to be monitored include detection of agent movement within a suppression system, indicating flow and operation. Other ancillary functions must also be monitored, including fuel control; power shutoff; control valve positioning and/or movement, indicating whether valves are in the open or closed position; trouble alarms; tamper alarms indicating a change in valve position; and changes in temperature, air, gas, air handling, and water pressure.

Fire Modeling

The field of fire investigation is ever changing and is similar to other analytical fields, particularly regarding the advancements and influences of technology. Fire investigation relies on scientific methodology, which must be performed in accordance with applicable science and engineering principles and processes. One technological tool using such principles to better understand and analyze fire phenomena is computer fire modeling.

Computer fire models use mathematical data and inputs to predict fire behavior and fire conditions, as well as to assist in establishing fire growth and propagation timelines based on the expected fire size given the available fuels. Fire models can provide additional details to explain system performance. Computer fire models can forecast system operation times and behavior and offer insights into how the system affected the fire and surrounding area, given the size and characteristics of the model fire.

Computer fire models can also assist you by supporting (or not supporting) your decisions and hypotheses. However, you should be careful not to exceed your abilities or skills when using computer fire models. Effective use of this investigation tool requires significant training, education, and experience, if you are to reach reliable and valid conclusions that can stand scientific challenges. If you do not possess such a background, you will need to employ outside resources to complete the model.

Collecting Documentation of Fire Protection Systems
Design Documentation

The initial approach to examine, document, and analyze a fire protection system is to recognize which codes and standards were used in laying out and installing the system and to acquire the design documents for review and comparison. Verify the editions of the codes and standards in force at the time of design and installation, as well as any additional design information, past or current, that played a role in the system's creation.

Typically, fire protection systems are designed and installed in accordance with NFPA standards. However, other sources for system design can be referenced and used. For example, FM Global publishes a large number of data sheets, best practices, and design standards for use at its insured facilities. Property insurance companies may also require a specific coverage or design, though their requirements are typically rooted in NFPA or FM Global design criteria. You can obtain the design information from building owners, system and building designers, system installers, or local regulatory personnel such as the building department or the fire marshal. When making these inquiries, it is important to understand both your role and the scope of the investigation (e.g., public or private) and to be aware of the tools at your disposal to acquire the needed documentation.

Permit History

Generally, fire protection systems are required and regulated by fire and building codes that reference design standards specific to the type of system proposed. The permitting process is critical to ensure that a review of the proposed system takes place and that the design meets the minimum code and standards. Normally, the fire and building codes require a permit application, plans, calculations, and manufacturers' data sheets to be submitted for review and approval by local or state fire or building officials.

It is important to understand the permitting process and to ascertain whether permitting is a requirement for system installation. It is rare that a jurisdiction would not require a permit; therefore, you will need to determine how and where to acquire permit information for a fire protection system involved in a fire scenario or other incident. Permit data can help you establish a true understanding of the system's functionality. An especially important consideration is whether system permitting and approvals, if required, were completed. The permitting files referenced may include initial permits, design drawings, additional requirements, rejections of the initial system design, and installation inspection and testing documentation.

Invoices and Contracts

Request and examine all records associated with fire protection systems, including, but not limited to, invoices and contracts relating to the installed system. Records may be available from the seller, installer, service provider, and/or building owner.

Codes and standards require the creation of design documents for review and approval by the approving authority. These documents serve as the initial platform for system construction. Upon completion of system installation, it is common to have as-built drawings created, documenting changes in the design and installation. These as-built drawings should be reviewed by the approving authority to ensure that any changes did not violate the code or referenced standard. In any case, you should acquire the as-built drawings for comparison to the installed system to confirm that the as-built drawings match the installed system.

Inspection and Maintenance Records

Make sure you acquire inspection, testing, and maintenance records, because these records provide a profile of ongoing system compliance and functionality. Proper documentation can be a good indicator of the likely performance of a system under fire conditions. These documents may be in the possession of several parties, including the building owner; a building representative; an insurance company; a professional engineer; the contractor who performed the inspections, tests, and maintenance; or local regulatory agencies such as the fire marshal's office.

The documents obtained may demonstrate a history of failing inspections, or of other defects that might have contributed to the fire you are investigating, particularly if you determine that some component of the system did not function or was compromised.

Product Literature

Whenever a fire protection system is proposed, designed, and installed, data and specification information should be obtained from the manufacturers of all parts and appurtenances utilized. This manufacturer data or product literature provides a variety of information and guidance related to system equipment, installation, and performance. Indeed, product literature can provide valuable insights and understanding regarding system components and operation. Moreover, the data in these documents can confirm

the equipment's capabilities and limitations and can explain and verify equipment testing and certification standards completed by certifying agencies such as FM Global, Underwriters Laboratories, or other nationally recognized testing companies. Acquire and review this kind of system literature, both current and referenced at the time of system installation, to gain a better understanding of the requirements when the system was installed versus any significant changes after installation. It could be critical to understand any flaws, failures, or needed corrections relating to system performance that may have been found and noted after system installation occurred.

Alarm/Activation History

Beyond notifying building occupants of fire, fire alarm systems are often monitored by an offsite monitoring company or proprietary, central, or remote supervising stations, where system data are recorded and maintained. If such data are available, retrieve this information as soon as possible through appropriate methods. The monitoring company should be notified as early as possible to preserve the system data for review.

You should have a basic understanding of the fire alarm system's capabilities found during the scene examination. For example, is it a zoned system (provides information about a general area of the building) or an addressable system (provides information about a specific device and location, time of incident, and condition of the device)? Addressable fire alarm panels are microprocessor-based units, so they hold considerable information and history about any events related to the fire protection systems that are monitored by the panel. Determine whether the fire alarm panel has any system information related to the fire that can be analyzed. In addition to verifying activation, confirm whether the system was operating on primary power or on a secondary power source, such as battery backup or a generator. As noted earlier, you must be aware that backup power may expire quickly after a fire, requiring rapid access to system data. Take a photograph of any data displayed by the panel as early as possible, before data are lost.

If the fire alarm system examination exceeds your knowledge, skills, and abilities, have appropriate resources available—such as outside engineering experts or systems technicians—to assist with the completion of this task.

Spoliation Issues

You must constantly be vigilant to avoid spoiling evidence during scene documentation and examination.

Fire protection systems within the fire scene can be considered physical evidence and must be properly protected and preserved as such. Fully document the systems with photos; however, avoid moving valves or switches, resetting alarms, or manipulating components, which can compromise the evidence. The event history in fire alarm systems should be retrieved by trained individuals who will not lose or compromise the data. Documentary evidence—for example, drawings, reports, contracts, and alarm data—should be preserved carefully. Documentation can include the use of photographs, evidence logs, copies, and proper collection materials and methods. Regardless of the evidentiary item, you must properly maintain a chain of custody for that evidence.

Internet Information

Much of the legwork and research that investigators once did outside of the fire scene is now being done by computer, and reliable computer and Internet search skills have become a critical tool in every fire investigator's toolbox. The Internet is the likely first stop for an investigator seeking information from governmental agencies and private entities, and an investigator should know how to access the websites of these agencies and entities.

Information related to fire investigations that is commonly obtained through the Internet includes property information and records, map and satellite data, weather data, business information, scientific and technical data, information about people, safety data sheets (SDSs) for substances and hazardous materials, and information about products and appliances. In fact, product and appliance specifications and users' manuals are now widely available online, even for products that are no longer being manufactured.

The investigator performing these searches must have the ability to discern which sites and which information are, in fact, reliable. Many sites on the Internet contain false information, and others provide unsupported and unreliable information that is relayed by persons acting in good faith but without good data. Any information gained from a site that is not a government or known private entity site, such as those listed later in this chapter, needs to be verified through other means. Indeed, the information found on Internet searches may simply serve as a lead for the investigator to seek more reliable data.

Internet-based information, if used as the basis for the investigator's conclusions, must not only be reliable, but also be captured and preserved in a manner that allows the investigator to refer back to it and potentially to present it in court.

Government Sources of Information

Government records provide a wealth of information. Street maps, building permits, blueprints, and property ownership records can contain vital data for a fire scene investigation. In the United States, municipal, county, state, and federal governments all maintain records that should be accessed as appropriate for background information. In other countries, the investigator should become familiar with the various levels and entities of government and the information that each may be able to provide.

Legal Considerations

The availability of information to the fire investigator is governed by legal considerations. The Freedom of Information Act provides public access to information held by the federal government. Most federal agencies have created procedures for the disclosure of information and appeals processes when a request is denied. Most U.S. states, provinces, and territories have also enacted similar laws that allow for the public disclosure of information concerning government operations and their work products. An investigator must be aware of these laws and understand that the laws, rules, and procedures may vary greatly between units of government. In addition, many laws and guidelines exist about access to social media and the uses that can be made of information posted by private individuals on these platforms.

Municipal Government

Townships, villages, and cities are all forms of local municipal governments. These entities create and enforce laws and regulations specific to their jurisdictional boundaries and provide various services funded through either levies or property taxes approved by their citizens.

TABLE 10-2 lists the positions, departments, and agencies typically found within municipal governments. Not every municipality will have each of the departments listed.

County Government

The next higher level of government is that of the county, borough, or parish. A county is a subdivision within a state that encompasses the various townships, villages, and cities within its boundaries. Like municipal governments, various laws and ordinances are created and enforced throughout the county as approved by the constituents.

TABLE 10-3 lists the positions, departments, and agencies typically found within county governments.

TABLE 10-2 Municipal Governments

Form of Municipal Government	Description/Benefit
Municipal Clerk	Maintains all records related to municipal licensing and municipal operations
Municipal Assessor	Maintains all public records related to real estate, including plot plans, maps, and taxable real property
Municipal Treasurer	Can provide public records related to names and addresses of property owners, legal descriptions of property, and the amount of paid or owed taxes on a property
Municipal Street or Public Works Department	Maintains records and maps of municipal conduits, drains, sewers, street addresses, and all old and current street names, including alleys and rights-of-way
Municipal Building Department	Has records related to building, electrical, and plumbing permits and archived building blueprints and files
Municipal Health Department	Maintains records of births and deaths and investigations related to health hazards
Municipal Board of Education	Contains records related to the school system and may assist with identifying and locating school-age offenders
Municipal Police Department	Can provide records related to local criminal investigations and evidence storage and retention
Municipal Fire Department	Maintains records related to fire and emergency medical services (EMS) incidents and life-safety inspections
Other Municipal Agencies	Varies, but may include public works, parks and recreation, and water distribution

TABLE 10-3 County Governments

Form of County Government	Description/Benefit
County Recorder	Responsible for recording legal documents that determine ownership of real property; maintains files of birth, death, and marriage records, as well as bankruptcy documents
County Clerk	Maintains public records related to civil litigation, probate records, and other documents related to county business
County Assessor	Maintains records related to property, maps and plats, including property owners, addresses, and taxable value
County Treasurer	Can provide information related to property owners, tax mailing addresses, legal descriptions, and the amount of either owed or paid taxes on property; also maintains all county financial records
County Coroner/Medical Examiner	Can provide information related to the identification of victims, manner and cause of death, and any items found either near or on the victim
County Sheriff's Department	Can provide both investigative and technical support for county criminal investigations; provides polygraph services, evidence collection, and evidence storage and retention
Other County Agencies	Various departments include parks and recreation, conservancy districts, Homeland Security, and the Emergency Management Agency

State Government

State, provincial, or territorial governments are subnational entities that function both independently and within the federal government (discussed later). For example, in the United States, state governments operate based on their own constitutions, but also adhere to the U.S. Constitution. Additionally, state governments may provide privileges such as driver's licenses, hunting and fishing licenses, marriage licenses, and birth certificates.

TABLE 10-4 lists the positions, departments, and agencies typically found within state governments.

Federal Government

Established by the U.S. Constitution, the federal government is the highest level of government in the United States. It includes three branches: executive, legislative, and judicial. These branches carry out the primary functions of government and are further divided into various departments and agencies.

TABLE 10-5 lists some of the positions, departments, and agencies found within the U.S. federal government.

TABLE 10-4 State Governments

Form of State Government	Description/Benefit
Secretary of State	Maintains records related to charters and annual reports of corporations, charters of villages and cities, and trade name and trademark registrations
State Treasurer	Maintains public records related to state financial business
State Department of Vital Statistics	Maintains records for births, deaths, and marriages
State Department of Revenue	May assist with locating tax records of individuals or corporations, both past and present, as well as locating individuals through child support records
State Department of Regulation	Source for information such as professional licenses, results of licensing exams, and regulated businesses
State Department of Transportation	May provide information about highway construction and improvements, motor vehicle accident investigations, and vehicle registration and operator testing and regulations
State Department of Natural Resources	Responsible for conservation and protection of water, lands, rivers, lakes, and forest areas; may also provide records related to hunting and fishing licenses, waste disposal regulations, and cooperation with the Environmental Protection Agency (EPA)
State Insurance Commissioner's Office	Can provide assistance related to licensed insurance companies, insurance agents (both past and present), and computer complaints
State Police	May provide information related to state criminal investigations; some agencies may conduct fire investigations or provide assistance to local agencies; some also operate their own forensic laboratory
State Fire Marshal's Office	Can provide information regarding fire incidents within the state, building inspection records, fireworks and pyrotechnics, and boiler inspections; most state fire marshals also have fire investigators who may provide assistance with conducting origin and cause investigations
Other State Agencies	May include the Department of Motor Vehicles, Department of Homeland Security, Emergency Management Agency, and liquor enforcement
County, District, and Superior Courts	May have information on civil or criminal litigation involving parties, witnesses, and/or property

TABLE 10-5 Federal Government

Form of Federal Government	Description/Benefit
Department of Agriculture	Maintains records related to food assistance programs, meat inspections, dairy products, and nutritional assistance; includes the U.S. Forest Service, which has law enforcement and fire investigation responsibilities and information on forestry and natural resources
Department of Commerce	Maintains records related (but not limited) to highway projects, names and addresses of ships fishing in local waters, trade lists, and patents; includes the National Institute of Standards and Technology (NIST), which engages in many activities related to fire safety, firefighting technology, and flammability, and does large-scale fire experimentation
Department of Defense	May provide public records related to the five military branches; all of these branches also maintain their own investigation units
Department of Health and Human Services	Maintains records related to the Food and Drug Administration and health care, and maintains an investigation unit; includes the National Institute for Occupational Safety and Health (NIOSH), which conducts firefighter fatality investigation and reduction programs
Department of Housing and Urban Development	Maintains records related to public housing and federal assistance
Department of the Interior	Maintains records related to fish and game activities; the National Park Service is part of this department, as is the Bureau of Indian Affairs (BIA)
Department of Labor	Can provide information related to labor and management, including overtime and pay; combats age discrimination
Department of State	Can assist with information on passports, visas, and import/export licenses, and foreign companies operating within the country and abroad
Department of Transportation	Can provide records related to transportation safety and hazardous materials
Department of Justice	Assists with records related to antitrust and civil rights violations. This department includes the Civil Rights Division, the Criminal Division, the Drug Enforcement Administration, the Federal Bureau of Investigation, and the Immigration and Naturalization Service. The Bureau of Alcohol, Tobacco, Firearms, and Explosives (ATF) within the Department of Justice may provide technical information on fire science and investigation, and has fire and explosion investigators including those assigned to the National Response Team (NRT). Its subject-matter experts include chemists and electrical engineers. ATF maintains the Fire Research Laboratory (FRL) for large-scale fire experiments and makes available to all public-sector investigators a web-based case management system known as BATS (Bomb Arson Tracking System), which provides information from across the United States. ATF also maintains information on license holders, manufacturers, and importers of firearms.
U.S. Postal Service	Postal inspectors may provide information on materials that have been routed through the mail system
Department of Energy	Provides information related to the nation's energy policies and programs
U.S. Fire Administration	Oversees the National Fire Incident Reporting System (NFIRS) and provides research, reference, and technical information related to fire investigation; also provides numerous fire service–based programs, training, education, and information databases

Form of Federal Government	Description/Benefit
National Oceanic and Atmospheric Administration	Maintains records of past and present weather data
Internal Revenue Service	Maintains public records related to compliance with all federal tax laws; may assist with matters related to the federal income tax.
Department of Homeland Security	U.S. Customs and Border Protection regulates importers and exporters, customhouse brokers, and truckers, as well as licensing vessels not licensed by the U.S. Coast Guard
U.S. Secret Service	Maintains public records related to counterfeiting and forgery of U.S. currency; conducts investigations related to threats against all current and former presidents and their families, as well as foreign heads of state
Federal Emergency Management Agency	Provides federal planning, response, and assistance related to federally declared disasters, both natural and human-made

Private Sources of Information

Numerous groups, professional services, and organizations may prove useful during a fire investigation. Test data, insurance records, and various standards may exist that provide valuable resources that are not available from governmental agencies. **TABLE 10-6** describes the private sources of information that are available to the fire investigator.

Myriad information sources may be accessed to assist in an investigation. As investigators continue to seek out all potential information sources, they may sometimes identify individuals or entities that normally would not be contacted during an investigation. An investigator should compile and frequently evaluate the **investigative file**. Doing so will help to identify areas for further investigation, evaluate the relationship between the information and documents that have been gathered to date, identify corroborating evidence, and identify any discrepancies that may exist within the information already obtained.

Relationship Between Public Agencies and Insurance Companies

Most states have some form of **arson reporting/immunity statute**—a legal provision that allows certain types of interactions between insurance companies and investigators whose duties include the investigation of arson and related crimes. (For more

on these laws, see Chapter 5, *Legal Considerations for Fire Investigators*.) Arson reporting/immunity statutes typically require insurers to report to authorities suspected acts of arson and provide for the release of some (or all) insurance claim information to these authorities in cases that may involve criminal acts. Requests from the authorities must be put in writing, and the information received must be kept confidential. These acts usually extend immunity from prosecution to insurance companies acting in good faith.

Public-sector investigators should become familiar with the arson reporting/immunity statute in the jurisdiction(s) in which they work, so that insurance information can be solicited when needed in an investigation. Investigators performing investigations for insurance companies should also review these statutes, as some of the statutes provide a mechanism for them to request records from public agencies as well.

Public- and private-sector investigators who are active in the field will inevitably have occasion to interact and network with each other at training events and conferences. These interactions provide opportunities for investigators to get to know and learn from others who have more experience and who have expertise in certain areas. Further, it is not uncommon for public and private investigators in a particular geographic area to be assigned to investigate the same fire(s). Communication between these two types of investigators can be informative, but both groups of investigators must be aware of any limits on the types of information they are permitted to share with regard to a case, and must avoid the temptation to request investigators from the other sector to take actions that they were not authorized to perform. For example,

TABLE 10-6 Private Sources of Information

Private Sources of Information	Description/Benefit
National Fire Protection Association (NFPA)	Develops, creates, and revises various standards and guides that may assist with an investigation
Society of Fire Protection Engineers	Works to advance fire protection engineering, including publishing various documents that may prove useful to the investigator
ASTM International	Develops voluntary consensus standards that may be used by architects during the design phase as well as procedures for fire tests
National Association of Fire Investigators	Provides training related to fire investigation topics; implemented the National Certification Board, which offers certificates such as Certified Fire and Explosion Investigator (CFEI), Certified Fire Investigator Instructor (CFII), and Certified Vehicle Fire Investigator (CVFI); presents the International Symposium on Fire Investigation (ISFI)
International Association of Arson Investigators (IAAI)	Dedicated to improving professional development of fire and explosion investigators; provides professional credentials including Fire Investigation Technician (FIT), Certified Instructor (CI), Evidence Collection Technician (ECT), and Certified Fire Investigator (CFI); offers training programs for fire investigators, including the CFITrainer.net online training platform
American National Standards Institute	Accredits the procedures of organizations that develop national consensus standards
Regional fire investigation organizations	May exist at either the state or town level; provide contacts that can be consulted during a fire investigation
Real estate industry	Maintains valuable records concerning structures and their owners, which can aid in the detection of fraud and arson
Abstract and title companies	Maintain records related to former and current property owners, as well as escrow account maps and tract books
Financial institutions	Maintain various records of both individuals' and businesses' financial records, as well as information of loan companies, brokers, and transfer agents
Insurance industry	Maintains valuable records concerning structures and their owners, which can aid in the detection of fraud and arson; maintains property loss and insurance claims databases; includes special investigation units (SIU) and performs fire investigations through employed or retained investigators
Educational institutions	Can provide information related to a person's background and personal interests
Utility companies	Maintain databases of customers, as well as documented problems on the status of their distribution equipment
Trade organizations	Act as a clearinghouse of information specific to their discipline; usually create and publish trade magazines
Local television stations	Often provide investigators with copies of videotape related to an incident
Lightning detection networks	Provide lightning and weather data, which can play an important part in causal analysis of a fire scene
Other private sources	Numerous other private sources exist that the investigator may find helpful depending on the specific needs of the investigation

a public investigator should not ask a private investigator to take an item from a scene if the public investigator's authorization to be on scene has expired. Asking the private investigator to take such an action can result in a finding that the government investigator acted impermissibly through an agent.

Data Analytics

Data analytics—that is, the process of analyzing raw data to find trends and answer questions—has emerged as a valuable tool for law enforcement and other government agencies in recent years. For example, persons performing data analytics may use existing crime reports and statistics to predict criminal activity, which in turn enables the jurisdiction to deploy police appropriately to prevent or intercede in crimes.

Other law enforcement data analytics programs gather information from many accessible sources to paint a picture of crime and behavior that would otherwise have been hidden. For example, a law enforcement analytics program might combine address information, business information, social media posts or statuses, geographic maps, device location data, and regulated business information (such as from banks, pharmacies, or firearms licensees) to reach conclusions or identify patterns that support investigations of known suspects or to identify suspects in criminal activity.

Fire investigators investigating serial arson or other repeated criminal behavior should determine whether applicable federal, state, or local agencies have access to data analytics that might assist in resolving these types of crimes.

After-Action REVIEW

IN SUMMARY

- The collection of data—a necessary element of the scientific method—goes far beyond the fire scene itself.
- The fire investigator should take whatever time and steps are necessary to ensure that all necessary data are collected, and should not feel rushed to make premature conclusions.
- Information obtained during interviews may prove to be the most valuable information gathered during a fire investigation.
- The fire investigator should attempt to conduct interviews as soon as possible to ensure that witnesses are located, identified, and interviewed in a timely manner.
- Interviews regarding fire location, causation, circumstances, and development should, when possible, be conducted by persons trained in the determination of fire origin and cause.
- A thorough understanding of the available facts for an incident can help the investigator create an interview plan and formulate relevant questions.
- Public-sector investigators should consider whether any Fifth or Sixth Amendment issues may exist with witnesses to be interviewed, and consult counsel as necessary.
- An investigation of a scene of any size may include a canvass or survey of people found in the vicinity of the fire.
- Almost any fire scene accessible or visible to the public likely has been recorded, at least in part, in the form of digital photos and/or video. The fire investigator should ask any eyewitnesses whether they have photos or videos of the fire.
- Fire scene video may also be available from the law enforcement agency with jurisdiction, as many agencies use dashboard cameras and body cameras that record the emergency scene.
- Reliable computer and Internet search skills are now a critical tool in every fire investigator's toolbox.
- Any information gained from a website that is not a governmental or known private entity site should be verified by the investigator through other means.
- Municipal, county, state, and federal governments all maintain records that can be useful and important as background information in fire investigations.
- Most states have some form of arson reporting/immunity statute, which allows certain types of interactions between insurance companies and investigators whose duties include the investigation of arson and related crimes.
- Law enforcement data analytics programs are being developed that gather information from many accessible sources to paint a picture of crime and behavior that would otherwise have been hidden.

KEY TERMS

Arson reporting/immunity statute State law requiring insurance companies to provide information on a fire investigation when requested by state authorities working on a potential arson or other crime.

Credibility The likelihood that a person is telling the truth.

Data analytics The process of analyzing raw data from various sources in an effort to find trends and

answer questions, including questions about criminal activity.

Investigative file The organized collection of all of the documentary information in a fire case, including verbal, written, and visual information from the scene; the reports of investigators and other professionals; and any other investigations or research that has been conducted.

On Scene

1. What are some general questions that you can plan to ask almost any eyewitness who says they saw the start or initial growth of a fire?

2. What is the difference between witness statements and scientific evidence such as fire patterns? Are there questions that you can ask that will test a witness's ability to accurately perceive and report what happened in a fire situation?

3. Does your state or other jurisdiction have an arson reporting/immunity statue? What prerequisites

must a public officer meet to be able to request information from an insurer under your jurisdiction's act? Does that act include reciprocal obligations for a public agency that receives a request for information from an insurance company?

4. What is the significance of the various types of evidence available from private sources? For example, what use would utility records, insurance claims history, or real estate records potentially be to a fire investigator?

Chapter Opener: © Jones & Bartlett Learning. Photographed by Glen E. Ellman; On Scene siren: © Bildgigant/Shutterstock.

CHAPTER **11**

Documenting the Fire Scene

KNOWLEDGE OBJECTIVES

After studying this chapter, you should be able to:

- Describe the use of photography in fire investigation. (**NFPA 1033: 4.3.2**, p. 274)
- Explain considerations for photographing a fire scene. (**NFPA 1033: 4.3.2**, pp. 274–275)
- Describe documentation of fire scene photos. (**NFPA 1033: 4.3.2**, pp. 274–283)
- Explain the considerations for video recording a fire scene. (**NFPA 1033: 4.3.2**, pp. 279–280)
- Describe the activities and items that should be documented at a fire scene. (**NFPA 1033: 4.3.2**, pp. 280–282)
- Describe the use of diagrams and drawings in fire investigation. (**NFPA 1033: 4.3.1**, pp. 283–286)
- Identify the elements that should be included in sketches and diagrams (**NFPA 1033: 4.3.1**, pp. 283–286)
- Describe the use of note taking in fire investigation. (**NFPA 1033: 4.3.3**, p. 286)
- Describe the use of reports in fire investigation. (**NFPA 1033: 4.3.3**; **4.7.1**, pp. 286–287)
- Explain the considerations for writing a fire report. (**NFPA 1033: 4.3.3**; **4.7.1**, pp. 286–287)

SKILLS OBJECTIVES

After studying this chapter, you should be able to:

- Document conditions at an investigation scene using various media. (**NFPA 1033: 4.3.1**; **4.3.2**; **4.3.3**; **4.7.1**, pp. 274–283)
- Diagram a fire scene. (**NFPA 1033: 4.3.1**; **4.7.1**, pp. 283–286)
- Write a written report documenting a fire investigation. (**NFPA 1033:4.3.3**; **4.7.1**, pp. 286–287)

You are the Fire Investigator

You receive a working fire notification for a single-family home. When you arrive, you meet with the incident commander. He tells you the fire had been confined to an upstairs bedroom. The investigation appears to be simple, and you take a few photos with your cell phone. You estimate the size of the room and make a quick sketch, although you do not obtain any measurements. You surmise that the fire "must be electrical" and leave the scene. The following day, you are told an occupant of the residence had been injured, and you are asked for the origin and cause of the fire. You return to the scene and discover the room has been cleaned out.

1. What information do you have to complete a proper investigation?

2. What documentation might you be able to obtain to assist with your investigation?

3. What are the advantages of completely documenting the scene?

JONES & BARTLETT LEARNING
NAVIGATE™ *Access Navigate for more practice activities.*

Introduction

In documenting any fire or explosion scene, the goal is to record the scene using media that will allow the investigator to recall, communicate, and document all observations later. The compilation of these data enables the investigator to recall the details of a fire scene investigation after substantial time has passed, and to support and verify opinions and conclusions in a court proceeding. The documentation types that are most often used in fire investigations include photographs, digital video, diagrams, maps, overlays, audio recordings, laser surveys, digital and handwritten notes, sketches, and reports.

Photography

The primary method of recording the scene is still photography, which can be supplemented with video photography. Photographs are effective reminders of what has been observed at the scene and allow a concise depiction of patterns, evidence, and other features for later presentation. They are acceptable for presentation in court, so long as they are objective and do not inflame the jury's emotions or exaggerate the situation. The presenter of the photographs to the court must be able to affirm that the photographs are relevant to the testimony and show a true and accurate depiction of the scene as they saw it.

Although both time and expense are considerations when taking photographs, it is always preferable to have too many photos rather than too few. Photographs should always include fire effects and patterns, artifacts, and any items of evidentiary value.

Every fire investigator should come into a fire scene with a fundamental understanding of photography, including familiarity with the equipment and accessories, lighting and movement, and common camera settings. Fire scenes are dominated by black and gray colors and shades, so taking photos with the proper lighting and exposure is critical to being able to review the scene later and demonstrate it to other persons.

To get the fullest representation of what occurred on the scene, it is best to prepare various forms of documentation. A digital video used in conjunction with still photographs, for example, will be more effective than photos alone. If you would like to become more familiar with cameras or video equipment, many courses are available that can provide this education. Training in crime scene photography may be offered at local or state criminal justice training academies or through college courses, but even basic training by camera clubs or supply stores may be useful.

Timing

Since firefighting is by nature destructive, the investigator should begin photography of a scene as early as possible, before the scene becomes altered, disturbed, or destroyed **FIGURE 11-1**. If it is possible to obtain photos during the firefighting operation, this can be very helpful. Similarly, if firefighting crews can pause the overhaul operation to allow the investigator to safely take photos, better documentation will result. Other cases in which timing is important include a building in danger of collapse, a building that must be demolished for safety reasons, and the presence of hazardous materials that are creating an environmental hazard. Further, when investigation activities begin

FIGURE 11-1 Since firefighting is by nature destructive, the investigator should begin photography of a scene as early as possible, before the scene becomes altered, disturbed, or destroyed.
© Hollyn Johnson/The Bay City Times/AP Images.

in earnest, the investigator should consider taking photos of the entire scene and fire area before investigators disturb anything further.

Photos should be taken anytime investigators discover items of significance or with potential evidentiary value. Even if removing and inspecting these items is appropriate, it is best practice to take the photos before touching them so that their exact location and orientation are documented. As the layering and debris removal process proceeds, periodic photos of the layers should also be taken, along with photos demonstrating the location of any specific items of interest in the area of origin. Finally, photographing a scene after it has been fully excavated and reconstructed is an important step not only to document the reconstruction itself but also to demonstrate the condition of the scene when investigators left it.

Basics

Cameras

Digital equipment for still photography and video has replaced film equipment as the technology of choice for scene documentation. Many different types of cameras are available at a wide range of prices. The camera choice is primarily predicated on financial resources and your personal skill level. Automatic cameras are the easiest to operate. They determine the primary scope of the photograph and then focus on the most obvious image in the viewfinder. These cameras can provide a sense of comfort to some investigators because they adjust the lens opening (f-stop), control the shutter speed, operate the flash, and focus the lens. These features remove many potential pitfalls for the inexperienced photographer.

Some skilled investigators prefer manual cameras because they allow for adjusting the focus and other settings to suit the immediate circumstances. The user can also obtain specialty photographs that the automatic camera's built-in options cannot perform. Most cameras have both manual and automatic settings.

The resolution of a camera affects the usable size of an image. A lower resolution will limit the size of an image for use as an exhibit at trial. Resolution is measured in pixels; the more pixels a camera has, the more detail it can capture and the larger the image that can be used for demonstrative purposes. Most digital cameras automatically display a digital image of the photo a second or two after the photo is taken, which allows the investigator to review the image immediately for proper exposure and focus. Digital cameras also provide the ability to enlarge the photo view to ensure that a sufficient level of detail is available. Most of these cameras have the capacity to take hundreds of photos, which can then be downloaded easily to a computer's hard drive. Depending on the agency's or company's procedures, the media card should be either stored and replaced with another, or erased and returned to service.

Many modern cell phones take photos and videos of excellent quality, and some investigators use their phones for these purposes on fire scenes. While such images can be acceptable for investigation purposes, some drawbacks exist. For example, cell phones usually have only a small, attached flash, which might not allow for taking an effective photograph, especially in dark conditions. By contrast, many digital cameras provide for stronger flash photography and can be used in tandem with remote flash units, which can improve photos. Further, unlike cameras, photography via cell phones does not allow for the use of conveniently removable disks and flash drives, so the investigator will not have a physical medium to introduce into an evidence accountability system. Finally, if the cell phone contains personal photographs as well as fire photos, the potential exists for a court challenge on the grounds that the investigator did not produce all photos, or a request for a subpoena or other examination of the phone to look for photos or to determine other circumstances surrounding the photography.

Image Authentication

To be admissible evidence in court, any photograph that is captured must be relevant to the testimony of the witness whose testimony is introducing the photo and must be a true and accurate representation of the scene or the item depicted in the photo. With readily available computer technology, digital images can be enhanced to correct brightness, color, and contrast. If an image has been enhanced, however, the investigator must keep the original photo for comparison if necessary and must document the enhancement efforts that went into creating the modified photo.

To best be able to authenticate an image in a legal proceeding, investigators should establish a process that preserves the original image and keeps track of all images and modifications. Procedures should also be established for image storage, such as placing the photos on an appropriate storage medium—or the use of a computer software program that does not allow the original image to be altered—and saving them with the original file name.

> **TIP**
>
> When using digital photography, do not delete any photos, even if they are not of the desired quality (e.g., if they are blurry or were taken accidentally). The case file should contain all photos taken to avoid any accusations that the investigator was trying to conceal evidence or information or portray the scene in an unrealistic light. Also, digital photos should not be renamed, although the investigator can make a copy of an original digital photo and then rename the copy for presentation purposes, as long the original is still being maintained and is available.

Lenses

A camera lens is used to gather light and to focus the image. Most cameras use compound lenses, meaning that multiple lenses are located in the same housing. A basic understanding of the lens's function is necessary to obtain quality photographs.

The lens aperture is an adjustable opening that controls the amount of light admitted. The adjustments of this opening are sectioned into increments called f-stops. The larger the f-stop number, the smaller the size of the opening and the better the depth of field will be. The depth of field is the range of distance in which an image will appear acceptably sharp—that is, the range of objects that are included within the focused area of the photo.

When taking photos, there is a trade-off between depth of field (f-stop) and adequate light (shutter speed). The photographer must balance the desire for more depth of field with the need for adequate light.

The focal length of a lens is an optical measurement, not a physical measurement of the lens; it determines the angle of view (how much of the scene will fit into the picture) and the magnification of items pictured in the photo. Digital cameras have variable or fixed focal length lenses that range from 20 mm to 1200 mm or

greater. Some digital cameras are equipped with optical zoom, digital zoom, or both. Macro lenses are useful for close-up photography. You should avoid changing camera lenses inside a fire scene to avoid contamination of the lens and camera.

Depth of Field

The area of clear definition, or depth of field, is the distance between the farthest and nearest objects in focus at any given time. The depth of field depends on the distance to the object being photographed, the lens opening, and the focal length of the lens being used. This factor can determine the quality of detail in the photographs; if the depth of field is too shallow, a photo of an object can appear focused only in a small area, with other parts of the object being blurry.

Filters

Use of a neutral UV filter on a camera is recommended for all fire scene photography. The UV filter does not alter the tone or color of the image, so court admissibility of the photo is not affected, but it provides protection for the delicate surface of a lens. In general, the use of color filters is not recommended, as the photos could turn out unexpectedly and could draw criticism in court. If a color filter is used, take a duplicate picture without the filter for reference.

> **TIP**
>
> Batteries will drain in cold weather. As a backup, always carry extra batteries for your camera. Keeping the camera in a warm vehicle until it is needed at a cold fire scene will help to extend battery life. To avoid potential damage to photography equipment from leaking batteries, do not leave batteries in your equipment for extended periods of time.

Shutter Speed

The shutter speed is the amount of time a shutter is open during an exposure (i.e., while taking the photo). Shutter speeds are typically denoted in fractions of a second. The higher the denominator, the faster the shutter works (e.g., 1/500 sec is faster than 1/100 sec) and the shorter the time the shutter is open. Faster shutter speeds are necessary to freeze fast-moving subjects. Although capturing a fast-moving subject is unlikely to be necessary in fire scene photography, faster speeds, when available, also reduce the risk of a photo being blurred due to movement or unsteadiness of the camera or the photographer.

However, a good exposure also requires a minimum amount of light. Thus, when the photographer chooses a faster shutter speed, a wider lens opening (decreased f-stop) will be necessary to get the same level of light in the photo. Conversely, if the investigator chooses a smaller aperture (by setting a higher f-stop), the depth of field is improved but less light is admitted, and a longer shutter speed is required.

If shutter speed is too low, there is a greater chance the resulting photo will be blurred due to the natural movement of camera and photographer. To avoid blurring the image, use of a tripod is required for shutter speeds less than 1/60 sec (60), although a flash can compensate for this problem under certain conditions. If a zoom or telephoto lens is used, the problem with movement and blurring gets worse, and the investigator will be required to set shutter speeds much higher to avoid blurred photos, or use a tripod to hold the camera steady.

Lighting

Lighting plays an integral part in photography. Although having sunlight available for scene photographs is usually ideal, even during the daytime the investigator may need to add flash or other alternative lighting to obtain good photos. Flash photography is usually necessary at night even when artificial scene lighting is available.

The flash attachment may be permanently mounted on the camera, temporarily mounted, or separate from the camera. Because of the large dark and light areas and the charred surfaces that are often found at fire scenes, these scenes are difficult to illuminate effectively for photography. A high-quality flash that can provide a vast area of light for multiple exposures is a necessary tool. Other supplemental light sources may include portable lights, such as floodlights and flashlights. However, floodlights can cause glare and distort appearances or distract from the subject of a photo, and flashlights can cause some areas of a photo to be bright while obscuring other areas.

Various "temperatures" of light are emitted from different light sources; for example, most individuals can distinguish the color of fluorescent light from that of incandescent light. A flash normally is designed to simulate natural sunlight. Another method to enhance light is through the digital camera's white balance setting, which can often be adjusted to address different temperatures. Become familiar with this setting and its use. For fire scene photography, an automatic or flash-white setting is recommended.

Ideally, the investigator will use a flash unit that can be separated from the camera system, allowing the ability to angle the light source as needed under the circumstances. This capability enables the investigator to use the built-in flash on the camera, with the separate flash unit increasing the amount of light available for a proper exposure and better demonstrating the details of the subject. Multiple flash units and remote operating devices called slaves can be used to illuminate large areas.

A **ring flash** is often used for close-up work. This specialized flash unit fits on the end of the lens and is often used when photographing a critical piece of evidence, such as an arc mark or tool mark. Specialized digital cameras are also available that take macro (close-up) photos with the assistance of a built-in ring flash.

Whenever concerns arise about the accuracy of an exposure, "bracket" it by taking a series of photographs that include a photo at the recommended f-stop, a photo at an f-stop setting below the recommended setting, and a photo at an f-stop setting above the recommended setting. Some digital cameras provide a setting that will automatically perform this function. When reviewing the photos, an investigator can select the best exposure of the three for presentation purposes.

Photo Technique

Many investigators take photos that follow a general sequence: They begin on the exterior perimeter of a building fire, progress to the interior, and then move from the least damaged areas to the most damaged areas. Another approach is to take photos in a sequence that generally illustrates the investigator's investigative process, starting with a general overview of the scene, then reflecting the exterior survey, followed by the interior survey, and finally layering and reconstruction in the area of origin. Critical evidence should be documented by photographing the subject from different angles and distances, including from above when possible. To show the degree of smoke spread or evidence of undamaged areas, it is important to photograph the entire fire scene, not just the hypothesized area(s) of origin. Evidence of alternative hypotheses should also be photographed.

Sequential Photos

Items of interest in a scene can be photographed sequentially to better illustrate their location and significance. For example, a photograph of a relatively small subject may be taken first from a distance (perhaps from the doorway into the room) to show the position of the subject in relation to other fixed objects, followed by a shot from a medium distance showing more detail, and ending with a close-up of the subject **FIGURE 11-2**. Sequential photography allows

FIGURE 11-2 Sequential photographs of a chair (**A–C**).
Courtesy of Jamie Novak, Novak Investigations Inc. and the St. Paul Fire Department.

the viewer to understand better the totality of the view and the relationship of the subject to the overall surroundings.

Mosaics

Another method for depicting the totality of the scene is to use mosaic photographs. **Mosaic photographs** are a series of photographs that cover a large area by overlapping the start of one photograph with the area where the previous photograph ended **FIGURE 11-3**. Mosaics can be useful when a panoramic view is desirable. To create a mosaic, first identify a landmark on the edge of the first photo you are taking. Each ensuing photograph should encompass a portion of the landmark in the previous photo, so that the entire area you wish to photograph is fully covered by one or more photos. A tripod will allow for a consistent

FIGURE 11-3 Mosaic photographs of a warehouse burn scene taken from an aerial truck.
© Jones & Bartlett Learning. Photographed by Glen E. Ellman.

mosaic pattern and alleviate movement and blurred photos. Many digital cameras have a preprogrammed feature that, when selected, automatically adjusts the camera for taking a seamless panoramic image.

FIGURE 11-4 A digitally stitched image.

Reproduced from *NFPA 921-2021, Guide for Fire and Explosion Investigations*, Copyright © 2020, National Fire Protection Association. This reprinted material is not the complete and official position of the NFPA on the referenced subject, which is represented only by the standard in its entirety.

If the camera does not include a panorama option, digital stitching computer programs can create mosaic images from a series of digital photos **FIGURE 11-4**.

Photo Diagram and Photo Log

When recording the scene, annotate a diagram of the site, identifying the point from which each photograph was taken, the direction of the photograph, the placement of the item, and the photo number. This annotated diagram is referred to as a **photo diagram FIGURE 11-5**. If a large number of photos are taken at the scene, the diagram can identify this information with respect to photos the investigator considers important or photos that are referenced in the report.

If the time of the photograph is important, it should be noted on the diagram. Investigators should identify their diagrams by affixing their initials, the date, the location of the scene, and any other pertinent identifiers. The resulting diagram will be helpful to individuals who did not visit the fire scene, as it gives them an overall picture of the condition of the scene and what was observed by the investigation team. Virtually all digital cameras record the date and time of a photo in the data that are saved with the photo, and many can imprint the date and time on the image itself.

Scene photos should be accompanied by a **photo log** that provides a list or table of photographs with a reference to each photo (such as the photo number assigned by the camera), and a brief description of the item or area photographed in each. The photo log should have a caption listing, at a minimum, the date of the photos, the scene involved, a case number if relevant, and the identity of the photographer. A photo log is essential for understanding and recalling what was depicted in the photos.

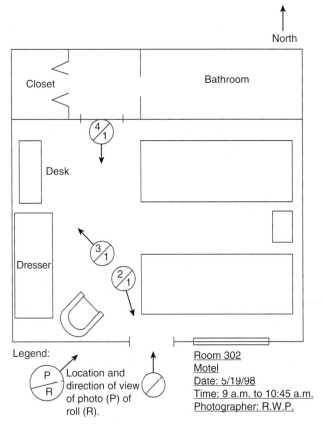

FIGURE 11-5 A diagram showing photo locations.

Reproduced from *NFPA 921-2021, Guide for Fire and Explosion Investigations*, Copyright © 2020, National Fire Protection Association. This reprinted material is not the complete and official position of the NFPA on the referenced subject, which is represented only by the standard in its entirety.

An investigator may work with another investigator to create a photo diagram or log while photos are being taken and even delegate the photography of the fire scene to others. However, take care that the photos that are being taken depict the items and views needed to fully document the scene. The investigator's needs should be clearly conveyed to the assistant taking the photos.

Photography and the Courts

Photographs play an important role for many reasons, including for presentation in court. Prior to any photo being admitted into the court, there may be several challenges to its use. The photo must be a true and accurate depiction of its subject and must be relevant to the testimony. A court may exclude photos that may be too inflammatory to jurors, such as photos that depict a gruesome death scene. Photos may also be rejected if they lack clarity or if they lack relevancy to the matter before the court. Most courts permit evidence in the form of printed photos as well as photos projected or depicted on screen.

Video Photography

Video photography has become an acceptable medium for orienting viewers to a fire scene. Newer digital video technology provides important low-light and self-focusing features. A big advantage of using video is the ability to orient viewers to the entire scene by recording complete views of important areas. Although video photography should not be the sole medium used to convey conditions at the scene, it can be an effective supplement to still photography, which remains important for documenting evidence and scene conditions. When documentation of testimony is needed, video cameras can be used to interview witnesses, suspects, occupants, and owners. It can also help orient the viewer to witnesses' activities and observations.

Most digital cameras also have video capabilities, and some can perform both functions at the same time. Investigators using this technology should record videos slowly and steadily so that a useful video results. Audio muting is recommended, although many cameras do not permit the microphone to be turned off. Some investigators choose to narrate while recording as a method of later refreshing their memory about the scene conditions, fire patterns, and location(s) of evidence. Whenever the microphone is working, investigators should limit background noises and distractions and should carefully avoid inappropriate comments or "off the cuff" analysis of cause, which may create problems or embarrassment later.

Suggested Activities to Be Documented with Photos and Video

It is important to document as much of the scene as possible. Some suggested activities to record are conditions on arrival, suppression activities, fire location and progression, overhaul activities, observers, patterns, and investigator work in origin and cause determination. The progression of the fire, its colors, its reaction to suppression activities, and the overhaul procedures employed are all important in helping the fire investigator determine the origin and cause. Photographs can also document the extent of injury to the victims or damage to the structure.

Photographs of the crowd observing the fire scene activity can help you identify individuals who may have knowledge beneficial to the investigation. These photographs can also help identify individuals who are seen at multiple fires or are known by the law enforcement or firefighting community.

All fire investigators should routinely determine whether initial witnesses to the fire, such as the 9-1-1 reporting party, documented what they saw with a camera phone or digital camera. Many times, media

FIGURE 11-6 A. Street signs are identifiable landmarks. **B.** A fire number is one possible address identifier.
(A) © Cindygoff/iStock/Thinkstock. (B) Courtesy of Rodney J. Pevytoe.

news outlets have staff respond to large fires. These professionals sometimes arrive on scene prior to fire apparatus and often have superior-quality photographic equipment. Reporters may be busy interviewing victims or witnesses prior to your arrival and should always be contacted.

Photographs of the suppression activities can help the fire investigator understand why the fire reacted in a particular manner. Also, when documenting suppression activities, the locations of hydrants, engine companies, apparatus, and hose placement should be noted.

Exterior photographs can be used to establish the location of the fire scene. To orient the viewer to a building, exterior photographs should include street signs or other identifiable landmarks that are likely to remain present for some time **FIGURE 11-6**. They should also include surrounding locations and all angular views of the exterior of the fire scene **FIGURE 11-7**. Exterior photos should be taken of the address numbers of affected buildings, which may be useful for documentation and search warrant purposes.

Structural photographs document the extent of damage to the structure after heat and flame exposure.

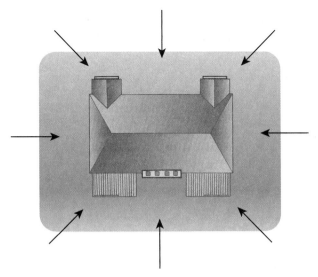

FIGURE 11-7 Photograph the scene from all angles and corners.

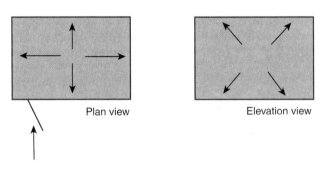

Plan view   Elevation view

FIGURE 11-8 Interior photographs should be taken of all four walls, the floor and ceiling, roof, and both sides of each door.

Reproduced from *NFPA 921-2021, Guide for Fire and Explosion Investigations*, Copyright © 2020, National Fire Protection Association. This reprinted material is not the complete and official position of the NFPA on the referenced subject, which is represented only by the standard in its entirety.

These photographs should be taken from multiple views to record heat and flame damage and fire spread patterns. Structural failures or deficiencies should be captured in photos because these can play a role in the fire. The resulting photographs should be useful in explaining the analysis of the fire scene to supervisors, prosecutors, the court system, and insurance representatives.

Interior photographs assist in documenting the scene and more thoroughly describing the conditions within the structure. All significant points that were accessed or created by the fire should be photographed, along with significant smoke, heat, and burn patterns **FIGURE 11-8**.

An investigator should photograph the condition of rooms within the fire scene to document thermal and smoke conditions. All heat-producing appliances or equipment should be photographed to document their condition, location, and role, if any, in the fire cause. The positions of controls on those devices or

FIGURE 11-9 The wick from a Molotov cocktail found at a fire scene.
© Jones & Bartlett Learning. Photographed by Jessica Elias.

equipment that is relevant to the investigation should be photographed, as well as all electrical cords and outlets pertinent to the fire's location.

Furniture and other contents involved and uncovered during the excavation and reconstruction should be photographed throughout the process and again after reconstruction. In addition, the investigator should photograph the protected areas and other evidence that is revealed by these items when they are moved.

The condition of entrance and exit routes, especially doors and windows, should be documented. The photographs should depict their condition at the time of the fire (e.g., whether doors were locked or open, whether windows were open or closed). The condition of interior fire detection and control devices should be documented as well, including whether smoke detectors or sprinklers were activated, and whether extinguishers were employed. Mechanical clocks may indicate the time at which power was cut off or fire or heat physically stopped their movement. Be cautious about interpreting the times on battery-operated clocks, however, as they could have stopped when the battery ran out instead of during the fire.

The utilities present at a structure should be thoroughly documented, including the condition of various controls and appliances such as transformers, panels, meters, regulators, and valves. Also document the point at which the utilities enter the structure, including their location and condition at the time of the fire. The condition of fuses and/or circuit breakers, the positions of all circuit breaker handles, and the panel's legend, when available, should be photographed.

During the scene examination, photographs should be taken of all items of evidentiary value. For instance, the photo shown in **FIGURE 11-9** documents the wick from a Molotov cocktail found at a fire scene. Photographs of potential evidence should allow the viewer to locate the item within the scene and perceive its general appearance. More detailed photographs of evidentiary items can be taken after they are moved if the photographer is unable to obtain detailed photos due to scene constraints such as lighting, location, or hazards. A highly visible ruler can be used to provide a scale to identify the size of items in photographs. In such case, two photos should be taken: one showing the evidence and another showing the ruler next to the evidence.

FIGURE 11-10 Documentation should include protected areas where a body was located.

Reproduced from *NFPA 921-2021, Guide for Fire and Explosion Investigations*, Copyright © 2020, National Fire Protection Association. This reprinted material is not the complete and official position of the NFPA on the referenced subject, which is represented only by the standard in its entirety.

TIP

Video recording can be used to document the examination, position, and condition of evidence.

Photographs should also be taken of any survivors' or victims' locations at the time of the fire, any indicators of actions taken by them during the fire, and any end result such as serious injury or death. Documentation should include marks on the walls and protected areas where a body was located **FIGURE 11-10**. If the scene includes a fatality, the position of the victim should be thoroughly recorded. If the full condition of the victim cannot be documented at the scene because of lighting, scene hazards, or other obstacles, additional photos should be taken as soon as possible after removal.

When a witness or victim reports that they saw a potentially significant event, an investigator should attempt to photograph the view from the position of the witness.

Aerial photographs can help clarify the scene's overall arrangement and the positions of large evidence items such as a vehicle or body **FIGURE 11-11**. The investigator may be able to take these photos from an aerial ladder or tower fire apparatus, or by means of an unmanned aerial system (drone), where permitted.

Satellite and street-level photography is available in many areas and viewable online in conjunction with mapping applications. These sources may provide both preincident and postincident photos of the scene.

Photos Obtained from Witnesses

Witnesses frequently take photos and videos of fires, and often appear at the scene long before the arrival of the emergency services. If an investigator receives photos or videos from a witness or other outside source, treat them as any other item of electronic evidence. Secure a copy using the same type of medium being used to secure other photos and videos, and

FIGURE 11-11 Aerial overview of a fire scene.

Reproduced from *NFPA 921-2021, Guide for Fire and Explosion Investigations*, Copyright © 2020, National Fire Protection Association. This reprinted material is not the complete and official position of the NFPA on the referenced subject, which is represented only by the standard in its entirety.

place them in evidence storage or with the evidence custodian, as appropriate. Gather details about the photos or videos such as who took them, when they were taken, and from what angle or perspective, and document this information in your report. Document when the photos were received and how they were secured as evidence items.

Presentation of Photographs

When displaying or presenting photography of a fire scene examination, you should choose the videos or photographs that depict the examination and causal determination with the greatest clarity. In addition, you should determine which media are most acceptable to the local court system, law enforcement agencies, fire departments, and insurance companies.

Computer-based and video presentations are effective means of conveying scene conditions and actions undertaken at the scene. Many law enforcement agencies, fire and police academies, school systems, and courts have the equipment necessary to support such computer or video presentations. Photograph enlargements and projected/screen images are commonly used as evidence in court hearings. Computer presentations are an effective and comprehensive medium for exhibiting large quantities of visual material, as they allow for integration of multiple formats into a single presentation. A hard copy of all material to be presented should always be made available to all parties. Also, ensure the screen presentation can be seen from all angles in the courtroom.

Diagrams and Drawings

Diagrams and sketches are used to support the fire investigator's memory and to document details of the scene. For example, a diagram could be used to gather details during an important witness interview or to provide a means for orienting photographs. Sketches can aid in documenting fire patterns, fire growth, and scene conditions. The accuracy of the data used to create these materials is essential.

Types of Drawings

A **diagram** is a formal drawing that is completed after the investigation. A **sketch** is generally a freehand diagram or a diagram drawn with minimal tools that is completed at the scene. The differences between the two types of drawings relate to the amount of detail that is incorporated in each. Comparing the less detailed sketch and the more detailed diagram, the differences between the two could include levels of detail relating to the type of construction of the structure, features of the structure, equipment, and other factors that may have played important roles in the origin, cause, and spread of the fire. Every investigation should include fire scene sketches, but especially investigations that are likely to lead to criminal or civil litigation.

When determining which type of drawing to use, first decide what needs to be shown on the drawing. If unable to create a detailed drawing, the fire investigator should create a rough sketch that depicts the area of fire origin, the arrangement of items within this area, and the locations of doors and/or windows. This rough sketch should also include accurate measurements of walls and ceilings; placement of doors, windows, and other fixed objects; and the positions of furniture and appliances within the area of fire origin. **FIGURE 11-12** includes an example of a rough sketch that shows the basic layout of the structure, as well as the finished diagram made from that sketch.

Several types of drawings can be used to illustrate fire scene conditions. Sometimes a detailed diagram is necessary for proper documentation. At other times, a rough sketch will suffice when the analysis and conclusions are simple. The type of drawing chosen depends on the degree of detail required.

When the scene or litigation is more complex, an investigator may need to develop or acquire detailed resources such as building plans or construction documentation. **TABLE 11-1** lists details that may be needed in such a case.

Drawing Tools and Equipment

The creation of a scene diagram is likely to vary depending on the specific situation. Often, the size and the complexity of the scene and the investigation dictate the type and degree of detail included. In large or critical cases, consider obtaining architectural plans or diagrams from equipment manufacturers. Surveyors may be useful in accurately mapping large-scale scenes.

Laser measuring devices are widely available and can assist the investigator in accurately measuring the dimensions of larger areas of a structure. Many larger municipalities and state and federal law enforcement agencies are able to perform computer-assisted mapping using laser devices.

When needed, three-dimensional (3D) imaging systems may potentially be available to borrow from law enforcement agencies, private investigation firms, builders, and contractors. Recent advances in these tools offer fire investigators the opportunity to create accurate 3D representations of fire scenes, which can greatly assist in better understanding the scene, including its dimensions and scale. 3D models can also help investigators test hypotheses by providing a more detailed picture of fuels, compartments, ventilation openings, and flow paths.

Recent advances in computers and software have allowed investigators to utilize various types of programs to develop diagrams. Computer-assisted design (CAD) programs may be used either alone or in combination with computer fire models. A good drawing package offers many useful tools. Some programs provide for multiple layers in the sketch, thereby enabling the investigator to depict both prefire and postfire conditions in the final diagram. No matter what type or brand of program is used, choose one that offers flexibility of design, automatic dimensioning, and a variety of "libraries" that contain predrawn objects such

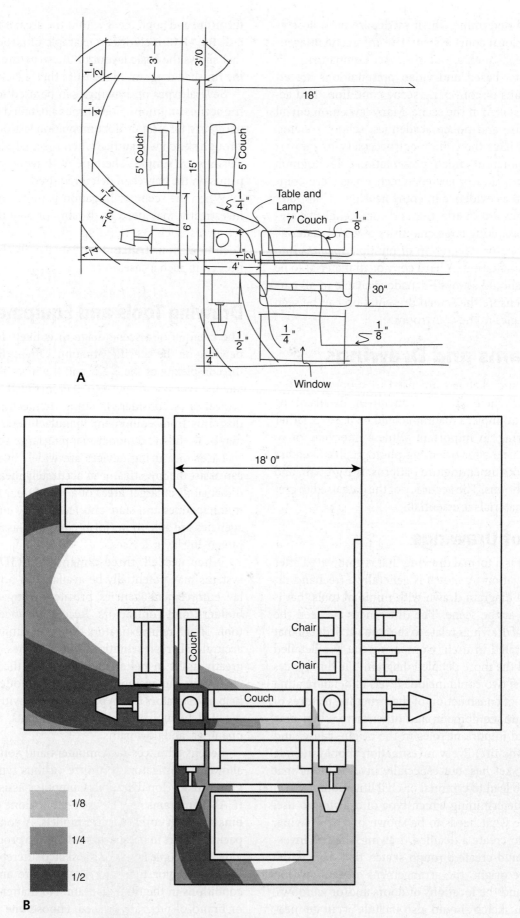

FIGURE 11-12 A. A rough sketch showing layout of a structure. **B.** A finished diagram of a structure.

TABLE 11-1 Detailed Documentation

Type of Documentation	Example
Site or area sketches	Show the location of apparatus, water supplies, or similar information; may also include inspection reports
Floor plans	Identify the locations of rooms, stairs, windows, doors, or other features
Elevations	Single-plane diagrams that show an interior or exterior wall and specific information about it
Details and sections	Show specific features of an item, such as the position of switches and controls or damage to an item
Exploded view diagrams	Used to show assembly of components or parts lists
Three-dimensional representations	Three-dimensional representation, regardless of whether these plans already exist. Thickness of walls; air gaps in doors; and the slopes of floors, walls, and ceilings are examples of the information that should be measured, obtained, and documented.

as furniture, electrical, plumbing, and HVAC (heating, ventilation, and air conditioning) components.

Diagram Elements

Depending on the complexity of the investigation, a wide variety of elements may be included on sketches and diagrams:

- General information: Identify the name of the person who created the diagram, the diagram title, and date of preparation.
- Identification of compass orientation: Most often the sketch will note north at the top of the page.
- Scale: Indicate on the drawing whether it is "not to scale" or "to scale." If the drawing is to scale, specify the scale being used.

- Symbols: Be consistent with the use of symbols, being sure to not use the same symbol for multiple purposes. It is recommended that an investigator use the same symbols as are used in engineering or architecture. Fire protection symbols can be found in NFPA 170, *Standard for Fire Safety and Emergency Symbols*.
- Legend: Create a legend for any drawing, indicating what the referenced symbols represent.

Prepared Design and Construction Drawings

Prepared design and construction diagrams are those that were developed for the construction or design of a building, equipment, or even an appliance. Care should be taken in the interpretation of these diagrams, however, because items or buildings may have been modified during their construction and/or use.

Architectural and Engineering Drawings

During the construction phase of a building, a variety of drawings are used by the contractors. These diagrams may depict only utilities, fire detection and suppression equipment, topography, or other unique aspects needed in the construction process. It can be helpful to be familiar with these drawings, and elements of the drawings may prove useful in the preparation of the investigative diagram.

Architectural or Engineering Schedules

In larger buildings, it often becomes necessary to detail the type of equipment used within the structure. A detailed list of this type is known as a **schedule**. Schedules, which may be broken down by equipment type, generally detail all of that particular type of equipment used within the structure. For example, a door schedule details the placement of all doors within the structure. Other examples where schedules are used include interior finish schedules, electrical schedules, HVAC schedules, and plumbing and lighting schedules.

Specifications

When an architect or engineer prepares a drawing, the types of materials used in construction are listed on the specification sheet. The specification sheet matches materials to their specific placement. For example, the specification sheet could annotate the use of R-17 insulation in the attic area, ⅝-inch plywood in a ground-floor bedroom, and so forth. These drawings can usually be obtained from the contractor or the building inspection department in the community in which the structure was built. They also include the architect's elevation drawings—that is, drawings of the outside of the building—which can be helpful if the building is a total loss. If possible, as-built plans should be obtained as well. These architectural representations indicate how the structure was actually built and may vary from the original plans.

Note Taking

Investigative note taking—that is, the construction of *field notes*—is important to supplement those items that cannot be photographed or sketched. Note taking may include names and addresses, model or serial numbers, witness interviews and statements, photo logs, identification of items, types of materials, data needed to produce an accurate computer model, or investigator observations such as fire effects and patterns and building conditions. Notes may also contain information that is obtained from witness interviews and document reviews that relate to the scene. Although they should not be used as an incident report, forms have been developed for note taking in the field in support of data collection. These forms may aid the investigator in gathering the data necessary to prepare the report.

Many investigators have found that digital recording devices are very useful for recording interviews with witnesses, victims, suspects, and others, as well as for recording notes during on-scene examinations. These electronic statements and notes can be easily downloaded into the case file on a computer along with digital photos and other digital information.

Be careful not to rely solely on one medium to record the scene. For example, the use of a portable recorder or another device should not be the only means of gathering data.

When completing the final report detailing the information about the fire scene, the documentation produced and recorded during the investigative process will be invaluable. A review of the notes, data forms, photos, and sketches made throughout the course of the investigation can help the investigator construct an accurate and detailed scene report.

Many departments have established a uniform policy on the retention of notes. Any notes maintained are subject to examination or "discovery" by interested parties, and this review may occur long after the information is collected or the report is written. Be aware of department/company policy on the retention of notes and adhere to it during every investigation.

Reports

The purpose of the fire investigator's written report is to document and communicate an accurate and concise record of the investigation's findings. Reports generally contain descriptive information, pertinent facts, and opinions and conclusions **TABLE 11-2**.

Every agency may have a different format for its reporting process. In addition, the format and content of the report will vary depending on the needs of the organization or request of the client on whose behalf the investigation was performed. NFPA 921, *Guide for Fire and Explosion Investigations*, does not prescribe a particular format of a report. When a case goes to court, rules of procedure may require certain report formats and/or the inclusion of specific categories of information.

All reports should be written in a clear, precise, and accurate manner. Such reports are often distributed to others outside of the investigator's department or company, and an organized, well-written report

TABLE 11-2 Report Information	
Type of Information	**Example**
Descriptive information	Date, time, and location of incident; date and location of examination; date the report was prepared; name of the person requesting the report; scope of the investigation; and nature of the report
Witnesses	All information necessary to identify and locate witnesses to the event and to other important facts
Interviews	Statements of witnesses to the event, and of other persons interviewed to gain data and information
Pertinent facts	Description of the scene, items examined, and evidence collected; data and observations that form the basis for hypothesis development, testing, and conclusions
Opinions and conclusions	The opinions and conclusions of the investigator; the foundation on which those opinions and conclusions are based; and the name, address, and affiliation of each person who expressed an opinion or conclusion

will create a favorable impression of the investigator's credibility and professionalism.

The report should describe the scene and state the times and dates when it was examined. It should list items of evidence examined and taken, and should detail the dates and locations of any subsequent examination of evidentiary items. All relevant information should be included, including all of the data and observations that support the investigator's working hypotheses. Sources, such as witnesses, should always be clearly identified, along with the information provided by each. Sometimes, it is important to relay specifically what a witness said; in such cases, the witness information should be related verbatim if possible, with quotation marks around the statement. In contrast, when witness statements are summarized or rephrased, quotation marks should not be used.

Be careful to avoid speculation and unsupported opinions. Where appropriate, technical definitions and information can include references to NFPA 921 and other relevant documents. Brief, cursory reports are discouraged because they are unlikely to meet the needs of the audience and may not be consistent with the requirements of NFPA 1033. NFPA 921 recognizes that run reports and incident reports such as those completed for the U.S. National Fire Incident Reporting System (NFIRS) are considered preliminary reports, and do not substitute for a fire investigation report. Run reports and NFIRS reports have many uses in the fire service, but usually are prepared quickly and lack the detail or depth that would be appropriate for expert investigative reports.

Reports are most effective when they are written in an organized fashion. The information in a report will be easier to understand if it is reported in categories; for example, assigning separate sections of the report to interviews, physical observations, outside research, conclusions, and other topics can be useful. It can also be helpful to develop a report template or use reporting software that provides such templates.

Opinions and conclusions should be placed together in a distinct section of the report, and this section should include the basis for each opinion. When a methodology is followed, the report should demonstrate how it was employed. In particular, the report should document how the scientific method was used in considering the fire's origin and cause. When possible, use the same format for each report.

In a team setting, different investigators may be assigned to document different parts of the investigation. For example, one investigator may be tasked with reporting on interviews they conducted (or for which they acted as a scribe), while another is asked to report on the physical findings in a scene. In such situations, only one investigator should report on any given topic, so as to avoid conflict and confusion among the contributors' reports. If the report discusses work or opinions of another investigator, that individual should be named in the report.

Every report should, at a minimum, be subjected to an administrative or technical review to ensure that the logic and conclusions can be followed by others and that typographical errors are avoided.

After-Action REVIEW

IN SUMMARY

- When documenting any fire or explosion scene, the goal is to record the scene using media that will allow for the recall, communication, and documentation of observations later.
- Photographs are effective reminders of what was observed at the scene and allow for a concise depiction of patterns, evidence, and other features for later presentation.
- To get the fullest representation of what occurred on the scene, it is best to obtain more than one form of documentation.
- The investigator should begin photography of a scene as early as possible, before the scene becomes altered, disturbed, or destroyed.
- Digital camera resolution is measured in pixels; the more pixels a camera has, the more detail it can capture and the larger the image that can be used for demonstrative purposes.
- To be admissible as evidence in court, photos must be relevant to the testimony of the witness whose testimony is introducing the photo, and must be a true and accurate representation of the scene or the item depicted.
- The focal length of a lens determines the angle of view (how much of the scene will fit into the picture) and the magnification of items pictured in the photo.

- The depth of field can determine the quality of detail in the photographs; if it is too shallow, a photo of an object can appear focused only in a small area, with other parts of the object being blurry.

- Do not delete any digital photos taken in a series at a fire scene, and do not rename them on the camera or computer.

- Shutter speeds are typically denoted in fractions of a second. The higher the denominator, the faster the shutter is, and the shorter the time the shutter is open. If shutter speed is too low, there is a greater chance of a blurred photo due to the natural movement of camera and photographer.

- Scene photos should be accompanied by a photo log that provides a list or table of photographs with a reference to each photo (such as the photo number assigned by the camera), and a brief description of the item or area included in each.

- Although video photography should not be the sole medium used to convey conditions at the scene, it can be an effective supplement to still photography.

- Photos or videos received from a witness or other outside source should be treated and secured like any other item of electronic evidence.

- Fire scene sketches and diagrams are used to support the investigator's memory and to document details of the scene.

- Investigative note taking is important to supplement those items that cannot be photographed or sketched.

- The purpose of a written report is to document and communicate an accurate and concise record of all factual findings, opinions, and conclusions.

- Brief, cursory reports of fire investigations are discouraged because they are unlikely to meet the needs of the audience and may not be consistent with the requirements of NFPA 1033.

KEY TERMS

Diagram A formal drawing that is completed after the scene investigation is completed.

Mosaic photographs A series of photographs taken when a sufficiently wide-angle lens is not available, and a panoramic view is desired. The mosaic is created by assembling a number of photographs in overlay form to give a more-than-peripheral view of an area.

Photo diagram A diagram indicating the directions from which photographs were taken.

Photo log A written chart or list of all photos taken at a fire scene, including photo numbers or other identifiers and a brief description of the item or area photographed.

Ring flash A type of flash that reduces glare and gives adequate lighting for the subject matter in close-up photography.

Schedule A list that details the types of equipment in a structure, which is typically used on larger projects.

Sequential photography A series of photographs that shows the relationship of a small subject to its relative position in a known area. The small subject is first photographed from a distant position, where it is shown in context with its surroundings. Additional photographs are then taken increasingly closer until the subject is the focus of the entire frame.

Sketch Freehand diagram or diagram drawn with minimal tools that is completed at the scene; it can be either three- or two-dimensional.

On Scene

1. What are some of the ways in which fire reports, photos, videos, and diagramming can be used to illustrate working hypotheses, and if selected, your final hypothesis, of fire origin and of cause?

2. Review your practice and procedure (or your department's written procedures) for handling digital photos and videos taken on a fire scene. How are they secured for later use? Are the originals saved and protected in their original format?

3. When would it be appropriate to develop visual documentation for a fire investigation that is more than a simple sketch? Which types of such documentation (e.g., floor plans, details and sections) would be appropriate in what circumstances?

4. Discuss the advantages of having proper spelling and grammar, as well as understandable organization, in an origin and cause report.

Chapter Opener: © Jones & Bartlett Learning. Photographed by Glen E. Ellman; On Scene siren: © Bildgigant/Shutterstock.

CHAPTER 12

Complex Investigations

KNOWLEDGE OBJECTIVES

After studying this chapter, you should be able to:

- Identify the interested parties in a complex investigation. (**NFPA 1033: 4.1.4**, p. 293)
- Describe the basic information and documents needed to manage a complex investigation. (**NFPA 1033: 4.1.4; 4.1.6**, pp. 293–294)
- Describe the coordination necessary when conducting a multiparty investigation. (**NFPA 1033: 4.1.4**, pp. 293–299)
- Describe the logistics involved in complex investigations. (**NFPA 1033: 4.1.4**, pp. 293–299)
- Describe the organization and management of a major loss investigation. (**NFPA 1033: 4.1.4; 4.2.1**, pp. 293–299)
- Describe the importance of communication among interested parties. (**NFPA 1033: 4.1.4**, pp. 294–295)
- Explain the considerations required for governmental inquiry into fire investigations. (**NFPA 1033: 4.1.4**, p. 298)
- Discuss site and scene safety considerations for complex investigations. (**NFPA 1033: 4.2.1**, pp. 295–297)
- Describe the special investigative considerations at a fatal fire scene. (**NFPA 1033: 4.4.1**, pp. 299–305)
- Describe the death-related pathological and toxicological examination of fire and explosion victims. (**NFPA 1033: 4.4.1**, pp. 301–304)
- Identify and describe postmortem tests and documentation. (**NFPA 1033: 4.4.1**, pp. 308–309)
- Identify and describe the important aspects of a death investigation. (**NFPA 1033: 4.4.1**, pp. 307–310)
- Describe the special investigative considerations for a fire that caused injury. (**NFPA 1033: 4.4.1**, pp. 307–310)

SKILLS OBJECTIVES

After studying this chapter, you should be able to:

- Assist in drafting protocols between parties when given incident and investigative information. (**NFPA 1033: 4.1.4**, pp. 293–294)
- Develop an investigative flow chart for conducting a large loss investigation when given incident information. (**NFPA 1033: 4.1.4**, p. 295)
- Evaluate a body for details about the fire scene. (**NFPA 1033: 4.4.1**, pp. 300–305)
- Secure a body as evidence. (**NFPA 1033: 4.4.1**, p. 305)
- Evaluate an injured victim for details about the fire scene. (**NFPA 1033: 4.4.1**, pp. 305–307)

You are the Fire Investigator

You are called to investigate a fatal fire in a single-family home that occurred during the early evening. The incident commander advises you that while the fire was largely contained to a laundry room, the residence was filled with smoke on the fire crew's arrival. A single victim was discovered deceased on the floor in a bedroom. The laundry was recently remodeled and contains a clothes washer and dryer.

1. What events or conditions may have kept the victim from escaping?

2. Given the existence of appliances in the room of origin, and the number of parties that might have been involved in the remodel, what precautions or steps should you take in the examination of these appliances?

3. Assuming that the appliance manufacturers and contractors request to be involved in the investigation of origin and cause, how can you proceed in a manner that protects the rights of all parties and allows for an effective investigation?

 *Access Navigate for more practice activities.*

Introduction

An investigation may be complex because of its size, scope, and/or duration. A fire investigator may become involved in a **complex investigation** whenever a fire occurs in a very large occupancy building, involves loss of life or high-value property loss, or requires coordination with various other investigators or team members. A public-sector investigator may be required to lead a team of investigators and other professionals in an occupancy with several floors or areas and may be required to coordinate the investigation with different types of professionals, such as machinery operators, coroner investigators, or building officials. A private investigator may need to coordinate with a number of interested parties and multiple investigations happening simultaneously.

This chapter provides recommended tools to manage and coordinate complex investigations involving large teams of different professionals, to lead and coordinate investigations encompassing large areas or multistory buildings, and to coordinate and manage investigations with multiple interested parties and investigators. Fatal fires usually involve multiple investigators and professionals, as well as their own complexities; they are also discussed in this chapter. Investigation methods and evidence collection techniques are discussed in other parts of this text.

Site and Scene Safety

Site and scene safety is paramount and must be addressed early in the investigation and planning process. The site should be evaluated for safety considerations, and safety information should be disseminated to all parties who are on scene or may arrive on scene. All parties should be reminded that structural and site safety are compromised in a fire scene. Fire scenes may not be familiar to people who are not fire investigators or firefighters; these individuals may include equipment operators, engineers, coroner personnel, and others. All parties bear responsibility for maintaining safety and for being trained to work with these hazards. It may be necessary to appoint a site safety person who has the authority to stop activities if safety concerns arise and until the issues can be adequately addressed. Hazardous materials provide additional, unique considerations for safety and hazard mitigation.

Safety concerns may be of such a magnitude that the retention of a company specializing in the evaluation of site hazards, the implementation of safety plans, and the provision of safety officers should be considered **FIGURE 12-1**. Site personnel must follow the safety officer's directives, although some may express objections to these orders and request reentry. A preliminary safety briefing specific to the particular site and scene may be required for anyone entering the investigation site. A record of those having attended such a safety briefing should be maintained. Unresolved objections, like disagreements, should be recorded and handled.

The site may need to undergo continued site monitoring for concerns such as structural conditions and airborne contaminants. During complex investigations, there may be times when no investigation activities are being performed because safety or environmental concerns need to be mitigated prior

FIGURE 12-1 A high level of site hazards (e.g., hazardous materials) may require specialized resources to ensure safety during site investigation.
© John Sartin/Shutterstock.

to conducting the scene investigation. Safety issues should be addressed in the protocol developed by the parties involved in the investigation, discussed later in this chapter.

Managing Investigations with Interested Parties

An **interested party** is "any person, entity, or organization, including their representatives, with statutory obligations or whose legal rights or interests may be affected by the investigation of a specific incident," according to NFPA 921. When interested parties are given notice of a fire event, they often will notify insurance companies of that event, and both the parties and insurers may engage and send their own investigators. When this happens, many investigators and attorneys may wish to participate in the investigation. Their roles are typically coordinated by the investigator representing the property owner or insurer.

Interested parties will include the owner and its insurance company and persons who may be parties to a potential civil case. In some circumstances, they may also include building component manufacturers, appliance systems providers, subcontractors and installers, protection and detection system installers, equipment manufacturers, and construction material and finish manufacturers. Other parties that should not be overlooked may include service companies that serviced the equipment in the area of origin, gas and

electric utilities, suppression or detection equipment representatives, and even some public agencies such as code enforcers or inspectors. The interested parties likely will retain their own experts, insurance companies, and lawyers, all of whom may participate in some aspect of the investigation.

Although coordination activities often happen with large-area and multiple-day investigations, they may also be needed with smaller fires. For example, if a large appliance may have been involved in a fire cause in a rented house, interested parties may include the landlord, tenant, product manufacturer, parts manufacturer, and installer, along with each party's insurance carrier.

The manager of the complex investigation will have the responsibility in most jurisdictions to identify the interested parties who should have access to the site. This process is highly dependent on the context and facts of the investigation, but to avoid spoliation of evidence issues (see Chapter 5, *Legal Considerations for Fire Investigators*), the manager should take an inclusive approach and make sure that no potential party is left out. For example, if the fire occurred in a newly constructed or remodeled building, interested parties would include not only the contractor, but also any subcontractors who worked in the area of origin. If a number of heat-producing appliances were present in an area of origin, interested parties may include the manufacturers and installers of each such appliance. Interested parties should be identified as early as possible in the investigation process.

Notification of interested parties is best coordinated with a legal advisor, who can help make the determination of whom to notify, help locate and identify parties, and create and send out the notices. Even when this is done, however, other parties may be identified during the course of the investigation. When this happens, these parties should be notified, and it may be deemed necessary to delay further investigation until they can participate. Interested parties that have become aware of the incident through their own sources and wish to participate in the investigation may also approach the investigator in charge.

Basic Planning and Protocol

A plan should be developed for the multiparty fire scene investigation. This plan, referred to as a **protocol**, lists the procedures to be followed during the investigation. It may be agreed upon by the parties or drafted by a party in charge (such as a property owner), and it should be provided to all interested parties prior to entry on the scene. The protocol may also be drafted with the assistance of legal counsel. Along with the procedures for the investigation, the protocol may designate responsibility for specific tasks, such as keeping the evidence that is collected, handling any disagreements that arise, and concerns and procedures surrounding scene safety. Other subjects that may be addressed in a protocol include costs, scheduling, communication, logistics, handling of evidence, documentation, and interviews. The protocol may include a **work plan**, which outlines "the tasks to be completed [during] the investigation, including the order [of tasks and the] timeline for completion," per NFPA 921. It should be developed with the input of the parties, where possible.

Sometimes more than one protocol may be developed for the fire investigation. For example, one protocol may be developed for the scene examination, and another for subsequent investigations or activities that occur outside the scene, such as interviews, evidence storage, lab examinations, and destructive testing.

Some interested parties may raise objections to the protocol. The handling of potential objections and disputes should be addressed in the protocol itself. As an example, it could state that a new suggestion will be evaluated and implemented if agreed to by the parties but that the property owner has the right to the final say. The protocol can, with agreement of or notification to parties, be amended to reflect new information or a new procedure.

The protocol may not be a true consensus because the scene manager, owner, or public authorities may need to override the decisions made on some issues. This may occur when there is a safety concern, a concern about criminal issues, or proprietary needs of the owner.

The following sections describe some of the issues that should be planned for in a complex multiparty investigation, and that can be addressed in the protocol and other agreements as necessary.

Meetings

A preliminary meeting should be held before starting the on-scene investigation to discuss safety considerations and set forth the ground rules for conducting activities. Preliminary information and discussions may also occur via telephone calls or e-mails. The following are some of the concerns to address during the preliminary meetings:

- Safety
- Protocols
- Planning and timing
- Access
- Organizing the investigative team

- Regular meetings
- Resources
- Preliminary information
- Lighting
- Securing the scene
- Sanitary and comfort needs
- Communications
- Interviews
- Plans and drawings
- Search patterns
- Evidence identification and handling

Meetings may be held as the investigation progresses to ensure that everyone is signed in, reminded of safety considerations, and informed about changes or scheduling. Only one person from each interested party should act as the spokesperson for that group. That person should also have the authority to make decisions on behalf of the group that they represent.

The investigator who manages the investigation site also needs to coordinate access to the site to ensure that only those persons who have been authorized gain access to the site. This access may depend on whether certain safety training has occurred, or environmental issues are maintained. More information on the handling of site-safety issues can be found in Chapter 6, *Safety*.

Communication Among Interested Parties

Communication among parties plays an important role in major loss investigations. Proper communication should be maintained to keep the parties informed of any safety issues, the progress of the parties, and the schedule for the scene work and examination. Lack of communication may leave parties with a sense that evidence is being hidden from them or that they are not being provided with information to protect their interests. This, of course, can lead to distrust and conflicts among the parties. Good communication and sharing of information, in contrast, can create a sense of cooperation among the parties and supports an uneventful, successful site investigation.

Information Dissemination and Sharing

The process for disseminating information among the parties needs to be addressed early; otherwise, a lack of information could lead to difficulties among the parties or even legal action. Without proper communication, the investigation may be plagued by attitudes of distrust, disagreements concerning the activities of the investigation, issues with timing of activities, and management issues. Routine communications may be conducted in various forms, such as through regular meetings, websites, e-mail, or bulletin boards.

Sharing of information among the parties can both expedite the investigation process and provide them with the data they need to develop an accurate analysis and hypotheses. It may have limits, however. In some cases, confidentiality agreements may need to be signed before interviews can be conducted or before information is exchanged to ensure that proprietary or secret information is held in confidence. Further, some information might not be shared at all, such as proprietary information on a manufacturing process or information that the owner is not willing to share with competitors. Concerns about confidentiality can sometimes be eliminated by having any party who receives the information sign a nondisclosure agreement stating that the information will not be shared with anyone else. Information sharing should be encouraged so that a complete data set is available for analysis.

If legally permissible, it is usually best to share information about which data are needed to conduct a complete analysis for hypothesis development, such as interviews with the first witnesses or the people who last used or maintained a suspect piece of equipment.

Website. When the investigation involves many parties and the scene work may take an extended period of time, establishing a secure website may be beneficial to keep the parties informed. Passwords can be used to limit access to the website or portions thereof. The website keeps interested parties informed about the investigation's progress during times when they are not required to be, or do not wish to be, present at the scene. It can include a schedule of activities, contact list, protocols, agreements, and safety instructions, and can also be a means to disseminate information or data.

Scheduling

It may not be feasible to allow access to all parties during all phases of an investigation if there would be too many people present for the allowable area, resulting in safety issues or lack of space to view the activities properly. A consensus or compromise needs to address these issues, which may include allowing access to only one person per interested party to view the activity or setting up audio and video feeds so that the investigation can be viewed from a safe location. It may be best if one person from each party is allowed to be present during important removal and documentation, if space and safety issues can be overcome.

Scheduling for a large number of interested parties may prove difficult. Not all parties may be available at the time scheduled for the activities. The scene investigation will likely have to move forward, anyway, and compromise on the schedule may be necessary when holding the site for extended times is not an option. As a compromise, a substitute person may need to attend for a party, or other parties may need to thoroughly document an activity. Scheduling activities that occur after the scene examination is usually not as time sensitive, such as the examination of artifacts that have been secured from the scene, and often a mutually agreeable time to perform the examination can be reached.

Good communication on scheduling will prove helpful in this respect because it allows parties to arrange their calendars as early as possible to accommodate the investigation schedule. Such communication should include the flow chart and activities that are occurring or are scheduled to occur, along with the best estimates of their timing.

The protocol and flow chart developed at the beginning of the investigation are important tools in scheduling activities and should address anticipated scheduling issues. Scheduling of the parties may still be a difficult task, however, and all parties should make an effort to accommodate the scheduling so that the investigation can move forward.

Cost Sharing

In some cases, the interested parties may agree to share the costs of some portions of the investigation activities. Such an agreement is especially likely to be reached if each of the parties has an interest in the results of the activity or needs copies of the result of the activities. Such cost sharing may include professional photography or videography during the investigative activities, activities related to developing drawings, or site-safety personal protective equipment (PPE). Other costs that may be shared include those for evidence storage, debris removal, personal comfort items, and specialized tests of the scene environment or evidence. However, some of the parties may not agree to cost sharing, especially if they think that they may have little involvement in responsibility for the fire. An agreement should be obtained for any cost sharing and put in writing, with the parties' signatures. Ideally, this step should happen early in the investigative process.

Issues in Site Management

Proper site management of a major investigation will provide the tools to limit safety-related problems, organize activities, handle evidence, and assist in other issues that may arise during the investigation. Implementing the management system prior to the start of the investigation will benefit the process as a whole. A site manager may be designated to manage the site and personnel; this manager may be different from the person managing the investigation itself.

Management of complex investigations may involve many different parties performing joint, yet independent, investigations simultaneously. Persons familiar with the National Incident Management System (NIMS) may be able to make use of the NIMS command model in the management of an investigation.

Control of the Site and Scene

Control of the site may change during the course of the investigation. Public-sector agencies may initially be in control. Once the public-sector investigators have completed their examination, they will often release the scene to the owner, another responsible person, or the insurance company. Following the public-sector examination of the site, the site owner or occupant will often have the authority to determine who has control of the scene.

Securing the Site and Scene

Maintaining security and controlling access are necessary at a complex fire scene to prevent contamination, safety concerns, and legal disputes. The **entity in control** is usually responsible for security and access control, and may need to consider installing fences and even assigning security personnel to assist with this task **FIGURE 12-2**. The owner should be consulted because there may be issues with proprietary information and valuable equipment remaining at the site.

Access to the site should be limited and monitored, with mandatory sign-in and sign-out sheets to account

FIGURE 12-2 Security personnel may be needed to monitor access to the investigation site.

© Alex Wong/Staff/Getty Images News/Getty Images.

for people who enter and exit the site. Coordination of access may involve name tags or escorts. At the end of an investigative day, the entity in control should confirm that all parties who entered the scene have exited and been accounted for.

Scene integrity is critical to a successful and defensible investigation and includes not allowing public access, not allowing debris or other contaminants on site, and keeping an accurate record of who was permitted on scene (or kept from entry). It must be maintained until the investigation is complete and the evidence is removed. Scene security may also be required to be maintained until the site has been made safe to passersby and others.

Restrictions or Requirements

Site-specific requirements to enter the fire scene may be established. Some common requirements to maintain scene security and safety include no smoking and no food, beverages, containers, or wrappers brought onto the scene. Scene safety often requires that food not be consumed while on the site because of the potential for ingestion of contaminants. Entering the site usually requires some form of PPE, such as the use of hard hats, safety shoes, and safety glasses. These requirements should be addressed in the protocol and site safety plan.

Delegation of Control and Transfer of Control

The choice of the person to be in control of the site is usually dictated by the circumstances at hand and often changes as the investigation moves forward. The party responsible for control immediately following fire suppression may be the public-sector investigator. If the public-sector investigator determines that no crime has occurred, the investigator will often turn the site and investigation over to others, likely the owner or the owner's representative. The owner likely has the legal authority to take control but may not have the resources or expertise to do so; thus, the insurance company and their representatives may take control of the site. Other issues of control may arise with tenant-occupied spaces and may require legal consultation to resolve.

The transfer of control should be planned ahead of time to ensure an orderly transition. The party relinquishing control should provide information on the changes that have been made to the site, evidence secured, and safety issues. This party may still participate in the investigation but is no longer in control. This often occurs when the public official determines

that the fire is likely accidental but wishes to continue to participate in the investigation.

Release of Information

Release of information to the public or media should be addressed early, especially with fires that lead to large losses. This activity is usually left to the public authorities and their public information officer (PIO). In incidents that lead to large losses and involve multiple public jurisdictions, this can be a sensitive issue. To reduce the potential for conflict during complex incidents, all public jurisdictions should provide a PIO, and these officers should operate a joint information center (JIC). The members of the JIC should manage information releases that are consistent for all jurisdictions. Private parties are unlikely to release information except through an owner representative. Generally, the investigator should not be involved in the release of information to the public.

Evidence Control

Handling of evidence is another key issue in management of complex investigations. An **evidence custodian** is designated to manage all aspects of this process. The protocol or a separate agreement should address how the evidence will be identified, documented, collected, preserved, transported, and stored (although if public-sector authorities are still performing a criminal investigation, this might dictate the method of collection and storage). Other factors to consider include identifying who will have access to the evidence once it is stored and how information will be distributed on proposed examination or testing of the evidence.

All interested parties should be allowed to document and see the evidence prior to its being changed or removed from the site. The evidence should be removed in an agreed-upon manner, usually following recognized practices identified in NFPA 921, *Guide for Fire and Explosion Investigations*. The removal should be carefully documented, as should the chain of custody for such evidence. See Chapter 5, *Legal Considerations for Fire Investigators*, and Chapter 9, *Identification, Collection, and Preservation of Physical Evidence*, for detailed discussions of documentation.

The evidence collected should be secured in a locked facility with fire-protection features. The cost for the storage should be agreed on by the parties, but usually is assumed by the owner's insurance carrier. The keys or security usually reside with the owner's representative, but multiple locks are sometimes allowed so that no one party has exclusive access to the evidence.

The interested parties should agree on access to view the evidence. When viewing occurs, this activity should be open to all parties. Nondestructive exams should be permitted to allow viewing and documentation of the evidence, and other interested parties should be notified of such an exam but need not be present. In contrast, when disassembly or destructive examination of evidence will occur, spoliation issues dictate that all interested parties participate in the exam and be permitted to have their own experts present. A specific protocol can be developed for the exam and should include identification of the laboratory where it will take place and the date of the exam.

Site Logistics

Numerous other coordinating activities may be necessary to permit the investigation to proceed safely and without conflict, and to identify and address the individual needs of investigators and other personnel on scene. Many of these require preplanning and may need to be covered in the protocol or in a work plan. Such logistics may include outlining in detail the scene investigation activities and environmental issues and obtaining trash collection services, toilets, equipment, and security personnel.

Identifying and obtaining the equipment needed to conduct the investigation is part of the preplanning. This may include large equipment to lift ceilings or walls or to remove debris or evidence. Laborers and small hand tools may also be needed to conduct the site investigation. The human resources required for conducting such an investigation may be obtained by retaining a construction company; some are equipped to provide such assistance.

In some large sites, transportation can be an issue. For example, parties may need transportation to the site from a remote parking area or to and from a checkpoint. Transportation of the evidence to the evidence storage facility should also be given consideration, especially if there is a need to transport large items.

The site may require an area for clean-up or decontamination, especially if hazardous materials are present. If materials are present that would pose a contamination hazard if brought outside the scene, decontamination personnel and equipment may be required. This may include a changing area, washing facilities, or even showers.

Other considerations include communication between parties on site, lighting or spotlights, ice and snow removal, heating of the scene, break and shelter areas, and food and water.

Public-Sector Considerations in the Complex Investigation

Public-sector investigators may be involved in large scenes with multiple investigators from a variety of agencies, which can include fire and police departments, multijurisdictional investigative teams, coroner or medical examiner personnel, building officials, and officials from local, state, and federal governments. In these situations, many of the factors discussed in this section apply, though there may not be the same tensions and issues between private entities with differing interests and needs. For example, many of the same logistical needs can come up with regard to group communications, safety notifications, logistics, and sharing of resources. This chapter provides a template to the public investigator managing such an investigation scene.

Public–Private Coordination

In most jurisdictions, public-sector investigators have control of a scene immediately after the fire. They are charged by statute with the investigation of fires and the determination of whether crimes were committed, and they may need to control the scene until that determination is made. However, private investigative interests should be kept in mind by the public-sector agencies, especially when criminal activity is ruled out or when it is not strongly suspected. Even though fire codes and other regulations state that the duty of the fire chief or other official is to determine the cause of all fires, under those circumstances serious consideration should be given to either bringing private-sector investigators into the investigation or turning the scene back to the person in control of the property for private investigation.

Fire officials may make this determination for several reasons. Often, the full investigation of a complex fire will require significant expenditures to obtain experts to work on scene, or to get laboratory or engineering testing, or to hire heavy equipment to remove and dispose of building components; the agency may be unwilling or unable to afford these items or to commit investigators to a multiple-day or multiple-week forensic investigation. Complex scenes may require the collection and transport of numerous noncriminal evidence items, and most public agencies are not equipped to transport or store noncriminal fire evidence.

Just as importantly, public officials recognize that private parties also have a strong interest in investigating fires and determining fire causes, as the losses in large fires can be catastrophic and the possibility of litigation significant. The resolution of civil fire cases, including determining the responsible party in the case of negligence or product liability, is important to society as a whole and to the parties and insurers involved. Therefore, in many instances it can be completely appropriate to involve private parties and allow them the opportunity to examine the scene and the evidence.

This cooperation can take several forms. Frequently, in cases where the public agency is satisfied that there is no significant public interest and that a criminal case is unlikely to result, the agency will leave the scene and hand it over to the responsible person or their insurance carrier or investigator. If no insurance carrier or investigator has arrived at the scene yet, the responsible person should be admonished to protect the scene and not to disturb or move any items in the fire area so that their insurer and/or investigators can effectively review the scene. The public-sector personnel should be aware, however, that the decision to conclude the scene examination and leave the scene can affect the agency's future ability to reenter the scene or maintain chain of custody of evidence, as discussed in Chapter 5, *Legal Considerations for Fire Investigators*. However, when there is the low likelihood of criminal charges in a given investigation, often the public agency will determine that it is safe to relinquish control. Public investigators in this situation must understand that if the fire origin and cause are undetermined when the scene is turned over, it is appropriate to put this interim conclusion in their reports, subject to further evidence, and to allow further conclusions to come from other expert parties. Moreover, the findings of private experts can always be incorporated into the public agency's eventual conclusions about the fire.

If a public agency turns over control of a scene, it can still leave one or more investigators on site to continue participating in, or observing, the investigation as it is being conducted by the private entities. This should be done with the understanding and agreement of the party responsible for the property. In the unlikely event that evidence of criminal activity is subsequently uncovered, the private entities must notify the public agency and permit it to resume control of the scene to resolve the criminal issue.

In other instances, a public agency in charge of a complex investigation may choose to remain on scene and in control of the investigation but invite private entities to observe or participate. This is an effective tool to ensure that all interests are represented at the scene and to avoid complaints that evidence was spoliated or was altered to the point that the private entities could not effectively conduct their own investigation.

The integrity of the scene is preserved as long as the actions of all personnel on scene are documented, and evidence is identified and collected in an appropriate manner. Public personnel who decide to take this route must decide what role each side will play in the examination. For example, investigations have been conducted at sites with the public-sector investigators taking the lead and doing the scene work, while the private investigators observe and photograph the process and the removal of debris and evidence. This can be an appropriate approach in a case when criminal activity is still a significant possibility.

Another choice, particularly when there is no strong indication of criminal activity, is to permit parallel investigations to proceed, with both public and private investigators performing tasks and observing the findings of the other parties. This can be an appropriate approach if a plan is established that ensures that both sides are independent and that neither side is taking instructions from the other. The public entity may continue to act as the party in charge, and a site plan for safety or on other topics may be developed. The parties should agree in advance how evidence will be collected and stored.

Public investigators should attempt to avoid altering evidence without consideration of other parties but, as NFPA 921 recognizes, it is not spoliation of evidence to perform normal scene sifting and debris removal. Even so, if it appears necessary to alter evidence, the public investigators should consider involving other interested parties. In any event, public investigators should document their alteration or removal of any evidence according to recognized standards and processes.

Fire and Explosion Deaths and Injuries

When a fire or explosion results in death or serious injury, the investigation becomes more complex and involved, and public-sector fire investigators will inevitably experience greater responsibility and scrutiny from their agency, other agencies, victims' families, and the press. As with any investigation, it is important that the correct procedures be followed from the beginning and that every fire and explosion involving serious injury receive a full investigation. Investigators should preplan and be prepared for the possibility of a fatal fire occurring in their jurisdiction, and should be able to call on other team members and professionals as needed to conduct these important investigations.

The investigation of a fatal fire or explosion is a two-part investigation, involving (1) the origin and cause of the fire or explosion and (2) the cause and manner of the death. The body of the victim is the most important piece of evidence in determining the cause and manner of the death. The following sections focus on the protection and proper treatment of a deceased fire victim, as well as the steps needed to correctly conduct the investigation.

Fire Suppression

Firefighters responding to a fatal fire scene must be made aware of the importance of preserving the scene as much as possible while still effecting appropriate fire suppression and essential rescue of potentially viable victims. Hose streams and overhaul activities can damage a body and other fragile evidence that may be present. If, on discovery of a victim, there is no question that the victim is deceased, then every effort should be made to leave the body in its original location. At a crime scene, it would be unthinkable to move a murder victim; the fire or explosion victim should not be treated any differently.

In the very uncommon event that it is determined that the body of the victim would be damaged further by allowing it to remain in its discovered position, it can be moved, if permitted by law, to preserve it for further examination, after it is fully photographed along with everything underneath and around it. Further, the body's physical location should be marked in a manner that allows the exact location to be recalled during debris removal, permitting the area to be further examined for evidence. Always remember that the body is a piece of evidence and should be treated as such.

Agency Notification

When a death occurs, most jurisdictions have legal requirements that certain parties must be notified; these parties can include law enforcement, medical examiners, coroners, and possibly others. Fire investigators must both understand these requirements and have the ability to notify the appropriate personnel in the event of a fire death. Ideally, the investigator will have already established contacts and developed relationships with these personnel and agencies prior to the fire event.

Documentation

To ensure complete documentation, the body and its surrounding area should be photographed before the scene is disturbed and throughout the process of debris and body removal. Attention should be given to photographing the body in relation to surrounding items and their condition, fire or explosion patterns, and other physical characteristics of the scene. Possible fire patterns and blast effects on the body should be well documented. All debris around, underneath, and on top of the body must be examined as possible evidence related to the death once the body is placed in the body bag, while the body is being moved, when the body is removed from the body bag, and during the time that clothing is removed from the body. Close-up and scale photos of burns and other injuries should be taken. The location in which the body was resting at the fire scene should be photographed as well.

In the event the deceased victim must be removed from the scene quickly, the investigator should still take a rapid series of photos if at all possible. If the victim's body is on scene but has been moved from its original position, document as much of the body as possible, including patterns on the body, and items around it. Attempts should be made to determine the path the body took as it was being removed. Further, if the body has already been removed from the scene prior to the investigator's arrival, the investigator can reconstruct its location for analysis purposes by knotting one corner of a sheet and positioning the sheet (with the knot representing the head) as the body was positioned.

Documentation of the scene should include sketches and diagrams that detail and record the physical dimensions of the scene, the contents of the scene, and the measurements of the body location. The sketches should include the outline of the body for reference purposes **FIGURE 12-3**. Often these sketches are useful in court when photos of a victim are deemed inadmissible due to their shocking or gruesome nature. At times, the investigator may choose to do a "victim sketch," which is used to detail burn and other injuries to the victim.

In examining and diagramming the scene, consider what evidence might exist that documents the activity of the victim during the fire incident, such as

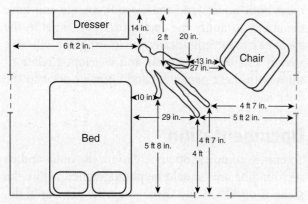

FIGURE 12-3 This diagram shows the location of the body and the dimensions of the surrounding area.

starting the fire, attempts to fight the fire, attempts to flee the fire or take refuge from it, or efforts to rescue persons, animals, or property.

> **TIP**
>
> In addition to the normal safety concerns that any fire poses, fatal fires present the risk of exposure to various biohazard materials. Appropriate PPE, decontamination, and disposal procedures should be addressed as part of the investigation planning.

Scene Examination and Evidence

Many fire and explosion investigations are performed by teams. In cases involving a death, the team may expand to include a police detective, the medical examiner or coroner, forensic laboratory personnel, and a forensic pathologist. If the body is badly burned, the expertise of a forensic anthropologist and a forensic dentist (odontologist) may be required.

While much of the investigative focus might concentrate on the area where the body is located, important evidence can often be found some distance from the body due to the victim's activities, the environment, and fire suppression efforts. The area where the victim may have traveled during the incident is important for examination by investigators, especially areas within arm's reach.

In many cases, the size of the fire scene may require more than one investigator to conduct the scene investigation: one investigator concentrating on the death investigation, while the other focuses on the determination of area of fire origin, point of origin, and the cause of the fire.

To aid in a thorough examination, you can use a grid system to divide the scene into sections **FIGURE 12-4**. Each grid section should be examined and documented, and evidence should be identified. If the scene does not lend itself to a grid search, a spiral pattern or other search method may be appropriate. Any method employed should provide overlap to allow for complete coverage.

Examine the area for all potential evidence. Small pieces of evidence can be located by removal of debris from a section and through careful screening of the debris. The use of multiple screens with different-sized mesh is helpful in this process.

Removal of Body

Removal of the body of a deceased fire victim should be done in the manner prescribed by the coroner, medical examiner (ME), or other official with authority. In most

FIGURE 12-4 This room has been marked off into sectors to ensure complete coverage.

jurisdictions, the body itself should not be disturbed without the coroner's or ME's approval. Thus, while the fire investigator is free to examine the victim's setting and surroundings and perform most of the tasks performed in any fire investigation, the investigator should avoid any activities that relate to the body itself without consultation and participation of the coroner or ME, including removal of any fire debris or building components positioned on top of the victim.

To facilitate the removal of the body from the scene, the coroner or ME and the fire investigator will often determine that it is necessary and appropriate to remove fire debris that might be present on the victim's body. This must be done gently and carefully and with a full examination and documentation of the debris being removed. Only debris that is not attached to the body should be removed in this fashion. Any item that adheres to the body should be left adhered, and any item that drops off of the body during its removal should be collected with the victim.

Once the body has been removed, examine the area where it was resting for potential evidentiary items. Note anything in the victim's hands. During all steps of the excavation and recovery process, take photos of the scene and the items uncovered. Treat the body not only with respect and care, but as the piece of evidence that it is, by ensuring proper consideration for maintaining the chain of custody and avoiding cross-contamination.

TIP

A body found at a fire or explosion scene must be treated as evidence. Not only do you need to maintain the chain of custody and avoid cross-contamination, but you also have to leave any material attached to the body intact for future examination.

TIP

The path and manner in which the victim was removed from the structure should be documented, because trauma to the victim's body may sometimes occur by accident during the removal process. In addition, the personnel who performed the removal should be interviewed and their responses should be documented.

Sometimes the victim's clothing may contain ignitable liquid residue. When possible, the clothing should be preserved for its evidentiary value. In cases where the body was transported away from the scene prior to the arrival of investigators, the investigators should contact the coroner or ME to request that they preserve the victim's clothing and any other evidence on the victim for collection. Victim clothing that will be tested for ignitable liquid residue must be promptly packaged like other fire debris, which usually means that a fire investigator or other person trained in fire debris collection should collect this evidence.

When the body has been badly burned or fragmented by the fire or explosion, take care to search the area for all human remains, no matter how small, and turn them over to the coroner or ME. Burned bones and tissue can blend in with fire debris and can easily be overlooked. In fact, the services of a forensic anthropologist are often used on fire scenes to ensure that all remains are identified and collected. These professionals can also provide advice on how to keep bones and bodies intact during the removal process.

Sometimes it is necessary to determine whether bones recovered at a scene are animal or human **FIGURE 12-5**. In instances in which animals die in a fire, it may be useful to conduct a postmortem examination on them to gather additional information.

Autopsy and Pathological/Toxicological Examination

Anytime a fire or explosion occurs and a body is discovered at the incident site, an autopsy should be performed by a competent forensic pathologist to determine the cause of death. A fire investigator should attend the autopsy, if possible, to point out specific evidence and take custody of evidence as it is recovered, and to share important investigative information with the medical personnel who are performing the tests. Artifacts of clothing, personal effects, bullet remains, chemical residues from ignitable liquids, various tissue and body fluids, and other items may

FIGURE 12-5 When examining remains at a scene, it is sometimes difficult to determine whether they are animal or human.
© Jason Edwards/Alamy Stock Photo.

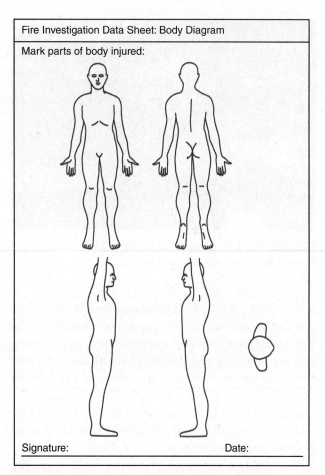

FIGURE 12-6 This chart can be used to diagram injuries to a deceased victim of a fire or explosion.

be recovered for future analysis. When ignitable liquids are suspected, the evidence must be handled in a specific way and, in most cases, is best collected by a trained fire investigator.

Fire investigators should become acquainted with the coroner's staff and medical examiners in their area and establish an investigative understanding with them. As part of this understanding, these professionals may give permission for the investigator to be present during examinations and other procedures such as photographic documentation.

Regardless of whether investigators are able to attend the autopsy, they should attempt to ensure that the pathologist performs the tests and examinations that are normally performed in a fire case, and any tests related to issues specific to the case. The following tests and examination areas are normally included in a fire-related autopsy:

- Blood: For levels of carboxyhemoglobin (COHb), hydrogen cyanide (HCN) concentration, drugs, alcohol, or poisons
- Internal tissue: For levels of volatile hydrocarbons, drugs, or poisons
- Stomach: For contents, including the presence or absence of soot
- Airways: For effects of the fire
- Internal body temperature: To assist in establishing a time and mechanism of death
- X-rays: To assist in identification of the victim, identification of injuries, and location of foreign objects in the body
- Clothing or personal effects: To assist in identification or check for the presence of ignitable liquids
- Recovery of foreign objects in or on the body (such as bullets, knife parts, explosive device components, and other items)
- Sexual assault evidence: To provide possible evidence of other crimes, identification of aggressor, and possible motivation for setting a fire
- Documentation by photography and sketching: To record injuries and burns to the body and evidence recovered from the body **FIGURE 12-6**

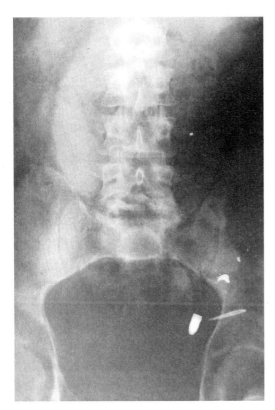

FIGURE 12-7 This full-body X-ray shows a bullet in the pelvic area of the fire victim.
© Jones & Bartlett Learning.

X-Rays

One of the first steps in examination of the body is to perform full-body X-rays. X-rays are useful in identifying any foreign matter present in the body, such as a bullet or knife tip **FIGURE 12-7**, which may provide evidence about the fire incident including any crimes committed against the victim. X-rays may also help to identify the victim through past injuries such as a prior bone fracture, surgical metal implant, or dental remains that can be compared with a known set of X-rays. In addition, this imaging may detect newly broken bones, which may be related to the fire incident and/or crimes committed against the victim.

> **TIP**
>
> Depending on the jurisdiction, X-ray examinations may not be routinely conducted on victims. Note that it may be necessary to specifically request this procedure be performed.

Carbon Monoxide and Other Toxic Products

Testing for the carbon monoxide (CO) levels in blood and tissue is one of the easiest and most common medical tests performed on the fire or explosion victim. This test often reveals much about the cause and sequence of events and should be performed whenever possible. Carbon monoxide is absorbed into the blood and tissue by breathing. It may cause a cherry-pink coloration of the skin, which may not be visible in victims with dark skin or in victims who are heavily covered in soot. The cherry-pink coloration could be visible in several areas on the body, including the lips, as well as in the blood that has pooled in the body (caused by postmortem lividity).

The medical examination process should also include a search for other toxic products. Products of combustion, such as hydrogen cyanide and hydrogen chloride, may be discovered in the blood and tissues. The same body elements may also reveal the presence of various drugs—medical and/or illicit—along with alcohol levels. This information may prove useful to investigators as they try to interpret the ability of the fire or explosion victim to deal with the events as they took place. Gas and liquid chromatography are two techniques that are used to determine the levels of various substances in the body.

Soot

The examination should look for and document smoke and soot in the breathing passages and in the stomach. While soot can be found on the exterior of the body, and in and around the nose and mouth, these deposits do not necessarily prove that the victim was breathing during the fire. In contrast, soot in the internal airways, such as the trachea and lungs, is usually consistent with the victim breathing in a smoke-filled environment, as is soot in the stomach. The presence or absence of soot in the internal airways may help establish a timeline for the events of the fire and the victim's death. An absence of soot in the internal airways may suggest that the victim was not breathing during the fire, because they died prior to the fire or in a different location.

> **TIP**
>
> In addition to being impaired from breathing carbon monoxide, hydrogen cyanide, and other fire gases, a fire victim may have been suffering from impairment due to alcohol and drugs. In such a case, the total impairment of the victim may be considered a combination of the prefire impairment and the exposure to fire toxins.

Burns

Burns to the body should be noted and documented. Burns may happen before or after the death of the victim. Blistering of the skin (second-degree burn) can happen to a more limited degree after death. The effects of heat after death include the dehydration of the muscle tissue. As the tissue loses hydration, a noticeable shrinkage of the muscle occurs, which in turn causes a constriction or tightening. This effect can be observed in the facial features of the victim and in the **pugilistic attitude**, a boxer-type stance observed in the hands and arms. This constriction of the muscles also takes place in the legs in the later stages of exposure. In the past, investigators seeing a victim in a pugilistic position often suspected—incorrectly—that the victim was in a defensive stance as if warding off blows to the chest and head. This constriction of the muscles can even fracture bones in the arms or legs.

Blood

During the examination of the body, the presence of any blood or suspected blood outside the body, on clothes, and/or on furnishings should be noted and documented. The locations of suspected blood should be correlated with information about the scene and information from the medical examiner in an effort to determine whether the blood is the result of the fire event or is an indication of non-fire-related trauma.

Consumption of the Body by Fire

Like many things in the fire scene, the body may be part of the fuel load. Investigators at a fire scene should examine the burn patterns and extent of burns on a fire victim as they would any other fuel item, to determine if fire patterns are present and how they correlate with other patterns in the area **FIGURE 12-8**. When exposed to sufficient heat, the body may be consumed to some degree. Although the skin and muscle tissues of the body are considered poor fuels, they dehydrate during fire and are consumed with time. The bones shrink and change color, then ultimately fracture. Body fat is combustible, and under some conditions, a victim's body fat may be absorbed by a wicking material, such as a cotton shirt, and serve as a fuel source for a small but concentrated flame. While the energy from the flame may be insufficient to ignite adjacent combustible materials, it can over time consume areas of the body with greater fat concentrations, such as the torso.

The skull may fracture or crack from the heat of the fire, most often along the suture lines. However, since gunshots and other trauma can also cause fractures, the ultimate determination of the cause of a fracture may need to be made by a skilled forensic pathologist or anthropologist.

FIGURE 12-8 The body is both part of the fuel load and a piece of evidence and must be examined accordingly.

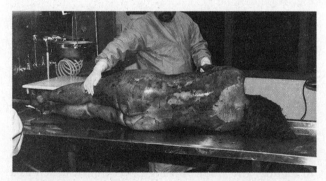

FIGURE 12-9 In the hours after death, blood pools in the lower elevations of a body. This pooling can help indicate how long the person has been deceased.
Courtesy of Mike Dalton.

No credible evidence exists to substantiate the theory of spontaneous human combustion. Thus, the investigator should not rely on this theory as a fire cause.

Postmortem Changes

After death, several changes begin to occur to the body. **Lividity**, the pooling of the blood in the lower elevations of the body caused by the effects of gravity, becomes fixed in the body 6 to 9 hours after death **FIGURE 12-9**. If the body is moved before lividity sets, the pooling of the blood can change.

Rigor mortis begins to stiffen the joints of the body a few hours after the death. The effects of the rigor mortis begin to leave the body after about 12 to 24 hours, proceeding from the body's extremities back to the torso and head of the body. Extreme muscular activities prior to death, as well as elevated surrounding temperatures, will speed the onset of rigor. Experienced forensic pathologists may use the status of the rigor on a body to estimate a time of death. Rigor should not be confused with the rigidity of the body's muscles caused by exposure to heat.

The rate of development of changes to the body after death may be affected by several factors, such as time and climate conditions. When possible, a victim timeline should be created. This timeline would include data such as the times of body discovery, removal, transportation, and examination, as well as environmental conditions such as temperature and manner of storage. All of this information could be helpful in the interpretation of medical data from pathological and toxicological findings.

Issues for Fire Investigators in Death Investigation

Issues that a fire investigator is likely to be required to address in a fatal fire investigation include the recognition of human remains, identification of a fire victim, investigating the prefire activities of a victim, and understanding cause and manner of death.

Remains and Victim Identification

As an initial matter, the fire investigator and forensic personnel may encounter some difficulty determining if a badly damaged body or body part belongs to a human or an animal. The remains of some large animals may resemble human remains. A trained anthropologist or other expert may be needed to determine with certainty whether remains are human.

The identification of a human fire victim is normally the job of the coroner and ME staff, but this information is also important to the fire investigator, who may be asked to assist in the identification process. If a body is only moderately exposed to heat and fire effects, it may be possible to make a visual identification of the victim. Even with moderate fire effects, however, facial skin can tighten and change the victim's appearance, and hair color may also change.

Clothing and personal effects may provide some circumstantial evidence that aids in identification. Fingerprints may be a possible source of identification. The pugilistic attitude causes the victim's fingers to curl into the palm, which may protect the fingertips sufficiently for printing and comparison. A victim's tattoos can provide a basis for identification as well.

X-rays of the dental remains with comparison to known X-rays of a victim are a reliable means of identification. The head and teeth are often well preserved in a fire. However, if the destruction to the head is extensive, use care when finding and recovering dental remains for reconstruction and comparison.

With the recent advances in DNA-related technology, its use for identification has improved. A comparison of the DNA of the victim to a family member may be necessary.

The bodies of children and infants pose special problems in identification, as their skeletal structure is less developed than that of an adult. It may be impossible to determine the gender of a child victim based on examination of skeletal remains alone. Further, a child has less development of the skeleton and less tissue mass, so in extreme cases a skeleton might be fully consumed in a fire, leaving few remains. Therefore, be especially cautious when examining a scene when there is a possibility of child remains.

Victim Activity

When trying to determine the cause and manner of death, it can be helpful to examine and interpret evidence to determine the victim's activities prior to or during the fire. The physical location of the body (e.g., in a bed), clothing, or items found around or near the body (e.g., a fire extinguisher) can provide information in support of conclusions about victim activity. Patterns of damage to clothing and/or the body should be considered in the context of the total scene. Closely examine any inconsistencies. Burn patterns to body or clothing may reflect attempts to extinguish the fire or to escape. Other evidence in the area may also be relevant, such as smoking materials or food being prepared for cooking. Knowledge of the victim's prefire physical abilities is helpful in understanding the victim's movement or lack thereof. Other types of investigational activities, including interviews of associated persons, may be of assistance in determining the victim's activities prior to death, including when the victim was last seen or heard from.

Combustion Products and Their Effects

Many products of the combustion process can affect the victim of a fire or explosion. These products may include carbon monoxide, carbon dioxide, nitrogen oxides, halogen acids, hydrogen cyanide, acrolein, benzene, particulates such as soot and ash, and aerosols. Inhalation of these products or direct skin contact can cause a variety of effects on the human body.

Carbon Monoxide

A product of all fires, carbon monoxide (CO) is generated by incomplete combustion of the fuel. The level of CO produced by each fire varies depending on the completeness of combustion and the type of the fuel package. Carbon monoxide is an anesthetic and an asphyxiant, so its inhalation can lead to dizziness and confusion during early exposure to levels on the order of 1000 to 1500 ppm. This can subsequently lead to death in as little as an hour. When inhaled, CO binds with the hemoglobin in the blood to form **carboxyhemoglobin (COHb)**. Because CO binds with the hemoglobin in the blood much more readily than oxygen does, it is possible for the blood to form dangerous levels of COHb with exposure to even low concentrations of CO. When the fire survivor is removed from the CO-exposure environment, the levels of COHb in the blood begin to decrease; however, this process can take many hours.

The stability of CO in the body is such that it can be tested for hours after the death. As a general rule, levels of COHb in the body of 50 percent or greater are considered fatal. The actual levels can vary, with death being possible in some cases at concentrations as low as 20 percent. The level of the COHb in the body may be helpful in making determinations about the death of the victim. When the level is less than 20 percent, the death most likely was caused by other factors, such as thermal injuries or physical trauma. When the level is 40 percent or greater, the toxic effects of the CO, either alone or in combination with other factors, likely contributed to the cause of death, with most victims dying in areas remote from the room of fire origin. Studies have shown that at least 60 percent of fire victims die from CO poisoning.

Be aware that other, non-fire-related factors can cause high concentrations of CO in the body. Smokers or people who have been exposed to automobile exhaust, for example, may have background CO levels of 4 to 10 percent.

The Coburn-Forster-Kane (CFK) equation or the Stewart equation can be used to estimate the quantity of CO the victim inhaled. Based on the concentration of CO produced by the fire over time, as well as the weight of the victim and the volume of air exchanged by breathing per minute, the CFK equation can provide a range of COHb values associated with the average exposure to CO with which to compare to the COHb value for a victim determined at autopsy. The Stewart equation estimates the CO inhaled based on the relationship among COHb, inhaled CO concentration, respiratory minute volume, and exposure duration. (For a detailed explanation of the CFK and Stewart equations, refer to NFPA 921, *Guide for Fire and Explosion Investigations*, Section 24.10.8.2.1–24.10.8.2.4.)

Cyanide

Hydrogen cyanide (HCN) is another toxicant produced during combustion. This gas is a by-product of the combustion of such common household items as wool, nylon, and various plastics. By itself or in combination with other toxic gases, HCN can incapacitate victims or cause their death. It is more rapidly absorbed into the blood through the respiration process than CO is, which explains HCN's rapidly lethal effects.

Cyanide affects the body's ability to use oxygen, rather than binding with oxygen, as CO does. Cyanide is most commonly measured in the blood but can also be measured in tissue samples of the liver, brain, lungs, and kidneys. Unlike CO, HCN does not always have good stability in tissues, so the distribution of cyanide to the organs may vary. Postmortem sample stability is related to time, storage (temperature), and preservation methods.

Other Toxic Gases

Just as a wide array of fuels are likely to be found in any common setting, so an equally wide range of gases may be produced during a fire. Many of these gases cause irritation and swelling or are toxic in nature. Hydrogen chloride and acrolein are two gases commonly found in fires. Hydrogen chloride is released during the burning of polyvinyl plastics, whereas acrolein is generated from the combustion of wood and cellulosic products. These and other combustion by-products should be considered when assessing the ability of a victim to act and move in a fire setting.

Soot and Smoke Exposure

Most fuel products will produce some form of smoke and soot material when burned—some more than others. Hot soot particles, when inhaled, may cause thermal injuries leading to edema (swelling). Soot may also transport toxins into the body or, in some extreme cases, block the victim's airway. The liquid mists of pyrolysis products are often acidic in nature and can cause systemic failures on inhalation.

Sublethal Inhalation Exposure

Narcotic gases such as CO and HCN or the effects of hypoxia may affect the responses of the person who is exposed to them by limiting both mental and psychomotor abilities. Additionally, some of these same by-products and smoke can interfere with respiration, irritate the eyes, and hinder the ability to see. The levels of these by-products in the body begin to decrease when the person withdraws from the affected area.

Irritant gases can alert people to a fire even when present in only low concentrations. These same irritant gases, when more concentrated, will cause irritation to the point that it may obscure vision or alter behavior. Postfire effects may produce respiratory edema or inflammation.

Smoke will obstruct vision and reduce the speed of travel during escape. The extent to which these effects influence the victim's attempts to escape the fire area may depend on several factors, including the victim's prior knowledge of the building/area, the extent of the smoke and heat conditions, obstructions to travel, and the condition or availability of exits.

Hypoxia

Hypoxia is caused by a victim breathing in a reduced-oxygen environment. As a fire burns in a confined area, it depletes the oxygen in the air. As the level of oxygen is reduced from its starting point of 21 percent to approximately 15 to 10 percent, a gradual increase in respiration occurs, followed by disorientation. Oxygen levels less than 10 percent cause unconsciousness and subsequent cessation of breathing. Tests conducted on the blood of a hypoxia victim after death will not provide reliable information because levels of oxygen and carbon dioxide begin to change after death.

Thermal Effects

Thermal effects of the fire can result in death or injury. In hyperthermia, the temperature of the body becomes greatly elevated. Depending on the time and exposure, hyperthermia may be classified as simple or acute in nature.

Simple hyperthermia is caused by extended exposure (15 minutes or longer) to hot environments. Over time, the body temperature increases, with internal temperatures above 109°F (43°C) being fatal in a few minutes. Humidity and moisture also compound the body's inability to shed excessive heat.

Acute hyperthermia is caused by exposure to high heat levels for a short period of time. Thermal burns also result from this exposure, but the cause of death is related to the elevated body temperature.

The inhalation of hot gases and various toxic gases causes edema and inflammation of the airway. The effects of hot gases are generally accompanied by facial

burns or singed facial hair. Inhalation of soot may produce thermal injury, introduce toxic compounds into the body, or cause physical blockage of the airway.

Skin Burns

When the temperature of the skin reaches about 110°F (45°C), pain will result. The transfer of thermal energy via conduction, convection, and radiation causes burn injury. Conductive heat transfer to the skin through clothing can occur even though the clothing may not show visible signs of heat. When skin is exposed to convective heat through air temperatures greater than 120°F (49°C), the result will be pain and injury. With radiant heat, the greater the heat flux will be, and the faster the tissues will suffer damage.

Inhalation of Hot Gases

The inhalation of hot gases may result in death or injury. Similar effects can be caused by chemical irritants. The inhalation of hot gases is often accompanied by burns to skin and facial hair. Research has shown that, in animals, hot gases of 932°F (500°C) result in larynx and trachea damage, and steam of 212°F (100°C) results in burns deep into the lungs.

Cause and Manner of Death

After the autopsy and other investigative steps are complete, the coroner or ME will provide an opinion about the cause and manner of the fire or explosion victim's death. The *cause* of death is the actual disease or injury that brought about the death, such as smoke inhalation, burns, or gunshot. The *manner* of death is how the injury or disease resulted in the death, and is usually classified in one of five ways: accidental, homicidal, suicidal, natural, or undetermined. Some of the ways that a victim can die in a fire setting were discussed earlier in this chapter.

Fire and Explosion Injuries

Deaths from fires and explosions can occur long after the event has ended. Serious-injury fires and explosions should be investigated like fatal investigations, to ensure a full investigation in the event that a victim subsequently dies.

In cases involving extensive injuries, investigators may not be able to obtain statements from the victim for an extended period. Nevertheless, they should follow all of the usual steps and leads to gather information—for example, locating surveillance footage and interviewing bystanders—and not delay these actions in hopes of interviewing the victim. Documentation of the nature of the victim's injuries should be conducted in a timely manner because, with time, these injuries may begin to heal and change in appearance.

Some jurisdictions have laws requiring medical personnel to report burn injuries much in the same manner as they report gunshot wounds. Such reporting laws can assist in identifying victims of abuse and assault, and can potentially identify a person who was burned while starting a fire. Be aware of the laws that apply within your jurisdiction.

Examination and Documentation of Injury Event

The examination of the serious-injury scene requires much the same work as the examination of the fatal scene. Physical evidence may be available that indicates the victim's activities and the source of victim injury. This evidence could include blood, clothing, and burned skin. The victim's location may be able to be determined from indicators such as soot lines of demarcation or protected areas. If reliable evidence of the victim's position is found, that position should be diagrammed just as it is in a fatal fire.

The victim's clothing may be removed by emergency or hospital personnel and should be collected as soon as possible to prevent its loss. Clothing may contain physical evidence (in the pockets, for example) or may have been a fuel source for the fire. If clothing is to be tested for ignitable liquid residue, it must be collected like fire debris so that the ignitable liquid, if any, does not evaporate.

Medical Evidence

Whenever possible, document the nature of treatment provided to a fire victim during transport and at the hospital. Obtain information on the time of toxicological testing and the results of those tests. The fire investigator may need to understand some of the following medical factors.

Burns

Many terms are used in medical reports to document evidence related to burns.

Degree of Burn. Burn injuries need to be documented and assessed in terms of their degree:

- First degree: Reddening of the skin; also called *superficial* burn
- Second degree: Blistering of the skin; also called *partial-thickness* burn

- Third degree: Full-thickness damage to the skin; also called *full-thickness* burn
- Fourth degree: Damage to the underlying tissue and charring of the tissue

Body Area (Distribution). The medical community estimates burn damage by using the "rule of nines." This rule is based on dividing the body into segments, each of which represents 9 percent of the total body area, and adding the injured portions of the body together to obtain an estimate of the body area affected by the burn injury **FIGURE 12-10**. The percentage of body area burned is sometimes used to predict survivability. Resources on body areas and burn survivability include Tables 24.8.1.3 and 24.8.1.4 in NFPA 921.

Mechanism of Burn Injury. The cause of a burn, whether by scalding, chemical exposure, or hot gases or flames, may not be distinguishable by appearance alone. However, there is a direct relationship between the formation of burns and the radiant heat flux. As an example, a radiant heat flux of $2 \ kW/m^2$ will cause pain after 30 seconds with no blistering of the skin, whereas a heat flux of $10 \ kW/m^2$ will cause pain after 5 seconds and blister the skin in 12 seconds. Because conducted heat brings the skin in direct contact with the heat source, it is more dangerous than heat transferred by radiation or convection. This conduction can also occur from heavy clothing that transfers enough heat to burn skin, even though the source does not burn the fabric.

Documentation of Thermal Injury. As with all elements in the investigation, thermal injuries should be well documented. This documentation may include sketches and color photography, preferably before treatment or healing of the wounds, which could change the appearance of the injury.

Inhalation

Fire produces various by-products that can affect the person who is exposed to those substances. Testing is needed to determine the levels of these by-products in the body and to understand the actions of the injured individual and the environment in which the victim was exposed.

The percentage of CO eliminated by the body is directly related to the concentration of oxygen available to the victim. The CO level of a victim who is subsequently placed in a normal atmosphere will decline by half in about 5 hours. This rate of decrease is greatly accelerated when the victim receives oxygen

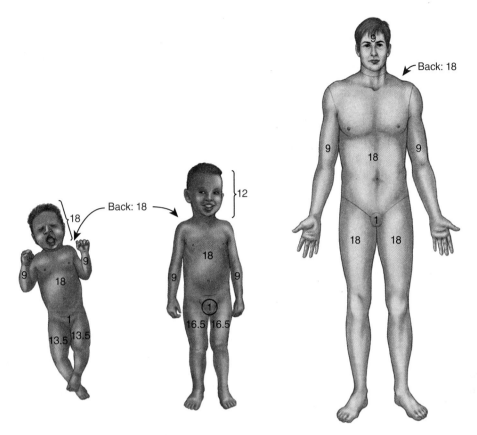

FIGURE 12-10 The rule of nines is a quick way to estimate the amount of surface area that has been burned.

during treatment. The half-life of CQHb in a normal environment is about 250 to 320 minutes, but when the victim receives 100 percent O_2, that half-life is reduced to 60 to 90 minutes. Knowledge and application of this information are useful when the COHb levels of people who received treatment are known and interpolated.

Hospital Tests and Documentation

On admittance to a hospital, a victim with fire-related injury should have a sample of blood taken and analyzed for level of COHb, HCN concentration, blood alcohol level, and presence of drugs, among other things. The blood sample should be obtained as soon as possible because the levels of most markers begin to decrease with time and treatment.

Access to Medical Evidence

Many federal and state laws affect an investigator's ability to obtain victims' medical records. Medical records may, in some cases, be obtained only with consent of the victim or through legal means, such as a court order. Knowledge of laws related to medical records is helpful in this process.

Obtaining the victim's past medical history may reveal conditions that could affect the individual's ability to comprehend, move about, or detect the dangers of a fire. It may also provide clues to deaths that were determined to have occurred prior to the fire, such as from a heart attack.

Explosion-Related Injuries

The location and distribution of injuries on the explosion victim's body may indicate the victim's location and activity at the time of the explosion and help establish the location, orientation, energy, and function of the exploding mechanism or device. The injuries related to an explosion scene are classified into four groups based on the explosion effect that caused them: blast pressure, shrapnel, thermal, and seismic.

Blast pressure injuries are caused by the concussion effects of the explosion. Damage to the internal organs is not uncommon, depending on the level of the blast pressure. The pressure wave may also propel the victim into objects and cause blunt trauma injuries, fractures, lacerations, contusions, and abrasions. With detonations, it may be possible for the body to experience severe injury or amputation. Small particles are also harmful if they are blasted into unprotected skin.

Shrapnel (solid fragment) injuries are caused by fragments from the blast center. These fragments may have serious effects when they penetrate the body, such as amputation, lacerations, or blunt trauma.

Thermal injuries may be caused by the explosive flame fronts, which usually produce first- and second-degree burns. Thermal injuries may also result from clothing: Synthetic fabrics may melt on the victim, and cotton fabrics may scorch. Thermal burns are mostly limited to first- and second-degree burns, but third-degree burns can also occur and may be fatal in nature. The brief exposure of the skin to the thermal event may cause skin damage on exposed surfaces, especially those surfaces in direct exposure to the source.

Seismic effects of the explosion may cause injury with the collapse of structures or of their structural elements. These injuries often take the form of blunt trauma, lacerations, fractures, amputation, contusions, and abrasions.

After-Action REVIEW

IN SUMMARY

- An investigator may become involved in a complex investigation whenever a fire occurs in a very large occupancy, involves loss of life or loss of high-value property, or requires coordination with various other investigators or team members.
- Site and scene safety is paramount in a complex investigation and must be addressed early in the investigation and planning process.
- Safety information should be disseminated to all parties and investigators who are on scene or who may arrive later.
- To avoid spoliation of evidence issues, the manager of a complex investigation should aim for inclusiveness when identifying and notifying interested parties.
- Notification of interested parties is best coordinated with a legal advisor, who can help make the determination of whom to notify, locate and identify parties, and create and send out the notices.
- An investigation protocol may outline investigative procedures, assign specific tasks, and specify how disagreements will be handled.
- A preliminary meeting should be held before starting the on-scene complex investigation to discuss safety considerations, address concerns, and set forth the ground rules.
- Proper communication should be maintained to keep all parties informed of any safety issues, the progress of the parties, and the schedule of the scene work and examination.
- When many parties are involved in the investigation and the scene work may take an extended period of time, establishing a secure website may be beneficial to keep the parties informed about the progress.
- Parties to an investigation may sometimes agree to share the costs of some portions of the investigation activities.
- Access to the complex investigation site should be limited and monitored, with mandatory sign-in and sign-out sheets to account for people who enter and exit the site.
- When public-sector investigators determine that no crime has occurred, they often turn the site and investigation over to others, likely the owner or the owner's representative.
- When disassembly or destructive examination of evidence is deemed necessary, spoliation issues dictate that all interested parties participate in the exam and be permitted to have their own experts present.
- Many logistics may need to be addressed in large investigations, such as security, transportation, trash collection services, toilets, equipment, food and water, and break areas.
- Private investigative interests should be kept in mind by public-sector investigators, especially when criminal activity is ruled out or when it is not strongly suspected.
- Complex fire scenes may be relinquished or shared by public agencies, especially in situations where there is not a strong indication of criminal activity.
- The investigation of a fatal fire or explosion is a two-part investigation, focusing on (1) the origin and cause of the fire or explosion and (2) the cause and manner of the death.

- Firefighters responding to a fatal fire scene must be made aware of the importance of preserving the scene as much as possible while still effecting appropriate fire suppression and essential rescue of potentially viable victims.

- To ensure complete documentation, the body of a fire victim and its surrounding area should be photographed before the scene is disturbed and throughout the process of debris and body removal.

- In examining and diagramming a fatal scene, consider which evidence might exist that documents the victim's activities during the fire incident.

- In addition to the normal safety concerns that any fire poses, fatal fires present the risk of exposure to various biohazard materials.

- Removal of the body of a deceased fire victim should be done in the manner prescribed by the coroner, medical examiner, or other official with authority.

- When a body has been badly burned or fragmented by the fire or explosion, take care to search the area for all human remains, no matter how small, and turn them over to the coroner or medical examiner.

- Anytime a fire or explosion occurs and a body is discovered, an autopsy should be performed by a competent forensic pathologist to determine the cause of death.

- A fire investigator should attend the autopsy, if possible, to point out specific evidence and take custody of evidence as it is recovered, and to share important investigative information.

- X-rays may help to identify a victim through past injuries such as a prior fracture, surgical metal implant, or dental remains that can be compared with a known set of X-rays.

- The absence of soot in the internal airways may suggest that the victim was not breathing during the fire, because the victim died prior to the fire or in a different location.

- A pugilistic attitude in a deceased fire victim is not an indication of self-defense, but rather a normal result of constriction of muscles in a fire event.

- No credible evidence exists to substantiate the theory of spontaneous human combustion, and the investigator should not rely on this theory as a fire cause.

- When trying to determine the cause and manner of death, it is helpful to examine and interpret evidence to determine the victim's activities prior to or during the fire.

- Combustion products that can affect fire victims include carbon monoxide, carbon dioxide, nitrogen oxides, halogen acids, hydrogen cyanide, acrolein, benzene, particulates such as soot and ash, and aerosols.

- Serious-injury fires and explosions should be investigated like fatal investigations, to ensure a full investigation in the event that a victim eventually dies.

KEY TERMS

Carboxyhemoglobin (COHb) The carbon monoxide saturation in the blood.

Complex investigation An investigation that generally includes multiple simultaneous investigations and involves a number of interested parties.

Entity in control The interested party who has or represents ownership of the scene or is in effective management of the site, scene, or evidence, and is organizing, directing, or controlling the joint actions of the other interested parties.

Evidence custodian Person who is responsible for managing all aspects of evidence control.

Hypoxia Condition caused by a victim breathing in a reduced-oxygen environment.

Interested party Any person, entity, or organization, including their representatives, with statutory obligations or whose legal rights or interests may be affected by the investigation of a specific incident. (NFPA 921)

Lividity Pooling of the blood in the lower elevations of the body after death, which is caused by the effects of gravity.

Protocol A description of the specific procedures and methods by which one or more tasks are to be accomplished. (NFPA 921)

Pugilistic attitude A crouching stance with flexed arms, legs, and fingers.

Work plan An outline of the tasks to be completed as part of the investigation, including the order or timeline for completion. (NFPA 921)

REFERENCE

National Fire Protection Association. 2020. *NFPA 921: Guide for Fire and Explosion Investigations*, 2021 ed. Quincy, MA: National Fire Protection Association.

On Scene

1. Consider some of the larger or more complex occupancies in your jurisdiction or service area. If a large-scale fire occurred there, what challenges might you face as an investigator tasked to determine the fire origin and cause?

2. What would be some effective ways to communicate or coordinate between investigators in a complex scene involving a dozen or more interested parties?

3. What are the factors that might influence the ability of a person in a fire building to escape the fire? Why might a person be overcome with fire or smoke even when trying to navigate a building or escape from it?

4. Research the agencies and personnel who might be involved in a fire-fatality investigation in your jurisdiction or service area. Identify the laws and requirements that would apply to them and to fire investigators in such an investigation.

Chapter Opener: © Jones & Bartlett Learning. Photographed by Glen E. Ellman; On Scene siren: © Bildgigant/Shutterstock.

CHAPTER **13**

Determining the Origin

KNOWLEDGE OBJECTIVES

After studying this chapter, you should be able to:

- Describe the recommended methodology for determining the origin of a fire. (**NFPA 1033: 4.2.6**; **4.6.5**, pp. 314–318)
- Describe how to analyze various data items collected during a fire investigation. (**NFPA 1033: 4.6.5**, pp. 318–324)
- Describe how to formulate hypotheses. (**NFPA 1033: 4.6.5**, pp. 322–323)
- Explain the process for testing hypotheses. (**NFPA 1033: 4.6.5**, pp. 323–324)
- Explain considerations for selecting a final hypothesis. (**NFPA 1033: 4.6.5**, p. 324)
- Describe the fire investigator's considerations when the origin cannot be sufficiently defined. (**NFPA 1033: 4.6.5**, p. 324)

SKILLS OBJECTIVES

After studying this chapter, you should be able to:

- Conduct an origin investigation using the scientific method. (**NFPA 1033: 4.6.5**, pp. 314–316)
- Formulate an opinion on the origin of a fire. (**NFPA 1033: 4.6.5**, pp. 318–323)
- Integrate findings from multiple modes of inquiry to create origin hypotheses. (**NFPA 1033: 4.6.5**, pp. 318–322)
- Test origin hypotheses. (**NFPA 1033: 4.6.5**, pp. 323–324)

You are the Fire Investigator

You are on the scene investigating a fire that has caused damage to four different businesses in an outdoor shopping center. After conducting an exterior examination, you note fire damage to only one rear exterior business door. Your interior examination of this business reveals significant smoke staining throughout, to a greater extent than you found in the other occupancies. The rear office inside the business has severe fire damage and deep wood char, and door jambs leading into this room show greater char toward the office than the rest of the business. Several desks and bookcases within this office have diagonal-shaped lines of discoloration.

1. What information may the occupants of the room be able to provide to identify the point of origin?

2. How would you review and correlate the fire patterns found inside and outside the business?

3. How would you document the area of origin?

4. How do the principles of fire dynamics figure into your analysis of the fire's origin?

Access Navigate for more practice activities.

Introduction

The fire area is the boundary of fire effects within a scene in which the area of origin will be located. It is determined by identifying the border between damaged and undamaged areas, which are distinguishable by fire effects and patterns created by flame, heat, and smoke.

According to NFPA 921 and NFPA 1033, the fire's **area of origin** refers to a structure, part of "a structure, or general geographic location within a fire scene, in which the point of origin of a fire or explosion is reasonably believed to be located." The **point of origin** is "the physical location within the area of origin where a heat source and a fuel first interact, resulting in a fire or explosion," per NFPA 921 and NFPA 1033. The goal of origin determination is to identify in three dimensions the location where the fire began. This task may be either easy or difficult, depending on the amount of evidence, facts, or data available; the reliability of the evidence; and the amount of destruction. In some cases, origin cannot be determined.

Determination of origin must be done prior to the analysis of cause. NFPA 921 states that if the origin cannot be determined, the cause usually cannot be determined, and if the correct origin is not identified, the investigator's determination of the cause will be incorrect. Even if a cause might seem likely in a given situation, the investigator must go through the steps for analysis of origin, using the scientific method, so that a full, rigorous analysis is completed. This adds reliability to the investigator's conclusions and demonstrates the investigator's understanding of appropriate investigative procedure. Further, it helps avoid the influence (or appearance of influence) of expectation bias. On occasion, the analysis can even result in a different finding than the investigator first expected to make.

In general, relevant information relating to origin can be obtained from one or more of the following sources:

- Witness information and/or electronic data
- Fire patterns
- Application of fundamental principles of fire dynamics

Past editions of NFPA 921 listed arc mapping as a fourth source of information. As of the 2021 edition, however, arc mapping has been included as a fire pattern, leaving just the three broad information categories listed here.

Overall Methodology

The overall methodology to be applied when determining a fire's origin is the scientific method. This section assumes that the investigator has reviewed the detailed discussion of the scientific method that appears in the methodology section of Chapter 1, *Introduction to Fire Investigation*. **FIGURE 13-1** depicts a flow chart in which the scientific method is adapted to the determination of origin. It includes recognition and definition of the problem, collection of data, data analysis, hypothesis development and testing, and selection, if possible, of the final hypothesis. In

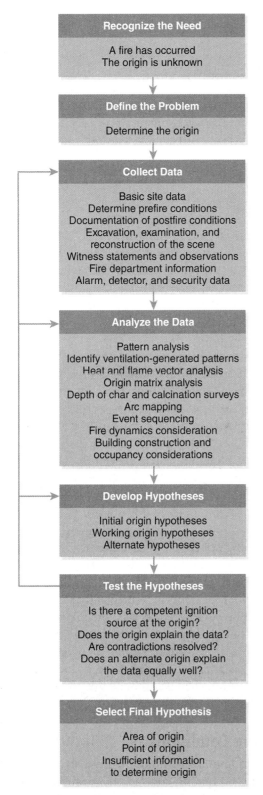

FIGURE 13-1 An example of applying the scientific method to origin determination.

Reproduced from *NFPA 921-2021, Guide for Fire and Explosion Investigations*, Copyright © 2020, National Fire Protection Association. This reprinted material is not the complete and official position of the NFPA on the referenced subject, which is represented only by the standard in its entirety.

this figure, unique steps are shown that assist in data analysis and hypothesis development, such as heat and flame vector analysis. The specifics of these steps are discussed in more detail in this chapter.

Within the overall methodology of the scientific method, NFPA 921 recommends the investigator follow this specific methodology for origin determination:

1. Initial scene assessment
2. Development of a preliminary fire spread hypothesis
3. In-depth examination and reconstruction of the fire scene
4. Development of a final fire spread hypothesis
5. Identification of the fire's origin

To determine origin, it is recommended that investigators use a systematic approach in examining each fire scene. This means employing similar techniques and processes on every fire, so that all steps will be covered each time and none will be omitted. Following this approach will help ensure that investigators do not overlook significant information and that they avoid jumping to conclusions or succumbing to expectation bias. Such systematic approaches include, for example, working from the area of least damage to the area of greatest damage, or from the highest point to the lowest point. Flexibility is important, though, because not all origins are located at the area of greatest damage or at the lowest point in a compartment.

To help ensure all proper steps are taken, checklists can be helpful. When using a checklist, make sure that all listed steps are followed, or that you can explain later why you did not take certain steps.

TIP

The purpose of determining the area and point of origin is to identify in three dimensions where the fire began. The ignition source is usually at or near the accurately determined point of origin.

The analysis of the various data an investigator collects will always yield at least one hypothesis (even if it is just a broad one, like "the fire started in the structure"), and usually multiple hypotheses, regarding the origin. These early hypotheses are considered "working" hypotheses. Testing of hypotheses should continue as new data are collected. Additional testing may result in revision or discarding of a hypothesis.

Testing of origin hypotheses requires inclusion of all available data and application of fundamental principles of fire dynamics. A fire investigator should not focus exclusively on identifying the first fuel ignited and a competent ignition source in the proposed area of origin. Indeed, failure to consider growth and spread can lead to erroneous conclusions. For example, a cigarette in a trash container may be a competent ignition source for paper in the container. However, the trash container must also be placed in proximity to secondary and subsequent fuels for the fire to spread. If the trash container were located next to a noncombustible wall with no other fuel in the area, it is unlikely the fire would spread. In contrast, if the trash container were placed under a wooden desk in an office cubicle, the fire would likely spread to adjacent combustibles. A **fire spread analysis** must be conducted to determine whether the physical damage and other available data are consistent with the origin hypothesis.

The body of data must be considered collectively to determine which of the hypotheses fit all of the available evidence. When an otherwise plausible hypothesis fails to fit some piece of data, try to reconcile the discrepancy and identify whether it is the hypothesis or the data that are wrong. In other instances, however, a single piece of evidence can be the basis for an accurate determination of origin. Such evidence may include a credible eyewitness, security video, or some piece of irrefutable physical evidence.

Many of the tasks performed by the fire investigator to determine the origin may occur simultaneously (e.g., photography, witness interviews). Likewise, application of the scientific method (data collection, analysis, hypotheses development and testing) occurs continuously during the investigation. These are not mutually exclusive processes.

Identification of the point of origin is not always possible, but an inability to determine the point of origin does not, in and of itself, rule out the possibility of developing a credible and defensible origin and cause hypothesis. For example, if the extent of damage is so great that the investigator is unable to pinpoint the exact point of origin, it may still be possible to put forward a supportable hypothesis relating to origin and cause.

Origin determination necessarily involves identification and interpretation of fire patterns, as described in Chapter 4, *Fire Effects and Patterns*. The surfaces in a fire scene record patterns continuously over the life of a fire. Importantly, as a fire progresses, previously generated patterns can be altered or obliterated. Determining the sequence in which these patterns were created, known as **sequential pattern analysis**, is the key to determining the origin. An investigator must collect and organize the fire pattern data into a sequential format to be able to comprehend and articulate the origin and spread of the fire.

Scene Examination and Data Collection

According to NFPA 921, the origin of the fire should be determined from information obtained from witness interviews and electronic data, fire patterns, and application of principles of fire dynamics. The process of planning for, responding to, and examining the fire scene (discussed in detail in Chapter 8, *Examining the Fire Scene*) is critical to collecting data that supports determination of the fire's origin. **TABLE 13-1** summarizes the scene examination process. Steps marked by an asterisk are directly related to the origin analysis.

Debris Removal and Reconstruction

Debris removal will likely be necessary to fully find and expose fire patterns and important material in the determination of origin; this process is described in Chapter 8, *Examining the Fire Scene*. Debris must be removed in a systematic and controlled manner and placed in a designated area where it may be examined by investigators for other interested parties. It should be documented layer by layer with photography and sketches.

Fire scene reconstruction normally takes place after the area or room of origin is identified. Furniture and other items that were moved or removed from the area are replaced after the layering process, so that fire patterns on these items can be seen without the interference of fire debris. The items are replaced in the area based on visual indicators that demonstrate where they were positioned before the fire occurred—for example, protected spots on the floor that show where a sofa sat before it was moved during firefighting operations. Witness statements are useful in this process as well, as are prefire photos if the investigator is confident that they accurately represent conditions that existed at the time of the fire. If there are several choices for the item's placement, all potential positions should be considered. If it is unclear where the item sat, it should not be part of the reconstruction.

If the investigator is able to reliably determine where structural components were located and how they were attached prior to the fire, these components can also be used in the reconstruction process.

Prefire Conditions

Persons familiar with the involved structure, such as owners, occupants, contractors, repair personnel, and neighbors, can provide information on such prefire characteristics as the type and placement of contents, general conditions or alterations to the structure, and fire protection systems. These people may be able to sketch the placement of furnishings or fixtures, or supply prefire photos or video.

Public records (e.g., county assessor or treasurer, fire department inspections, and building department permits and inspections) will also help to determine

TABLE 13-1 Scene Examination and Data Collection

Step	Data Gathered/Tasks Performed
* Gathering basic incident information	Location, date, time, weather, complexity and nature of incident, injuries
Safety planning and preparation	Plan for and obtain personal protective equipment and respiratory protection; assess building hazards and integrity
Organization of investigation team	Assemble team, assign team functions, consider specialized personnel
Pre-investigation team meeting	Introduce personnel, safety briefing
Tools and equipment	Be prepared with basic and specialized investigation tools
Specialized personnel and technical consultants	Consider the need for specialized and technical personnel
Management of documents and information	Develop a system to retain and organize the information learned
* Initial scene assessment	Overall assessment of the fire occupancy and surrounding area to determine the scope of investigation and human resources needs; perform site safety assessment
* Identify the scope	Define the overall scene, identify the fire area, examine the full site for physical limits, perform a walk-through of the exterior and interior, evaluate damage, establish building identifiers and weather
Secure the scene	Secure the scene using personnel and materials as necessary; contact the incident commander to learn about activities and alterations by fire personnel
* Exterior survey	Perform a detailed exterior surface examination to document its construction, state of repair, utilities, openings, and fire damage
* Interior survey	Perform a detailed examination of interior rooms; identify safety hazards, damage levels, openings, maintenance, smoke and heat movement, and prefire conditions
* Evaluate building systems	Identify and evaluate prefire conditions and involvement of electrical, fuel gas, HVAC (heating, ventilation, and air conditioning), and fire protection systems
* Identify and protect physical evidence	Locate, protect, and document physical evidence; limit suppression and overhaul, and identify their effects; protect knobs, switches, and valves
* Interpret fire patterns	Identify and interpret fire patterns to assist in locating the area and point of origin
* Debris removal and examination	Remove and examine debris in a systematic manner, avoiding contamination and spoliation (this step may also be used in the determination of cause)
* Fire scene reconstruction	Reconstruct the scene as appropriate by replacing items into their prefire position to further identify patterns and fire movement and intensity

prefire conditions. Internet searches can sometimes lead to photographs of the interior of the scene from real estate listings. Internet and commercial sources may be able to provide aerial or satellite images.

Computer Fire Modeling

Mathematical calculations and computer fire models are sometimes used in reconstructing the fire dynamics or fire development; however, these methods require a certain level of expertise. They are also highly dependent on the correct information being entered to produce a valid solution. If investigators are not comfortable using these tools, they should seek another professional who is proficient in computerized fire modeling. For the model to work correctly, careful measurements and documentation of all building levels and dimensions, penetrations such as windows and doors, and fuel packages are needed.

Witness Observations and Electronic Data

Witness observations and electronic data collectively form one of the three bases of information used in a determination of origin. Many questions may be appropriate for a given witness relating to the occupancy and to the events surrounding the fire, but the exact questions are highly dependent on the situation and the witness's role. With regard to origin, a witness may have seen the fire start or observed it in its very early stages. That information, of course, can be very helpful if the witness is determined to be credible, has a vantage point that permits adequate observation, and makes statements that are supported by other evidence. Other witnesses may also have origin information. Consider asking the following origin-directed questions:

- Where in the occupancy was the fire when you first saw it?
- How big was the fire when you first saw it, and did it change in size over time?
- What time did you first see the fire?
- Did you see the fire directly, or did you see only a glow or smoke?
- What actions were taken to extinguish or fight the fire?
- Did you see the fire spread? If so, describe the spread of the fire.
- From what location were you observing the fire?
- Describe whether windows and doors were open or closed.

- Did any windows break, or were any windows or doors opened or closed during the fire's progress? If so, how did the fire react?
- Did you take any photos or videos of the event or the fire?

All witness statements must be evaluated not only for the credibility of the witness but also as to the witness's location during the fire and their ability to clearly perceive what was occurring during the event. Consider also whether facts supplied by witnesses were offered in a descriptive narrative or were provided in response to specific questions.

As described elsewhere in this text, electronic data are often available and may be very useful in understanding origin and fire spread. Members of the public in populated areas frequently take videos and photos of fires. These images can vary widely in their usefulness but are important for the investigator to review and compare if possible. Videos are also routinely taken by surveillance cameras, including those at banks and automated teller machines (ATMs), as well as by other electronic means such as dashboard cameras, police body cameras, and even doorbell cameras, and should be sought out by the investigator. These recording devices may be located far from the scene, and the investigator should consider canvassing the neighborhood to find any security cameras with relevant footage. Because the devices recording these videos may have a limited storage capacity, the videos should be acquired as soon as possible.

Electronic data that can prove helpful in origin determination may also be available from fire alarm and suppression systems in the fire building. In addition, some burglar and security alarms may be activated by flame, by heat, or by movement in general (which might correspond to the movement of flame or smoke within a protected area); data from these systems should be obtained as well.

Origin Analysis

The scientific method dictates that once available data have been collected from the scene and other appropriate sources, one or more hypotheses should be developed regarding the area and point of origin of the fire. Any such hypothesis must offer a credible explanation of the fire's origin and must be supported by the data that have been gathered throughout the fire investigation process. All hypotheses that are appropriate given the facts should be considered and tested; the investigator must take care not to simply formulate a single preferred theory without regard to other credible explanations.

As mentioned, three information pillars support origin determination: witness observations and electronic data, fire patterns, and fire dynamics. The first two comprise data gathered on scene and with other methods; the last, fire dynamics, involves application of physics and chemistry principles to the data acquired. The investigator may wish to develop hypotheses of area of origin based on witness observations and electronic data separate from those that can be formed from fire patterns, then compare the two sets to see which hypotheses are supported by the others and which information needs to be reconciled.

With regard to scene data, some investigative and analytical techniques that may be useful are fire pattern analysis, heat and flame vector analysis, depth of char surveys, and depth of calcination surveys. The origin hypothesis should explain not only where the fire started but also how it spread to the next available fuels, and throughout the compartment and structure.

Any evidence that contradicts the proposed origin hypothesis must be resolved. When new data are received, the investigator must evaluate it against the hypotheses that are being considered. Also, as part of the testing phase, the investigator must take steps to disprove each hypothesis that is developed. If the hypothesis is not supported by the facts, it must be rejected or modified, and testing must begin again. The scientific method is an iterative process, meaning that it may be necessary to return to one of the earlier steps and repeat the process from that point to ensure that a sound conclusion is reached. For example, if no hypothesis survives the testing, more data may be necessary.

Fire Pattern Analysis

The general purpose of analyzing fire patterns is to determine fire growth and movement, and to trace the movement of the fire back to its origin. To do this, an investigator must understand the principles of fire dynamics, heat transfer, and fire pattern generation and apply them correctly during the analysis. Chapter 1, *Introduction to Fire Investigation*, covers the basic methodology, and Chapter 4, *Fire Effects and Patterns*, describes the patterns and the science underlying them.

It is essential that all observed patterns be considered in the analysis. Reliance on a single pattern, except in cases of extremely small fires, may result in inaccurate origin hypotheses. Analyze whether each pattern represents fire movement or whether it is explained by a nearby fuel package. Determine whether a pattern "points" to an area generally or to a specific direction within the compartment. Consider whether it suggests a heat source at floor level, or at some other level within the room, and whether intervening objects might have changed the appearance of the pattern.

Remember that the fire patterns found after suppression efforts represent what is left of all patterns recorded during the life of the fire. As the fire progresses, it can obliterate or alter earlier patterns. This sort of progression is particularly prevalent with full room involvement, post-flashover fires, or structural collapse. Also, pattern interpretation can be significantly complicated if rekindling occurs, creating new patterns.

Fire patterns are generated by either the spread of the heat, flames, and gases (movement patterns) or the intensity of burning (intensity patterns). Importantly, factors such as fuel form, type, and geometry; heat release rate; and ventilation may lead to intensity patterns that may not correlate with the origin. Fire movement patterns may ultimately be better indicators of origin than intensity patterns are, simply because an intensity pattern can be created by any large fuel package burning, regardless of whether it is related to the origin of the fire.

Fuel Considerations

Many factors affect fire pattern generation. Identical fuel packages will burn differently depending on their placement in the room. The size, geometry, and heat release rate of the fuel, as well as the location of ventilation, will have at least as much effect on fire pattern generation as the length of time the fire burned. A fire investigator should not assume that the point of greatest damage was where the fire burned longest and, therefore, is the origin. The possibility that flames spread from a smaller fuel package to a larger package must also be considered.

Ventilation Considerations

Ventilation can have a dramatic effect on fire behavior and heat release rate. Identification of ventilation factors is crucial when evaluating fire patterns. For example, ventilation-controlled fires will burn with greater intensity (and cause significant damage) near openings that provide access to oxygen. In these fires, the original fire may diminish or stop burning when oxygen is depleted in an area of the compartment, but continue to burn vigorously near the ventilation openings where air continues to be available. Full room involvement will also change, and may mask or obliterate, some patterns made during the fuel-controlled (early) phase of the fire.

When performing sequential pattern analysis, the investigator must determine whether a pattern could be accounted for by ventilation effects. This includes a determination whether the compartment went to flashover and/or was ventilation controlled. If a compartment fire becomes ventilation controlled, dramatic patterns may form near and around any ventilation openings. If other areas exist where fuel-rich smoke can mix with fresh air, such as on a wall opposite a doorway opening, ventilation patterns may appear there as well.

Common patterns seen in this situation include large surface areas of damage near airflow or on surfaces across from the airflow opening, greater damage in these areas, lines of demarcation angled away from ventilation openings, large areas and increased magnitude of damage under a window, and increased damage around the seams of gypsum wallboard.

Patterns that cannot be accounted for by ventilation should be closely evaluated. When conducting a sequential pattern analysis, it is important to recognize that full room involvement does not obliterate all patterns. As a fire progresses toward sources of ventilation, some patterns, such as mass loss–type patterns on furniture in or around the area of origin, may be preserved. Further details are discussed later in this chapter in the "Origin Matrix Analysis" section.

Flashover Considerations

Because flashover results in generalized burning of horizontal surfaces (when sufficient air exists to permit such burning), patterns may be created on these surfaces, including on the floor of the compartment and on horizontal surfaces underneath furniture. Although these flashover-related patterns can reflect low burning, they are not related to the origin of the fire. Further, as discussed earlier, flashover can cause a compartment involved in fire to become ventilation controlled as the oxygen is depleted in the compartment, resulting in the formation of dramatic ventilation patterns that are unrelated to the fire's origin. When analyzing fire patterns, the investigator should attempt to determine if flashover occurred in the compartment, and take this possibility into account in the analysis.

Effects of Suppression Activities

Fire department activities may have an effect on fire patterns and fire movement and should be fully understood by the investigator. The investigator should ask firefighting crews where they advanced into the compartment, if they did, and which extinguishment activities took place there. The investigator should also determine which ventilation activities took place and how they affected the fire, and which vent openings might have been created, or sealed off, during firefighting (including the simple opening and closing of doors). Also, fire crews should be asked if any hose streams were observed to move or "push" the fire. All of these factors may influence the patterns that the investigator sees in the compartment, and knowing

about them will help the investigator understand the fire patterns and movement.

Heat and Flame Vector Analysis

In a structural fire, the investigator often finds numerous fire patterns on different surfaces. The investigator should inventory all patterns seen in a given room or compartment and take steps to resolve them into a heat and flame movement theory and potentially an origin theory. However, a fire pattern might not point directly at the source of heat or fire movement that created it. For example, loss of mass on one arm of a chair indicates a greater heat flux from that side, but does not necessarily point more specifically to the source of that heat. Similarly, a partial "V" pattern in the form of a diagonal line on the side of a dresser might indicate heat coming from inside the room, but does not necessarily indicate how far the heat source was from the dresser, nor does it indicate whether the heat source was to the left, center, or right.

Nevertheless, by reconciling these various patterns as a group, the investigator can frequently form a picture of the fire's movement and/or intensity within a compartment. If certain patterns demonstrate that greater heat was present in the west side of a room than in the east side, for example, and other patterns show that greater heat was present in the north side of the room than in the south side, the investigator may be able to reconcile the two findings to conclude that the northwest quadrant of the room needs further examination. Although it is not uncommon to find conflicting patterns, in general the reconciliation of patterns within a compartment can be an effective way to trace heat movement and find a fire's origin.

One of the analytical tools that an investigator uses is **heat and flame vector analysis**, which seeks to create a visual overview of fire patterns. To use this method, first prepare a diagram of the scene. This diagram should be reasonably detailed and include the locations of doors, windows, and pertinent contents showing identifiable patterns.

Next, draw arrows on the diagram to depict the interpretation of heat and/or flame spread based on the patterns in the scene. These arrows can point in the direction of fire travel or back to the heat source, as long as the direction is consistent throughout the diagram. The length of the arrow should reflect the actual size or magnitude of the pattern. In a legend accompanying the diagram, the investigator may also provide details about the pattern—for example, pattern geometry, height above the floor, height of the pattern vertex, and the fire effect that constitutes the pattern. Once the diagram is complete, the various vectors can be viewed together to show actual heat and flame spread directions.

The point of the heat and flame vector analysis is to illustrate graphically the interpretation of the fire patterns. Such a diagram can also be used to identify conflicting patterns that need to be explained **FIGURE 13-2**.

Depth of Char Analysis

Sometimes fire patterns are not visually obvious. Depth of char or depth of calcination surveys may help a fire investigator identify areas of greater and lesser heat damage and define the associated lines of demarcation defining the patterns. Depth measurements are plotted on a diagram. Points of equal or nearly equal depth, called **isochar**, can be connected by lines, thereby revealing patterns.

With **depth of char analysis**, measurements of the relative depth of char on identical fuels are plotted on a detailed scene diagram to determine locations within a structure that were exposed to a heat source and were part of the fire spread. The key to appropriate use of this tool is documenting the relative depth of charring from point to point and locating places where the most severe damage occurred. Investigators may then be able to deduce the direction of fire spread. Remember, though, that severe damage may be the result of

many factors, including fire movement, nearby ventilation points, and nearby high-energy fuel packages.

Several factors can influence the validity of char depth analysis:

- Single versus multiple heat or fuel sources
- Ventilation factors influencing the rate of burning or fire intensity
- Consistency of the measuring technique and method
- Suppression and overhaul efforts, both of which can modify or destroy char patterns

Depth of char analysis is most reliable when applied to fire spread evaluation. It is not intended to determine specific burn times or to quantify the intensity of heat from any particular fuel package.

Consistency in the method of measuring char depth is critical to generating reliable data. Thin, blunt-edge probes are best used for this purpose; sharply pointed objects are not suitable. Certain types of calipers with round depth probes and tire tread depth gauges are excellent choices. The same tool should be used on comparable materials. Consistent pressure of insertion is also necessary.

When measuring char depth, note whether any of the material has been consumed or broken off during the fire or suppression/overhaul. The depth of the missing material should be added to the measured remaining depth of char **FIGURE 13-3**.

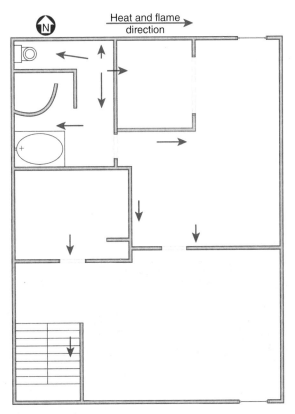

FIGURE 13-2 A simplified example of a heat and flame vector diagram.

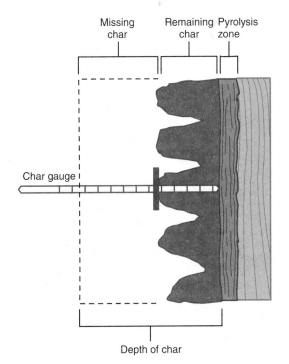

FIGURE 13-3 The depth of the missing material should be added to the measured remaining depth of char.

Reproduced from National Fire Protection Association. 2020. *NFPA 2021: Guide for Fire and Explosion Investigations*, 2021 ed. Quincy, MA: National Fire Protection Association.

Where fugitive fuel gas was the first fuel ignited, char depths are likely to be relatively uniform. Variances in char depth may exist in locations where the gas became pocketed or ignited other fuels. Deeper charring may exist at the point of the gas leak if there was continued burning of a pressurized gas jet.

Depth of Calcination Analysis

Depth of calcination surveys measure the relative depth of calcination (physical changes in gypsum wallboard) and are plotted on a diagram to determine locations within a structure that were exposed to a heat source for the longest durations. The relative depth of calcination on gypsum wallboard can indicate differences in total heating. Deeper calcination findings indicate longer exposure to a heat source or exposure to greater intensity or temperature during the fire.

The same cautions regarding depth of char measurements apply to depth of calcination measurements. Additionally, the finish on the wallboard (e.g., paint, stucco) may influence the heat effect, particularly if the finish is combustible. Water applied during fire suppression can also soften gypsum wallboard to the point where no reliable measurements can be obtained.

When measuring depth of calcination, conduct a vertical and lateral survey of the affected area, at regular intervals of 1 foot (0.3 m) or less. A depth of calcination diagram can be prepared from these data in the same manner described for depth of char measurements.

Origin Matrix Analysis

When a compartment becomes fully involved, the availability of oxygen there is limited. Thus, it is possible that the area of origin will not be actively flaming and ultimately may be less damaged than other areas located near ventilation sources. In this situation, simply relying on the areas of greatest damage to form hypotheses may lead to incorrect conclusions. An **origin matrix analysis** is a tool that can assist the investigator in analyzing fully involved compartment fires, as it takes into account the changes stemming from full fire development **FIGURE 13-4**.

Origin matrix analysis is a cognitive testing tool that may be used to develop and evaluate origin hypotheses. It explicitly recognizes the fact that conditions change over the life of a fire. Once a room becomes fully involved in fire and then burns for what can be considered a long duration, available oxygen can be depleted from some areas of the room, including the area where the fire originated. In Figure 13-4, a theoretical room is divided into four quadrants, only one of which has a ventilation opening (Quadrant 3).

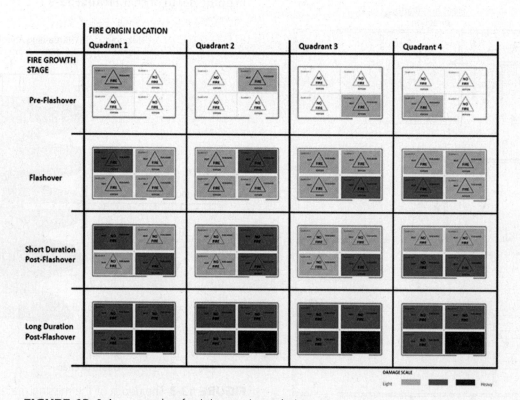

FIGURE 13-4 An example of origin matrix analysis.

Reproduced from NFPA 921-2021, *Guide for Fire and Explosion Investigations*, Copyright © 2020, National Fire Protection Association. This reprinted material is not the complete and official position of the NFPA on the referenced subject, which is represented only by the standard in its entirety.

A fire is ignited in each of the quadrants in turn, and observations are made of damage levels as each fire progresses through and past flashover.

If any of the fires is extinguished at flashover, light damage will be observed around the room, with heavier damage in the quadrant where the fire started. As each fire progresses post flashover, however, oxygen is depleted and heavier damage is seen not only in the quadrant of origin but also at the ventilation opening: There, the fire continues to burn fuel gases at the point where oxygen is still available. (Note that pyrolysis of solid fuels will continue in oxygen-deficient portions of the room, so damage such as char and loss of mass will continue in these areas at a slower rate.) If the fire is allowed to progress for a long duration post flashover, the theoretical room will display heavy damage throughout, with the most distinct damage being found at the ventilation point and not at the origin.

Conditions will, of course, vary from fire to fire. However, the origin matrix demonstrates that burning at the ventilation opening may create its own damage unrelated to the fire origin. Indeed, if the fire continues to burn to a significant extent after flashover occurs, the overall room damage can look similar regardless of which quadrant held the origin of the fire. The investigator must, therefore, consider the length of burning for each fire, which ventilation openings might have supplied oxygen, and whether flashover occurred. At times, it may not be possible to determine the correct origin of a fire that burned significantly post flashover or if the fire became ventilation controlled for any reason.

As the preceding discussion suggests, ventilation-generated patterns can sometimes obliterate origin-related fire patterns. If, however, the origin of the fire was located away from those ventilation-generated patterns, the origin patterns may remain identifiable. Therefore, all patterns must be considered in the data analysis and must be documented by the investigator. Identification of origin-related patterns is usually more difficult when multiple ventilation openings exist in a fully involved compartment.

While the origin matrix in Figure 13-4 applies to a theoretical room with only one vent opening, an origin matrix analysis may be a helpful analytical tool in any involved room. In such an analysis, the involved room is divided into sections. Smaller compartments can usually be divided into quadrants (although there is no specific requirement for size or geometry of the sections), and fire patterns and ventilation openings can then be located on the sketch. Next, each section is evaluated for overall thermal damage. This damage can be described as "light," "medium," or "heavy." Using a convenient color scheme, the investigator places the correlating color to each section. Heavy damage by ventilation openings can then be properly attributed to ventilation effects, not origin patterns.

Ultimately, if the fire originated near a ventilation opening, origin patterns may be obliterated by ventilation-generated intensity patterns. While the point of origin might not be independently identifiable, the area (section) of origin may still be able to be identified.

On the basis of analysis of all of the data, an investigator can now develop one or more hypotheses to explain the origin and spread of the fire. A hypothesis may be developed early in the investigation and not be disproved by the additional data. Ultimately, a fire investigator must be able to articulate, no matter the order of the investigative and analytical steps taken, how the steps taken conform to the scientific method.

Testing the Hypotheses

Once hypotheses have been developed, the scientific method requires that each be tested. Such testing can be done cognitively as well as by experiment, as described in Chapter 1, *Introduction to Fire Investigation*. Deductive reasoning tests are based on the premise that if the hypothesis is true, then the fire scene should exhibit certain characteristics (assuming the fire did not grow so large the characteristics were obliterated). Techniques for hypothesis testing include timeline analysis, origin matrix analysis, fire modeling, and experimental testing. Application of these techniques to hypothesis testing can help identify gaps or inconsistencies in the proposed explanations of the fire's origin.

TIP

A critical question to answer is this: "Are there any other origin hypotheses that are consistent with the data?"

Timelines can be used to understand the relationship in time among significant events or circumstances identified or known to have occurred during the fire.

Fundamentals of fire dynamics can be applied to a hypothesis through various computer programs. These fire models use incident-specific information (collected by the investigator) to predict fire development at given points in time.

Physical experiments, including live burn exercises, can also be used to test a hypothesis. If the experiment produces results consistent with the evidence and data from the fire scene, the experiment is said to support the hypothesis. Conversely, if the experiment produces different evidence, the hypothesis may need to be discarded or new information developed and considered in the analysis.

The following questions should be answered during the process of hypothesis testing:

- Could a fire starting at the hypothesized origin result in the observed damage?
- Are the growth and development of a fire starting at the hypothesized origin consistent with the established timeline?
- Is the proposed hypothesis consistent with the physical damage observed and the witness statements and electronic data?
- Is there a competent ignition source at the hypothesized origin?
- Are other origin hypotheses consistent with the data?

The lack of a competent ignition source in the area of origin may require the investigator to reanalyze the validity of the origin hypothesis. Contradictory witness observations must be explained through physical evidence, reliability, or weight of the information. The origin should not be "determined" just because a readily ignitable fuel and a potential ignition source are found together. When testing the hypotheses, the investigator must actively attempt to disprove them, not simply find data that support them. This helps avoid expectation and confirmation biases as described in Chapter 1, *Introduction to Fire Investigation*.

Selecting the Final Hypothesis

Once the origin hypotheses have been tested, only one hypothesis should survive the testing for the investigator to be able to select it as the final hypothesis. A technically valid origin determination is one that is uniquely consistent with the available data. The investigator should then review the entire process to ensure that all credible data are accounted for, and credible alternative hypotheses have been tested and eliminated. The investigator should document the facts that support the origin hypothesis to the exclusion of all others.

It is unusual for a hypothesis to be totally consistent with all of the data collected. However, not all data collected have the same value. Thus, the data must be evaluated by the investigator to determine their reliability and relative weight. Contradictions must be recognized and resolution attempted. Incomplete data can make it impossible to achieve this kind of resolution. Ultimately, if contradictions cannot be resolved, the origin hypothesis should be reevaluated and may need to be rejected. If no hypothesis survives the testing phase, the original determination may need to be specified as inconclusive.

After-Action REVIEW

IN SUMMARY

- The goal of origin determination is to identify, in three dimensions, the location where the fire began.
- Determination of the fire's origin must be done prior to the analysis of its cause. Usually, if the origin cannot be determined, the cause cannot be determined; moreover, if the correct origin is not identified, the investigator's determination of the cause will be incorrect.
- In general, relevant information related to fire origin can be obtained from witness information and/or electronic data, fire patterns, and the application of fundamental principles of fire dynamics.
- The overall methodology to be applied when determining a fire's origin is the scientific method.
- As a fire progresses, previously generated patterns can be altered or obliterated.
- Persons familiar with the involved structure, such as owners, occupants, contractors, repair personnel, and neighbors, may be able to provide information on prefire characteristics of the building.
- All witness statements must be evaluated not only for the credibility of the witness but also in regard to the witness's location during the fire and their ability to clearly perceive what was occurring during the event.
- Once available data have been collected from the scene and other appropriate sources, one or more hypotheses should be developed regarding the area and point of origin of the fire. As part of the testing phase, the investigator must take steps to disprove each hypothesis that is developed. If the hypothesis is not supported by the facts, it must be rejected or modified, and testing must begin again.

- Fire pattern analysis is done to determine fire growth and movement, and to trace the movement of the fire back to its origin.

- Reliance on a single pattern, except in cases of extremely small fires, may result in inaccurate origin hypotheses.

- The size, geometry, and heat release rate of the fuel, as well as the location of ventilation, will have at least as much effect on fire pattern generation as the length of time for which the fire burned.

- Identification of ventilation factors is crucial in evaluating fire patterns. When performing sequential pattern analysis, an investigator must determine whether a pattern could be accounted for by ventilation effects.

- Because flashover results in generalized burning of horizontal surfaces, patterns may be created on these surfaces, including on the floor of the compartment and on horizontal surfaces underneath furniture.

- When analyzing fire patterns, the investigator should attempt to determine if flashover occurred in the compartment, and take the effects of flashover into account in the analysis.

- Tools to help evaluate fire patterns include heat and flame vector analysis, depth of char analysis, depth of calcination analysis, and origin matrix analysis.

- Once hypotheses have been developed, the scientific method requires that each be tested. Testing can be cognitive or by experiment.

- Physical experiments, including live burn exercises, can be used to test a hypothesis.

- Once the origin hypotheses have been tested, only one hypothesis should survive the testing for the investigator to be able to select it as the final hypothesis.

KEY TERMS

Area of origin A structure, part of a structure, or general geographic location within a fire scene, in which the "point of origin" of a fire or explosion is reasonably believed to be located. (NFPA 921 and NFPA 1033)

Depth of calcination surveys Measurements of the relative depth of calcination (observable physical changes in gypsum wallboard) plotted on a detailed scene diagram to determine the locations within a structure that were exposed to a heat source for the longest durations.

Depth of char analysis Measurements of the relative depth of char on identical fuels plotted on a detailed scene diagram to determine the locations within a structure that were exposed to a heat source for the longest durations.

Fire scene reconstruction The process of removing debris and replacing contents or structural elements in their prefire positions. Reconstruction can also include re-creating the physical scene at a different location to more clearly assess damage and patterns.

Fire spread analysis The process of identifying fire patterns related to the movement of fire from one place to another, along with the sequence in which the patterns were produced, so as to trace the fire back to its origin.

Heat and flame vector analysis A method used to assist with fire spread analysis and origin determination, in which arrows representing the investigator's assessment of the direction of heat/flame spread are placed on a detailed diagram of the fire scene; the resulting diagram provides a visual illustration of the movement of the fire from its origin through the building.

Isochar A line on a diagram connecting equal points of char depth.

Origin matrix analysis A cognitive testing process used in the analysis of fully involved compartment fires to determine their origin, and to account for the effects of ventilation on the fire patterns.

Point of origin The physical location within the area of origin where a heat source and a fuel first interacted, resulting in a fire or explosion. (NFPA 921 and NFPA 1033)

Sequential pattern analysis Application of principles of fire science to the analysis of fire pattern data (including fuel packages and geometry, compartment geometry, ventilation, fire suppression operations, and witness information) to determine the origin, growth, and spread of a fire.

REFERENCES

National Fire Protection Association. 2020. *NFPA 921: Guide for Fire and Explosion Investigations*, 2021 ed. Quincy, MA: National Fire Protection Association.

National Fire Protection Association. 2021. *NFPA 1033: Standard for Professional Qualifications for Fire Investigator*, 2022 ed. Quincy, MA: National Fire Protection Association.

On Scene

1. Why does NFPA 921 state that without correct determination of origin, the investigator's determination of cause will likely be incorrect?

2. Fire investigators routinely gather data in support of origin determination even before they arrive on scene. Which types of information could you gather from the moment of dispatch for a wildland fire? A commercial building fire?

3. Review the information on origin matrix analysis, and consider how the damage levels, and corresponding patterns, depicted on the matrix might change with the addition of one or more ventilation openings at various locations and levels of the room depicted.

4. What are some physical experiments that a fire investigator could perform to help determine the area of origin or point of origin of a fire?

Chapter Opener: © Jones & Bartlett Learning. Photographed by Glen E. Ellman; On Scene siren: © Bildgigant/Shutterstock.

CHAPTER 14

Cause Determination

KNOWLEDGE OBJECTIVES

After studying this chapter, you should be able to:

- Describe the use of data analysis in cause determination. (**NFPA 1033: 4.6.5**, p. 228)

- Explain methods for identifying the ignition source and sequence. (NFPA **1033: 4.2.6**, pp. 328–330)

- Describe identification of the first fuel ignited. (**NFPA 1033: 4.2.6**; **4.6.5**, p. 329)

- Describe the recommended methodology for determining the cause of a fire. (**NFPA 1033: 4.6.5**, pp. 330–333)

- Describe the process of elimination and negative corpus.

- Describe levels of certainty in investigative opinions. (**NFPA 1033: 4.6.5**, p. 333)

- Describe considerations when evaluating a potential electrical cause. (**NFPA 1033: 4.2.6**; **4.2.8**; **4.4.3**; **4.6.5**, pp. 334–336)

- Describe considerations when evaluating appliances as a cause. (**NFPA 1033: 4.2.6**; **4.2.8**; **4.4.3**; **4.6.5**, pp. 336–339)

- Identify and describe common residential appliances and their operation.

- Identify and describe common appliance components.

- Describe how to record the fire scene when an appliance is involved. (**NFPA 1033**; **4.3.3**, pp. 354–356)

SKILLS OBJECTIVES

After studying this chapter, you should be able to:

- Identify the ignition source. (**NFPA 1033: 4.6.5**, pp. 328–329)

- Identify the first fuel ignited. (**NFPA 1033: 4.6.5**, p. 329)

- Conduct a cause investigation using the scientific method. (**NFPA 1033: 4.2.6**; **4.6.5**, pp. 330–333)

- Integrate findings from multiple modes of inquiry to create cause hypotheses. (**NFPA 1033: 4.6.5**, pp. 330–333)

- Test cause hypotheses. (**NFPA 1033: 4.6.5**, p. 333)

- Formulate an opinion on the cause of a fire. (**NFPA 1033: 4.6.5**, pp. 331–333)

- Assess an appliance for its potential as a cause. (**NFPA 1033: 4.2.6**; **4.2.8**; **4.4.3**; **4.6.5**, pp. 336–354)

You are the Fire Investigator

You are investigating a fire in the kitchen of a single-family residence, where you have determined the origin to be the kitchen counter. Appliances in the area of origin include a coffee maker, toaster, electric slow cooker, and radio. You observed that all of these items were plugged in during your inspection.

An interview with the occupants of the structure reveal they had left for work approximately 15 minutes prior to the fire. They had used the coffee maker, toaster, and radio during their preparations for the day. They reported no issues or troubles with these appliances.

1. How would you apply the scientific method to this investigation?

2. Which potential sources of ignition will likely need to be considered in the area of origin?

3. Which data would you collect to identify the ignition source?

 JONES & BARTLETT LEARNING **NAVIGATE**™ *Access Navigate for more practice activities.*

Introduction

Cause is defined as "the circumstances, conditions, or agencies that bring together a fuel, the ignition source, and the oxidizer, resulting in a fire or explosion," according to NFPA 921 and NFPA 1033. The **determination of cause** is the process of identifying the factors that allowed a fire to occur, including the type and form of the first fuel, the competent ignition source, and the circumstances and human actions that brought the ignition source and the fuel together. Fire cause determination generally follows origin determination, and a determination of cause generally can be considered reliable only if the fire's origin has been correctly determined.

As with fire origin, the cause of the fire is determined using the scientific method. Origin and cause are distinct determinations. Although some of the data gathered during the origin determination will likely apply in the determination of cause, the fire investigator must essentially start the analytical process over when applying the scientific method to cause.

The nature of the investigator's assignment may affect the amount of work devoted to cause determination. For example, if a government investigator narrows the point of origin to an appliance, further cause determination may require expert or engineering examination that the agency does not wish to arrange and pay for. If there is no evidence of an incendiary involvement, public agencies will frequently decide that it is not appropriate to continue the public-sector investigation under these circumstances, and will turn control of the scene over to the owner and private fire investigators to finish the cause determination. This will preserve the scene and the appliance for the party that can conduct a more comprehensive examination, and

the public agency will by necessity consider the fire undetermined pending the further examination. In contrast, if the origin suggested criminal intent, then the public agency would likely continue the investigation.

Where the Fire Started

Ignition Source

The source of the ignition energy will be at or near the point of origin. Often, the investigator can find the source of the ignition, or its remains, at the point of origin **FIGURE 14-1**. If present, the ignition source may be damaged or even unrecognizable. It may have been destroyed by the fire or moved by suppression operations, or it may have been moved or transported away from the origin before the fire. Because there are many ways in which the ignition source may have been moved or removed from the point of origin, an investigator should

FIGURE 14-1 Often, the source of the ignition will remain at the point of origin.
© FirePhoto/Alamy Stock Photo.

always consider the possibility of ignition sources that are no longer found at the point of origin.

Any identified ignition source must be "competent," meaning that it has sufficient energy to ignite the first fuel, and it is capable of transferring that energy to the fuel long enough to bring the fuel to its ignition temperature. This process proceeds through three steps:

1. **Generation:** The competent ignition source must be able to generate energy in the form of heat and must be generating heat at the time of ignition.
2. **Transmission:** The heat source must be able to transmit enough energy to the fuel so that the fuel reaches its ignition temperature. Transmission may occur by conduction, convection, radiation, or direct flame contact.
3. **Heating:** The target fuel must have heated to the point of ignition. This depends on several factors, including the material's thermal inertia—that is, the properties of a material that characterize its rate of surface temperature rise when exposed to heat. Chapter 2, *Fire Science for Fire Investigators*, provides more information on thermal inertia.

When the investigation shows that the fuel ignited was a vapor, such as gasoline vapor, liquefied petroleum (LP) gas, or natural gas, a competent ignition source must have been present when the fuel gas was within the range of its flammable limits at the ignition source. The ignition source may be difficult to determine because it may have been an arc from a switch or electric motor or an open flame such as a pilot light.

The ignition source for ignitable vapors may also be more difficult to identify because of the ability of the vapor cloud to migrate throughout multiple compartments. Within the area of the vapor cloud, multiple competent ignition sources may have been present. Detailed interviews with individuals who have sufficient knowledge of the contents of the room of origin and the activities that took place there may assist with determining which ignition sources were present, and whether they could have been competent and generating heat at the time of ignition.

First Fuel Ignited

The first material ignited that sustains the combustion process beyond the actual ignition source must be identified. This means, for example, that if a wood match was used to light paper in a waste can, the first fuel was the paper in the waste can—not the wood of the match.

Although the first fuel ignited may have an ignition temperature that is well within the temperatures

produced by the ignition source, the configuration of the first fuel ignited is important to consider. In case of like fuels, where one has a high surface-to-mass ratio and the other has a low surface-to-mass ratio, the fuel with the high surface-to-mass ratio will require less thermal energy to ignite. An example is a 2-inch by 4-inch (5-cm by 10-cm) block of wood and the same block of wood shaved into a pile of thin shavings. With the same overall mass but a much higher total surface area, the pile of shavings will ignite at the same temperature but will require less energy to bring the fuel to the ignition temperature. In all instances, the heat source must have sufficient heat and duration of exposure to heat the fuel source to its ignition temperature.

Oxidizing Agent

In most fires, the oxygen in the air serves as the oxidizing agent, or oxidant. Fire intensity can be enhanced by the introduction of other oxidants into a fire. Commonly encountered supplemental oxidants include medical oxygen (e.g., compressed gas cylinders or oxygen generators/concentrators) and chemicals such as pool sanitizers. Residue of the chemical oxidants may remain in the area after the fire. If such residue is found, it should be collected for laboratory analysis. If an oxidant other than atmospheric oxygen is identified at or near the area of origin, document it and consider its role in fire development.

Ignition Sequence

The mere presence of a fuel and a competent ignition source does not alone result in a fire. Thus, a key consideration in the investigation is to identify the **ignition sequence**, or the events that brought the ignition source and the fuel together, thereby establishing the fire cause **FIGURE 14-2**.

As an example, an investigator may conclude that a kitchen fire originated in a toaster. But what event caused this fire to occur? Several possibilities exist for consideration—a design defect, alterations (or even sabotage) to a component, malfunction of an internal component, or improper usage by the consumer. Investigators should be prepared to address the initial event, possibly a failure mode, as long as it is within their respective scope of expertise. In some instances, an additional expert may be required to more thoroughly unravel the event and the failure mode of a suspected appliance or component. It may also be useful to develop timelines to organize and analyze collected data.

A fire investigator should also be prepared to describe the way in which the first fuel ignited was

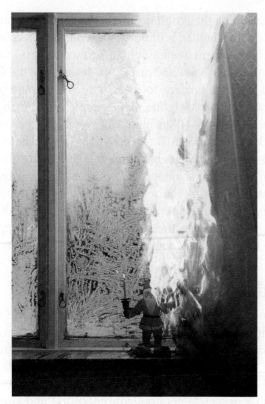

FIGURE 14-2 The cause of a fire includes three elements: an ignition source, the first fuel ignited, and the event that brought the ignition source and the first fuel together. Here, the candle is the ignition source, the curtain is the first fuel ignited, and the event that brought them together was a placement of the candle in proximity to the curtain.

© Johner Images/Getty Images.

then able to ignite subsequent fuels, causing the fire to spread. If the hypothesized location of the ignition is not in the main area of fire destruction, it will be necessary to demonstrate a mechanism of fire spread that allowed the fire to propagate along the path it was believed to have taken.

The Scientific Method in Cause Determination

As mentioned, the determination of origin must occur before the determination of cause is made. Once origin is identified, the process to determine cause begins in earnest with a reapplication of the scientific method. While the first two steps of the scientific method will be similar for the origin and cause investigations, the rest of the analysis in the latter case will be specifically directed at the cause. The data collected for the origin determination will be supplemented with further cause-specific data collection, and the formulation and testing of hypotheses are done anew **FIGURE 14-3**.

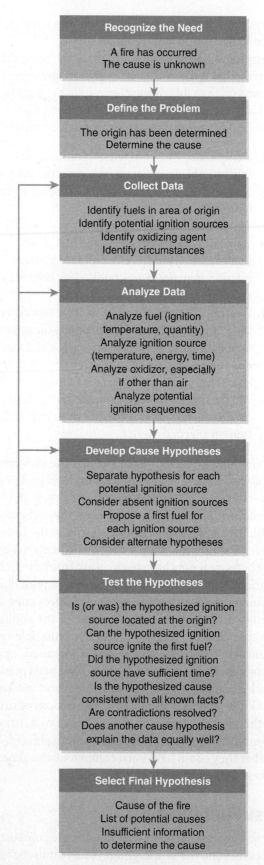

FIGURE 14-3 Example of applying the scientific method to fire cause determination.

Reproduced from *NFPA 921-2021, Guide for Fire and Explosion Investigations,* Copyright © 2020, National Fire Protection Association. This reprinted material is not the complete and official position of the NFPA on the referenced subject, which is represented only by the standard in its entirety.

Data Collection

Essential data to be collected to determine cause include identification of all fuels and fuel packages, potential ignition sources, and unusual oxidants in the area of origin. Sources of this information include results of the examination of the scene and debris, information from persons with knowledge of the events or the occupancy, and prefire photographs, videos, and plans or diagrams.

All potential ignition sources in the area of origin must be identified. No matter the size of the area of origin, an investigator should examine every potential ignition source within it to determine whether it was or could have been competent. Information should be collected about all appliances and electrical items in the origin area, as well as all other heat-producing items, including those that would normally not produce enough heat to ignite a first fuel but might do so due to malfunction or misuse.

Evaluation of the competence of a hypothesized ignition source requires identification of the initial fuel, even though it may not have survived the fire in recognizable form. The initial fuel may consist of building materials, interior furnishings, or a component part of an appliance. When diffuse fuels (i.e., gas, vapor, or dust) are the initial fuel, the point of origin (ignition) may be remote from the location of other fuels that sustained the combustion. A flash fire may ignite light combustibles, such as window treatments, in multiple locations, complicating the effort to identify the point of ignition. Thus, the investigator should collect as much information as possible about fuels and their prefire locations, including furniture, wall coverings, window treatments, plastic and paper items, and other possessions.

Data may also need to be collected from outside the area of origin—for example, if a comparison sample is needed of unburned fuel or carpet, or if an **exemplar** ignition source is available. Also, consider ignition sources that do not correspond to a physical device that has been identified on scene, such as open flame and static electricity.

Data Analysis

Once all of the potential initial fuels, potential ignition sources, oxidants, and relevant circumstances/ activities have been identified, analysis of the whole of the data can be conducted. Avoid jumping to a conclusion about a fire cause just because a fuel and an ignition source happen to coincide with the area of origin. A separate hypothesis must be developed and tested for each possible initial fuel/ignition source/

oxidant combination identified. All potential ignition sources that are present or believed to have been present should be considered in separate, alternative hypotheses. As discussed earlier, consider whether every potential ignition source was competent in light of the energy of the ignition source and the physical properties of the fuel ignited. This process is known as the **ignition source analysis**.

The form and geometry of the initial fuel must be such that the ignition source is capable of generating and transferring sufficient heat energy over the allotted time to result in ignition. Heavier fuels such as structural lumber are less likely to ignite as a first fuel without a very intense heat source. The first fuel ignited must also be capable of igniting the subsequent fuels for the fire to spread. Identification of this information is known as a **fuel analysis**.

It may be helpful to create a written table or matrix that lists all ignition sources on one axis and all potential first fuels on the other axis. Then, each ignition source may be considered with each first fuel to determine whether that combination could constitute a cause hypothesis and if not, why not. Notes can be made in each square of the table as to whether that specific combination could have produced ignition. This analysis can be done for all items in the full area of origin.

Some caveats apply, however. First, understand that the cause is likely to be located in the point of origin unless it was moved. So, if the fire investigator's analysis finds a viable cause in an area outside of the point of origin, more analysis is needed. Second, many of the first fuels and ignition sources will obviously not have ever had the opportunity to come together, so those combinations can be easily eliminated. Despite these limiting considerations, creating an ignition fuel matrix is still a valid exercise to perform, as it demonstrates the number of potential hypotheses that can be considered within a given area.

The investigator must also establish the ignition sequence. In conducting this analysis, it is important to establish—based on evidence—the events that occurred or that were logically necessary for the fire to have started. Factors to consider in developing an ignition sequence should include the fuel analysis and ignition source analysis just mentioned. Consider the following points:

- How and when the initial fuel came to be present at the origin
- The fuel's ignition temperature, configuration, orientation, quantity, and other properties
- How and when the oxidant came to be present (if an unusual oxidant was involved)

- How and when the competent ignition source came to be present

- How and when the competent ignition source transferred heat to the initial fuel, including the temperature of the ignition source and the amount of time for which it could have been applied to the first fuel

- How safety devices and features designed to prevent fire from occurring operated or failed to operate

- Which acts or omissions, and in which sequence, brought together the fuel, oxidizer, and ignition source

- How the initial fuel ignited subsequent fuels, resulting in fire spread

If the hypothesized ignition is not located within the main area of fire destruction, then the investigator must be able to show that some mechanism or process would have allowed the fire to propagate into the areas that it burned.

Logical Inference of Ignition Source

At some scenes, no physical evidence of the ignition source may be found after the fire, yet the ignition sequence can still be logically inferred. In such a case, the ignition source must still be established with evidence—not by the absence of it. That is, through testing of alternative hypotheses of potential ignition sequences, an investigator can identify the one sequence that is consistent with all known facts.

For example, it may be possible to develop a cause hypothesis when the first fuel is known but multiple competent ignition sources were present at the scene. Suppose a fire or explosion occurred in an area containing vapors, such as from a fuel gas leak, and multiple pilot lights or other appliances that could produce arcs were located in the area. Even if a specific point of origin is not identified, it may still be appropriate to hypothesize that one of the pilots or arcs acted as the ignition source, assuming the hypothesis is supported by the other data.

The following situations, among others, can support a determination of an ignition scenario even in the absence of an identified ignition source:

- Diffuse fuel explosions and/or flash fires occurred.

- Ignitable liquid residue, as confirmed by laboratory analysis, is found within a scene in a location(s) that does not have a non-incendiary explanation.

- Multiple fires are confirmed to have occurred. (The appearance of multiple fires can be explained

by other fire spread factors, as explained in Chapter 2, *Fire Science for Fire Investigators*, and Chapter 4, *Fire Effects and Patterns*.)

- Trailers are observed (but note that other phenomena can create the appearance of trailers).

- The fire was observed and/or recorded at its inception or before it spread to a secondary fuel.

Keep in mind, though, that any hypothesized ignition sequence must always have data to logically support it.

Testing of Hypotheses

As prescribed by the scientific method, an investigator should develop and test alternative hypotheses of cause to identify the one that is consistent with all known facts. Hypothesis testing may reference scientific and trade literature, which can contain not only useful information and data but also descriptions of experiments and testing performed by others. Investigators may also choose to do their own physical tests or experiments, while keeping in mind that such tests or experiments must be performed in a way that will produce reliable and applicable results for the investigation at hand.

In addition, cognitive testing of hypotheses is commonly and appropriately used by fire investigators to test ideas and hypotheses against the known data. The goal is to rigorously test all hypotheses in an effort to disprove each, as described in Chapter 1, *Introduction to Fire Investigation*. When examining each hypothesis, ask the following questions:

- Is the hypothesized ignition source a competent ignition source (temperature, duration) for the proposed first fuel?

- Is the time required for ignition consistent with the timeline associated with the facts and the hypothesis?

- What circumstances brought the ignition source together with the first fuel, and do these circumstances make sense in light of common sense and the data collected?

- What failure modes would have had to occur to result in heating and ignition?

Other tools to analyze the incident are discussed in Chapter 15, *Analyzing the Incident*, including timelines and fault tree analysis.

Process of Elimination and Negative Corpus

The process of elimination is crucial in the analysis of cause and in the efforts to test and evaluate hypotheses, and is an essential part of the scientific method. In many investigations, investigators will be successful in testing and disproving a range of hypotheses, to the point that one hypothesis remains that has not been refuted. However, as mentioned earlier, this hypothesis must not be adopted as the final opinion simply because it is the last one standing. Rather, it must be supported by, and consistent with, the evidence. Thus, investigators must not accept an ignition source for which no supporting evidence exists solely because they believe they have eliminated all other ignition sources known or suspected to have been in the area of origin.

The flawed technique just described is termed *negative corpus*. Negative corpus is inconsistent with the use of the scientific method that is required of fire investigators in NFPA 1033 because it involves making conclusions without the collection and analysis of data. Use of negative corpus has typically been seen in fires classified as incendiary, where an investigator would propose and eliminate a number of accidental ignition sources, then rely on such elimination as proof of an incendiary (human hands or open flame) ignition. To avoid any suggestion that the conclusions reached in the cause investigation rely on negative corpus, base all hypotheses and determinations on the analysis of facts, or logical inferences drawn from the facts, that are derived from observations, physical and other evidence, science, calculations, and experiments.

Selection of the Final Hypothesis

If all viable cause hypotheses have been considered and tested, and only one remains after testing, then it may be selected as the cause of the fire if it encompasses all three elements of cause (ignition source, first fuel ignited, and circumstances uniting the two). Once the final hypothesis is known, however, the investigator should still review the entire process to ensure that all credible alternative hypotheses have been eliminated. A critical question to ask is whether other hypotheses remain that are consistent with the known data. If so, further data are needed to clarify the cause.

Selecting a final hypothesis may not always be possible if insufficient data are available to the investigator. Further, that step may need to be deferred until the investigator has sought the assistance of a specialized professional to examine equipment or appliances. If two or more viable hypotheses remain, or if none has survived the testing process, the investigator must consider the fire undetermined until further conclusions can be made.

Levels of Certainty

It can often be appropriate to establish the level of certainty for an opinion that is formed about cause. That is, the investigator must assign a weight to, or the level of confidence in, any particular piece of data considered and the final determination of cause. Two potential levels of certainty can be assigned to a cause, based on the investigator's confidence in the data collected:

- Probable: A level of certainty that corresponds to the hypothesis being more likely true than not, or a level greater than 50 percent. This level of certainty is necessary for any hypothesis of cause to be selected as a final hypothesis (after full testing of all hypotheses). Because the "probable" level of certainty is greater than 50 percent, only one hypothesis can rise to this confidence level. The term "most probable" suggests there are multiple probable hypotheses, of which one is more likely than the others; such a statement is illogical and should not be used.
- Possible: A level of certainty that applies when a hypothesis is viable but has not been demonstrated to be probable. In the event that two viable hypotheses remain after testing but neither rises to the level of being probable, each must be considered possible. When the level of certainty in all hypotheses tested is "possible," the cause of the fire should be considered undetermined.

Do not confuse the terms "possible" and "probable" here with the term "probable cause" as used in the legal context of a criminal investigation. Likewise, do not equate these terms with the burdens of proof required in civil or criminal cases. (See Chapter 5, *Legal Considerations for Fire Investigators*.)

Classification of Fire Cause

While the fire cause is the specific event giving rise to the fire, the **classification of cause** is simply an effort to categorize the cause for purposes of assigning responsibility, reporting purposes, or statistics. From its inception through the 2017 edition, NFPA 921 contained a full chapter on cause classification. In the 2021 edition, however, the NFPA 921 Committee eliminated this chapter from the guide, after determining that the classification of cause created more problematic issues than benefits.

Cause classification is not a substitute for the cause determination itself; indeed, over the years confusion has arisen about how a given fire cause should be classified. At times, the same set of events could potentially be classified in different ways, depending, for example, on the mental state of the actor involved. Just as importantly, the cause of fire speaks for itself, whether or not it is placed into a classification by the fire investigator. If a fire is classified by an investigator as incendiary, for example, and a criminal arson charge results, the defendant's guilt or innocence at trial will depend not on the category in which the fire was placed, but rather on the specific facts as to how the fire started. In other words, the judge and jury will seek to determine whether the defendant's act met the elements of the crime charged.

Notwithstanding the elimination of the cause classification chapter from NFPA 921, many agencies still rely on classification of fire causes for statistical and reporting purposes, and some national data systems include cause classification. For reference, the pre-2021 NFPA 921 contemplated that fires could be classified into the following four categories:

1. Accidental: Fires that are not the result of an intentional act, including friendly fires that are ignited deliberately but become hostile.
2. Natural: Fires that ignite without human action or intervention, such as from lightning, wind, and earthquakes.
3. Incendiary: Fires that are intentionally ignited in an area or under circumstances where and when there should not be a fire. Note that "arson" is a crime as defined by statutory law and may have different elements than an incendiary fire classification.
4. Undetermined: Fires that have not yet been investigated, or are under investigation, or have been investigated but the cause is not proven to an acceptable level of certainty.

TIP

Suspicious is not an accurate description for a fire cause. Avoid using this term.

Electrical Fire Causes

It is common to find electrical wiring and/or components within the fire debris in a structural fire of any magnitude. Electric appliances and devices are everywhere, and electrical wiring and connections are found in walls, ceilings, attics, and crawlspaces. However, the presence of electrical wiring or components in fire damage, or even in the origin of a fire, does not by itself indicate that an electrical event was the cause. Even if electrical activity is evident on these components, the activity might be due to the component being attacked by the fire while energized. Electrical components can be, and often are, the victims of a fire that started from nonelectrical causes.

Because of the presence of electricity and electrical components in most buildings, it is a key skill for the fire investigator to be able to understand how electrical systems can cause fire and under what circumstances; to distinguish between situations in which an electrical component may have contributed to the cause of a fire; and to recognize situations in which a component found in a fire, while damaged, did not contribute to its cause.

Wiring and electrical installation performed in the United States and Canada in compliance with applicable codes, such as the National Electric Code (NEC), are generally safe under normal use. The energy supplied by an electrical system, however, is sufficient to cause fire under a variety of circumstances, including improper installation, abuse and misuse, impact and movement, and defective devices, among others. The investigator is cautioned to refrain from the temptation to name electrical components as fire causes without a full and appropriate investigation. See Chapter 7, *Building Electrical and Fuel Gas Systems*, for more information.

Examination Procedure

The examination of the building's electrical system may occur throughout the investigation, and good practice in all building investigations includes thorough documentation and review of the electrical system. The arc mapping process, if used, can aid in determining the origin of the fire (see Chapter 4, *Fire Effects and Patterns*). Once origin is determined, the investigator will focus on any relevant electrical components and devices that may be found in the area of origin, with the goal of determining whether they provided the heat source for the fire or otherwise contributed to its start or propagation.

Before analyzing an electrical circuit or equipment, evaluate the safety of the installation and make sure that all components have been de-energized. See Chapter 6, *Safety*, for further information, including a description of the lockout/tagout procedure. Also, keep in mind the legal principles surrounding spoliation of evidence. Remember that electrical wiring and componentry are very fragile after a fire. Do not move knobs, switches, or breakers, and refrain from disassembling electrical parts or equipment. Refer examinations for which you are not qualified to the appropriate expert, such as an electrical engineer.

The investigator should fully document the building's electrical system, and specifically identify and diagram any circuit components that may be involved in the fire event. Note the circuit breaker or fuse status. Examine and document exposed wiring and electrical damage during the debris removal process. If wiring becomes exposed during layering, document its position, along with the area or level of the building or room it appears to have come from. Carefully trace any wiring that may be involved in the fire to determine which appliance or connection it leads to, and where it comes from. If wiring leads back to an outlet, document its condition but do not unplug it.

In the debris, inspect items that look like electrical wiring to determine whether they are truly wiring, or whether they are metal from other objects. When wiring is found, document whether the wire is solid or stranded; whether it has insulation and what type; whether it is severed, melted, or crushed; where it ran in the compartment; and whether it matches other wire in the room. Inspect wire visually, and then run a gloved hand along it to feel for anomalies. Attempt to recognize electrical parts that may be found—for example, the metal components of an outlet or switch (some familiarity with these components is useful; items found on scene can also be compared to new exemplars). Outlets involved in fire should be examined to determine whether the contacts show any localized deterioration.

Heat-producing electrical equipment in the area and at the point of origin should be fully documented and examined, while taking care not to damage it or take it apart. Look for fire patterns on the equipment and note whether they appear to be interior to the device or exterior. Document the switches and knobs, if any, and determine whether it was connected to electricity and energized. Document any evidence of the manufacturer's identity.

Remember that even when an electrical device or appliance is turned off, if it is plugged in (or otherwise connected to the power supply), it can be assumed that there is electric potential in the cord leading up to the device. Furthermore, it is common for today's appliances and electrical devices to be consuming power and using electricity internally even when shut off. For example, a coffee maker with a time display, or any device with a working digital readout or indicator light that remains on when the device is switched off, is active and using electrical power while it is plugged in.

Analysis

Electrical wiring and appliances in the area of origin should be inspected for any anomalies that are different in kind or degree than the damage that might be expected in a fire. Chapter 7, *Building Electrical and Fuel Gas Systems*, discusses numerous types of anomalies and their causes. Anomalies are reviewed here, along with a discussion of what they might mean, and what they might not mean. Note that if the investigator finds evidence of arcing or other electrical activity in wiring, the only conclusion that is permissible from this finding alone is that the device or wiring was energized at the time of the fire. It is critical for the investigator to be able to examine an anomaly in wire and form a hypothesis as to how it was formed.

Arcing

A parting arc only melts the metal at the point of initial contact. Thus, copper wire that has melted due to arcing will show round, smooth "beads," with clear lines of demarcation between damaged and undamaged areas. If the arcing occurred between the wire and an adjacent wire, there may be a corresponding area of damage on the other wire—for example, a bead on one wire and a notch or divot in the other. General melting that does not show these characteristics was likely caused by heat from the fire. If an arcing event occurred in the wire but the wire was melted by the fire afterward, it may be difficult to discern evidence of the initial arcing event.

High-current arcing signifies that the circuit was energized at the time the arc occurred. Such an event happens when a high amount of current runs between conductors or from a conductor to ground. It can be caused by a short circuit or ground fault, or when insulation between conductors is burned away in a fire, allowing conductors to touch. A short circuit, particularly in wiring, is not usually expected to last for a significant duration because it should be stopped either by the activation of circuit protection or by the melting of the conductors until they are too far apart to permit the flow of electricity. A ground fault should be stopped either by a ground-fault circuit interrupter (GFCI) or by a standard circuit breaker or fuse.

As described elsewhere, arcing may result from numerous causes, some of which could indicate the fire's cause and many of which do not. When arcing is found in electrical components, further analysis is needed within the context of the facts uncovered by the investigator to determine whether the arcing was related to the fire's cause or resulted from heating by the fire. Anytime arcing is believed to be cause related, the investigator must identify the failure mode or event that resulted in the arcing and determine the ignition scenario. Arcing is generally a poor ignition source for all but the most diffuse fuels, particularly because many arcs will result in conductor melting or the activation of circuit protection, thereby stopping the current and eliminating the heat source.

Overload

An overload results from persistent overcurrent, which causes resistance heating in the conductor or other equipment. In a conductor, the fire investigator may see damage along the wire, including sleeving, defined as the softening and sagging of the thermoplastic insulation on the conductor, which may also be found where the conductor enters the circuit breaker; and offsets, in which sections of the conductor are offset from each other. If the overload is severe and persistent, the overheating of the conductor may ignite nearby combustibles. For example, when a conductor runs over several wood ceiling joists, charring or loss of mass might be observed on each.

Overload to wire conductors happens when the current in the conductor is greater than its ampacity, causing the conductor to become heated. Current can rise to a level far in excess of ampacity in a short circuit or ground fault situation, but these faults typically cause the overcurrent protection to operate before the conductor becomes excessively hot. Thus, for overcurrent in this situation to cause overload and overheating of a conductor, there must be a failure of the circuit protection device. Overload can also happen when the current in a conductor exceeds its ampacity but is less than the current rating for the circuit breaker or other protection. For example, if a 30-A circuit breaker is improperly installed to protect a circuit with an ampacity of 15 A, then 30 A could be drawn through the circuit, causing overcurrent without activating the circuit breaker. In addition, an overload situation can happen when the amount of current drawn in a circuit is equivalent to, or less than, the circuit's ampacity, but an undersized conductor is part of the circuit—for example, if equipment is drawing current through an extension cord that is not adequate for that amount of current.

Note that a moderate increase in amperage in a circuit that exceeds its ampacity may not cause overheating sufficient to cause ignition. However, the results could be more severe if the conductor experiencing overcurrent is protected by insulation.

Resistance Heating

A poor connection at a terminal of a duplex receptacle (outlet) or other equipment can cause rising temperatures and the build-up of oxide on the connection, further raising resistance and increasing temperature. When this occurs, a glowing connection can result and can ignite nearby combustibles. In this event, the investigator may see oxidation, pitting, damage, or destruction of metal or of the conductor at screw terminals and other connections. Also, if the device is damaged by fire, the damage from resistance heating may be recognizably greater than the damage to the rest of the device. The investigator, upon seeing such damage, should still develop and test hypotheses relating to whether the poor connection was related to the fire, how it occurred, which combustibles were ignited, and how the fire spread from the connection.

Static Electricity

Determining whether static electricity contributed to a fire cause requires the investigator to collect as much data as possible about the setting where the fire occurred, and operations that may have been underway when it started. For example, the investigator should be alert for situations involving tank filling and discharge, spraying, and conveyor belt operations. Similar to other cause hypotheses, a hypothesis regarding static electricity discharge must address the circumstances in which the heat source was generated and discharged, the fuel was ignited, and the heat source was united with the first fuel.

Appliances

Appliances are often considered as potential ignition sources when they are found in an area or point of origin in a fire. As with electrical equipment, though, the mere presence of an appliance at the site of the fire's origin does not establish that it contributed to the fire's cause. The investigator should consider an appliance at the origin equally plausible as, but not more likely than, any other potential ignition sources.

If the appliance is believed to be the ignition source, the investigator must be prepared to explain how the appliance produced the heat, whether it did so as part of its ordinary operation or due to a failure mode, and whether and how the first fuel came into proximity with it. Note that it is possible that an ignition within an appliance could ignite the casing or other parts of the appliance, which would then be regarded as the first fuel ignited. If the heat source was within the appliance and the first fuel is the appliance's casing, the investigator still must be able to explain what circumstance or failure led to ignition. If unable to do so, the investigator has not explained one of the three essential elements of fire cause—that is, the act or circumstances that brought the heat source into contact with the first fuel—and therefore must regard the fire as undetermined.

In a fire scene, the following indicators may support the theory that an appliance caused the fire:

- Fire patterns that point to the area of origin being near the appliance
- Severe fire damage to the appliance

In addition, if the appliance operates on electrical power:

- Evidence of arcing damage found either inside or near the appliance

Procedure

Before an appliance is considered as a possible ignition source, fire patterns should be used to determine whether the appliance was located at the point of origin. Evaluate the fire patterns on the appliance in relation to the remainder of the fire scene. If the appliance shows more severe damage than surrounding items, this could indicate that the fire originated at the appliance. Keep in mind that plastic appliance parts may have severe damage or be missing in a fire, but that this is not necessarily an indicator that the appliance was the point of origin. All other causes for the fire damage that occurred at or near the appliance must be considered, including debris drop-down or other ignition sources that happen to be next to or on the appliance. In addition, the type of enclosure on the appliance must be considered in determining the origin area. Appliances with non-fire-resistant plastic housings can suffer severe fire damage as a result of exposure to fire. An X-ray may be required to see inside melted or collapsed masses to determine the condition of key components.

Verify that the appliance was connected to an electrical power supply and determine whether it was energized at the time of the fire. It is important to trace and document the power source all the way back to the service panelboard. Branch circuit conductors should be recovered along with switches and, if possible, circuit protection devices, disconnects, and fuses.

Reconstruction of the fire scene and replacement of the appliance in its prefire position may be necessary to document the fire patterns and provide supporting evidence that the fire originated at the appliance.

To determine how an appliance located at the origin could generate sufficient heat energy to start a fire, answer the following questions:

- Was the appliance attached to a power source or fuel source at the time of the fire?
- Was it energized (and/or operating) at the time of the fire?
- Had the appliance recently been operating poorly or abnormally?
- Were repairs or maintenance recently performed on the appliance?
- Had the appliance recently been moved?
- Have there been any recalls on the appliance?
- Does the appliance appear to have been modified in any way?
- Could significant heat be generated under recent operating conditions?
- Was moisture present or a factor?
- Did nearby combustible materials come into close contact with a heat-emitting part of an appliance?
- How old is the appliance?

The goal in answering these questions is to determine whether sufficient heat existed, which material was first ignited, and how the material was ignited by this heat. Note that an appliance does not necessarily have to be operating to supply an ignition source to cause a fire. As mentioned previously, many appliances continue operating and using electricity even when they are in standby mode.

It is helpful to consider issues that have arisen with specific types of appliances. For example:

- Water-using appliances, such as dishwashers, washing machines, water heaters, and outdoor appliances: Consider fire caused by water/moisture or the lack thereof.
- Heat-emitting appliances, such as heaters, transformers/motors, and clothes dryers: Consider fire caused by ignition of a nearby combustible material, such as clothes or lint/dust that has built up.
- Appliances that draw a large amount of current for an extended period of time, such as heaters and air conditioners: Consider fire being caused by a poor connection or an overloaded circuit.
- Heavy appliances, such as freezers or refrigerators: Consider the possibility of a fault in the power cord caused by recent moving of the appliance, as the appliance may have been

left sitting on the cord or may have crimped or frayed the cord while moving across it.

- Appliances with controls, timers, or electronics and cabling in them: Consider poor power or circuit board connection problems; relay, power switching, or control component failures; or contamination of sensor connections.
- Fuel gas–powered appliances, such as stoves and dryers: Determine whether all gas valves were on or had been turned off, and whether gas service was supplied to the building.

Appliance Operation

It is important to understand how an appliance or device operates during normal operation, so as to help explain how a failure might have occurred. Appliance repair books can be a reference source for the design and operation of appliances. Sometimes the manufacturer will provide, on request or online, design documentation such as schematics, part lists, operational guides, or software listings that are not normally available to the average consumer.

Disassembly

Usually, disassembly of an appliance should not occur in the field, as this procedure can result in claims of spoliation of evidence. Before beginning any disassembly, identify the specific reason disassembly is needed and create a protocol with the objectives, ground rules, and extent of disassembly identified based on the specific reason. Trying to reveal or clarify the condition or status of some aspect of the evidence is generally the objective sought. For example, some disassembly may be needed to identify the manufacturer of a product. Going beyond what is necessary to do this is not recommended. Investigators should disassemble appliances only if they have proper expertise, have studied the appliance information, have considered the implications of evidence spoliation, and have given notification to interested parties, including the appliance manufacturer and the insured.

Each step of the disassembly process should be documented. Take notes and photographs or videos documenting the entire disassembly process. X-rays may be considered if disassembly is not possible. Fuses, heating elements, relay contacts, and windings can be examined in this manner **FIGURES 14-4** and **14-5**. When handling appliances, stay alert to the possibility of sharp edges, moving parts, and even chemicals or other harmful substances associated with the appliances, and wear appropriate personal protective equipment (PPE). Further, note that many appliances

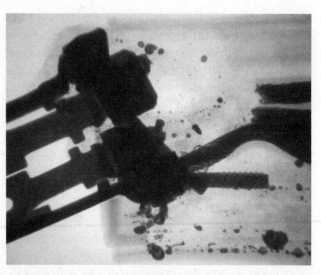

FIGURE 14-4 X-ray of a receptacle with internal damage.
Reproduced from *NFPA 921-2021, Guide for Fire and Explosion Investigations*, Copyright © 2020, National Fire Protection Association. This reprinted material is not the complete and official position of the NFPA on the referenced subject, which is represented only by the standard in its entirety.

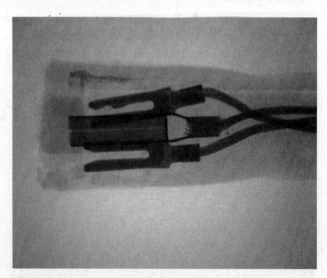

FIGURE 14-5 X-ray of the female end of an extension cord depicting internal damage.
Reproduced from *NFPA 921-2021, Guide for Fire and Explosion Investigations*, Copyright © 2020, National Fire Protection Association. This reprinted material is not the complete and official position of the NFPA on the referenced subject, which is represented only by the standard in its entirety.

continue to draw electricity even when shut off, and some electric and electronic parts may remain charged with electricity for some time after disconnection of power.

Exemplar Appliances

An exemplar is an exact duplicate of the appliance in question. An exemplar can be used to help understand how an appliance operates and can be used in testing a proposed ignition scenario. The model and serial numbers on the appliance in question can help you obtain an exact duplicate. If the exemplar is not exactly the same, the investigator must determine whether it is suitable to use in testing. Testing should verify that the appliance could not only generate the necessary heat or electrical discharge but also ignite the fuel load.

The Consumer Product Safety Commission is a good resource regarding any recall notices on the appliance.

Appliance Components

Many different appliances exist, each of which has a different use and a different construction. The following sections describe the components commonly found in appliances.

Housings

The **housing** of an appliance is the outer shell of the appliance that contains the working components. Although most housings are made of metal or plastic, they may also be made of wood, glass, or ceramic.

Steel

Most metal housings are fabricated from steel or stainless steel because of their strength and durability. Steel housings may be coated with enamel or plastic, which may be a factor when the appliance is exposed to fire. An example would be a refrigerator or washing machine.

Postfire patterns and colors depend on many factors. Ordinary steel housings can have a mottled appearance with colors that include blue-gray, white, black, and reddish brown. Bare galvanized steel could have a whitish coating. Protective coatings on the steel can cause many varied colors when exposed to heat or flames **FIGURE 14-6**.

> **TIP**
> A steel housing does not shield internal components of the appliance from reaching high temperatures.

Aluminum

Aluminum housings are generally made from formed sheets or castings. Pure aluminum has a melting temperature of 1220°F (660°C). Aluminum alloys have slightly lower melting temperatures.

Other Metals

Other metals, such as zinc or brass, may be used for appliance housings, often for decorative purposes.

FIGURE 14-6 Varying degrees of color change to the side of a stove as a result of heat exposure.

Reproduced from *NFPA 921 2021, Guide for Fire and Explosion Investigations*, Copyright © 2020, National Fire Protection Association. This reprinted material is not the complete and official position of the NFPA on the referenced subject, which is represented only by the standard in its entirety.

Zinc melts at 786°F (419°C) and appears as a lump of gray metal when it melts. Brass is an alloy and softens over a range of temperatures; however, it generally has a melting temperature of 1740°F (950°C). Although brass is sometimes used as a housing material, its primary use is for electric terminals.

Plastic

Plastic housings are used in appliances that do not normally operate at high temperatures. Plastics can melt at low temperatures; they char and decompose at high temperatures. Some plastics will continue to burn on their own once ignited, whereas others may not, due to added fire retardants or the chemical composition of the plastic.

Following a brief fire, the plastic housing could be melted and partially charred. An investigator must determine whether the heat source was inside or outside the appliance. X-rays can provide a view of the internal metal components. In a severe fire, the entire plastic housing may be consumed, but this does not necessarily indicate that the fire started within the appliance.

Phenolic plastics are highly resistant to heat and are often used to make coffeepot handles and circuit breaker cases. These plastics do not melt and do not support combustion. Phenolic plastics form a thin, gray ash layer when moderately heated and turn to a gray ash in a sustained fire. A thin, gray ash layer on the inside, and not the outside, of a phenolic plastic component may indicate that the heating was internal.

In general, there are two types of plastics: thermoplastics and thermosets. Thermoplastics will melt and flow when heated in a fire. In contrast, thermosets will char but generally retain their shape when heated.

Wood

Wood is occasionally used for appliance housings. It may be completely consumed in a fire or may have indicative fire patterns. These fire patterns may help to identify whether the fire originated inside or outside the appliance.

Glass

Glass is most commonly used for transparent covers and doors and for decorative purposes on appliances. It readily cracks and may soften and drip under fire conditions.

Ceramics

Ceramics are generally employed for use as a novelty housing and do not melt in fire; however, the decorative glaze that often coats ceramics could possibly melt. Ceramic is also used to support or house the electrical components.

Power Sources

The power source for most appliances is the alternating current (AC) supplied by the electrical utility company. Information in this chapter is limited to single-phase power that is 240 V AC or less. Power in the United States is supplied at 60 Hz, 120/240 V AC. Most appliances operate on 120 V AC. However, appliances that require 240 V AC (e.g., dryers, ranges, hot tubs) can also work on the same system.

Electrical Cords

Electrical cords can comprise two or three conductors. These conductors are stranded to provide flexibility. Cords with two conductors are found on appliances made before 1962 and on some newer double-insulated appliances. Cords with three conductors are found on newer appliances, with the third conductor being used as a ground for the appliance. These cords can be used for 120- and 240-V AC appliances.

Stranded conductors usually survive a fire event, but may be brittle following exposure to a fire. Electric cords are sometimes trampled on or end up under furniture or under the appliance itself, resulting in the potential for poor connections or overheating. With a high enough load current, poor connections increase the risk of fire due to resistive heating at the point of connection. Contact clearance spacings may become compromised due to mechanical damage, especially with higher voltages, resulting in faulting. Moisture or contamination can cause a failure of insulation or of a connection.

Plugs

Plugs made prior to 1987 and rated for 20 A or less have two straight prongs of equal width. Plugs made after 1987 and rated for 20 A or less have a neutral prong that is wider than the hot prong—an arrangement known as a **polarized plug**. Some newer plugs have a third prong for grounding. Plugs may have the conductors attached to the prongs inside a molded plastic housing. Some plugs have information stamped on the blades, which is useful for identifying manufacturing information such as date and location of manufacture.

Loose-fitting plugs can create a resistive heating connection, potentially causing a fire that can be identified by missing or partially melted away prongs and receptacles. Whereas the conductors and prongs usually survive a fire, other brass parts may melt. The receptacle does offer some protection to the face of the plug, which can then support a hypothesis regarding whether the appliance was plugged in.

Step-Down Transformer, Power Supply (Adapters)

Some appliances operate at lower voltages, such as 6, 12, or 24 V AC or DC. A **step-down transformer** reduces the 120 V AC provided at the receptacle to the required voltage. The transformer may be a part of the appliance, or it may be separate from the appliance and either plug into the wall receptacle or be in line with the power cord. The wire running from the adapter to the appliance is usually a thinner, two-conductor cable. Shorting of the lower-voltage wire is not likely to cause a fire; however, some low-voltage transformers can produce significant enough energy to make resistive heating connections an issue. Low-voltage (12–48 V DC/V AC) loads are especially vulnerable because the power of a resistive connection varies with the square of the current (the current multiplied by itself). For a 12 V AC or DC load, the current is thus 100 times higher than for a 120 V AC load.

For example, to provide 120 W of power to a low-voltage lamp, 1 amp of current is required, but at

12 V AC or DC, this becomes 10 amps of current. If the load connection has a 0.3-ohm resistive connection, the 120 V AC connection dissipates 0.3 watt of power, whereas the 12 V AC or DC connection dissipates 30 watts of power.

If embedded with a combustible material, unprotected connections can cause a fire. Therefore, if used outdoors, splices and connections need to be sealed and approved for outdoor use to avoid moisture intrusion.

Appliance power supplies typically have protection for loads. They can use internal transformers or power switching/transforming electronics to isolate the load from the line voltage. These devices can overheat if covered with insulating material. Power supplies should be considered potential ignition sources because they typically convert high-energy power sources to low-energy loads, but the circuits that do this may malfunction and overheat. Many appliances and electronic devices contain unfused internal power supplies, which can overheat if an internal fault occurs. Some larger entertainment equipment power supplies feed remote devices such as speakers or monitors, which should also be examined for possible faulting. Cables for connecting peripheral components could be considered as ignition sources if too small or not the right type.

Batteries

Batteries are energy storage devices and possible ignition sources. They are used for remote and portable devices, as well as some stationary ones. A "primary" battery is one that is designed to be discharged only once, whereas a "secondary" battery can be discharged and recharged multiple times. Batteries come in a variety of designs and chemistries, and at times a particular chemistry is more suitable for a given application. Batteries can also be arranged into battery packs, which act as a single unit but consist of multiple batteries connected either in series or in parallel.

Devices that have larger batteries include computer backup systems, security panels, hand tools, and some larger equipment such as floor cleaners, lawn tools, and electric scooters. The style of batteries used with larger devices includes auto, marine, and motorcycle batteries, and the types of cells used are typically lead-acid, although lithium and nickel–metal hydride (NiMH) cell packs are increasingly being included with these devices. Laptop computer batteries and power tools typically use NiMH or lithium-ion cell packs because of their lightweight construction. Smaller batteries,

including AA/AAA/C/D-sized lithium and NiMH cells, contain enough energy to potentially cause a fire if a direct short occurs. In addition, 9-V batteries are well known for providing enough energy to start a fire under some circumstances.

The remains of batteries are usually found after the fire, and it is important to determine what they were connected to. Examine cable connections from batteries to loads for faulting and heating (sleeving or melting of insulation). On systems with large batteries, such as cars or vans, cables should be protected by a fuse installed close to the battery. Recover fuses and/ or breakers to determine proper sizing.

Under normal conditions, small battery-powered circuits generally do not allow sufficient heat build-up to cause ignition. In certain conditions, however, even one battery can provide sufficient power to ignite materials. Lithium batteries have been known to cause fires due to faulty wiring in their electronic components. In some cases, the solution in the lithium cells will physically leak, indicating the possibility that the battery served as an ignition source.

The voltage and energy density for a battery are thus important considerations and are determined by its chemistry. **TABLE 14-1** identifies several different types of rechargeable batteries and their characteristics.

Lithium batteries have significantly greater energy density per volume (W-hr/L) compared to lead-acid batteries. Compared to a lead-acid battery, the lithium battery has more than three times the energy density.

FIGURE 14-7 shows the discharge curve for alkaline versus lithium batteries. As this curve indicates, lithium is a more potent energy source than alkaline in terms of its ability to maintain full voltage during discharge. In a direct short situation, it is possible, but unlikely, that a few alkaline cells could ignite very fine combustible materials.

Battery chargers are also possible ignition sources, so cables running from the battery to the charger must be examined for evidence of failure. Battery chargers control and limit the current required to charge the battery. Typically, they protect the load with fuses or circuit breakers. Some chargers use timers to control the charging rate and time; others use the voltage and/or current supplied to the battery to control the charging rate. Sometimes they control the charging power path based on battery presence.

Battery chargers are similar to low-voltage power supplies, which have line voltage (120 V AC) input and a controlled DC output. Misconnection or poor connection of the charger to the battery terminals is a common error to look for during the cause investigation. Because of resistive heating connections, nearby

TABLE 14-1 Battery Energy Characteristics

Manufacturer	Type	Chemistry	Nominal Cell Voltage	Nominal Capacity (A-hr)	W-hr/L
Mfgr 1	AA	Li ion	3.6	0.75	365
Mfgr 2	AA	Li metal	3.0	0.8	324
Mfgr 3	AA	NiCad	1.2	1.0	200
Mfgr 4	AA	NiCad	1.2	1.0	200
Mfgr 5	AA	NiMH	1.2	1.5	246
Mfgr 6	AA	NiMH	1.2	1.5	246
Mfgr 7	AA	Rechargeable alkaline	0.9–1.4	1.6	220
Mfgr 8	Box	Sealed lead acid	12	12	103
Mfgr 9	Box	Sealed lead acid	12	12	99

Abbreviations: Li, lithium; NiCad, nickel–cadmium; NiMH, nickel–metal hydride.

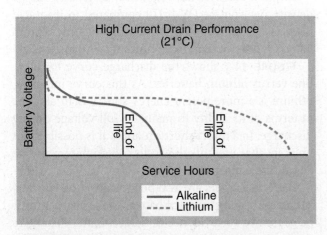

FIGURE 14-7 Battery voltage-during-discharge curves for alkaline (red) and lithium (green) cells.

combustible materials can become heated to ignition temperatures. Some batteries, if improperly charged, can themselves become overcharged and explode or violently expel hot electrolytic contents, becoming an ignition source. Further, it may be possible that the user was using the incorrect charger.

While charging, batteries are often (but not always) disconnected from the equipment that they power during normal operation. If so, this may eliminate the equipment from being a candidate ignition source.

Protective Devices

Overcurrent protection devices employed in appliances and electrical service panels often take the form of fuses and circuit breakers. The remains of the device after a minor fire may show whether the overcurrent device operated. The remains of the appliance after a major fire, however, may show only that the protective device was present.

In the case of a fuse, if the current is moderate (less than twice the rating), the fuse metal simply melts, which breaks the circuit. If the current is excessive, the metal vaporizes, leaving an opaque deposit on the glass tube or window.

Switches

Many different switch designs are used in common appliances. A **switch** is used to turn the appliance on or off or to change the operating conditions. Postfire examination of a switch can sometimes determine its state and, therefore, whether the appliance was energized at the time of the fire. Do not operate the switch. Instead, attempt to check the electrical continuity of the switch while it is in place, while recognizing that temperatures, arcing, and mechanical damage can affect the resistance reading. The remains of the switch can be very delicate. Carefully document the positions of the knobs, levers, or shafts while the switch is in place. It is important to make sure that the power to the appliance is off or that the appliance has been unplugged before testing. An investigator should not disassemble the switch unless qualified to do so. Fire impinging on a switch may destroy the housing, but the switch contacts can still show abnormal operation and should be preserved as valuable evidence.

TABLE 14-2 identifies several different switches used in appliances and how they operate. Many of these switches may also act as sensors. **FIGURE 14-8** shows the typical thermal cutout switch and thermostat in a coffee maker. Both are normally closed and if either one operates and opens the circuit, the appliance will stop functioning.

Fluid Pressure Switches (Capillary Tube)

A fluid pressure switch operates through a fluid in the sensing bulb, which is located in an area that can get hot. The fluid expands in the bulb when heated and applies pressure to a bellows device, which in turn opens a set of contacts, breaking the flow of electricity. High temperature can cause the fluid to undergo a phase change, which forces it out of the bulb. The resulting pressure increase of the fluid may crack the bulb or split the tube.

Bimetal Switches

Bimetal switches are much more common. They are composed of two dissimilar pieces of metal that are joined together to form a flat piece. The metals expand

TABLE 14-2 Switch/Sensors Commonly Used in Appliances			
Switch Type	**Property Sensed**	**Actuation Method**	**Mechanism**
Manual switch	None	Push button	Mechanical contacts
	None	Throw lever	Mechanical contacts
	None	Rotate knob	Mechanical contacts
Circuit breaker	Overcurrent	Current heats a bimetal element and energizes the coil	Mechanical contacts
Fuse	Overcurrent	Temperature melts the fuse	Fuse
Relay	None (computer-controlled input)	Current energizes the coil	Mechanical contacts
Solid-state relay	None (computer-controlled input)	Low-voltage command from computer	Solid-state transistor, MOSFET, Triac
Bimetal thermostat	Temperature	Temperature coefficient of expansion of different metals actuates the switch	Mechanical contacts
Thermostat	Temperature	Temperature expansion of a single metal actuates the switch	Mechanical contacts
Capillary thermostat	Temperature	Temperature expansion of a liquid actuates the switch	Mechanical contacts
Thermal cutoff	Temperature	Temperature melts the fuse, actuates the switch	Mechanical contacts
Tip-over switch	Orientation	Switch contacts displaced by gravity	Mechanical contacts
Door close switch	Open/close position	Switch contacts are displaced by the door in place	Mechanical contacts
Pressure switch	Water level or air pressure	Switch contacts enabled/displaced by pressure in the line	Mechanical contacts

FIGURE 14-8 Thermal cutoff and thermostat on a coffee maker.
Courtesy of Ray Franco.

at different rates when heated, causing the combined metal structure to bend. This bending motion can actuate a set of electrical contacts, blocking or permitting the flow of electricity. After a severe fire, the bimetal device can be distorted far beyond its operating position. This effect can be a result of heat from the fire and does not necessarily indicate the component was defective.

Expanding Metal Switches

An expanding rod switch employs a long rod that is exposed in the heated area. As the temperature rises, the rod expands, opening a set of electrical contacts.

Melting Switches

Some cutoff devices are based on a material that melts when the normal operating temperature is exceeded. When the material melts, it releases a spring that opens a set of contacts, which stop the flow of electricity. Note that these devices (and fuses in general) may be deliberately bypassed to allow the appliance to be operated without any thermal protection.

Motion Switches

Appliances such as portable electric heaters usually employ a motion (tip-over) switch to ensure that they are operated in their correct, designed position. A typical motion switch uses a weighted arm that hangs down and opens the contacts if the heater is tipped over. Another type of motion switch uses a vial of mercury (or other conductive fluid) that flows away from contacts when the appliance is tipped.

Contact Damage

Mechanical switches can fail because of an overload that overheats internal parts or welds the contacts. Poor internal connections can also cause destructive heating and failure. Switches normally show damage created by a parting arc on their metal contact surfaces. Contacts designed to be operated in normal use, such as in a thermostat, are normally pitted because of frequent opening and closing. Safety cutoff switches, however, should not have pitting on the contact faces. If they do, this is a sign of frequent abnormal operation and possibly evidence of a defect. Most switches are designed to snap open or closed to avoid surface pitting, erosion, or possible welding. The presence of damaged or welded contacts does not necessarily prove that the switch caused the fire. Electrically welded contacts typically have normal shapes and have their faces stuck together. Contacts that are melted together in one lump were probably exposed to external heating.

Contact misalignment may be a function of fire heat damage rather than indicating operational failure of the switch.

Solenoids and Relays

Electromechanical solenoids and relays control high-power circuits by using a low-power circuit. The remains of these devices are normally found after a fire, and their contacts should be inspected to determine whether they were stuck together during the fire.

Electronic Relays/Switches

Some appliances use electronic switches to turn on loads. Typically taking the form of semiconductor components, these switches are called *solid state* because they have no physical moving parts and thus no contacts. Therefore, it is not possible to determine their switch status at the time of the fire. They typically have current limit and overtemperature protection, and sometimes overvoltage protection. A small current or a voltage is applied to the gate of the switch, which causes it to allow current flow to a load in the appliance. This type of switch is often used where control of the switches occurs via computer or a special algorithm based on low voltage logic. A data sheet describing the operation of the switch can be obtained from the switch manufacturer.

Additional elements are sometimes added upstream of the switch. Examples include a surge protection device such as a metal oxide varistor or a series PTC (positive temperature coefficient) resistance to control overcurrent. Some equipment, such as motor drives, may use soft-start circuits to reduce inrush current to loads sensitive to current surges. Solid-state relays usually have added features such as zero-voltage or zero-current turn-offs to reduce further electrical noise caused by switching.

An advantage of electronic switching is that, depending on how the switch is used in the control circuit, contact arcing can be eliminated. Unfortunately, this makes discovery of abnormal operation and safety issues difficult, because the evidence of failure cannot be seen on an X-ray scan. In many cases, the manufacturer of the switch may be able to generate a failure analysis report for the device, which may give information about the failure mechanism of the switch.

Electronically controlled loads, such as motors, are controlled by semiconductor switches, as described previously. These devices require power-switching electronics called *drives*. Several different types of motor drives are available, which use different types and ratings for their electronic switches. A common control method used by all motors, however, is the controlled timing of electronic switches to deliver current to the motor windings in the proper sequence to produce the torque required to move the motor rotor. Motors with greater torque require greater current and those with faster speed require greater voltage, which means it is possible to calculate (approximately) the amount of power consumed by the motor and, in turn, delivered by the drive. Depending on the amount of power consumed during the motor's operation, if a fault occurs in the motor winding or in the electronic switch feeding it, the resulting energy discharge can be sufficient to ignite combustibles nearby, inside the motor, or on the drive circuit board. Various types of switches are used to control motors, though a description of them is beyond the scope of this text. However, this information is available in application manuals from various power semiconductor and motor drive manufacturers.

Other types of electronically controlled loads require timed electronic switches, which act in a similar manner to motor drives. These may be lighting, heating, or cooling devices or subassemblies within an appliance controlled by precise application of power to the load. A commonly seen example is a phase-fired Triac controlling a lighting or heating element. The exact control and operation of the Triac are beyond the scope of this text but can be explored further by examining circuit diagrams and data sheets.

Transformers

Transformers are devices that can increase or decrease an AC voltage. They are used to reduce line voltage, usually 120 V AC or 240 V AC, to a lower voltage for a specific appliance or device.

The winding insulation in the transformer may deteriorate after extended use at high ambient or internal temperatures. As this insulation deteriorates, the impedance drops and more current flows, which in turn generates more heat. This can lead to severe heating, which can cause the windings to fail by melting or can create a ground fault. The heat that is generated may ignite the winding insulation or combustibles in the vicinity of the transformer. Components of a transformer often survive a fire. Internal damage of the windings may be shown by a pattern of internal heating or arcing from turn to turn.

A thermal cutoff switch may be mounted on the windings to remove power when the maximum rated temperature is exceeded. If the transformer is encased in steel, it is unlikely that temperatures could become hot enough to ignite nearby combustibles. However, if there is no steel covering and if paper or plastic winding forms are part of the transformer, these can themselves ignite and cause flames to impinge nearby combustibles.

Motors

The power provided by motors ranges from ⅓ to 1 horsepower in major appliances to small, fractional horsepower levels in smaller appliances.

Small motors that drive cooling fans are generally not sources of ignition. Typically installed in bathrooms, small shaded pole motors are open-frame construction and are routinely exposed to moisture. Dust and lint tend to accumulate on these motors, which can cause overheating, resulting in a fire. These motors are required to have thermal cutoffs to prevent overheating.

AC (induction) motors in major appliances usually have starting windings with a centrifugal switch to disengage these windings after start-up. The starting windings contain a start capacitor that should be examined during the fire investigation. The capacitor may have failed or decreased in value with age, causing the motor to take longer or even stall during start-up, resulting in overload current in the windings. This can cause the windings to heat sufficiently to ignite the insulation and plastic materials often found around the motor. Protection for the motor often takes the form of a fuse link or a thermal cutoff switch that may be mounted on the windings. Motor bearings can over-heat and ignite nearby combustibles because of loss of lubrication or increase in friction.

In recent years, the permanent magnet motor has become a more popular choice for appliances. In these motors, the rotor windings are replaced with permanent magnets; in addition, the commutation brushes, which can be a source of arcing, are sometimes removed. This arcing is controlled and is low energy and capable only of igniting combustible gases, vapors, and dusts.

Stepper motors are used in small-movement de-vices, usually electronics such as printers, clocks, and copiers. Because these are typically low-voltage de-vices, they are not likely to produce the energy needed to ignite a fire.

Heating Elements

Heating elements can ignite combustibles that are in contact with the element itself. Appliances with heat-ing elements are designed to maintain a distance be-tween the element and surrounding combustibles. Sheathed elements are found in ovens and ranges. They are made of a high-temperature, high-resistance wire, usually nickel–chromium–iron, which is some-times called nichrome wire. This wire is surrounded by an electrical insulator and encased in a metal sheath, sometimes made of steel. Baseboard and space heaters have sheaths made of aluminum, which generally melt from external fire exposure. Open elements are com-posed of wires or ribbons constructed from nichrome. Some appliances use a fan to remove heat from the element and to distribute it into the room.

Heat tape, which is used to warm pipes near ap-pliances, consists of two specified lengths of a known high-resistance element that are electrically isolated except where they are connected at the end of the wires. If a poor connection occurs at the crimp con-nection point, nearby combustibles may be ignited by the generated heat. One element method uses steel wire, and another uses conductive plastic. The investi-gator should determine the required resistance of the heating element, and test it for resistance by using a volt-ohmmeter.

Due to the high amount of current required by electric heating appliances, resistance heating can occur if poor connections exist. The investigator can refer to the installation guidelines to see whether elec-tric power cords should be used with the particular heating appliance.

Lighting

Appliances often employ lighting to illuminate work ar-eas, dials, or internal cavities. This lighting is normally of low wattage and is not prone to ignite combustibles. The lighting types normally found are incandescent and fluorescent. Light-emitting diode (LED) lighting is becoming more popular with the advent of high-in-tensity LEDs, but their power level is still usually lower than that required for ignition. Some types of high-er-wattage incandescent lighting may ignite combusti-bles if they come in contact, such as if an incandescent droplight were taken into an attic and fell into cellulose insulation. Fluorescent lighting operates at a higher voltage, but the tubes normally do not get hot enough to ignite combustibles.

Recessed lights may also generate significant heat, and if the top of the fixture is covered by insulation, the insulation could ignite. Recent building codes specify the use of thermal cutoff switches to interrupt the circuit above a certain temperature. Fixtures with thermal cutoffs and with protective separation of hot bulbs from insulation are marked with an "IC" nota-tion, which means "insulation contact."

Any lights installed in bathrooms are susceptible to moisture and dust accumulating at the terminals. Over time, this can lead to poor connections and po-tential resistance heating.

Fluorescent Lighting Systems

Fluorescent lighting systems are commonly em-ployed in office settings. They use one or more glass tubes filled with a starting gas and low-pressure mer-cury gas. An electrical discharge is sent down the length of the glass tube, exciting the mercury gas. Two methods of starting fluorescent lights are used: the ballast discharge system, which is more common, and the preheated filament system. Older ballasts were magnetic and typically contained transformers. Electronic ballasts are more popular for newer light-ing systems, especially in compact fluorescent lamps (CFLs), and work on a switched transistor princi-ple. The ballast is used for creating the high start-up discharge voltage required to establish a starting arc across the lamp electrodes. This process is enabled by the starting gas, which results in the mercury vapor creating ultraviolet light that is converted to visible light by the coating on the inside of the tube, known as the phosphor or fluorescent powder. Once the current has started flowing and the lamp is op-erating, the ballast acts to limit the current passing through the tube.

There are two main types of fluorescent light bal-lasts: magnetic and electronic ballasts. The ballast in fluorescent lights can overheat because of internal short circuits and ignite combustible ceiling materi-als. Magnetic ballasts incorporate either a reactor or a transformer. Interior fluorescent light fixtures manufactured after 1968 are required to have thermal protection in the ballast.

A "P" notation, which is found on most metal housings of light fixtures, indicates that the ballast has thermal protection. All fixtures, for both indoor and outdoor use, manufactured after 1990 are required to have thermal protection in the ballast; however, thermal protection does not ensure that a failure cannot occur. Both electronic and magnetic fluorescent light ballasts contain pitch or potting compound within the ballast. This provides for better heat transfer, reduces noise, and holds the internal parts in place. This pitch can ooze out because of either internal heating or fire exposure. Unless it is already burning, the pitch will not usually ignite other materials.

The electronic ballasts in CFLs use semiconductor switches (transistor/MOSFET/Triac) and relatively small passive circuits (capacitor/inductor) switching at high frequency to create the electrical arc required for their operation. Some electronic ballasts employ self-resetting thermal protectors, whereas others use fuses for thermal protection. The ballast control ICs usually contain overcurrent protection, which must work correctly for the switching action to occur and for these ballasts to work correctly. For an unsafe condition to occur, the IC must malfunction (many ICs have protection against this possibility as well), and the semiconductor switches must short circuit to generate significant heat.

Failures of magnetic ballasts that frequently initiate fires include arc penetrations into combustible ceiling materials or nearby combustibles and extreme coil overheating that conducts heat into nearby combustibles. If a ballast failure is suspected as the cause of a fire, the ballast, along with the fixture and the wiring, should be preserved for examination by qualified experts. The fixture itself, along with the conductors feeding power to it, should be examined for failures, such as arcing nearby or inside the fixture and failure of the lamp holders.

High-Intensity Discharge Lighting Systems

High-intensity discharge (HID) lighting systems are often employed in warehouses, manufacturing facilities, and "big box" retail stores. They utilize a lamp that has a short tube filled with a metal vapor such as sodium, mercury vapor, or metal halide. An electrical discharge is created along the length of the tube, exciting the metal vapors in the tube, which in turn creates light. HID lights operate at higher pressures than common fluorescent lighting systems. These systems also employ a ballast and a capacitor for the starting voltage and for limiting the current flowing through the lamp. The ballasts may be electronic or magnetic, and some are protected by fuses or thermal protectors. The ballasts are typically mounted inside the fixture enclosure above the lamp holder and reflector assembly.

Most mercury and metal halide lamps use a cylinder composed of fused silica/quartz with electrodes at either end, called an **arc tube**. The arc tubes of metal halide lamps are designed to operate under high pressure, from 5 to 30 atmospheres, and at temperatures up to 2012°F (1100°C). They can unexpectedly rupture due to internal or external factors such as a ballast failure or misapplication or if finger oils are present. Mercury vapor arc tubes can operate in the range of 1112°F to 1472°F (600°C to 800°C) and at pressures from 3 to 5 atmospheres. High-pressure sodium lamps operate at lower pressures, typically close to 1 atmosphere. If the arc tube ruptures for any reason, the outer bulb may break. In such a case, pieces of extremely hot glass and quartz arc tube may be discharged into the surrounding environment, igniting combustibles where they land.

If an investigator suspects an HID lighting fixture as the cause of the fire, the whole fixture must be preserved for laboratory examination. The investigation of a suspect HID fixture requires the evaluation of the fixture's power supply, wiring, ballast, capacitors, lamp, and any lenses present. The fixtures should not be disassembled at the fire scene, but rather carefully preserved for laboratory examination. It is important to find as much of the lamp remains as possible, preserving all pieces of the arc tube and shielding them from additional damage. The edges of the glass jacket and of the arc tube remains should not be handled, disturbed, or cleaned prior to the laboratory examination. The lamp frame and lamp base pieces should also be collected and preserved.

Owing to the damage associated with building collapse, firefighting, and overhaul, the mere presence of a fractured arc tube is not definitive proof that the arc tube ruptured and ignited the fire. It can be helpful to gather additional exemplar lamps and fixtures from the scene for comparison and to create a usage history of the lamps in the structure.

Miscellaneous Components

Dimmers and speed controllers are examples of miscellaneous components found in appliances. Older appliances may contain rheostats or wire resistors. Components in newer appliances tend to be solid-state devices and are often destroyed in a fire unless the fire is of brief duration. The components that survive the fire may be displaced from printed circuit boards because most solders melt at temperatures as low as 400°F (204°C).

Timers can be built into the appliance or stand alone as separate devices; they can be driven by small motors. Often, timers are badly damaged in a fire. Failure of the timer usually results from gears wearing out or losing teeth, but is generally not the cause of the fire.

Thermocouples and thermistors measure temperature. If they are damaged or if their connections are contaminated, they could compromise the operation of the appliance.

A thermopile is a series of thermocouples. This type of device is used in gas appliances to keep a valve open when the pilot is burning. Newer appliances use electric igniters instead of standing pilots.

TABLE 14-3 lists some sensors commonly used in appliances. The sensors mentioned earlier in Table 14-2 are not included here.

Common Residential Appliances

The following sections discuss the components and operation of common appliances.

Ranges and Ovens

In the common household range or oven, heat is provided by electricity passing through resistance heating coils or by the burning of natural gas or propane **FIGURE 14-9**. The temperature in the oven is controlled by a thermostat and a valve or switch on the fuel or power supply. In a gas range, a burner fuel supply valve that is manually operated controls the fuel flow rate. Ignition is by a standing pilot flame or an electric igniter. In an electric range, the temperature control cycles the flow of electricity through the heating coil to maintain the selected temperature of the electric range. The timing device controls the ON and OFF periods of electric current; it is typically a knob, dial, or digital keypad that can be manually adjusted. Thus, selecting a higher setting will leave the current ON longer and OFF less.

Although stoves and ovens are typically made of noncombustible materials, any combustible material left on the stove or in the oven while the stove or oven is turned on can catch fire. Thus, the postfire settings of the controls are important to document.

Air Conditioners and Heat Recovery Ventilators

Air conditioners and heat recovery ventilators (sometimes called HRVs or ERVs) are often operated continuously. Over years of operating in a dusty environment, the dust or contamination in the airflow path can settle over electrical components such as capacitors and cause tracking faults, leading to a high-resistance fault that could ignite nearby combustible material. Compressor and fan motors in these units can overheat

TABLE 14-3 Sensors Commonly Used in Appliances			
Sensor Name	**Property Sensed**	**Sensor Type, Output**	**Typically Controls**
Thermocouple	Temperature	Dissimilar metal contact produces a millivolt signal	Heat-producing appliances
Thermistor	Temperature	Thermal-affected change in resistance	Heat-producing appliances
Pressure sensor	Differential or absolute pressure	Strain measurement, change in resistance	Appliances controlling water level or air flow
Flame sensor	Presence of flame	Infrared detector, microvolt signal	Gas valve, blower motor
Current sensor	Current	Current passes through series resistance causing a voltage drop, or magnetic measurement causes a change in resistance	Current-controlled heaters or cooking appliances, variable-speed motors
Proximity sensor	Presence or absence	Capacitance sensor, magnetic sensor, or optical sensor	Door sensors, touchpad, motor-speed sensor

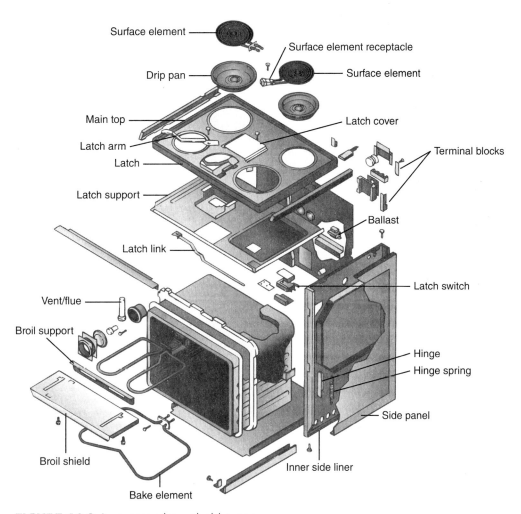

Surface element
Surface element receptacle
Surface element
Drip pan
Main top
Latch cover
Latch arm
Latch
Terminal blocks
Latch support
Ballast
Latch link
Latch switch
Vent/flue
Broil support
Hinge
Hinge spring
Side panel
Broil shield
Inner side liner
Bake element

FIGURE 14-9 A common household range.

because of bearing friction, increased load, or capacitor degradation, causing excessive temperature rise. If properly designed, the motor's thermal protector will open, and the motor will stop running before it reaches dangerously high temperatures; however, this may not prevent a fire if this condition is allowed to persist for a long time. Motor capacitors can develop internal resistance over time, which affects their performance and, if they are encased in plastic, can cause sufficient heating to self-ignite.

Water Heaters

One cause of fires from electric water heaters is moisture contacting the electrical controls. Leaks can occur in water heaters because of corrosion, particularly around the elements or the water inlets and outlets. Additionally, electrical circuits within the heater may come in contact with the polyurethane insulation in the heater jacket, allowing a fire to start and smolder, delaying detection. A lack of water also creates the potential for overheating and damage to

the control circuitry, which can malfunction and cause a fire.

Fires caused by gas water heaters are usually the result of combustibles being placed too close to the appliance. Blocking the flow of combustion air into the appliance may allow flame rollout, in which the burner flame extends outside of the combustion chamber. Evidence of such a failure will include patterns emanating from the combustion chamber. The heater may also be installed without adequate clearance from other combustibles, or the exhaust flue may be blocked.

Coffee Makers

Standard coffee maker components consist of a water reservoir, heating tube, carafe, and housing **FIGURE 14-10**. The heating tube heats the water flowing through it from the water reservoir and forces the water over the ground coffee. The coffee then drips into the carafe. In most models, the carafe sits on a warming plate that is heated by the same

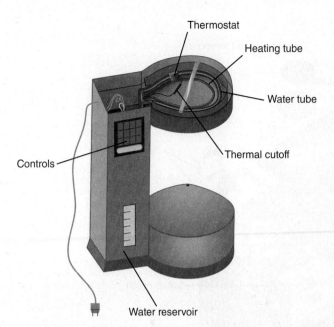

FIGURE 14-10 Coffee maker components.

resistance-heating element that heats the heating tube. The resistance heater is controlled by a thermostat that cycles the flow of electricity as needed. Two thermal cutoffs are usually installed adjacent to the heating element and serve as high-limit safety devices. Some coffee makers include automatic timers that turn off the coffee maker after a selected period of time and turn the appliance on at a predetermined time. If a coffee maker is suspected as the cause of the fire, the investigator will have to determine how three safety devices failed and allowed the fire event to occur.

Toasters

The common toaster uses resistance heaters in combination with a sensor that controls the toasting time. The sensor typically consists of a bimetal strip that can be adjusted for different levels of toast darkness. Sometimes, the bimetal strip may become blocked by a piece of toast, causing the toaster to stay on and overheat. For this reason, a separate heater is sometimes used to heat the bimetal strip and remove it from the area where food can affect it. When the toasting timer indicates completion, a mechanical latch partly releases the food and turns off the bimetal heater. When the bimetal switch opens, after cooling, the food is allowed to rise to the release point. Newer designs employ an electronic timer to set the toasting time and utilize an electromagnetic latch to release the tray and turn off the heating element.

Electric Can Openers

An electric can opener uses a geared electric motor and generally can operate only when the lever is depressed; this lever holds the cutting wheel in place and acts as the power switch to the motor. Because they operate only when the user presses the lever, these devices usually do not cause fires. Some can openers may have thermal protectors for motor overheat protection; however, as with other appliances, electrical arcing or high-resistance faults can occur due to poor connections or damaged cords.

Cooling Fans

The capacitors in box fans and oscillating fans have been known to overheat and cause the plastic casing near the capacitor to ignite. The ventilation available around the motor makes ignition from the motor difficult but not impossible. When installed in bathrooms or other damp environments, however, fan motors are exposed to moisture and can accumulate dust and lint, which can cause overheating, resulting in a fire. These motors are required to have thermal cutoff switches to prevent overheating.

Refrigerators

A household refrigerator contains the following components, among others **FIGURE 14-11**:

- Evaporator
- Condenser
- Compressor (motor driven)
- Heat exchange medium (fluorocarbon or Freon)
- Tubing to connect the components

Warm air inside older appliances evaporates the heat exchange medium (Freon), turning it to a vapor and transferring the heat to the Freon from the enclosure. The coolant vapor moves to the compressor, where it is compressed. As the coolant condenses to a liquid in the condensing coil, it gives off the heat picked up in the enclosure. The air around the evaporator is cooled, and the air around the condenser is heated. The cool air is circulated in the refrigerator, and the warm air is dissipated into the room. This cycle is controlled by a thermostat or a timing device. The compressor is typically powered by an electric motor equipped with a thermal cutoff switch contained in a sealed container that functions as a heat sink. Recalls have been issued for refrigerators that have experienced fires caused by electrical compressor relay failures and faulty wiring.

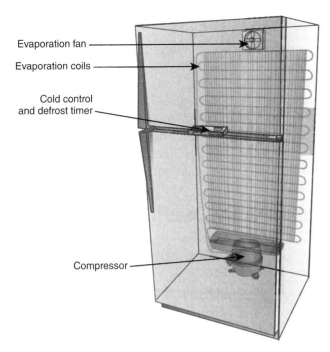

FIGURE 14-11 Refrigerator components.

Other systems that may be found in refrigerators include the following:

- Lighting
- Ice maker/water fill valve
- Ice and water dispenser/water fill valve
- A fan for the condenser and possibly a fan for the evaporator
- Heating coils for:
  - Automatic defrosters
  - Prevention of water condensation
  - Drain pans and ice makers

Freezers and refrigerators can cause a fire if their power cords or plugs are damaged. For this reason, it is important to ask the occupants if they recently moved these appliances, as such movement might possibly cause the appliance to roll over and rest on the power cord.

Some older refrigerators have a plastic tray with elements to evaporate the defrost water. If the drain valve becomes blocked, this tray can overheat, crack, and expose the elements. When the blockage clears, water fills the tray, and a short circuit and fire can occur because of the exposed elements.

Electric Water Dispensers

Failure of the thermal cutoff device or thermostat on water dispensers is a possible cause of fire with these appliances. As the reservoir empties, water can reach high temperatures and evaporate. If there is a leak, water may reach the electrical components located underneath the water reservoir, possibly causing a fire.

Deep Fat Fryers

Deep fat fryers are typically found in restaurants and fast-food outlets, though smaller models can be found in the home. The thermostat on this appliance controls the temperature of the cooking oil; however, it could fail (most commonly when the capillary line connecting the temperature probe in the vat to the thermostat breaks), allowing the oil to overheat and ignite. When this happens, a charred or blackened residue is often found in the cooking vat.

Thermostat contacts have also been known to weld closed with these appliances.

Dishwashers

The dishwasher is a household appliance that employs both a pump and a motor. The pump motor may or may not have thermal protection. An electric resistance heater may be used to dry the dishes after the water has been drained, and may be used to heat the water, subjecting it to submersion.

A major cause of dishwasher fires is moisture contacting the conductors, especially at the top of the door where the controls are located. The plastic components that release the detergent could weaken over time, becoming brittle, cracking, and then leaking. When the door is opened, the water inside the door can reach the controller at the top of the door and cause a fire.

Other potential causes of dishwasher fires are as follows:

- Repeated opening and closing of the door over time can stress the wiring harness. If the insulation breaks down, a short circuit can result.
- A combustible material contacting the electrical elements could cause a fire.
- Some control modules in dishwashers contain relays that have failed and ignited the plastic housing.

Microwave Ovens

A microwave oven uses a magnetron to generate radio-wave (i.e., microwave) energy that heats the food. Microwaves are usually equipped with a thermal cutoff switch. The appliance is typically equipped with internal lighting and a food rotation tray, and it may have thermal cutoff switches above the enclosure. Radiation that is emitted will excite water molecules; if there is no moisture available, other, less volatile molecules will absorb the radiation and start to heat. Therefore, any material that has dried out can overheat and ignite in a microwave oven. Also, any metal objects inside a microwave oven can cause severe arcing, which could lead to a fire.

Space Heaters

According to the *Ignition Handbook*, there are two types of space heaters: convective and radiant. A **convective heater** relies on natural or forced ventilation to transmit heat throughout a room. An example of a convective heater is the oil-filled portable heater. A **radiant heater** has a heating element that glows. These appliances often are equipped with tip-over switches or other protective devices.

Although heaters are often blamed for fires, these fire events may have more to do with the large electrical currents involved or with misuse of the heater rather than with the basic safety of the appliance. Fires may be caused by overloading of an extension cord that is not rated for the current drawn by the heater. Combustibles may be placed too close to the heater or safety devices bypassed to ensure continued operation in cold weather. When designed and used properly, electrical heaters are a safe form of heating. A failure of most modern space heaters would require the failure of multiple safety devices, which must be taken into consideration.

The following conditions are known to cause fires in portable electric space heaters:

- Combustible material coming into contact with the heating element(s), or combustible material placed inappropriately close to a heating element and igniting from radiant or convected heat
- Degradation of wiring, leading to the potential for a high-resistance connection
- The heating element detaching because of rough handling and later igniting the plastic casing
- Restriction of the inlet air, but not the overtemperature sensor, causing the heating element to overheat and ignite the plastic casing
- Failure or blockage of the fan, thereby stopping the removal of heat from the heater
- Accumulation of dust, lint, or hair in internal spaces of heaters that have air flow
- The heater falling over and igniting nearby combustible items (such as items on the floor)

Electric Blankets

An electric blanket employs an electric heating element inside the blanket. The controls are typically located on the blanket's power cord. Thermal cutoffs are located near (and usually sewn onto) the elements in the blanket—sometimes as many as 12 to 15 thermal cutoffs. The blanket is designed to be laid flat and used on top of another blanket. Covering or folding the electric blanket could lead to a failure.

Window Air-Conditioner Units

Window air-conditioner units are designed to be placed in a window to cool a room. They function and operate similarly to refrigerators and use similar components and principles for cooling. The air from the room is circulated through the unit, past the evaporator (which cools the air), and the cool air is then discharged into the room by a fan. A thermal cutoff prevents overheating of the fan. These appliances may be powered by 120 or 240 V AC. Fan or compressor motor overheating can cause a fire.

TIP

It is important to interview the property owner about how the appliance was used or possibly misused. This includes inquiring about how the appliance was used, whether there were any repairs or problems with the appliance, whether it was energized at the time of the fire, and how it was controlled.

Hair Dryers and Hair Curlers

Hair dryers employ a high-speed fan to force air across a heating element and thereby heat air. They usually have a thermostat to control the temperature and a thermal cutoff in the event of an overheating event.

Hair curling irons use an electric resistance heater inside a wand that is equipped with a thermal cutoff switch. Some models allow water to be added to generate steam.

Clothes Irons

Clothes irons use an electric resistance heater that can heat in both the vertical and horizontal positions. They also include one or more thermostats and thermal cutoff switches. The condition of these switches should be checked if the iron is suspected of causing a fire. Water can be added to many models for generating steam. Controls on these appliances include temperature selection and an on/off switch.

Clothes Dryers

Clothes dryers work by circulating heated air through a rotating drum that contains the wet clothing. The air is discharged from the drum through a filter (lint trap) and into a duct that is typically discharged to the exterior of the house. The dryer uses either an electric heating element or the combustion of fuel gas to heat the air. The components in a clothes dryer include timing controls, humidity sensors, heat source selectors, intensity selectors, thermal cutoffs, a blower motor, and heating elements. **FIGURE 14-12** shows the internal components of a clothes dryer.

A clogged lint filter can blow lint back into the interior of the dryer, which can then settle across the base, eventually reaching the element and igniting. Frictional heating from a piece of clothing caught between the moving parts may also cause a fire, and the heating of contents such as vegetable oil–soaked rags may cause them to spontaneously combust. Check the contents of the dryer after a fire has occurred during the fire investigation.

The clothes dryer design includes a flexible bearing, usually made of a composite of polyvinyl chloride (PVC) and other synthetics or cotton, on which the drying drum rests. If this bearing moves out of position, it can contact the element and ignite. Because of PVC's fire-retardant properties, these bearings can smolder for very long periods undetected before igniting the clothes inside the drum. A fire from a clothes dryer can therefore occur hours after the dryer has been disconnected from power.

Consumer Electronics

Consumer electronics are very common household appliances. Examples include laptop and desktop computers, cell phones, tablets, gaming systems, radios, music systems, Blu-ray or DVD players, and video cameras. All of these appliances have a power supply, circuit boards, and a housing.

Some electronics with circuit boards (including television sets, stereo components, and computers) have experienced recalls in the past due to poor solder joints and faults in the transformers. When components in a high-energy circuit (such as a power supply) are poorly soldered, the connections may loosen with time and use, resulting in carbon

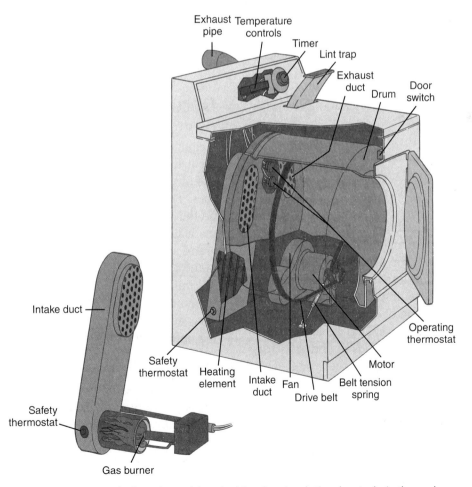

FIGURE 14-12 A clothes dryer dries clothing by circulating heated air through a rotating drum.

tracking on the board. Over time, resistive heating can cause a fire.

Solid-state switches are used to drive speakers, displays, and motors. If these devices come loose from the tight bonding to their heat sinks, they can short circuit and explode. The resulting sparks can ignite nearby combustible material.

Portable equipment, such as laptop computers, may use lithium batteries that can fail if overheated or overcharged.

Entertainment system/television enclosures are manufactured using fire-rated plastics that do not easily ignite. However, because so many different plastics are used in the manufacturing process and their fire ratings vary, it is still possible for electrical faults to cause a television set to catch fire.

Older fax machines equipped with thermal print heads can generate sufficient heat to start a fire if covered with combustible material.

Recording the Fire Scene When an Appliance Is Involved

The material in this chapter expands on the information in Chapter 16, *Documentation of the Investigation*, and Chapter 26, *Appliances*, of NFPA 921, *Guide for Fire and Explosion Investigations*. The documentation methods outlined in those chapters should be used to record details and findings in investigations of fires involving appliances.

Appliance evidence is vulnerable to damage or alteration due to many factors, including the activities of fire crews, restoration crews, and occupants; adverse weather conditions; and general deterioration of fire-damaged materials. Investigators locating an appliance in a fire scene should not disturb it until it has been thoroughly documented. When it becomes necessary to alter the scene or to move evidence, an investigator should consult NFPA 921 for guidance on handling of evidence. Improper handling can result in claims of spoliation of evidence. To avoid such claims, all fire investigators should be familiar with and practice the required methods of notifying other parties and documentation prior to scene or evidence alteration. See Chapter 5, *Legal Considerations for Fire Investigators*, for more information on legal concerns when moving or altering evidence.

After an origin area has been identified and one or more appliances have been found in the origin area, the appliance(s) should be documented through photographs, diagrams, and measurements. Ascertaining the positions or settings of any controls, securing the appliance nameplate data, and gathering all component parts of the appliance must also be accomplished as part of the documentation of the scene.

Photographs

An investigator should photograph the entire scene, taking care to photograph the appliance from many different angles, including any electrical and fuel lines supplying the appliance. Record the entire area, including the appliance, to establish its location in relation to combustibles and landmarks in the room. Take close-up photographs to provide detailed information such as positions of controls and thermal protection devices. The appliance should not be moved until all other fire scene documentation and appliance documentation have been completed. Witness information can often be helpful to establish the appliance's position and proximity to combustibles as well as the power state and any recent problems with operation of the appliance.

Diagrams and Measurements

Diagram the fire scene and locate the appliance on the diagram. Make sure to include measurements locating the appliance in relation to fixed landmarks, combustible fuels, power sources, and other elements in the fire scene.

Documenting the Appliance

On a sketch or photograph of the appliance, with descriptive annotations if warranted for clarity, document the information outlined in **TABLE 14-4** to the extent possible, albeit without creating spoliation issues. This information may require product research on the appliance. Some appliance manufacturers have manuals or online databases that include helpful diagrams and schematics of their appliances. An exemplar of the component(s) in question will provide guidance by similarity, enabling the investigator to understand the function or placement of components in the appliance, and will often include descriptive design or layout information. Because this exemplar-comparison analysis usually does not occur at the scene, you should still document the information in Table 14-4 at the site before moving and transporting the appliance for storage or laboratory inspection.

The investigator should also obtain the following identifying information from the appliance:

- Manufacturer
- Model number
- Serial number

TABLE 14-4 Appliance Documentation

Item	Description	Examples
Primary power source type	Describe the type of primary energy used in the appliance.	Electric, gas (natural or propane), fuel oil, solar, or wind
Power source, energy storage	Describe the service origin or storage or generation site of the power.	Utility electrical power supply, battery, propane storage tank, natural gas utility, generator with uninterruptable power supply
Energy source adapter	If one is present, describe the adaptation method (converts the energy source to a usable form).	Transformer, power supply, pressure regulator
Power/fuel source feed and connections	Describe feed wire and interface connections to the power source.	Electrical wire distribution = 14/2 AWG solid wire with standard duplex outlets; extension cord = 12 ft 12/2 AWG stranded wire with 4-tap power strip; natural gas 3/8-inch ID quick connect male plug with 3/8-inch ID copper pipe
External overload protection	Describe the method of external protection from energy overload.	Ground-fault circuit interrupter (GFCI) protection, circuit protection (breakers or fuses), regulator
Bonding and grounding	Describe how the appliance is grounded.	Two or three wire with chassis connected ground, two wire insulated, line–line with earth ground, transformer isolated
Power for controls for the appliance	Describe the energy source for the controls used.	Electrical controls—110 V AC, mechanical switches, pneumatic switches, hydraulic servo valve
Internal circuit protection and disconnects	Describe mechanisms used to protect internal functions of the appliance and describe their "as found" state.	GFCI, surge protectors, in-line fuses/circuit breakers, disconnect switches, thermal cutout switches
Operational controls	Describe the appliance control components and their settings as found.	Dials, switches, power settings, thermostat settings
Feedback devices and sensors	Describe the control feedback condition as found.	Temperature sensors, pressure/flow sensors, level sensors, position feedback
Movable parts	Describe the condition and position of important moving parts.	Doors, vents, valves
Cleaning/cooling/heating components	Describe the condition of filters and cooling/warming portions of the appliance.	Filters, cooling fans/pumps, heaters
Clocks and timers	Describe the state of timing mechanisms (if available) in the appliance.	Clock hand position, timer position

- Date of manufacture
- Name of product
- Warnings and caution notes
- Recommendations and ratings
- Additional data located on the appliance, such as installation diagrams

This information is sometimes found on labels or plates located behind service panels or hidden behind or under the appliance. Often the owner will keep all appliance manuals together.

TIP

It is essential to document the status and condition of the key control and protection components listed in Table 14-4, which may be found on or in the vicinity of the appliance, before moving it. Although moving a control knob or switch is highly discouraged, if one needs to be moved, mark the position first so it can be returned to its original position.

Recovery and Reconstruction of Appliance Components

If possible, gather together the various components that may have been moved during the fire or firefighting operations, and reconstruct the appliance in its prefire location. Make every effort to recover the property in its original state (as found), with the least damage possible. If necessary, plan for a truck or trailer to transport the appliance to a secure location. The transportation plan should also include the items necessary to secure and protect the device, such as covers, padding, bags, chains, straps, or other securing devices. Take care to keep the appliance upright if possible and to avoid wind or vibration disturbance of sensitive portions of the appliance as the vehicle is driven.

Avoid testing or operating the appliance at the fire scene, as this could further damage the appliance, and because it may be unsafe to do. At this point, all testing must be nondestructive, such as testing for electrical continuity or resistance using a volt-ohmmeter **FIGURE 14-13**. X-rays may be useful to determine continuity of fuses, thermal protectors, heating elements, and relay contacts. Transformer and motor windings can sometimes be examined with X-rays.

FIGURE 14-13 This ohmmeter is being used to check the electrical continuity of a coffee maker.
Courtesy of Ray Franco.

After-Action REVIEW

IN SUMMARY

- Fire cause determination generally follows origin determination, and a determination of the cause generally can be considered reliable only if the origin has been correctly determined.
- The cause of the fire is determined using a renewed application of the scientific method.
- Usually, the source of the ignition energy will be found at or near the point of origin.
- Any identified ignition source must be "competent"; that is, it must have sufficient energy to ignite the first fuel and be capable of transferring that energy to the fuel long enough to bring the fuel to its ignition temperature.
- The ignition source for ignitable vapors may be more difficult to identify because of the ability of the vapor cloud to migrate throughout multiple compartments.
- To be a viable part of an ignition scenario, a first fuel must be in a configuration that permits it to be ignited by the ignition source.
- In most fires, the oxygen in the air serves as the oxidizing agent.

- The investigator should be prepared to describe the ignition sequence, and the way in which the first fuel ignited was able to ignite subsequent fuels, causing the fire to spread.
- All potential ignition sources in the area of origin must be identified.
- The investigator should collect as much information as possible about fuels and their prefire locations, including furniture, wall coverings, window treatments, plastic and paper items, and other possessions.
- The investigator should avoid jumping to a conclusion about a fire cause just because a fuel and an ignition source happen to coincide with the area of origin.
- Sometimes, physical evidence of the ignition source may not survive the fire, but the ignition sequence can still be logically inferred. This information must be established with evidence, not by the absence of it.
- Hypothesis testing for origin determination may make reference to scientific and trade literature, physical tests or experiments, and cognitive testing.
- "Negative corpus" reasoning is inconsistent with the use of the scientific method—the method that fire investigators are required to apply, as stated in NFPA 1033.
- Selecting a final hypothesis may not always be possible if insufficient data are available or if expert examination is needed.
- The classification of causes has been removed from NFPA 921 and is no longer supported by this document, although certain agencies may still engage in classification of causes for statistical or other purposes.
- The presence of electrical wiring or components in fire damage, or even at the origin of a fire, does not by itself indicate that an electrical event was the cause.
- Electrical components can be, and often are, the victims of a fire that started from non-electrical causes.
- Before analyzing an electrical circuit or equipment, the investigator should evaluate the safety of the installation and confirm that all components have been de-energized.
- The investigator should not move knobs, switches, or breakers, and should refrain from disassembling electrical parts or equipment until it is appropriate to do so and appropriate personnel are involved.
- When evidence of an arc is found in an electrical circuit, without any further data the investigator can conclude only that the circuit was energized at the time the arc occurred.
- In an overloaded conductor, the fire investigator may see damage along the wire, including sleeving and offsets.
- An overload can happen when the current in a conductor exceeds its ampacity but is less than the current rating for the circuit breaker or other protection.
- When resistance heating occurs at an electrical junction, a glowing connection can result, which then ignites nearby combustibles.
- As with electrical equipment, the mere presence of an appliance in the fire's area of origin does not establish that it had any contribution to the fire's cause.
- The investigator should consider an appliance at the origin to be equally plausible as, but not more likely than, any other potential ignition sources.
- With any appliance under examination, the investigator should verify whether the appliance was connected to an electrical power supply and whether it was energized at the time of the fire.
- It is important for the investigator to understand how an appliance or device operates during its normal operation, as this information may help explain how a failure might have occurred.
- Usually, disassembly of an appliance should not occur in the field, as it can result in claims of spoliation of evidence.
- An exemplar appliance can be used to help understand how an appliance operates and can be used in testing a proposed ignition scenario.
- Overcurrent protection devices such as fuses and circuit breakers may be found in appliances.
- Investigators who locate an appliance in a fire scene should not disturb it until it has been thoroughly documented.
- Although moving a control knob or switch is highly discouraged, if one needs to be moved, mark the position first so it can be returned to its original position.

KEY TERMS

Arc tube A cylinder of fused silica/quartz with electrodes at either end that is used in most mercury and metal halide lamps.

Cause The circumstances, conditions, or agencies that bring together a fuel, the ignition source, and the oxidizer, resulting in a fire or explosion. (NFPA 921 and NFPA 1033)

Classification of cause An effort to categorize the fire cause for purposes of assigning responsibility, reporting purposes, or statistics.

Convective heater A heater that relies on either natural or forced ventilation to move air across a hot surface, heating the air and dispersing it throughout the room.

Determination of cause The process of identifying the factors that allowed a fire to occur, including the type and form of the first fuel, the competent ignition source, and the circumstances and human actions that brought the ignition and the fuel together.

Exemplar An exact duplicate of an appliance.

Fuel analysis Identification of the first fuel item or package that sustained combustion beyond the ignition source as well as subsequent target fuels beyond the initial fuel.

Housing The outer shell of an appliance.

Ignition sequence The sequence of events and circumstances that allowed the initial fuel and the ignition source to come together and result in a fire.

Ignition source analysis Process through which all potential ignition sources in the area of origin are identified and then considered in light of the physical properties of the first fuel ignited, fundamental scientific principles, and other available data.

Polarized plug A plug made after 1987 for 20 A or less, which has a neutral prong that is wider than the hot prong.

Radiant heater A portable electric heater that utilizes a glowing heating element.

Step-down transformer Device that reduces the 120 V provided at the receptacle to the required voltage.

Switch Device used to turn an appliance on or off or to change its operating conditions.

Transformers Devices that can increase or decrease an AC voltage.

REFERENCES

Babrauskas, V. *Ignition Handbook: Principles and Applications to Fire Safety Engineering, Fire Investigation, Risk Management, and Forensic Science.* Issaquah, WA: Fire Science Publishers/ Society of Fire Protection, 2003.

National Fire Protection Association. 2020. *NFPA 921: Guide for Fire and Explosion Investigations,* 2021 ed. Quincy, MA: National Fire Protection Association.

National Fire Protection Association. 2021. *NFPA 1033: Standard for Professional Qualifications for Fire Investigator,* 2022 ed. Quincy, MA: National Fire Protection Association.

On Scene

1. Why does NFPA 921 strongly disapprove of the use of "negative corpus" reasoning in the analysis of fire cause? What could be the result if an investigator incorrectly draws a conclusion when there are no data to support it?

2. Why does NFPA 921 instruct the investigator to determine the origin of a fire before the cause? If an occupant who did not witness a fire is nevertheless certain what caused the fire, should you still follow all of the steps recommended by NFPA 921?

3. If an investigator discovers electrical wiring and equipment at a point of origin, what steps can the investigator take to analyze whether the electrical equipment and wiring were involved in the fire's cause?

4. In what ways could an exemplar of an appliance assist an investigator in analyzing fire cause, when the original appliance is not available for testing or examination?

Chapter Opener: © Jones & Bartlett Learning. Photographed by Glen E. Ellman; On Scene siren: © Bildgigant/Shutterstock.

CHAPTER 15

Analyzing the Incident

KNOWLEDGE OBJECTIVES

After studying this chapter, you should be able to:

- Identify analytical tools used in fire investigation. (**NFPA 1033: 4.6.2**; **4.6.5**, pp. 360–368)
- Explain the use of timelines in fire investigation. (**NFPA 1033: 4.4.3**; **4.6.2**, pp. 360–362)
- Explain the purpose and implementation of system analysis in fire investigation. (**NFPA 1033: 4.4.3**; **4.6.2**, pp. 362–364)
- Identify the components of mathematical and engineering modeling, and explain how they can be used to investigate fire incidents. (**NFPA 1033: 4.4.3**; **4.6.2**, pp. 364–365)
- Explain the use of graphic representations, including computer modeling, in fire investigation. (**NFPA 1033: 4.4.3**; **4.6.2**, pp. 364–365)
- Explain the process of fire testing and its importance in fire investigation. (**NFPA 1033: 4.4.3**; **4.6.2**, p. 368)
- Identify and describe incendiary fire indicators. (**NFPA 1033: 4.2.3**; **4.2.6**, pp. 370–380)
- Describe the assessment of fire growth and damage in incendiary fires. (**NFPA 1033: 4.2.3**; **4.2.6**, pp. 374–375)
- Identify and describe potential incendiary fire indicators not directly related to combustion. (**NFPA 1033: 4.2.3**, pp. 375–377)
- Describe factors to consider after a determination is made that a fire was incendiary. (**NFPA 1033: 4.6.4**, pp. 378–380)
- Recognize the motive(s) for a fire based on the known list of motives for fire-setters. (**NFPA 1033: 4.6.4**, pp. 378–380)
- Describe the general characteristics of human response at a fire incident. (**NFPA 1033: 4.6.4**, pp. 380–381)
- Describe how specific features of a fire and fire environment can impact human behavior. (**NFPA 1033: 4.6.4**, pp. 380–381)
- List ways that human action or inaction can impact fire occurrences. (**NFPA 1033: 4.6.4**, pp. 382–385)
- Describe the three recognized age categories of youth fire-setters, and summarize the reasons that youths are drawn to setting fires. (**NFPA 1033: 4.6.4**, pp. 384–385)
- Describe the ways in which an occupant may react once a fire threat is identified. (**NFPA 1033: 4.6.4**, p. 385)

SKILLS OBJECTIVES

After studying this chapter, you should be able to:

- Reference standards, guidelines, and regulations dealing with safety warnings and product design.
- Use analytical tools as part of a fire investigation. (**NFPA 1033: 4.6.5**, pp. 360–368)
- Use models to analyze fire incident data. (**NFPA 1033: 4.4.3**; **4.6.2**, pp. 364–365)
- Examine fuel loads and their composition and location to understand fire growth.
- Recognize the various motives of fire-setters. (**NFPA 1033: 4.6.4**, pp. 378–380)
- Examine fire protection systems to determine whether they have been altered or sabotaged. (**NFPA 1033: 4.6.4**, pp. 375–378)
- Analyze fire-related human behavior. (**NFPA 1033: 4.6.4**, pp. 382–385)
- Integrate knowledge of human behavior into the total fire investigation. (**NFPA 1033: 4.6.4**, pp. 382–385)
- Gather evidence of motive and opportunity related to an incendiary fire. (**NFPA 1033: 4.6.4**, pp. 378–380)

You are the Fire Investigator

You are called to investigate a fire that has damaged a furniture refinishing business. As you go through the debris and evidence in the area of origin within the building, you find a plastic can that has mostly been consumed by fire. However, some of its contents are still present; they consist of cotton rags showing evidence of spontaneous ignition.

Employees of the facility inform you that the business has strict rules about the use and disposal of cotton rags that are used to apply linseed oil to furniture. They state that they always put used cotton rags outside of the building or in a special drying area where they are not piled up. Your observations appear to contradict the employees' statements about their handling of the rags.

1. What potential events may have led to the cause of the fire in the business?

2. What role may the company's employees have played in the cause of the fire?

3. How might you establish the cause and determine potential responsibility for these fires?

Access Navigate for more practice activities.

Introduction

The overall methodology used to determine the origin and cause of a fire is the scientific method. As part of the scientific method, the fire investigator can use a number of techniques to assist in the analysis of the cause of fire, its development, and other factors that may have contributed to its spread or its severity. The following sections describe some of the techniques that investigators use during the analytic phase of the fire incident investigation.

Analytical Methods

During the investigation of a fire or explosion event, investigators will often be faced with a multitude of facts and data. To formulate hypotheses about the event, they will need to be able to organize and analyze this wealth of information. Several analytical tools are available to assist in interpreting the information obtained, including the following:

- Timelines
- System analysis
- Mathematical/engineering modeling
- Graphic representations
- Scene reconstruction
- Fire testing

Timelines

A **timeline** is a graphic or narrative representation of events related to the fire incident, arranged in chronological order. Understanding the timeline of any incident is fundamental to performing failure analysis, because it helps in determining the sequence of events that occurred. A timeline is a comprehensive listing of specific relevant events that can be verified to a stated degree.

Timelines may include events that occurred before, during, and after the fire. Estimates of fire size or conditions are often valuable in developing these representations. By applying known information about fire dynamics, fire conditions can be related to specific events. Information that is gathered from detection and suppression systems may also be useful in determining the spread of the fire as well as establishing the times at which specific events occurred. In all cases, the value of the timeline is directly related to the accuracy of the information it includes, and the reliability of a timeline is a function of the level of confidence that can be placed in its elements. Not all information can be traced to a specific time; some may have to be listed as a time interval during which the event occurred.

A variety of components are used to develop timelines. These include incidents described as either hard times or soft times.

Hard Times

When developing a timeline, incidents that can be related to a known exact time are generally referred to as **hard times TABLE 15-1**. This means that the time of occurrence is specifically known. For example, a fire department's incident history records can provide the specific times when units were dispatched, arrived on scene, and so forth. Such data serve as benchmarks in developing the timeline.

TABLE 15-1 Hard Time Sources

Dispatch telephone or radio logs
Emergency medical service reports
Alarm system records (on-site, central station, fire dispatch, etc.)
Inspection reports (building, health, fire)
Utility company records (maintenance, emergency, and repair records; cell phone records; power company monitoring data)
Private videos and photos; video from surveillance cameras
Media coverage (newspaper photographer, radio, television, magazines)
Timers (clocks, time clocks, security timers, water softeners, lawn sprinkler systems)
Weather reports (weather service, airports, and lightning tracking services)
Maintenance records (current and/or prior owner/tenant)
Interviews establishing observations and activities of witnesses
Computer-based fire department alarms, communication audio tapes, and transcripts
Building or systems installation permits

When collecting information on hard times, the investigator must verify whether the time recorded by the source, or source device, is accurate. For example, if an investigator finds that a surveillance camera has the incorrect time at the point when the investigator is obtaining times and recordings from it, likely the time shown on the recording will also be inaccurate. The investigator may be able to use the current time discrepancy to make conclusions about the recorded time. If the surveillance camera displays the time of 4:30 PM at 3:30 PM, for example, the investigator may be able to conclude that a recent video that is time-stamped for 10:00 AM is displaying activities that actually occurred at 9:00 AM.

Soft Times

Other times can be considered soft times. **Soft time** is either estimated or relative and is generally provided by witnesses. For example, a witness may not be able to state exactly when a certain event such as a flashover occurred or when an occupant was told to leave a building, but may be able to relate such observations to another event, such as the arrival time of the fire department or the moment when a bus or train went by, a television program was on, or another event occurred. Through interviews and further gathering of data, you may be able to narrow down the range for such soft times to relative time periods. **Relative time** can be subjective in nature and varies with the witness

providing the information. Witnesses should refer to their actions and observations in relation to each other and in relation to other events, and they should be as specific as possible.

Estimated time is an approximation based on information or calculations that may or may not be relative to other events or activities.

Benchmark Events

Some events are particularly valuable as a foundation for the timeline or may have significant relation to the cause, spread, detection, or extinguishment of a fire. These events are referred to as **benchmark events**.

For example, Mr. Jones ends his workday and punches out of work at 6:00 PM. He begins his 25-mile drive back home in heavy traffic. As he turns onto a side street, he passes a furniture store and notes that the store is closed and appears quiet. Mr. Jones continues his travel and passes the fire station, seeing fire trucks responding to a fire call at the furniture store. The trucks radioed in their departure for the fire at 6:32 PM. Using the benchmark events established with the work time clock and the fire department dispatch records, an estimated relative time may be able to be established for when Mr. Jones passed the store.

Multiple Timelines

A variety of timelines may be required to effectively evaluate and document a sequence of events precipitating the fire, events during the actual fire incident, and postfire activities. Depending on complexity, events may be evaluated in a macro or micro timeline.

A macro evaluation of events can cover months or even years and may incorporate activities that occurred long before the fire, such as those related to building construction, modification of codes, or code enforcement activities. Conversely, a micro evaluation looks at small or narrow segments of the macro timeline in detail. Examples of micro incidents could include travel times for a witness to walk between rooms or the dispatch and on-scene times.

Parallel timelines can be used to look at multiple events that occur simultaneously. Graphic timelines can be very helpful in putting together the sequence of events that transpired, and approaches involving matrices may be helpful when multiple events occur simultaneously **TABLE 15-2**.

The actual creation of the timeline may be accomplished by applying any of several methods. A simple paper-and-pencil timeline may be all that is needed. Alternatively, a variety of computer software programs can be used to construct timelines and investigative charts. Some larger law enforcement agencies

TABLE 15-2 Example of an Abbreviated Matrix Used for Developing a Timeline

Time	Engine 1	Engine 2	Ladder 1	Battalion 1
0604	Dispatched	Dispatched	Dispatched	Dispatched
0605	Responding	Responding	Responding	Responding
0608	On scene, establishing command			
0610	Making entry with hose line			
0611		On scene	On scene	
0613	Attacking fire in northwest bedroom	Advancing to second floor with hose line for primary search and rescue	Extending ground ladder to roof for ventilation	
0614				On scene, assuming command
0615	Fire knocked down in northwest bedroom	Victim located in hallway, second floor	Accessing roof	Requesting ambulance
0616	Conducting primary search, first floor	Victim removal	Making ventilation opening	

The investigator may want to add columns for the police department or for non–fire service witness accounts.

may have an intelligence section whose personnel will assist with the gathering of the information and then construct reference charts and reports, such as timelines.

Scaled Timeline

A **scaled timeline** is useful in displaying the time and event and its chronological relationship to other events. Such a timeline will show the time of each event with the spacing between the time events being scaled in a manner that indicates the elapsed time between each event **FIGURE 15-1**. The scaled timeline may also list hard times above the line and soft times below the line.

System Analysis

System analysis is an analytical approach that takes into account the characteristics, behavior, and performance of a variety of elements—including human activities and mechanical features of equipment—and integrates them to provide as complete a picture of events surrounding an incident as possible. Simple

system analyses include evaluations of incidents that investigators make every day. An informal system analysis can be used, for example, with a properly functioning stove left on by a homeowner, which leads to a kitchen fire: One aspect of such an incident involves human factors, and the second aspect involves the properties of the stove and its surroundings. Together, these make up the system whose properties led to the incident.

A variety of tools can be used in analyzing such incidents. Some of these tools (such as fault trees and failure mode and effects analysis) are discussed in the following subsections, and they provide a systematic method for analyzing systems to determine hazards or faults.

Fault Trees

Fault trees (also known as decision trees) illustrate the series of events and decisions that must take place for a specific outcome to occur. When this type of graphic logic or reasoning is applied to a situation, the solution may become more readily apparent, and incorrect

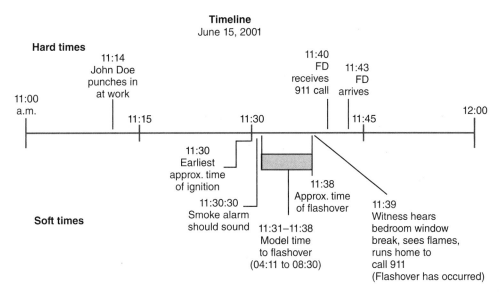

FIGURE 15-1 A scaled timeline.

Reproduced from NFPA 921-2021, *Guide for Fire and Explosion Investigations*, Copyright © 2020, National Fire Protection Association. This reprinted material is not the complete and official position of the NFPA on the referenced subject, which is represented only by the standard in its entirety.

solutions may be eliminated from consideration. A fault tree diagram places, in logical sequence and position, the conditions and chains of events that are necessary for a given fire or explosion to occur.

Similar decision trees are used in electronics or computer programming, with "and" and "or" gates being placed at the points where decisions must be made **FIGURE 15-2**. For an "and" gate, all events or conditions must be present. For an "or" gate, any one of several events or conditions must be present.

Fault trees identify the conditions and chain of events involved in a fire or explosion. If the conditions or events did not happen in an order that would lead to the event, then the proposed scenario is not possible. The investigator can apply probabilities to the events to identify their likelihood. In some cases, the end result may include multiple plausible scenarios for the fire scene.

Software for developing fault trees is readily available. The data and conditions to be used in fault tree analysis or failure mode and effects analysis (discussed in the next section) can be obtained from the following places:

- Operations and maintenance manuals
- Maintenance records
- Parts replacement and repair records
- Design documents
- Services of an expert with knowledge of the system
- Examination and testing of exemplar equipment or materials
- Component reliability databases
- Building plans and specifications

- Fire department reports
- Incident scene documentation
- Witness statements
- Medical records of victims
- Human behavior information

Failure Mode and Effects Analysis

Failure mode and effects analysis (FMEA) is another graphic method used to determine the causes and effects involved with an event or a subevent leading to a fire. By identifying specific components associated with potential ignition sources or fire spread, it may be possible to identify specific predecessor events that occurred or activities that preceded an incident.

In FMEA, the investigator fills in a table with various column headings to address the needs of the particular investigation. Although the column headings are flexible, each FMEA contains at least the following items:

- The item or action being analyzed
- The basic fault (failure) or error that created the hazard
- The consequences of the failure

The development of an FMEA table can be either highly involved or quite simple, depending on the complexity of the incident or the depth of analysis required. When compiling the information for FMEA, consider the environmental conditions and the process status for each item or action. Probabilities or degrees of likelihood can be assigned to each occurrence. Also, remember that the accuracy of the table depends on identifying the system components and human actions that are relevant to the incident.

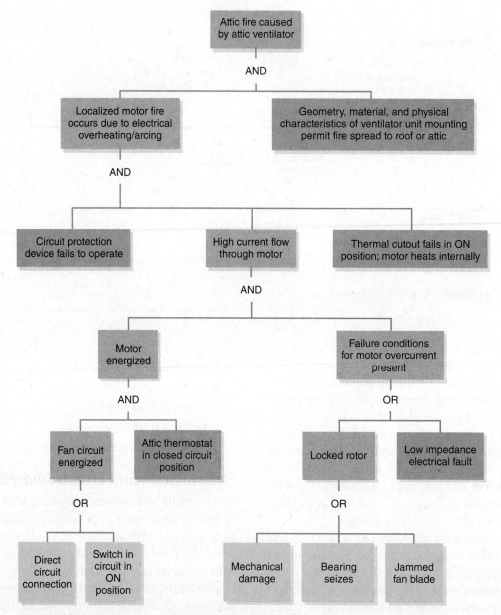

FIGURE 15-2 Fault tree showing the combination of "and" and "or" gates.

Reproduced from NFPA 921-2021, *Guide for Fire and Explosion Investigations*, Copyright © 2020, National Fire Protection Association. This reprinted material is not the complete and official position of the NFPA on the referenced subject, which is represented only by the standard in its entirety.

TIP

The minimum information for FMEA includes a list of all system components and human actions that may have led to the incident, possible failure modes for each component and action, and the immediate consequences of each failure.

Mathematical/Engineering Modeling

Mathematical (engineering) modeling of fire incidents is generally conducted by using calculations or formulas to evaluate specific issues or by using computer-aided analysis to examine more complex fire dynamics and fire progression issues. Fire modeling software is widely available.

One group of modeling techniques involves the application of engineering models. Data are developed that incorporate both known and approximated properties of materials and systems, specific features and components of a fire incident, and physical property estimates defined to a stated degree of certainty. Accepted engineering and analytical techniques are then applied to these assumed "fact sets" or "modeling input data" to develop descriptions of what may or may not have occurred under a clearly stated set of conditions.

Mathematical (engineering) models can be as simple as calculation of flammable gas concentrations with a manual calculator to determine the upper and

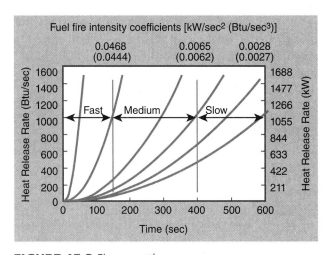

FIGURE 15-3 Fire growth curves.
Reproduced from NFPA 72-2016, *National Fire Alarm and Signaling Code*, Copyright © 2016, National Fire Protection Association. This reprinted material is not the complete and official position of the NFPA on the referenced subject, which is represented only by the standard in its entirety.

lower explosive limits for a given gas in a given space. Other mathematical models may be quite complex, such as those used for fire growth modeling; these involve the application of zone or finite element techniques that must be run by a qualified modeler on a well-equipped personal computer. As an example of a mathematical model, **FIGURE 15-3** depicts graphics showing heat release for National Fire Protection Association (NFPA) standard fire growth rates for slow, medium, fast, and ultrafast fire growth rates.

It is important to understand that these techniques are merely tools used to test a hypothesis, and that repeated calculations should be carried out to bracket the conditions that most likely existed with a given incident. This type of modeling can be used to determine the most probable scenario given a specific set of boundary conditions. Models, however, do not necessarily provide a definitive solution. As with all evidence, their results should be considered in light of other data that have been gathered during the fire investigation. Models can frequently be used to help support or disprove a particular hypothesis. Even in investigations in which the origin and the cause of the fire are not in question, the fire model may be useful in the understanding of fire damage or injury.

The use of these analytical tools may depend on the scope of the fire investigator's role in a specific investigation and its practical purpose in the question at hand. The services of a special expert may be needed to complete the analysis.

Various disciplines use forms of mathematical models to simulate or predict events using established scientific principles and empirical data. In some cases, these models have also provided useful information for fire or explosion investigations.

Limitations of Mathematical/ Engineering Modeling

Mathematical (engineering) modeling, whether it involves hand calculations or computer-based fire models, has some limitations and relies on certain assumptions that need to be considered by the investigator. These models are useful in testing hypotheses but should never be used as the sole basis of a fire origin and cause determination.

The results of mathematical models may be affected by several factors. Inputs to the model, for example, are subject to uncertainties—though the use of standardized methods of input for the selected model can narrow these uncertainties. Often, generic data from fire science literature or exemplar models are used in the models and may also contribute to the uncertainties. In addition, results from the mathematical models are subject to uncertainties from the approximations made within the model itself.

When using a model, retain the original input and output files as part of the investigative record. This process will allow other interested parties to examine the full content of the work product.

Mathematical modeling must be conducted by an individual who has sufficient training and experience to carry out the type of analysis required. The sorts of engineering modeling techniques that can typically be used effectively in fire investigation work are discussed later in this chapter.

Heat Transfer Analysis

Heat transfer models allow the investigator to determine how heat was transferred from a source to a target by one or more of the three heat transfer modes: conduction, convection, or radiation. Such models can be used to test hypotheses about, for example, the competency of a given heat source to act as an ignition source in a given fire causation scenario. Other useful applications of heat transfer modeling are to explain damage or ignition to adjacent buildings via fire spread, ignition of secondary fuel items, and transmission of heat through building elements.

Flammable Gas Concentrations

By determining the concentration of a gas within a given space, an investigator can support or disprove a hypothesis about whether the presence of a flammable gas during the incident was a contributing factor to the sequence of events that occurred. In addition, flammable gas concentration results can either support or refute hypotheses about the location and size of a leak or the competency of an ignition source.

Hydraulic Analysis

When a fire is not controlled by an operating sprinkler system, it is important to determine whether the sprinkler system functioned as designed and intended, and whether the design was adequate for the particular occupancy and furnishings present.

The analysis of a sprinkler system and its water supply should determine whether the system and water supply were matched to the hazard being protected. Questions should be asked to determine whether any faults were present in the system, such as closed valves or faulty sprinklers. It may also be necessary to evaluate why an apparently functional system did not control a given fire. This involves an analysis of the water supply characteristics, water flow through the piping network, and sprinkler discharge characteristics. In addition, the investigator may need to develop a fire growth or heat release model to assess the potential effectiveness of the sprinkler system under various circumstances.

Hydraulic analysis can be used on a variety of fire protection systems, such as wet and dry, antifreeze, carbon dioxide, gaseous suppression agent, dry chemical, and fuel distribution systems.

Thermodynamic Chemical Equilibrium Analysis

Fires and explosions that are believed to be caused by reactions of known or suspected chemical mixtures can be investigated by conducting a thermodynamics analysis of the probable chemical mixtures and potential contaminants. Thermodynamics analysis can be used in incidents involving chemical reactions and to determine the feasibility of a given scenario. This analysis may be useful in evaluating a hypothesis about chemical reactions, the role of contamination, the role of the ambient conditions, the potential for overheating, and other related scenarios. Computer programs can be used to make these predictions based on input data detailing the properties of the chemicals in question. Such an analysis might show, for example, that the chemicals would not create a thermodynamic reaction, allowing this hypothesis to be eliminated. Alternatively, the analysis might show the chemicals are capable of being thermodynamically favored, in which case further investigation would be required to determine whether their reaction would be sufficient to cause ignition.

Structural Analysis

Structural analysis can provide clues to why a building collapsed at a given point during a fire. This type of analysis can be critical in pointing the investigator toward the area where the fire was able to have the most significant impact on the strength of the building. Structural analysis normally requires the services of an engineer with failure analysis training.

Egress Analysis

The reason a fire victim did not escape from a given fire scene is a critical question for a fire investigator to answer. The cause of such a failure may be related to the egress design features (or lack thereof) of a building or to maintenance issues. The investigator should obtain data such as the locations of the exits, egress routes, travel distance, and egress route widths to help in this analysis. Some computer-based fire models, such as the National Institute of Standards and Technology's (NIST's) Consolidated Model of Fire and Smoke Transport (CFAST), have modules that can provide egress data for the specific fire situations they can be used to model.

Fire Dynamics Analysis

A number of methods are available to assist in the analysis of fire dynamics and fire growth, including specialized fire dynamics routines (hand calculations) and computer models (zone models and field computational fluid dynamics [CFD] models). These analytical methodologies can be used to test and evaluate origin and cause hypotheses, and can assist in the evaluation of physical and eyewitness evidence used as part of the hypothesis development process.

Computer models can assist an investigator in attempting to determine the growth and development of the fire. These models, which are based on established mathematical equations, may be useful in the prediction of several fire-related factors and/or events, such as the following:

- Time to flashover
- Gas temperatures or concentrations
- Flow rates of fire-related gases
- Temperatures of interior surfaces
- Time of activation of fire detection and suppression devices
- Effects of events such as opening doors and windows

As noted earlier, computer models can be useful tools in the evaluation of the hypothesis formed by an investigator. Using known input data, the results from the model can be compared with witness accounts and physical evidence as a test of the investigator's hypothesis or the witness's account.

A note of caution is warranted, though: These models are only tools, and are subject to certain limitations based on uncertainties that must be assessed and analyzed. Although they can provide additional data on a fire, modeling results must be weighed in relation to other information gathered during the investigation and the reliability of the input data and assumptions made.

TIP

Fire dynamics and fire growth can be analyzed using several methods, including the following:
- Specialized fire dynamics routines, such as hand calculations
- Computer models, such as zone models and field CFD models

Variables or uncertainties that can influence fire modeling results include the following:
- Fire load characteristics
- Ventilation openings (size and open or closed)
- Heating, ventilation, and air conditioning (HVAC) flow rates
- Heat release rates

Specialized Fire Dynamics Routines

Specialized fire dynamics routines are simplified procedures that require minimal data to run a computer model and can often answer a narrowly focused question. These procedures often involve the use of specific algebraic hand-calculation equations and generally require much less data and information. The equations included in these routines have been derived through research and experimentation and require the input of variables derived from information developed from the fire scene and obtained from appropriate reference materials.

The equations and hand calculations in fire dynamics routines can be utilized to evaluate specific issues, including the following:

- Time to flashover
- Heat flux
- Heat release rate
- Time to ignition
- Flame height
- Detector activation
- Gas concentration
- Flow rates of smoke, gas, and unburned fuels

Computer Fire Models

The two primary types of computer models currently used to assess fire growth are zone models and CFD models. Computer models allow for a more complex and detailed analysis of fire growth and behavior that is useful in both predictive analysis and testing and postincident investigation and hypothesis testing. These models also incorporate certain assumptions that must be considered when analyzing results and are subject to limitations.

Zone models usually divide a compartment into two zones: a hot upper zone and a cooler lower zone. They assume universal conditions prevail throughout the zone and allow for the expansion of the zones during the course of the fire. These models are often run on personal computers and are generally well accepted and validated by peer review.

CFD models are more complex than zone models and divide the compartment into many small cells. Numerous calculations occur in each cell, and activity in one cell affects the surrounding cells. These models allow for a more detailed analysis of a fire event, but also require a much greater level of expertise to use them correctly. They require larger computer capabilities and can be time consuming to operate. CFD models are useful in evaluating fire progression in areas with irregular geometry or where very fine detail is required. The use of these models in fire investigation and litigation is increasing.

Graphic Representations

Another useful tool to represent what has occurred during a given fire incident is graphic representations. These include drawings of all sorts, as well as physical models and computer animations, which are increasingly being used today. Some computer models also work with a graphic interface that allows for a visual representation or interpretation of the data derived from the model results. For instance, NIST's Fire Dynamics Simulator is a CFD model that works with NIST's SmokeView program to provide a visual animation of the fire progression. Fire Dynamics Simulator is a large-eddy simulation code for low-speed flows, with an emphasis on smoke and heat transport from fires.

Graphic representations are frequently used by both the investigation team and forensic evaluation personnel. Among the reasons that an investigation team might use graphics are to understand an incident location better, to assist in interviewing witnesses, and to define and identify materials and systems and their involvement in an incident. In forensic applications, graphics can be used to help a judge and/or jury to better understand important features of a fire scene and the underlying scientific and engineering principles that caused a particular outcome in a given fire.

Graphic representations should not be confused with mathematical models, described earlier. Mathematical models are based on calculations, whereas graphic representations are based on geometric representations of a scene or a set of facts. Modern modeling programs permit the program operator to open and close doors and windows, change furnishings, obstruct flow paths with victims, incorporate weather conditions, and make myriad other changes to test their effect on fire behavior.

Guidelines for Selection and Use of a Fire Model

Numerous factors should be considered when selecting the fire model to use for a particular fire incident. First, identify the hypothesis or hypotheses to be tested or other questions to be answered by the modeling procedure. The model that is chosen on a preliminary basis should be validated (this information should be available from the model developer and available in the model's documentation). If possible, consider the degree of uncertainty that is expected to result from the use of the model, both from the model itself and from the data that will be entered **FIGURE 15-4**.

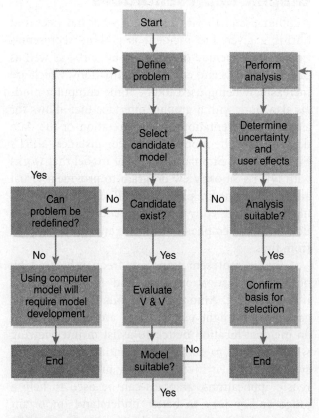

FIGURE 15-4 Fire model selection flowchart.

Reproduced from NFPA 921-2021, *Guide for Fire and Explosion Investigations*, Copyright © 2020, National Fire Protection Association. This reprinted material is not the complete and official position of the NFPA on the referenced subject, which is represented only by the standard in its entirety.

Fire Testing

Fire testing is a process that can help check the data collected or test a specific hypothesis. Testing can be conducted in the field and/or in a controlled environment and may range from bench tests to full-size re-creations of the event. Such tests are useful in determining several factors or events in the fire, such as origin and cause, fire spread, combustion characteristics, and effects of fire on materials.

As with fire modeling, when used in a proper manner, fire testing can be useful in the evaluation of a hypothesis. However, the investigator must be mindful of potential differences and inaccuracies between the test conditions and the conditions at the time of the actual fire. For example, differences may exist in weather conditions, missing windows or loss of glass, fuel loads associated with contents, and ventilation effects. Fire testing may be used as a tool to examine a hypothesis, but the data obtained should not be relied on as absolute. In other words, this testing can provide useful information, but it is not possible to re-create all conditions of a given fire perfectly. To the extent possible, the methods, procedures, and instruments used in testing should follow accepted norms of practice. Following these established practices will help ensure the credibility of the results.

Examples of fire testing include determining the burn-through time of a wall or door using recognized fire endurance testing techniques, such as those found in ASTM E119-20, *Standard Test Methods for Fire Tests of Building Construction and Materials*, and determining the heat release characteristics of a piece of cushioned furniture to assess whether it could have been responsible for the fire growth and damage patterns noted. Application of the types of data that are developed in testing can provide invaluable information to support or refute various hypotheses being evaluated. Nevertheless, fire testing does have its limitations, in that it is not possible to re-create all aspects of a specific fire.

FIGURE 15-5 shows a series of images of 4-minute fire growth progression prior to extinguishment. The fire was initiated with a single match igniting a single sheet of newspaper.

FIGURE 15-6 shows temperature development during ASTM E1537-16 fire testing of a couch. Measurements were made at locations 4 ft (1.2 m) above the couch and directly above the couch back.

Data Required for Modeling and Testing

To conduct valid modeling and testing, it is important to gather data that are as accurate and complete as possible. As in other endeavors, the "garbage in,

FIGURE 15-5 A fire's progression over a 4-minute period.
© Jones & Bartlett Learning. Photographed by Glen E. Ellman.

garbage out" concept applies here. It should be anticipated that during court proceedings, any test results will be subjected to a Daubert review, and the results that are presented will only be as valid and accurate as the information from which they were derived and the care that was taken to develop them. (See Chapter 5, *Legal Considerations for Fire Investigators*, for more information on the Daubert rule.)

Important information to be obtained includes structural information, materials and contents, and ventilation information.

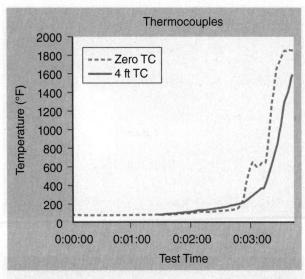

FIGURE 15-6 Temperature development during ASTM E1537-16 fire testing of a couch.

Structural Information

Structural information, such as the room dimensions and the sizes of structural components as they apply to the fire scenario, should include the following:

- Length, width, and height of rooms and buildings
- Wall thickness
- Slopes of floors and/or ceilings
- Construction materials, including wall coverings
- Construction features present, such as types of doors, windows, HVAC, and stair locations

Materials and Contents

A meaningful analysis of a fire requires understanding of the heat release rate, the fire growth rate, and the total heat released. This analysis should be documented and include the following items:

- Type of contents, including materials
- Location of contents
- Configuration and condition of contents

Ventilation

Understanding ventilation conditions is important to the validity of a fire test or model. Data on the positions and conditions of openings should be included, as should the following:

- Locations of openings
- Sizes of openings
- Status (open or closed)

- Area of usable opening (fully open, partially open, closed)
- Ventilation effects, including wind and HVAC
- Fire department operations

Incendiary Fires

An **incendiary fire** is a fire that is intentionally ignited in an area or under circumstances where and when a fire should not occur. The following sections cover the indicators that point to incendiary fires, including indicators not directly related to combustion and other evidentiary factors such as suspect development and identification.

Incendiary Fire Indicators

The indicators discussed in this section should be studied for their possible support of the hypothesis that the fire is incendiary.

Multiple Fires

Multiple fires are fires with no obvious connection that would have allowed one fire to ignite the fuel in (or spread to) another area. A fire in the basement that spread through the walls of a balloon-frame house to the attic would not be classified as an incident with multiple fires, because the fire spread naturally through a vertical opening and ignited combustible fuels in a location that was remote from the initial fire. An investigator must determine that a separate fire was not the natural result of the growth and spread of the initial fire.

Other "natural" means of fire spread that could cause multiple fires, or the appearance of multiple fires, include the following:

- Conduction, convection, or radiation
- Flying brands
- Direct flame impingement
- Falling flaming materials (drop-down), such as curtains
- Fire spread through shafts, such as pipe chases or air-conditioning ducts
- Fire spread within wall or floor cavities within balloon construction
- Heat from a fire accumulating at the ceiling level, and igniting other materials near the ceiling
- Overloaded electrical wiring
- Utility system failures
- Fuel gas or dust explosions
- Lightning
- Rupture and launching of aerosol containers

Further, an investigator should not confuse the site of a previous fire with that of a more recent fire. This could cause the investigator to develop the erroneous hypothesis that multiple fires occurred simultaneously. The fire history of the structure should be obtained and examined.

Apparent multiple points of fire can result from sustained burning or smoldering during or after fire suppression or overhaul. Full-room involvement or post-flashover conditions can make identifying multiple fires more difficult or impossible. It is important to conduct a full scene examination to determine whether separate fires took place, and to establish areas of lesser or greater damage and fire spread patterns. If this full scene examination is not conducted, valuable evidence may not be identified, and an investigator may not be able to determine accurately whether separate fires occurred. Any investigation should include, where permitted, examination of areas in a subject structure that do not appear to have been damaged by fire.

Trailers

A trailer is a deliberately introduced fuel or manipulation of existing fuel(s) used to aid the spread of a fire from one area to another. One fire will, in turn, ignite other areas via these trailers **FIGURE 15-7**.

FIGURE 15-7 Here, paper towels were used (unsuccessfully) as a trailer.
Courtesy of Joseph Whittaker, Nassau County Fire Marshals.

Trailers can leave distinctive patterns on horizontal surfaces such as floors. When the floor is cleared of debris, the pattern may be easily discernible; however, it is important to determine that these patterns are not the result of other mechanisms or materials, such as open areas bordered by protected areas or the effects of flashover and full-room involvement. When a room becomes fully involved, radiant heat can create burn patterns on floors that can be misinterpreted as trailer burn patterns (see Chapter 4, *Fire Effects and Patterns*, for additional information).

Materials that can be used as trailers include the following:

- Ignitable liquids
- Clothing
- Paper
- Straw

Many common household goods (cleaning fluids, gasoline, etc.) are ignitable liquids, but their presence on the fire scene is not necessarily indicative of use as a trailer. It is not the fuel that constitutes a trailer, but rather the manner and location in which the fuel was used.

Lack of Expected Fuel Load

When the observable fire damage is not consistent with the observable fuel load, further investigation is warranted. An investigator should attempt to quantify the fire damage that would be caused by the known or reported fuel load in relation to the observable fuel load. However, the absence of expected fuel loads is not enough to classify the fire cause as incendiary.

Examples of areas and spaces that routinely have low or limited fuel loads include corridors, stairways, hallways, and vacant homes. If the origin for a fire is in a low- or limited-fuel area, the investigator should look for physical evidence of fuels such as an ignitable liquid and take samples; however, burning in these areas may not be unusual if it represents fire extension or movement from another area, particularly if the adjacent space has developed past flashover.

Lack of Expected Ignition Sources

Another indicator that deserves further investigation is lack of a readily apparent competent ignition source at the fire origin. Investigators may have to look closely through the debris in the search for ignition sources that have burned, melted, or been consumed. Closets, crawl spaces, and attics are typical areas, rooms, and spaces in which a limited number of heat sources are present. Investigators should be careful to examine all fire-burned areas in their totality and not focus on one particular area where the fire damage appears to be greatest. Areas of heavy burn can be influenced by products with a high heat release rate, delayed extinguishment, and ventilation effects in that particular area and may not be the specific area of origin.

Exotic Accelerants

Mixtures of fuels containing Class 3 or Class 4 oxidizers and thermite mixtures may be considered exotic accelerants. Some oxidizers are capable of self-ignition. These types of accelerants can cause exceedingly hot fires and generally leave residues that may be visually or chemically identifiable. Indicators of these substances, sometimes termed **high-temperature accelerants (HTAs)**, include the following:

- Rapid rate of growth
- Brilliant flares
- Melted steel or concrete

Other reasons for dramatic fire effects and patterns should always be considered. These might include ventilation effects, delays in fire suppression, effects of particular fire suppression tactics, or type and configuration of fuels.

Forced Entry

Broken door frames and locks, broken windows, and pry marks can be evidence of forced entry. Points of entry in a building should be carefully examined and documented for potential physical evidence. Since first responders and suppression personnel may force entry as part of their response to the fire, the fire investigator should interview these persons to determine whether they forced entry. In some cases, forced entry may be evidence of burglary. In others, it may be staged by the owner or resident in an attempt to mislead the investigation. Finally, it is also possible that signs of forced entry originated from an earlier event and are not related to the fire.

Unusual Fuel Load or Configuration

A fire-setter may seek to create a fire that will burn more aggressively or effectively by moving contents or materials into a configuration that allows for more rapid fire growth or fire spread than would be expected if the contents were spaced farther apart. This could also be done in an attempt to provide more complete burning of the fuels. Witnesses may be able to provide information about the positions or locations of contents prior to the fire.

The types of fuels can be evaluated to determine whether they would ordinarily be expected in a given occupancy, as fuels may be added to an area to assist in fire growth or spread.

An unusual fuel load or configuration should not be assumed to be related to the fire cause. If a fuel load appears truly unusual, the investigator should seek further information from the occupant and/or first-arriving engine company to determine whether the load or configuration is considered abnormal or has a rational explanation.

Burn Injuries

Anytime victims experience burn injuries in a fire, those burns should be analyzed to determine whether they provide information about the cause of the fire. Determine the circumstances of the injury, and if possible, determine whether the injury itself provides information about how it was received (e.g., a burn resulting from a hot object or open flame). Because burn injuries may be sustained while setting an incendiary fire, local hospitals should be contacted for identification of recent burn victims. Some jurisdictions require the reporting of burn injuries.

A detailed interview of the burn victim may be helpful to determine the origin, cause, or spread of the fire, as well as the victim's activities prior to and during the fire. This information can then be compared to physical findings at the scene to determine whether they are consistent. All burn injuries should be documented, and samples of the victim's clothing may be taken in appropriate cases for laboratory testing for ignitable liquids.

Incendiary Devices

Incendiary devices include a wide range of mechanisms used to initiate an incendiary fire. If they are used, remains of the device can often be found at the scene.

Almost any appliance or heat-producing device can be used as an incendiary device. If there are no other obvious ignition sources, then efforts should be made to determine whether a device was used. Some examples include the following:

- Combination of cigarette and matchbook
- Candles
- Wiring systems
- Electric heating appliances **FIGURE 15-8**
- Fire bombs/Molotov cocktails **FIGURE 15-9**
- Paraffin wax–sawdust incendiary device (fireplace starters)

Delay Devices

Some incendiary devices are constructed as delay devices to allow the fire-setter time to leave the area safely

FIGURE 15-8 This heater was used as an incendiary device.
© Jones & Bartlett Learning. Photographed by Glen E. Ellman.

FIGURE 15-9 Fire bombs are also referred to as Molotov cocktails.
© Jones & Bartlett Learning. Photographed by Jessica Elias.

FIGURE 15-10 An example of a time-delayed incendiary device.
© Jones & Bartlett Learning. Photographed by Glen E. Ellman.

or to establish an alibi. If, during the investigation, the investigator finds a device that was not activated, they should not move it **FIGURE 15-10**. Adequate precautions and safeguards should be taken, including evacuation of the area and notification of trained explosive ordnance disposal personnel if an active or live device is found.

A fire-setter may attempt to mask the true cause of the fire by placing an appliance within the scene, so that it can be discovered as the "obvious" cause of the fire. For example, a fire-setter may pour an ignitable liquid into a coffeemaker, causing a fire to occur. An investigator should not assume that the fire was caused by the appliance's malfunction. Further testing or evaluation may be warranted to determine the true cause of the fire.

It is important to remain aware of spoliation issues when conducting this type of investigation. Moreover, if investigators do not have the necessary engineering training, they should seek qualified assistance. As an example, a pot-on-the-stove fire may cause heavy fire damage to some or all of the control mechanisms on the panel at the upper rear portion of the cooktop surface. Because the components are largely composed of a phenolic plastic material, much of it may be heavily damaged, and removing the back panel of the control panel could cause the contacts to collapse and fall apart. Caution should be taken to protect these mechanisms until all interested parties have been put on notice, and to ensure they are present before any destructive testing is done. An alternative to destructive testing is the use of an X-ray machine to document the contact positions of the controllers without taking the panel apart.

Because arsonists may set multiple fires, the investigator should inspect the fire building fully to determine whether other fires occurred that have separate origins. The arsonist may have used similar devices in these other fires that can provide valuable clues. Examine these areas for debris that may contain delay or other incendiary devices or methods that would contribute to the ignition sequence. In cases of multiple fires, often at least one incendiary device has failed to operate, leaving an investigator with valuable evidence. Look for trailers from one burn area to another.

TIP

During investigations of possible arson fires, you should work closely with law enforcement and prosecuting authorities.

Presence of Ignitable Liquids in the Area of Origin

The presence of ignitable liquids in an unusual or un-expected place may indicate that the fire was incendiary. Take care to document the location where the ignitable liquid was found. On the one hand, the presence of a perceived ignitable liquid found on the floor area of a garage may not be unusual considering that vehicles, lawn equipment, and fuel cans may be stored there. On the other hand, an ignitable liquid found on the floor area in a kitchen in a random linear pattern should be considered as the possible result of the application of an ignitable liquid.

It is the investigator's responsibility to determine whether the liquid's presence is the result of an intentional placement to start, accelerate, and/or spread the fire. Samples of the debris should always be taken for laboratory analysis. Laboratory confirmation of the presence of an ignitable liquid is a critical step if an investigator's cause hypothesis includes application of such liquids and can help substantiate a criminal case.

Use of Ignitable Liquid Detection Canines

An ignitable liquid detection canine (termed "IGL canine" in NFPA 921) is specially trained in the detection of various ignitable liquids and can provide useful assistance in identifying areas at the scene that might contain such liquids **FIGURE 15-11**. While these dogs have been referred to as accelerant detection canines, the term "accelerant" should be avoided, as it suggests the intention of a person in placing an ignitable liquid in a specific location; in reality, the IGL canine simply detects the presence of such a liquid, but does not identify the intent behind it.

FIGURE 15-11 An ignitable liquid detection canine may be a useful addition to a fire investigation.
© Jones & Bartlett Learning. Photographed by Glen E. Ellman.

The IGL canine's abilities and reliability depend on both its training and its handler, and the investigator should be familiar with the handler and the training the dog has had and continues to receive. An IGL canine–handler team should be certified by an appropriate certifying body and must undergo periodic proficiency assessments.

The fire investigator should use an IGL canine in conjunction with other investigative and analytical methods, training and experience, and the information uncovered in the case. If the investigation indicates that a sample should be collected from a certain area, this collection can and should be done even if the area did not receive a positive canine alert. Conversely, even if an IGL canine alerts in a certain area, this is not conclusive proof of the presence of an ignitable liquid there.

Although the IGL canine has the reputation of having a powerful nose, its ability to distinguish between ignitable liquids and background materials commonly found in buildings is far more important. This sets the IGL canine apart from other types of law enforcement dogs that detect unusual substances, such as drug- and bomb-detecting dogs, as the IGL canine must be trained not to alert on common petroleum-based substances such as burned plastic and carpet.

To include an IGL canine in an investigation, perform all of the normal investigative steps, identifying areas where an ignitable liquid may have been used. Brief the canine–handler team about the areas they are requested to examine, and any safety hazards that might be found within those areas. The handler will then enter the identified area(s) to seek any places where the canine alerts to a potential ignitable liquid. The handler identifies these places with a token or other item, and the investigator (with the input of the handler, where appropriate) collects fire debris or other evidence for laboratory examination. If possible, before sealing the evidence containers, place the cans of collected fire debris, along with cans of collected comparison sample materials, out for review by the IGL canine to see if the dog continues to alert to the evidence after it has been collected.

All substances collected based on an IGL canine alert must be validated through laboratory analysis. Without such validation, it should not be used as evidence of the presence of such substance in criminal or civil court.

Assessment of Fire Growth and Damage

If a fire spreads more quickly than can be explained by the expected fuel load or beyond the area where it would normally be expected to be confined, the investigator should look more closely at which factors might have contributed to the fire growth. Fire growth depends on a large number of variables, including the volume of the compartment, the height of the ceiling, the heat release rates of fuels, the location of the initial fuel package within the compartment, the location of subsequent fuel packages, and ventilation. In the absence of physical evidence, an investigator is cautioned against using subjective terms such as *excessive*, *abnormal*, *unusual*, or *suspicious* to support an incendiary fire cause determination. The proper use of a fire model can provide assistance in determining the accuracy of an investigator's observations.

Potential Indicators Not Directly Related to Combustion

Certain indicators can assist an investigator in developing an ignition hypothesis, questions for witnesses, or avenues for further investigation. These indicators typically are not related to the determination of the fire cause, but rather tend to show the possibility that somebody had prior knowledge of the fire.

Remote Locations with Blocked or Obstructed View

A fire-setter may start a fire in a remote location or one obscured from public view. Most important are actions taken just before the fire to obscure the view, such as newly covered or painted windows, or furniture placed in front of exterior openings. These findings can provide information that is helpful in establishing timelines and questioning witnesses. However, remember that accidental fires can also start in remote locations, so no cause determination conclusions should be made solely based on the location of origin.

Fires Near Service Equipment and Appliances

To make a fire appear to be accidental, a fire-setter may set a fire near appliances, hoping that the appliance will be assumed to be the ignition source for the fire. Each appliance should be carefully evaluated to determine whether it was truly the ignition source. Keep in mind the issues related to spoliation before any destructive evaluation.

TIP

Always strive to perform your investigations in a systematic manner. Establish a normal routine for each fire you investigate so that eventually you will automatically perform all of your investigations in that same manner, consistent with NFPA 921, *Guide for Fire and Explosion Investigations*, Sections 4.2 and 4.3.

Removal, Replacement, or Absence of Contents Prior to the Fire

Contents are sometimes removed or replaced with items of lesser value prior to a fire. Careful documentation of the remains and the debris may be useful in establishing a fraudulent insurance claim, even in situations in which the fire cause cannot be determined.

The determination that contents have been removed or replaced requires verification that the items were present at some time prior to the fire. Verification may be made through corroborated witness statements, photos, inventory, or sales receipts. It may be necessary to document all of the occupancy's contents, not just those near the origin.

Sometimes personal items or irreplaceable items may have been removed prior to an incendiary fire. Examples include jewelry, photographs, pets, tax records, business records, and firearms.

Items and contents that may be replaced can include the following:

- Residential: furniture, clothing, appliances, jewelry, guns
- Industrial/commercial: machinery, equipment, stock, merchandise
- Vehicles: tires, batteries

If the contents of the building are abnormal given the occupancy, this could be an indication that further investigation is necessary.

Blocked or Obstructed Entry

To allow the fire more time to grow, the fire-setter may place obstructions to hinder or slow the firefighting operations. Any unusual obstructions that deny fire vehicle access should be noted and evaluated. Obstructions may include fallen trees, barricades, or larger obstructions that have the potential to block the access of emergency vehicles. To prevent firefighter access to the structure, doors and windows may be obstructed as well.

TIP

Document each step of your investigation photographically. You can never take too many photos. The one photo you will need will invariably be the one that you did not take.

Sabotage

The term **sabotage** refers to intentional damage or destruction. Fire-setters often try to set up conditions that lead to rapid and complete destruction of a building and its contents. To accomplish this goal, the

FIGURE 15-12 Fire-setters may sabotage fire protection systems to delay notification of the fire department.
© DWImages/Alamy Stock Photo.

fire-setter may sabotage the fire protection systems so notification to occupants and the fire department is delayed **FIGURE 15-12**.

Damage to Fire-Resistive Assemblies

Fire-resistive assemblies are installed to separate a structure into compartments, thereby confining fire and smoke from spreading to other parts of the structure. The most common method of fire travel through a structure is open doors, with open stairwell doors having an even greater impact on fire spread. Fire-setters may penetrate fire-resistive assemblies (walls, floors, or ceilings) or prop open doors to facilitate fire and smoke movement. Be aware, however, that penetrations in these assemblies may also be the result of firefighting activities or poor construction, and doors may be propped open prior to the fire by occupants to improve access or ventilation. Given these possibilities, it is important to conduct a thorough investigation and not to assume sabotage.

Damage to Fire Protection Systems

Numerous types of fire protection and suppression systems may be found in an occupancy. If they are operating normally, they should help detect or suppress the fire before it has a chance to cause any significant damage. Some might also be designed to alert the fire

department in case of fire. These systems can include heat, smoke, and flame detection systems and security cameras, sprinklers and standpipes, and extinguishing systems.

If a fire suppression system or fire detection system failed, inspect the system for any signs that may indicate the cause of the failure, such as the following:

- Improper installation
- Tampering
- Lack of maintenance
- System shutdown
- Equipment or structural assembly failure

It is also important to determine whether these conditions existed prior to the fire. Inspection and maintenance records could provide valuable clues regarding the system's prefire status.

Some methods of disabling these systems include the following:

- Removing or covering smoke detectors
- Obstructing sprinklers
- Shutting off control valves
- Damaging threads on standpipes, hose connections, and fire hydrants
- Placing debris in pipe splitters and fire hydrants or disconnecting alarm bells or sirens
- Starting multiple fires to overload the suppression system

An investigator should always examine the fire protection systems to determine whether any tampering or disabling of the systems occurred, and should check with alarm-monitoring facilities to determine the times at which trouble or alarm signals were received. This information may assist the investigator in defining the timeline for the fire.

TIP

Some fire-setter actions can create additional safety concerns for firefighters and investigators alike. For example, a fire-setter may intentionally weaken supportive structures in hopes of delaying firefighting efforts and obscuring evidence. Investigators must be attuned to the potential for sabotage to a building and take the necessary safety precautions.

Damage to Buildings

A fire-setter may intentionally damage the building for several reasons, including (1) to hinder firefighters' ability to fight the fire effectively and (2) to provide avenues for the fire to spread beyond its area(s) of origin. Examples of such activities could include the following:

- Cutting openings in the floors that firefighters could fall through, or breaching other fire-rated assemblies, such as walls or doors, to permit the spread of fire
- Jamming or barricading doors and windows to make it difficult for firefighters to enter
- Sabotaging fire-rated doors, fire dampers, and the like so that they are in the open position during a fire and will not close automatically

Exercise due care during the fire scene examination to determine whether the failure of any particular fire protection system or component was the result of an intentional act and not an outcome of lack of maintenance. For example, fire- and smoke-rated doors are sometimes propped open or disabled by occupants as a normal practice. Other reasons for failure should also be examined, such as manufacturer or installation defects.

Opening Windows and Doors

To provide ventilation to a fire, the fire-setter may open exterior windows and doors that would not normally be open (during cold weather, for example). The objective could be to have the fire spread outside of the compartment of the area of origin through creation of artificial avenues of fire propagation. The investigator should determine whether the occupants normally propped the doors open or whether this action may have been done to spread the fire. Wind direction may also be a factor that the fire-setter attempted to exploit for spreading the fire.

TIP

Examine all available court, real estate, business, maintenance, code enforcement, and any other records and documents for the vehicle, property, structure, or person you are investigating. This may help establish motive for the fire.

Other Evidentiary Factors

Other evidentiary factors include those indicators that can be analyzed after the fire has been determined to be incendiary, in an effort to develop a potential profile of the suspect.

Serial Crimes

Patterns in fire-setting are important in seeking to develop the profile of a suspect, and eventually to identify them. **Serial fire-setter** and **serial arsonist** are the terms used to identify individuals or groups who are involved in three or more fire-setting incidents. In a series of fires, the investigator should analyze which fires may be attributable to a single suspect or group by considering the following factors:

1. Geographic location or cluster: Fire-setters tend to act within the same geographic location or neighborhood.
2. Temporal frequency: A serial arsonist may choose the same time period or day of the week.
3. Materials and methods: Repetitive fire-setting behavior not only occurs within the same geographic location but also uses similar fire-setting materials and methods.

Further, a perpetrator can at times be identified through the analysis of crime attributes that suggest a connection to a single individual:

- **Modus operandi (MO):** The method of operation used by the offender. This predominantly applies to serial arsonists. The offender's actions during the perpetration of a crime form the MO. The offender develops and uses the MO over time because it works, but it also continuously evolves. MOs are a learned behavior and may be modified over time as the offender becomes more sophisticated and confident.
- **Personation** (the signature): Unusual behavior by an offender, beyond that necessary to commit the crime. The offender may add personal meaning or ritualistic behavior to the crime, potentially creating commonalities that can be used to identify the offender.
- **Staging** (alteration of the crime scene): When someone purposely alters the crime scene prior to the arrival of police. There are generally two reasons associated with staging:
  - To redirect the investigation away from the most logical suspect
  - To protect a victim or the victim's family from a gruesome or demeaning scene by changing the victim's or body's position

Timed Opportunity

Fire-setters sometimes take advantage of conditions or circumstances that add to the chances of successful destruction of the property. A timed opportunity can also increase their chances of not being apprehended. Examples of a few timed opportunities are as follows:

- Natural conditions such as hurricanes or floods, snowstorms, or high winds

- Civil unrest
- Fire department unavailability due to response to another alarm (potentially called in by the suspect), parades, or other calls

Motives for Fire-Setting Behavior

Motive is the inner drive or impulse that is the cause, reason, or incentive that induces or prompts a specific behavior. Motive indicators in the fire investigation process can be used to help identify potential suspects, but they should not be used to determine or classify the fire cause. Whereas **intent** is generally necessary to show proof of a crime and refers to the state of mind that exists at the time a person acts or fails to act, motive is the reason that an individual or group may do something and is not generally a required element of a crime.

The classifications discussed in this chapter are based on Douglas et al., *Crime Classification Manual* (CCM), 2013. This manual can help investigators gain as much information as possible by identifying essential elements of analytical factors used to classify the motive of a murder, arson, or sexual offense. The behaviors listed in the CCM may identify a possible motive, leading investigators to possible suspects. These behaviors apply whether the fire is the result of a one-time occurrence or the action of a serial fire-setter.

There are three classifications of repetitive fire-setting behavior:

1. **Serial arson** involves an offender who sets three or more fires, with a cooling-off period between fires.
2. **Spree arson** involves an arsonist who sets three or more fires at separate locations, with no emotional cooling-off period between fires.
3. **Mass arson** involves an arsonist who sets three or more fires at the same site or location during a limited period of time.

Motive Classifications

The National Center for the Analysis of Violent Crime (NCAVC) has identified the following six motive classifications as the most effective in identifying offender characteristics for fire-setting behavior:

1. Vandalism
2. Excitement
3. Revenge
4. Crime concealment
5. Profit
6. Extremism

Vandalism. Vandalism is defined as mischievous or malicious fire-setting that results in damage to property. Common targets include educational facilities and abandoned structures, but incidents also include trash fires and grass fires. Vandalism falls into two categories:

- Willful and malicious mischief: Such incendiary fires that have no apparent motive or are seemingly set at random. These fires are often attributed to juveniles or adolescents.
- Peer or group pressure: Recognition or pressure from peers can inspire vandalism, and is a predominant motive among juveniles. Juveniles often are swayed more by the people they are with than by the potential consequences of their actions.

Excitement. Some people set fires because they seek excitement. This kind of perpetrator generally does not intend to hurt anyone in the fire, but unplanned injuries and deaths can occur. There are four subclassifications to this motive:

- Thrill seeking: Offenders are filled with feelings of excitement and power and are often repetitive fire-setters. A psychological need or desire drives them to set these fires.
- Attention seeking: Fire-setters in this classification have a need to feel important.
- Recognition: These offenders are sometimes referred to as vanity or hero fire-setters. Firefighters (paid or volunteer) and security guards may be candidates for this classification. The need for recognition, praise, and/or reward can be a driving motive for their behavior. If they are successful, they will most likely repeat the offense multiple times. Such offenders are often still present at the scene and provide extensive details about the location of the fire and how to reach that location. They are usually very forthcoming with information and helpful to firefighters and police.
- Sexual gratification or perversion: Although considered rare, there are documented cases of offenders who set fires as a means for sexual release. They may remain in the area to permit them to observe the fire, although they may be hidden from view.

Revenge. An offender may set a fire as a means of revenge for a real or imagined injustice done to them or to someone they care about. These fires are sometimes premeditated and well planned. They are usually one-time events, but in the case of serial arsonists will occur multiple times. This category includes four subcategories:

- Personal retaliation: Something commonly occurs that triggers a retaliatory response. The event may have been a fight, an argument, or a feeling of being taken advantage of, involving either the offender or others close to them. Common targets are the victim's home, personal possessions, or vehicle.

- Societal retaliation: A serial offender may fall into this subclassification. The perpetrator may have feelings of loneliness, rejection, persecution, abuse, or inadequacy that drive them to set fires. As these feelings recur, the offender may continue to set fires as a release and for a feeling of gratification. Look for a similar MO in these fires.

- Institutional retaliation: This motive usually arises with individuals who have a grudge against the institution. The offender may be a former employee, customer, patient, or student. Common targets are religious, medical, governmental, and educational institutions or corporations.

- Group retaliation: Targets may involve fraternal, religious, racial, or other groups, including gangs. Evidence of graffiti, symbols or markings, and other vandalism may be found at these scenes.

Crime Concealment. In crime concealment, arson is generally a secondary or collateral criminal activity. The purpose is to conceal the initial criminal activity that occurred. This category includes three subcategories:

- Murder concealment: When a murder has been committed, the fire is usually set to disguise the fact that a death occurred prior to the fire or to destroy forensic evidence that could identify the victim. The fire may also be set with the aim of destroying evidence that could link the offender to the crime.

- Burglary concealment: These fires are usually set to hide the fact that a burglary occurred or to destroy any evidence that could help identify the burglar/fire-setter.

- Destruction of records or documents: The general target in this category is records or documents associated with the business, institution, or corporation. Fires may be started in the file cabinets, which often are found with the drawers left open, or in folders of the files or other documents. Investigators should consider employees or owners of the location as potential suspects.

Profit. Fires set for profit involve those set for material or monetary gain. The monetary issue may be either directly or indirectly linked to the fire. Profit-motivated arson is a commercial crime that involves the least amount of passion of any of the motives for arson. Direct gain may come from insurance fraud, elimination or intimidation of business competition, extortion, removal of unwanted structures to increase property values, or escaping financial obligations. Further subclassifications include fire-setting to liquidate property, to dissolve a business, to conceal a loss, or to liquidate inventory. Other categories include employment, parcel/property clearance, and competition.

Many tools exist to help the investigator determine financial motives in a fire. The examination of bank records, insurance policies, code enforcement or fire inspection complaints or sanctions, National Insurance Crime Bureau records, tax records, and so forth can often uncover potential motives for the fire. Some indicators identified by an investigator could include the following:

- Financial stress
- History of code violations
- Fires at additional properties owned by a single individual or group
- Over-insured property

Liens, attachments, unpaid taxes, mortgage payments in arrears, real estate for sale, poor business location or competition, economic decline, outdated or overstocked product, loss of jobs, and over-insurance or recent changes in policies to inflate the values can all establish or support the motives for the fire. The investigator should also examine past insurance histories to determine whether the claimant may have had previous claims for fires in other buildings or vehicles.

Extremism. Extremism as a motive comes into play when the fire-setter aims to further a political, social, or religious cause. These fires may be set by individuals or groups. The fire-setters generally have a great degree of organization, as reflected in their use of more elaborate ignition or incendiary devices. Terrorism and riot/civil disturbance fall into the extremist category:

- Terrorism: The targets chosen by terrorists are rarely random structures, but instead are typically chosen based on a specific significance that will provide the greatest impact for the message that they want to deliver. Common objectives may be of political or economic significance. Political targets can include government offices, newspapers, universities, political party headquarters, animal research facilities, abortion clinics, and military or law enforcement installations. Economic targets may include business offices, distribution facilities, banks, financial institutions, or companies thought to have an adverse impact on the economy. Fire, explosives, and other weapons may be used in these assaults.

- Riot/civil disturbance: Fires that occur during riots or civil disturbances are usually intentional and are generally accompanied by looting and vandalism. Investigators need to ascertain whether the fire was the act of a rioting crowd or of building owners who want to benefit financially from having the fire be attributed to the ongoing riots.

Human Behavior and Fire

A fire's origin, development, and consequences are all related—either directly or indirectly—to the actions and/or omissions of human beings. Understanding the behavior of witnesses or occupants is integral to the fire scene examination. By understanding why the occupants or witnesses behaved in a particular manner, an investigator can better evaluate fire development, cause, and the evidence found on scene.

Research conducted over the past several decades indicates that an individual's or group's behavior before, during, and after a fire can provide valuable insights for the investigator. Factors that affect this behavior include characteristics of the individual, characteristics of the group or population to which that individual belongs, characteristics of the physical setting where the fire occurs, and characteristics of the fire itself.

Characteristics of the Individual

An individual's physical limitations, such as age (as it relates to mobility), physical disabilities, injuries, medical conditions, and chemical impairment, can adversely affect the person's ability to take appropriate actions before and during a fire. The very young and the very old are most susceptible to these limitations. Cognitive limitations can hinder an individual's ability to recognize and react appropriately to a fire or explosion. Some factors that can limit a person's cognitive ability are age (as it relates to mental comprehension,

such as a child hiding instead of escaping), level of rest, alcohol use, drug use (legal or illegal), developmental disabilities, mental illness, and inhalation of smoke and toxic gases.

Greater familiarity with the setting can make a person's escape during a fire more likely, although physical limitations and cognitive impairments can minimize the advantages of such familiarity. In larger and unfamiliar structures, individuals tend to leave by the same route they took to enter, likely because the existence of an emergency causes their minds to focus on known, familiar exits and limits their ability to seek or consider alternative ones.

Characteristics of the Group or Population

The size of a group that a person is with can temper the person's response to a threat or purported threat. An individual's tendency toward delayed or inappropriate reaction is increased as the size of the group increases. Research has shown that this phenomenon occurs because individuals in groups will delay their responses to sensory cues until the other group members also acknowledge and react to those cues. When a group has a formalized structure, with defined and recognized leaders, the group tends to react to fire incidents more quickly and in a more orderly fashion; however, the reaction is not always appropriate. Examples of this type of group include school, hospital, nursing home, and religious facility populations.

Research has indicated that permanence, or the degree of familiarity among the individuals in a group, also affects response times. If the group is established and its members know one another well, such as in a family, sports team, or club, the individuals react and notify each other in a more timely manner than they do if the group is newly formed or its members are unfamiliar with each other. Further, a group's roles and norms in terms of gender, social class, occupation, or education can affect its response to threats. For example, in a fire or explosion situation, studies have shown that males are more likely to engage in activities to suppress or defuse the threat, whereas females are more likely to engage in reporting of the threat.

Characteristics of the Physical Setting

The characteristics of a burning structure will affect the response of the individuals within it. If occupants are unfamiliar with the building, they often react with increased levels of stress, which in turn can result in unpredictable behavior. Furthermore, as mentioned

FIGURE 15-13 A fire alarm system can include lights and horns to notify occupants.
© Jones & Bartlett Learning. Courtesy of MIEMSS.

FIGURE 15-14 Sprinklers indicate the presence of a fire suppression system.
© Jones & Bartlett Learning. Courtesy of MIEMSS.

previously, occupants who are unfamiliar with a building often try to exit by the door that they used to enter, even if that behavior moves them in the direction of the threat. Therefore, it is important that exits be clearly marked. Moreover, the number of exits available for escape drastically affects the occupants' behavior. If there are too few exit routes or if the exits are blocked or restricted, occupants are exposed to additional danger.

Fire alarm systems can help occupants recognize a threat **FIGURE 15-13**. Research has also shown that voice or directive messages may generate better responses than the use of strobes and horns alone. However, if the building has experienced a number of false alarms, people tend to delay their response until the actual fire emergency is confirmed, resulting in greater danger of being trapped. Although fire suppression systems provide greater time for occupants to react to a threat, occupants may overestimate the amount of extra time available **FIGURE 15-14**. Further, some systems, once they have discharged, can create reduced visibility, impacting the occupants' ability to escape.

Characteristics of the Fire

An individual's or group's response to a threat is tempered by their perception of the hazard or threat. For example, most individuals do not understand the

threat presented by flames and the ability of fire to grow exponentially and release toxic products into the air. In consequence, they may tend to disregard small flames as a source of danger. Similarly, many people lack knowledge about the toxic and incapacitating effects of smoke or believe that lighter-colored smoke is less dangerous than darker smoke.

Many people do not understand that fire, because it consumes oxygen, may cause a reduction of available oxygen in enclosed spaces. Similar to the effects of combustion by-products, depletion of the oxygen level to less than 15 percent of the air results in impairment of motor and mental skills and can ultimately be fatal. The behavior displayed by victims in these scenarios may appear inappropriate but may be explained by confusion resulting from inhalation of toxic gases or oxygen deprivation. Note that an individual may be impaired by more than one factor; in particular, an individual may have had an impairment before the fire (such as from alcohol) and suffer from additional impairment from the effects of the fire or the products of combustion.

Examples of Human Behavior Contributing to Fire Fatalities

Over the course of history, many catastrophic fire events have resulted in significant loss of life. Although many factors usually contribute to the scale of the fire event—such as criminal activity, inappropriate use of heat-producing devices, and uncorrected code violations—the response of the occupants also plays a factor. Several examples are highlighted here.

The Station Nightclub Fire

On February 20, 2003, the band Great White was engaged to play before a packed nightclub in West Warwick, Rhode Island. At approximately 11:07 PM, the band began to play its opening song, "Desert Moon." As part of the performance, the band set off three pyrotechnic "gerbs," which shot fountains of sparks against the walls of the stage area. The ignition of foam material on the walls of the stage area led to a fast-developing fire that took the lives of 100 people. That night, a TV news crew was filming the event to produce a segment on nightclub safety. Ironically, their video captured the tragic event that is one of America's deadliest fires.

Watching the video footage, viewers are offered a horrific accounting of fire dynamics and human behavior. Spectators and band members failed to see the wall covering ignite or realize its danger. First recognition of the fire by patrons occurred 24 seconds after ignition, with most patrons beginning to evacuate the building 30 seconds after ignition, when the band stopped playing. Some of the patrons had been drinking alcoholic beverages. The majority of the patrons attempted to exit the nightclub through the main entrance—the entry point they were familiar with—even though there were three other exits for the structure.

The resulting mass exodus through the main doors led to congestion of people. The doorway soon filled with a tangled mass of victims. In the end, 96 people died in the fire, and another 4 died in local hospitals. An additional 230 people were injured, with only 132 escaping unharmed.

MGM Grand Hotel Casino Fire

On November 21, 1980, a fire broke out in a vacant restaurant in the MGM Grand Hotel and Casino in Las Vegas, Nevada, resulting in the death of 85 people and the injury of 650 others. The fire originated in a wall soffit in an area where a sprinkler exemption applied, which contributed to its rapid development.

The fire spread was aided by wallpaper, polyvinyl chloride (PVC) piping, and plastic mirrors and eventually traveled through the lobby, sending a fireball out the hotel's front entrance. Most of the fatalities occurred as a result of toxic fumes that were circulated through the hotel's HVAC system. A significant number of the fatalities occurred in the upper floors of the hotel from smoke inhalation and carbon monoxide poisoning, many of them in the stairwells that served as chimneys transporting the toxic smoke and gases to the upper areas of the structure. Once in the stairwells, the occupants became trapped as the doors

behind them locked and the only open exits were on the ground floor and roof.

A review by the NFPA indicated that the hotel occupants did not panic, and many took rational steps to preserve their lives, including warning other occupants, placing towels at the base of their doors, and covering their faces with wet cloths.

Several months later, on February 10, 1981, an arson fire occurred at the Las Vegas Hilton. Using the knowledge they gained from the investigation of the MGM Grand Fire, firefighters used local television networks to notify people to stay in their rooms and out of the hallways and stairwells. Although eight people died in this fire, the steps taken by firefighters are believed to have prevented a much greater loss.

Kiss Nightclub Fire

On January 27, 2013, a fire caused by the inappropriate use of an outdoor pyrotechnic device inside the Kiss Nightclub in Santa Maria, Rio Grande do Sul, Brazil, resulted in the death of 242 people and the injury of 168 others. Much like the Station Nightclub fire, a pyrotechnic device ignited acoustic foam lining the ceiling of the stage area. Following the breakout of the fire, a stampede occurred as people tried to escape. These efforts were significantly hampered by the lack of exit signs and emergency exits. Most of the victims succumbed to smoke inhalation, and many people died as they either attempted to hide in bathrooms or mistook their doors for exits. More than 150 people were injured by the crush at the front door and the rapidly accumulating smoke within the nightclub.

Other factors contributing to the significance of this event include the furnishing of false information regarding the number of emergency exits during the permit-issuance process and the lack of or false information provided regarding functioning fire extinguishers.

Following this catastrophic event, Brazilian authorities conducted safety inspections of similar nightclubs, resulting in the closure of 58 facilities for safety concerns.

Factors Related to Fire Initiation

Fire and explosion incidents frequently occur as a result of an act or omission by one or more individuals occurring before or during the fire. Moreover, individual actions can encourage (or prevent) the spread of fire. An investigator should assess these actions, which may include the opening or closing of doors, operation of protection systems, and rescue efforts.

Common examples of human actions relating to fire initiation and spread include improper maintenance and operation of equipment or appliances; careless housekeeping; failure to follow product labels, instructions, warnings, and recalls; and violations of fire safety codes and standards.

Improper Maintenance and Operation

During the lifetime of most equipment, it is subject to a prescribed maintenance and cleaning schedule, which is usually provided by the manufacturer and should be followed to prevent malfunction. When the equipment is capable of explosion or starting a fire, the required maintenance becomes critically important. Likewise, the operating procedures for equipment or appliances are designed to ensure safety. If either of these two areas is neglected or improperly performed, the lapse can lead to a fire or explosion. Examine all maintenance records and operating instructions carefully to determine whether the correct procedures were followed.

Housekeeping

Household equipment that is capable of initiating an explosion or fire typically has instructions identifying the recommended clearance distances between the equipment and combustibles. For example, paper products or ignitable liquids should not be stored adjacent to the pilot light of a water heater. Carelessly discarded smoking materials, such as cigarettes or matches, can ignite a fire. Grease build-up in cooking areas or improperly stored cleaning solutions are other hazards. Note any housekeeping irregularities that may have contributed to the fire or explosion.

Manufacturer's Product Labels, Instructions, and Warnings

Labels, instructions, and warnings are placed on products to prevent their misuse or abuse **FIGURE 15-15**. Manufacturers' labels inform the user of the product's capabilities. Instructions provide the user with all pertinent information about how the product is intended to be used. Warnings alert the user to the dangers that can occur if the product is not used as intended and remind the user about the hazards of a product. A proper warning contains four key elements: an alert word to signal danger, a statement of the danger, a statement of how to avoid the danger, and an explanation of the consequences of the danger.

FIGURE 15-15 The investigator should read any labels, instructions, or warnings on products.
© Michael Ledray/Shutterstock.

According to ANSI Z535.4, *Product Safety Signs and Labels*:

- CAUTION: Indicates a potentially hazardous situation that, if not avoided, may result in minor or moderate injury.
- WARNING: Indicates a potentially hazardous situation that, if not avoided, could result in death or serious injury.
- DANGER: Indicates an imminently hazardous situation that, if not avoided, will result in death or serious injury.

Government Standards on Labels, Instructions, and Warnings

The government, like many manufacturers, also implements its own standards, guidelines, and regulations dealing with safety warnings and safe product design **FIGURE 15-16**. While manufacturer's warnings are common and helpful when they accompany a product, government standards serve a crucial role when it comes to defining parameters of a given product or substance.

Recall Notices

A recall notice is a method of notifying consumers of a product defect that was identified after the product was released for use by consumers. For the most part, recalls result from the identification of a dangerous situation that could arise even when the product is used in its intended manner. If an individual disregards a recall notice and continues to use the product, the outcome can be a fire, explosion, or other catastrophic event. Recall notices can be found through the Consumer Product Safety Commission.

(1) ANSI standards on labeling:
    (a) Z400.1/Z129.1 Hazardous Workplace Chemicals - Hazard Evaluation and Safety
       Data Sheet and Precautionary Labeling Preparation
    (b) Z535.1, Safety Colors
    (c) Z535.2, Environmental and Facility Safety Signs
    (d) Z535.3, Criteria for Safety Symbols
    (e) Z535.4, Product Safety Signs and Labels
    (f) Z535.5, Safety Tags and Barricade Tapes (for Temporary Hazards)

(2) UL standard on labeling: UL 969, Standard for Marking and Labeling Systems

(3) United States Federal Codes and Regulations:
    (a) "Consumer Safety Act" (15 USC Sections 2051–2084 and 16 CFR 1000)
    (b) "Hazardous Substances Act" (15 USC Sections 1261 et seq. and 16 CFR 1500)
    (c) "Federal Hazards Communication Standard" (29 CFR 1910)
    (d) "Flammable Fabrics Act" (15 USC Sections 1191–1204 and 16 CFR 1615, 1616, 1630–1632)
    (e) "Federal Food, Drug and Cosmetic Act" (15 USC Section 321 (m) and 21 CFR 600)
    (f) OSHA Regulations (29 CFR 1910)

(4) Industry standard: FMC Product Safety Sign and Label System Manual

FIGURE 15-16 Standards on labels, instructions, and warnings.

Modified from NFPA 921-2021, *Guide for Fire and Explosion Investigations*, Copyright © 2020, National Fire Protection Association. This reprinted material is not the complete and official position of the NFPA on the referenced subject, which is represented only by the standard in its entirety.

TIP

Both government and industry have developed accepted guidelines and standards for labels, warnings, and product instructions. Institutions such as American National Standards Institute (ANSI), Underwriters Laboratories, FM Global, and various United States Codes and regulations within the Code of Federal Regulations (CFR) have addressed these requirements.

Violations of Fire Safety Codes and Standards

Noncompliance with fire safety codes and standards can result in a fire or explosion. Noncompliance may be either deliberate or unintentional. When an investigator conducts an origin and cause determination, it is frequently difficult to determine whether deliberate misuse or abuse of the product, carelessness, or some other factor contributed to the event. Examine training records, maintenance records, and other documentation and look for a pattern that might point toward the potential cause of the event. Check with the local fire prevention bureau to see if the business had been cited for any fire code violations.

Children and Fire

Children's curiosity about fire and experimentation with fire play have been determined to be relatively normal activities in childhood. Besides curiosity,

children can be motivated to set fire out of frustration, anger, revenge, or a need for attention. The location and motive for setting fires will vary according to the age and developmental stage of the child.

Preschool-age children understand the world mostly as it affects them (egocentrism), so they may not understand cause and effect, such as a small fire growing into a big one, unless they have experienced the event personally. They are motivated by pleasure, and typically set fires out of curiosity. Early elementary school–age children may be interested in the process of fire, such as consumption of materials, and tend to observe and imitate the actions of adults. They still have difficulty comprehending cause and effect.

Children between approximately 8 and 12 years old understand cause and effect. Emotional and crisis fire-setting behavior may be seen in this group, due to inability to cope with trauma and change. Fires may also result from the child taking on too much responsibility and allowing a fire to grow without requesting help so as to not be perceived as incompetent.

Finally, adolescent children may set delinquent fires. They may experience the need for acceptance

TIP

Although the fire prevention and investigation communities use the term "fire-setters" in discussing youths involved in starting fires, you should avoid terms such as "fire-setter" or "arsonist" when discussing youth-involved fires with the youths involved, their families, or the public during your investigation.

by peers, a need to test limits, and issues with boredom. While they have more responsibility and freedom than younger children, they may be unwilling to think a situation through and take responsibility for negative outcomes.

Investigation of Fires Involving Youths

The fire investigator plays a very important role in the investigation of fires where youth involvement is suspected. Information and evidence obtained or observed at the fire scene may be critical in allowing the fire investigation team and other specialists to adequately identify and address issues involving a youth involved in fire.

The investigator should know the legal rules that apply in the jurisdiction in question, including the age at which a youth can be charged in the criminal law system (and whether the juvenile or adult system would be involved), and the role of parents and guardians.

The investigator should take note of where the fire began, and whether the fire involved planning or easily accessible lighters and fuels. With younger children, the investigator may find evidence of fire where the child sleeps or plays, using readily accessible fuels and ignition sources and potentially involving toys, bedding, or clothing. Older children may set fires in easily accessible public areas, and these fires may involve the use of lighters, matches, fireworks, or devices. Vegetation may be involved, as may ignitable liquids.

Recognition and Response to Fires

In a fire situation, an occupant's ability to recognize the danger is critical to their survival. The occupant must be able to respond appropriately to the perceived and actual dangers associated with the event. Sensory perception can be affected by the person's physical or mental state or by ingestion of alcohol or drugs.

The occupant's actions are based on four factors:

- Sight: Direct view of flames, smoke and visual alarms, or flicker
- Sound: Crackling of flames, failure of windows, audible alarms, a dog barking, children crying, voices, or shouts
- Feel: Temperature rise or structural failure
- Smell: Smoke odor

FIGURE 15-17 Interviews can lead investigators to important information.
© Glen E. Ellman.

When the occupant identifies a threat such as a fire or explosion, that person must make a decision about how they will react. The occupant has several choices, including ignoring the problem, investigating, fighting the fire, signaling an alarm, rescuing or giving aid to others, fleeing the fire, remaining in place, or reentering the structure after successfully escaping. The final decision will be influenced by the individual's state of mind.

The ability to escape is affected by the identifiability of escape routes, distance to the escape routes, fire conditions (such as smoke, heat, or flames), the presence of dead-end corridors, the presence of obstacles or people blocking the escape path, and the individual's physical disabilities or impairments.

Interviews with event survivors can provide information that will be beneficial in determining how people behaved before and during the fire or explosion **FIGURE 15-17**. These interviews can help establish the following points:

- Prefire conditions
- Fire and smoke development
- Fuel packages and their location and orientation
- Victims' activities before, during, and after discovery of the fire or explosion
- Actions taken by individuals that resulted in their survival (e.g., escaping or taking refuge)
- Decisions made by survivors and reasons for those decisions
- Critical fire events such as flashover, structural failure, window breakage, alarm sounding, first observation of smoke, first observation of flame, fire department arrival, and contact with others in the building

After-Action REVIEW

IN SUMMARY

- Several analytical tools are available to help the investigator organize and analyze multiple facts and data, including timelines, system analysis, mathematical and engineering modeling, graphic representations, scene reconstruction, and fire testing.
- Timelines may include events that occurred before, during, and after the fire.
- Hard times can be provided by public safety dispatch, devices that record time, private photos and videos, and alarm system data.
- When collecting information on hard times, the investigator must verify whether the time recorded by the source, or source device, is accurate.
- Soft time may be estimated and may relate to other events observed by a witness.
- A variety of timelines may be required to effectively evaluate and document a sequence of events precipitating the fire, events during the actual fire incident, and postfire activities.
- Timelines may be created with pencil and paper or with computer software; they can also depend on the assistance of law enforcement intelligence personnel.
- Failure mode and effects analysis allows the investigator to identify important predecessor events to an incident, through identifying specific components associated with potential ignition sources or fire spread.
- Mathematical (engineering) models can be simple or very complex; more complex modeling may require the use of powerful personal computers.
- Mathematical and engineering models have several potential limitations, including quality of input, generic data, and approximations made within the model.
- Hydraulic analysis helps the investigator determine whether a sprinkler or suppression system and water supply were matched to the hazard being protected.
- Structural analysis can provide information about why a building collapsed at a given point during a fire.
- Fire dynamics analysis can help the investigator test and evaluate origin and cause hypotheses, evaluate witness evidence, and determine fire growth and development.
- Graphic representations help investigators understand an incident location better, assist in interviewing witnesses, and help define and identify materials and systems and their involvement in an incident.
- Fire testing may be done in the field or in a controlled environment; this process is performed to help check data collected or test a specific hypothesis.
- Numerous data points must be gathered about a building and its contents to ensure that a fire model relies on accurate information.
- The presence of multiple fires may support an incendiary determination, although several other phenomena may give the false impression of multiple fires.
- The appearance of trailers can be significant to the analysis of a fire, although other phenomena may give the false appearance of the remnants of a trailer.
- Points of entry in a building should be carefully examined and documented for potential physical evidence.
- The fire investigator who sees signs of forced entry should interview first responders and suppression personnel to determine whether they forced entry as part of their response to the fire.
- Almost any appliance or heat-producing device can potentially be used as an incendiary device.
- A fire-setter may attempt to mask the true cause of the fire by placing an appliance in the scene to be discovered as the "obvious" cause of the fire.
- Because arsonists may set multiple fires, an investigator should inspect the fire building fully to determine whether other fires occurred that have separate origins.
- The presence of ignitable liquids in an unusual or unexpected place may indicate that the fire was incendiary, but may also have numerous non-incendiary meanings.

- The ignitable liquid detection (IGL) canine's abilities and reliability depend on its training and its handler.
- An IGL canine–handler team should be certified by an appropriate certifying body and must undergo periodic proficiency assessments.
- The investigator may collect debris samples from an area even in the absence of a canine alert there.
- A canine alert in a certain area is not conclusive proof of the placement of an ignitable liquid, and a canine alert is not appropriate evidence in court unless the material on which the canine alerted is confirmed by laboratory analysis to contain ignitable liquid.
- A fire-setter often starts a fire in a remote location or one obscured from public view.
- Contents are sometimes removed or replaced with items of lesser value prior to an incendiary fire. Careful documentation of the remains and the debris may be useful in establishing a fraudulent insurance claim, even in situations in which the fire cause cannot be determined.
- Sabotage of detection systems and fire-resistive assemblies may support a determination of an incendiary fire.
- Fire-setters sometimes take advantage of conditions or circumstances that add to the chances of successful destruction of the property, such as emergency conditions, civil unrest, or the unavailability of the fire department due to parades or other alarms.
- Understanding the behavior of witnesses or occupants allows the investigator to better evaluate fire development, cause, and evidence found at the scene.
- Physical and cognitive/mental limitations can affect an individual's ability to recognize and react appropriately to a fire or explosion.
- The behavior of children with fire, and the reasons for fire-setting behavior, will depend on the age and maturity level of the children.

KEY TERMS

Benchmark events Events that are particularly valuable as a foundation for the timeline or may have significant relation to the cause, spread, detection, or extinguishment of a fire.

CFD models Analytical models for fire behavior based on computational fluid dynamics.

Estimated time An approximation based on information or calculations that may or may not be relative to other events or activities.

Failure mode and effects analysis (FMEA) A technique used to identify basic sources of failure within a system and to follow the consequences of these failures in a systematic fashion.

Fault trees Logic diagrams that can be used to analyze a fire or explosion; also known as decision trees.

Hard times Specific points in time that are directly or indirectly linked to a reliable clock or timing device with known accuracy.

Heat transfer models Models that allow the investigator to determine how heat was transferred from a source to a target by one or more of the common heat transfer modes: conduction, convection, or radiation.

High-temperature accelerants (HTAs) Mixtures of fuels with Class 3 or Class 4 oxidizers and thermite mixtures.

Incendiary devices A wide range of mechanisms used to initiate an incendiary fire.

Incendiary fire A fire that is intentionally ignited in an area or under circumstances where and when there should not be a fire. (NFPA 921)

Intent Normally an element of the proof of a crime; the state of mind that exists at the time a person acts or fails to act.

Mass arson The setting of three or more fires at the same site or location during a limited period of time.

Modus operandi (MO) The method of operation used by the offender.

Personation Unusual behavior by an offender, beyond that necessary to commit the crime.

Relative time The chronological order of events or activities that can be identified in relation to other events or activities.

Sabotage Intentional damage or destruction.

Scaled timeline Timeline in which the spacing between the time events being scaled is depicted in a manner that would show the elapsed time between each event.

Serial arson Multiple arsons by an offender who sets three or more fires, with a cooling-off period between fires.

Serial arsonist A serial fire-setter.

Serial fire-setter An individual or group involved in three or more fire-setting incidents.

Soft time Estimated or relative point in time.

Spree arson The setting of three or more fires at separate locations with no emotional cooling-off period between fires.

Staging Purposeful alteration of the crime scene prior to the arrival of police.

System analysis An analytical approach that takes into account characteristics, behavior, and performance of a variety of elements.

Timeline A graphic or narrative representation of events related to the fire incident, arranged in chronological order.

Vandalism Mischievous or malicious fire-setting that results in damage to property.

Zone models Computer fire models that divide a compartment into two zones: a hot upper zone and a cooler lower zone. (Douglas et al., 2013)

REFERENCES

Douglas et al. *Crime Classification Manual: A Standard System for Investigating and Classifying Violent Crimes*, Third Edition. Hoboken, NJ: John Wiley and Sons, 2013.

National Fire Protection Association. 2020. *NFPA 921: Guide for Fire and Explosion Investigations*, 2021 ed. Quincy, MA: National Fire Protection Association.

On Scene

1. Have you used any of the analytical methods discussed in this chapter in a fire investigation or in another context? When do you believe each would be most useful to the fire investigator?

2. During witness interviews, if given a soft time by a witness, are there ways in which the investigator could further inquire or investigate to help convert these times into hard times, or at least to provide a more accurate time frame for the witness's observation?

3. Review the arson statutes and related laws in your jurisdiction. How do they differ from the definition of "incendiary" provided here and in NFPA 921? If your jurisdiction's laws provide for different levels or grades of arson offenses, what separates each offense from the others?

4. After reading about the major fires described in this chapter, list some codes and laws you are familiar with that address life safety in public buildings, and explain how they relate to the scenarios described. What role does the fire investigator play in determining which codes and laws should be passed to protect the public?

Chapter Opener: © Jones & Bartlett Learning. Photographed by Glen E. Ellman; On Scene siren: © Bildgigant/Shutterstock.

CHAPTER **16**

Explosions

You are the Fire Investigator

You are assigned to conduct an origin and cause investigation at a residence. While inspecting the kitchen, you note damage on the walls and ceilings indicating that they were pushed outward; however, they contain very little thermal damage. You determine that the residence is connected to a natural gas line provided by the local utility company. Numerous potential heat sources are seen within the residence, including a water heater and a furnace, both of which have flames as part of their normal operation.

1. What safety issues may exist at this scene?
2. How will an assessment of the damage assist with determining the point of origin?
3. How will you identify the ignition source?

Access Navigate for more practice activities.

Introduction

In fire and explosion investigations, an **explosion** is defined as a sudden conversion of potential energy, either chemical or mechanical, into **kinetic energy** (the energy possessed by a system or object as a result of its motion). The conversion into kinetic energy produces and releases gases under pressure, which then cause mechanical damage to materials, such as movement, shattering, and other changes. Whether an explosion has occurred is determined by examining the evidence. For example, **blast overpressure** can damage containers, structures, and equipment, and cause injury to persons. Although an explosion is usually accompanied by a loud noise, owing to the production and violent release of gases during the event, that sound is not the primary criterion for identifying an explosion.

The ignition of a flammable vapor–air mix in a closed container that pops off the lid or damages the can is an explosion, whereas the same ignition occurring in an open field may not be an explosion as defined by NFPA 921. A failure and bursting of a container caused by hydrostatic pressure, such as water pressure, is not an explosion because it was not created by gas. Factors controlling the explosion effects and damages include the type, configuration, and amount of fuel, as well as the size, shape, and material of the container or construction and the type and amount of venting present.

Flash fires are a unique form of burning that can lead to investigative confusion because of the fire patterns they create. A **flash fire** is a fire that spreads rapidly by means of a flame front through a diffuse fuel such as dust, gas, or ignitable liquid vapor, but without production of damaging pressure. The ignition of these fuels does not necessarily result in an explosion.

Whether an explosion occurs will depend on numerous factors, including the location and concentration of the fuel, the location of any obstacles, and the strength and geometry of any confining vessel.

The *Explosions* chapter of NFPA 921 primarily discusses the terms and techniques pertaining to explosions of diffuse fuels in residential and commercial framed structures, so investigating explosions in other types of structures (for example, concrete structures or portable buildings) may require additional research and specialized expertise. Further, solid and liquid explosives require their own expertise and are not covered at all in NFPA 921. Be aware of the possibility of secondary explosions and/or devices. If you are unsure about the possibility of secondary devices, do not proceed; request qualified personnel to conduct the investigation. It is best to have mutual aid or specialized resources identified prior to the time of need. These resources will vary depending on the jurisdiction but are often available from local, county, state, regional, and federal law enforcement, fire, and other agencies.

Types of Explosions

The two major types of explosions, mechanical and chemical, are distinguished by the source and the mechanism that produces the explosion **FIGURE 16-1**. Several subtypes of explosions are also identified under these major types.

Mechanical Explosions

A **mechanical explosion** does not involve changes in the basic chemical nature of the substance(s) in the vessel; it is a purely physical reaction. A mechanical explosion entails the rupture of a vessel or container

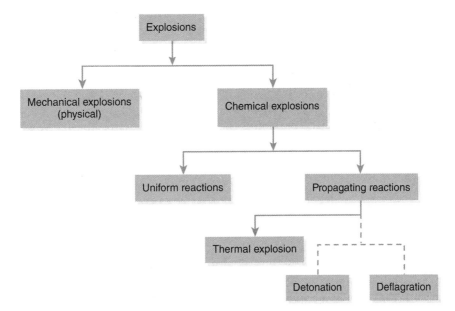

FIGURE 16-1 Types of explosions.

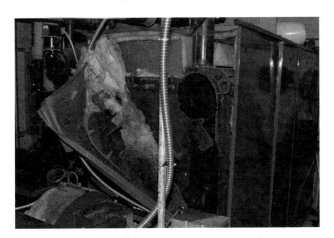

FIGURE 16-2 BLEVE damage.

(such as a cylinder, tank, or boiler) that results in the release of pressurized gas or vapor. The pressure leading to the mechanical explosion is not due to a chemical reaction or change in chemical composition of the substances involved.

A **boiling liquid/expanding vapor explosion (BLEVE)** is a frequently seen subtype of mechanical explosion. A BLEVE occurs when containers of liquids are under pressure and reach temperatures above their atmospheric boiling point. The vessels involved may range from a small butane lighter to a railroad tanker, and an ignitable liquid need not be involved. Indeed, a common example of a container that may experience a BLEVE is a steam boiler, where the heated steam provides the pressure to rupture the vessel **FIGURE 16-2**.

For the container to become heated to the point of explosion, the internal temperature of the vapor or liquid in the container must be raised. This can occur when the vessel is exposed to fire. In such a case, the heat increases pressure within the vessel, leading to rupture. Because the liquid's pressure drops dramatically when it is released, the liquid may vaporize almost instantly. If the vessel contents are ignitable, a fire of the contents is likely to be ignited either from the original heat source that caused the BLEVE or by another electrical or mechanical source.

Nonfire causes of BLEVEs include overfilling, runaway reaction, vapor explosion, or mechanical failure.

Chemical Explosions

A **chemical explosion** is one in which an exothermic chemical reaction is the source of the high-pressure gas, and the fundamental chemical nature of the fuel is changed. Although chemical explosions can involve solid combustibles or explosive mixtures of fuel and an oxidizer, the most typical events are propagating reactions that involve gases, vapors, or dust mixed with air.

Combustible Explosions

The most common chemical explosions are those caused by the burning of combustible hydrocarbon fuels. A **combustion explosion** is characterized by the presence of a fuel (such as dust), with air acting as an oxidizer. The elevated pressures are created by the rapid burning of the fuel and the production of combustion by-products and gases. The velocity of the flame front's propagation through the fuel determines whether the combustion reaction is classified as a deflagration (slower than the speed of sound) or detonation (faster than the speed of sound).

Uniform reactions occur more or less equally throughout the material and include ordinary chemical reactions that form gaseous products at a rate faster than they can be vented. Propagating reactions initiate at a specific point in the material and continue through the unreacted material.

Thermal explosions are the result of exothermic reactions occurring within confined areas without any means of dissipating their heat of reaction. This can accelerate to the point at which high-pressure gases are generated and an explosion occurs.

A **deflagration** is a reaction that travels through the air (propagates) at subsonic velocities. A deflagration can be vented successfully.

NFPA 921 defines a **detonation** as a combustion reaction that propagates at supersonic velocities, more than 1100 ft/sec (340 m/sec). Other authorities define the minimum detonation speed as 1000 m/sec (3250 ft/sec) or higher, and explosives typically have propagation speeds many times the speed of sound.

The difference between a detonation and a deflagration focuses on the magnitude of pressure versus time for the system involved in the combustion reaction. A significant difference between deflagration and detonation is the time over which the event occurs: A detonation is faster and cannot be vented because of the speed of the reaction.

Subtypes of combustion explosions are classified as flammable gases, vapors of ignitable (flammable and combustible) liquids, combustible dusts, smoke and flammable products of incomplete combustion (such as in a backdraft explosion), and aerosols.

Electrical Explosions

Electrical explosions occur when high-energy arcing generates sufficient heat to cause an explosion. Thunder accompanied by a lightning bolt is an example of an electrical explosion effect. Such explosions require a special expertise to investigate and are not covered in NFPA 921.

Nuclear Explosions

In a nuclear explosion, the high pressures within the primary system and secondary system (such as steam generators and boilers) are created by the enormous heat produced by the fission or fusion of atoms. NFPA 921 does not cover the investigation of nuclear explosions; investigations of such events should be referred to experts in this area.

Characterization of Explosion Damage

The terms *low-order damage* and *high-order damage* are preferred to characterize explosion damage. The differences in damage are influenced more by the rate of pressure rise and the strength of the confining vessel or structure than by the maximum pressures within the system.

Low-Order Damage

Low-order damage is produced by pressure rising at a slow rate. The characteristics of low-order damage include walls bulged out or laid down, roofs lifted slightly, windows dislodged with the glass intact, and thrown-out debris that is generally large and found within a short distance from the structure **FIGURE 16-3**.

High-Order Damage

High-order damage results from a rapid rate of pressure rise **FIGURE 16-4**. This kind of damage is characterized by walls, roofs, and structural members shattering, which results in small, pulverized debris that is often thrown great distances from the structure (hundreds of feet).

FIGURE 16-3 The windows are dislodged as a result of low-order damage to this dwelling.
© Chris Pole/Shutterstock.

FIGURE 16-4 Shattered remains as a result of high-order damage.
© Urbancow/iStockphoto.com.

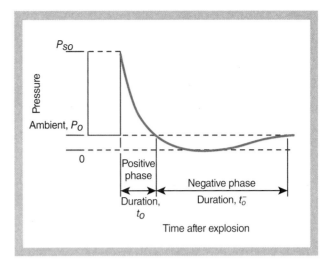

FIGURE 16-5 Typical pressure history from an idealized detonation, measured at a point away from the point of detonation.
Reproduced from NFPA 921-2021, *Guide for Fire and Explosion Investigations*, Copyright © 2020, National Fire Protection Association. This reprinted material is not the complete and official position of the NFPA on the referenced subject, which is represented only by the standard in its entirety.

Dynamics and Effects of Explosions

An explosion is a gas dynamics phenomenon that, under ideal theoretical circumstances, will manifest itself as an expanding spherical heat and pressure wave front. The heat and pressure waves produce the damage characteristic of explosions. The effects of explosions can be classified into four major groups:

- Blast overpressure and wave effect
- Projected fragment effect (shrapnel)
- Thermal effect
- Seismic effect (ground shock)

Blast Overpressure and Wave Effect

The blast overpressure and wave effect results from the production of large quantities of gas by the explosion of the material. The gases move outward at high speed from the point of origin and then return. The blast overpressure and wave effect creates a positive-pressure phase characterized by the outward movement of gases and displaced air, followed by a negative-pressure phase in which the air rushes back toward the area of origin.

Positive- and Negative-Pressure Phases

The positive-pressure phase, in which the expanding gases move outward from the point of origin, is more powerful than the negative-pressure phase and is responsible for most of the pressure damage, including weakening of the structure. The negative-pressure phase follows the rapid outward movement of the positive-pressure phase, which leaves low air pressure behind it **FIGURE 16-5**. Air then moves into the low-pressure area toward the point of origin. The negative-pressure phase can cause significant damage and result in further weakening of an already-compromised structure. It can also propel artifacts, including items of physical evidence, toward the point of origin, potentially contributing to concealment of the point of origin.

Shape of Blast Wave Front

The shape of the blast front from an idealized explosion is spherical, expanding evenly in all directions from the epicenter. Factors such as confinement, obstruction, ignition position, cloud shape, or concentration distribution at the source of the blast pressure wave may all change and modify the direction, shape, and force of the front.

When the containers, structures, or vessels that contain or restrict the blast overpressures rupture, they often break into fragments that may be thrown over great distances, depending on their size and shape. These fragments, which are also called missiles, shrapnel, debris, or projectiles, can cause significant damage.

Confinement changes the shape and force of the front. The direction of the blast wave front may be altered as a result of either the venting path or redirection when the front is reflected off a solid object. The force decreases with distance from the epicenter, assuming no propagating reactions occur. The correlation between the rate of pressure rise and the damage effects of the explosion is shown in **TABLE 16-1**.

TABLE 16-1 Rate of Pressure Rise Versus Damage	
Rate of Pressure Rise	**Damage**
Slow	Pushing or bulging type of damage Weaker parts of the structure will rupture first Characteristic of low-order damage
Rapid	Shattering of the confining vessel or container Debris will be thrown over great distances Characteristic of high-order damage

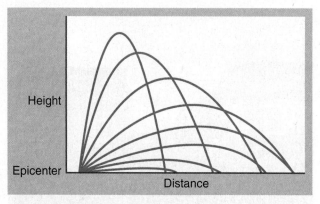

FIGURE 16-6 The distance shrapnel is propelled depends on the initial direction as well as on the weight and aerodynamic characteristics of the projectile.

TABLE 16-2 Thermal Effects from Various Explosions	
Explosion Type	**Effect**
Combustion	Releases heat energy Can cause secondary fires
Chemical	Releases great quantities of heat
Detonating	Produces extremely high temperatures of short duration
Deflagration	Produces lower temperatures of longer duration

Where the pressure is less rapid, the venting effect has an important impact on the maximum pressure that develops. For example, in fugitive gas explosions in residential or commercial buildings, the maximum pressure is limited to a pressure that is slightly higher than the pressure that the components of the building enclosure can sustain without rupture. In a well-built residence, this pressure seldom exceeds 3 psi (21 kPa).

Rate of Pressure Rise Versus Maximum Pressure

Damage caused by the explosion blast pressure front is not only a product of the total amount of energy, but also, and sometimes more significantly, depends on the rate of energy release and the resulting rate of pressure rise.

Damage from slower rates of pressure rise is similar to low-order damage: Windows and other weaker structural members will fail, allowing the blast pressure to vent directly out, thereby mitigating the explosive effects. Damage from a rapid pressure rise is similar to high-order damage because venting does not occur, there is more shattering of the confining structure components, and debris is thrown over great distances. Essentially, a slow rate of pressure rise decreases the maximum pressure through venting. (See NFPA 68, *Standard on Explosion Protection by Deflagration Venting*, for a discussion of calculating the theoretical effects of venting during a deflagration.)

Projected Fragment Effect

The projected fragment effect, also called the shrapnel effect, refers to the pieces of debris that result from the rupture of vessels or containers as a result of the blast pressure fronts. The shrapnel may cause personal injury or damage a great distance from the source of the explosion. It may also sever electric utility lines, fuel gas lines, or storage containers. The distance over which the shrapnel will be propelled largely depends on the initial direction as well as on the weight and aerodynamic characteristics of the projectile **FIGURE 16-6**.

Thermal Effect

The thermal effect may occur because the explosion releases quantities of energy that can be sufficient to ignite nearby combustibles. As noted in **TABLE 16-2**, detonation produces extremely high temperatures of limited duration, whereas deflagration produces lower temperatures of longer duration. **Fireballs** (momentary balls of flame during or after an explosion, which may present high-intensity, short-term radiation) and **firebrands** (hot or burning fragments propelled from the explosion) are possible thermal effects, especially with BLEVEs of flammable vapors. Secondary fires may increase the damage and injuries from the explosion. Unfortunately, it is often difficult to determine which occurred first—the fire or the explosion.

The thermal damage from a chemical explosion depends not only on the nature of the explosive fuel but also on the duration of the high temperature. Table 16-2 provides an analysis of explosion types and their resulting thermal effects. An investigator should note the part that time plays in thermal effects. Because these effects may ignite fires away from the explosion center, they must be investigated thoroughly to determine the reason for them and to prevent them from misleading the investigation.

Seismic Effect

The transmission of tremors through the ground is known as the **seismic effect**. This effect is a result of the blast wave's expansion, which causes structures to be knocked to the ground. Small explosions result in negligible seismic effects, whereas larger explosions may produce damage to structures, underground utility services, pipelines, tanks, and cables.

Factors Controlling Explosion Effects

The following factors determine the effects of an explosion:

- Type and configuration of fuel
- Nature, size, volume, and shape of the containment vessel
- Level of congestion and obstacles within the vessel
- Location and magnitude of the ignition source
- Venting of the containment vessel
- Relative maximum pressure
- Rate of pressure rise

The nature of these factors and their variables can produce a variety of physical effects. The variables that may affect the characteristics of a blast pressure front as it travels from the source include blast pressure front modification by reflection or refraction and blast focusing.

The blast pressure front may be amplified because of its *reflection* from objects in its path, causing overpressure to increase to an extent that depends on the angle of incidence and the overpressure involved in the incident. This phenomenon is negligible with deflagrations.

A blast pressure front that encounters a layer of air at a significantly different temperature may bend or *refract*. This occurs because the speed of sound is proportional to the square root of the absolute temperature and, therefore, to the density of the air. This effect is also negligible with deflagrations.

Reflection in corners may *focus* the blast pressure in the corner, and refraction from a temperature inversion can focus pressure on the ground adjacent to the center of the explosion.

The make-up of the fuel has a significant effect on the results of an explosion. When the fuel mixes with air and creates turbulence, the flame speed, rate of combustion, and rate of pressure rise all increase. The shape, size, and location of obstacles within the container affect the turbulence and consequently the severity of the explosion. The fuel's container has a great effect on the explosion as well: Its size, shape, construction, volume, materials, and design all impact the fuel's reaction.

In addition, both the location and the magnitude of the ignition source influence an explosion. The closer the ignition source is to the wall of the container, the sooner the flame will reach the wall and be extinguished, leading to lower pressure and a smaller explosion. The closer the ignition source is to the center of the container, therefore, the more pressure that can build within the container and the greater the explosion will be.

The damage created by the explosion is greatly affected by the ventilation of the fuel prior to the explosion, particularly if the fuel is diffuse, such as gas, vapor, or dust. The more ventilation there is, and the closer that ventilation is to the ignition source, the weaker the explosion will be. A weaker explosion creates less damage, although the damage will be greatest in the path of the venting—for example, a window or a door. Ventilation has less of an effect when the explosion is caused by a detonation because there is not enough time for the venting to relieve sufficient pressure.

Seated and Nonseated Explosions

The crater or area of greatest damage may be characterized as the **seat of the explosion**. The seat may be of any size, depending on the amount and strength of the explosive material. A seated explosion is generally characterized by high pressure and rapid rates of pressure rise. Types of fuels that may generate seated explosions include explosives, steam boilers, highly confined fuel gases and liquid fuel vapors, and BLEVEs in small containers.

Explosives generally have a highly centralized epicenter, or seat. Accordingly, because these explosions have a high-velocity positive-pressure phase, they usually produce craters or localized areas of great damage. Boiler and pressure vessel explosions exhibit effects similar to explosives, albeit with less localized overpressure adjacent to the source. Fuel gases are ignitable vapors that, when confined in small vessels, may also produce seated explosions. Finally, a BLEVE produces a seated explosion if the confining vessel is of a small size (such as a can or barrel) and the rate of pressure release is sufficiently rapid.

A **nonseated explosion** occurs when the fuels are dispersed or diffused at the time of explosion, and demonstrate moderate rates of pressure rise and subsonic explosive velocities (supersonic detonations may also produce nonseated explosions, depending on the conditions). Fuel gases, such as natural gas and liquefied petroleum (LP), usually produce nonseated explosions because the gases involved are found in large containers such as rooms. Explosions associated with the vapors of flammable fuels or combustible liquids are usually nonseated explosions. A subsonic explosion speed, as well as the magnitude of the area that the explosion covers, means that small, high-damage seats will not occur.

Dust explosions most often occur in confined areas with wide dispersal, such as grain elevators, processing plants, and coal mines; these large areas of origin also preclude the development of explosion seats. Similarly, smoke explosions or backdrafts usually involve a widely diffused volume of combustible gases and particulate matter. Accordingly, the explosive velocities are subsonic, which limits the production of pronounced seats.

Gas and Vapor Explosions

Explosions from fuel gases or the vapors of ignitable liquids are the most commonly encountered explosive events. Explosions involving lighter-than-air gases occur less frequently than do explosions involving gases or vapors with a vapor density greater than 1.0 (heavier than air). Ignition temperatures in the range of 700°F to 1100°F (370°C to 590°C) are common. Thus, fuel gas–air mixtures are the most easily ignitable fuels that may result in an explosion. Minimum ignition energies begin at 0.25 mJ (an extremely low level of energy). **FIGURE 16-7** shows the damage resulting from a vapor explosion.

FIGURE 16-7 Low-order damage caused by a propane vapor explosion.
Courtesy of Rodney J. Pevytoe.

Interpreting Explosion Damage

The explosion damage to structures is related to several factors:

- Fuel–air ratio
- Vapor density of the fuel
- Turbulence effects
- Volume of the confining space
- Location and magnitude of the ignition source
- Venting
- Strength of the structure

Fuel–Air Ratio

The entire volume of air need not be occupied by a flammable mixture of gas and air for an explosion to occur. Indeed, relatively small volumes of explosive mixtures capable of causing damage may result from gases or vapors collecting in a given area.

Mixtures at or near the lower explosive limit (LEL) or the upper explosive limit (UEL) of a gas or vapor usually produce less-violent explosions than do those near the optimal explosive concentration. The optimal concentration is usually just slightly richer than the stoichiometric level. Because this **stoichiometric mixture** (slightly fuel-rich environment) produces the most efficient combustion, the result will be the highest flame speeds, the quickest rates of pressure rise, and the maximum pressures, and consequently the most damage.

By contrast, explosions of mixtures near the LEL do not tend to produce large quantities of post-explosion fire, because most of the available fuel is consumed during the explosive propagation. Explosions of mixtures near the UEL tend to produce larger post-explosion fires because fuel remaining from the explosion of the fuel-rich mixture may continue to burn after the initial explosion.

Flame speed can vary dramatically, depending on the temperature, pressure, confining volume, confining configuration, combustible concentration, and turbulence. The **burning velocity** is the rate of flame propagation relative to the velocity of the unburned gas ahead of it. The fundamental burning velocity is an inherent characteristic of a combustible material; that is, it is a fixed value. Defined as the velocity at which a flame reaction front moves into the unburned mixture as it chemically transforms the fuel and oxidant into combustion products, the burning velocity is only a fraction of the flame speed. The **flame speed** is the product of the velocity of the flame front (created by the Z value of the combustion products caused by the increase in temperature), any increase in the number of moles (a measurement of the amount of a substance based on the number of particles it contains), and any flow velocity caused by the motion of the gas mixture prior to its ignition. The burning velocity of the flame front can be calculated from the fundamental burning velocity as reported in NFPA 68, at standardized conditions of temperature, pressure, and composition of unburned gas.

An explosion can occur even if the fuel has not spread throughout the entire container. In such a case, the fuel's properties help determine the reaction that occurs. Similarly, it cannot be assumed that because no explosion occurred, the container is not full of fuel. A delicate balance exists among the various elements involved.

The fuel–air ratio becomes of further importance in investigating an explosion when the following concepts are applied:

- Adiabatic flame temperature: The theoretical temperature of a complete combustion process that loses neither energy nor heat to its outside environment.
- Laminar burning velocity: The flame propagation rate relative to the unburned gas and the movement of the front into the unburned gas as it chemically combines the fuel and oxidizer into a combustible.
- Expansion ratio: The rate of expansion of a fuel–air mixture's products of combustion after ignition behind the expanding flame front.
- Laminar flame speed: The speed of a freely burning flame relative to a fixed point without turbulent effect.
- Turbulent flame speed: The speed of a turbulent combustion, which is generally relevant to a real explosion. Turbulent flame speeds may exceed the speed of sound in the unreacted mixture.

Vapor Density

Air movement, from both natural and forced convection, is the dominant mechanism for moving gases within a structure. Vapor density is the ratio of the average molecular weight of a given volume of gas or vapor to that of air at the same temperature and pressure; this value is also referred to as specific gravity. The vapor density of the gas or vapor may affect the movement of the fugitive gas as it escapes from its container or system. Heavier-than-air gases and vapors (vapor density greater than 1.0, such as ignitable liquids and LP gases) have a tendency to flow toward lower areas. In contrast, lighter-than-air gases (such as natural gas) tend to rise and flow toward upper areas, so their explosion can have an epicenter at a level or floor that is actually higher than the gas source or leak. Where lighter- or heavier-than-air gases are involved, the operation of heating and air conditioning systems and temperature radiance may cause mixing and movement that can reduce the effects of vapor density. Field tests have confirmed that vapor density effects are minimized by most air exchange rates that exist in older homes. Thus, vapor density effects are greatest in still-air conditions.

Full-scale testing of flammable gas concentrations has shown that near-stoichiometric concentrations of gas develop between the location of the leak and either the ceiling (for lighter-than-air gases) or the floor (for heavier-than-air gases). A heavier-than-air gas that leaks at floor level may initially concentrate to a greater extent at floor level, but with time will slowly diffuse upward. The inverse relationship is true for a lighter-than-air gas leak at ceiling height.

When fuels mix with air to their flammable/explosive ranges, the mixture usually consists of more air than fuel; thus, the vapor density is closer to that of air (1.0) than that of the fuel. Hydrogen, acetylene, and ethylene are exceptions to this rule, however.

Turbulence

Turbulence within a fuel–air mixture increases the flame speed, which in turn increases the rate of combustion and the rate of pressure rise. The level of turbulence may vary according to the size and shape of the combining vessel and, therefore, may affect the severity of the explosion. Be especially aware of any mixing or turbulent sources, such as fans and forced-air ventilation, and watch for chevrons or other structural components that could create turbulence.

Nature of the Confining Space

The size, shape, construction, volume, materials, and design of the confining space greatly affect the nature of the explosion damage. In a confining space, a smaller space for the same amount of fuel will increase the rate of pressure rise, causing a more optimal explosion. In other words, the same fuel amount in a steel drum will result in a more violent explosion as compared with an explosion in a house. Similarly, obstructions such as walls, columns, and/or other large objects increase flame speed and, therefore, also increase the rate of pressure rise.

Location and Magnitude of the Ignition Source

The highest rate of pressure rise will occur when the ignition source is located in the center of the confining structure. Although the energy of the ignition source has a minimal effect on the course of the explosion, unusually large ignition sources such as explosive devices can significantly increase the speed of pressure development, thereby causing the event to transition from a deflagration to a detonation.

Venting

The nature of the venting of the containment vessel for gas, vapor–gas, or dust-fueled explosions will affect the damage inflicted by the explosion. A length of pipe may rupture in the center if it is sufficiently long, because venting is insufficient for the explosion energy. A room may experience destruction or merely movement of the walls and ceiling depending on the number, size, and location of doors and windows **FIGURE 16-8**. Furthermore, the venting of a vessel structure can cause damage to materials in the path of the venting. Notably, if the explosion is classified as a detonation, venting effects will be minimal because of the high speeds of the blast pressure fronts—the movement of the pressure fronts is too fast to allow any venting to relieve the pressure.

Strength of the Structure

A structure, for the purposes of this chapter, is defined as anything that can be used to contain something else. The characteristics of a structure impact its strength against an explosion. The strength of the structure will depend on the type of structure; the materials

FIGURE 16-8 Damage to a wall from a gas explosion.
Courtesy of Nina Scotti, NMS Investigations, Inc.

that compose it; and potentially, in the case of a building, the services or utilities which with it is equipped. The characteristics of a structure impact its strength against an explosion. For example, a glass jar (simple structure) would react differently from a five-story building made of steel and glass and serviced with natural gas and electrical service (complex structure).

Underground Migration of Fuel Gases

Both lighter-than-air and heavier-than-air fuel gases that have escaped from underground piping systems can migrate underground to enter structures, resulting in fires or explosions there. The fugitive gases may sometimes permeate the soil, migrate upward, and dissipate harmlessly in the air. In contrast, if the surface of the ground is obstructed by rain, snow, freezing, or paving, the gases may migrate laterally and enter structures through disturbed soils.

Fugitive gases may enter buildings through any opening into the structure, such as sewer lines, electrical and telephone conduits, or drain tiles, or through basement and foundation walls. The fire investigator should note that natural gas and propane have no natural odors. Foul-smelling compounds (such as ethyl mercaptan) are routinely added to gases, but some individuals may lack the ability to smell these compounds; further, under some conditions, the effectiveness of the odorant can be reduced. Odorant verification should be part of any explosion investigation. Gas detectors may be helpful in locating sources of gas leaks.

Multiple Explosions

Secondary explosions or cascade explosions may result when gas and vapors have migrated to adjacent stories or rooms of a building, resulting in the formation of pockets of fuel. Thus, when ignition or explosion takes place in one story or room, subsequent explosions may occur in adjoining areas or stories. The migration or pocketing of gases may produce areas with different air–fuel mixtures. Accordingly, the dynamics of air–fuel mixtures may result in a series of vapor–gas explosions depending on the ratio of the air–fuel mixture in each area.

Dust Explosions

Violent explosions can be fueled by dust that is dispersed within the air. This is true both for combustible materials as well as for materials not normally considered to be combustible. For example, dust explosions may occur in agricultural products, carbonaceous materials, chemicals, dyes and pigments, metals, plastics, and resin. Published values for dust explosion properties should be treated cautiously, as numerous factors may affect the explosiveness, including (but not limited to) chemistry, composition, particle size, and moisture content. When needed for an investigation, a sample of the dust suspected to be involved in the incident should be retained for analysis and testing.

Significance of Particle Size

The rate of pressure rise generated by combustion largely depends on the surface area of the dispersed dust particles **FIGURE 16-9**. The finer the dust particles are, the more violent the explosion will be. The total surface area, and consequently the violence of the explosion, increases as the particle size decreases. In general, an explosion hazard can exist when the particles of the dust are 500 microns or less in diameter.

Concentration

Minimum explosive concentrations (MECs) may vary with the specific dust, but unlike with most gases and vapors, there is generally no reliable maximum limited concentration. The reaction rate is controlled more by the surface-to-air mass ratio than by a maximum concentration. MECs may range from as low as 0.03 oz/ft³ to 2 oz/ft³ (30 g/m³ to 2000 g/m³), with the most common concentrations being less than 0.25 oz/ft³ (250 g/m³).

The combustion rate and maximum pressure will decrease if the mixture is either fuel rich or fuel lean. The rate of pressure rise and the explosion pressure are low at the LEL and with a high, fuel-rich concentration. Thus, as in gases and vapors, the pressure

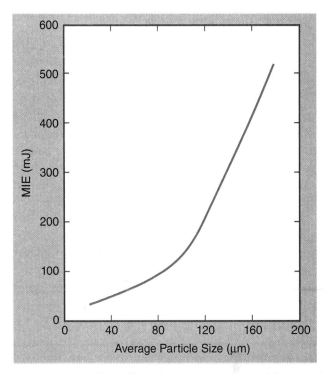

FIGURE 16-9 The effect that an average particle diameter of a typical agricultural dust has on the minimum ignition energy (MIE).

Reproduced from NFPA 921-2021, *Guide for Fire and Explosion Investigations*, Copyright © 2020, National Fire Protection Association. This reprinted material is not the complete and official position of the NFPA on the referenced subject, which is represented only by the standard in its entirety.

rise and the maximum pressure that occur in dust explosions will be high if the dust concentration is at or close to the optimal mixture. The rate of combustion, and in turn the rate of pressure rise, is greatly increased when turbulence occurs within the suspended dust–air mixture. Furthermore, the size and shape of the confining vessel affect the nature of turbulence and, therefore, can have a great effect on the severity of the dust explosion.

Moisture

Generally speaking, increasing the moisture content of the dust particles increases the minimum energy required for ignition and the ignition temperature of the dust suspension. The initial increase in ignition energy and temperature is low, but as the limiting value (or boundary) of moisture concentration is approached, the rate of increase in ignition energy and temperature becomes high. When the moisture value exceeds the limiting value, most dust suspensions will not ignite. However, some water-reactive materials, such as metals, may be more reactive and burn more rapidly at moderate moisture content levels than when they are dry. Note that the moisture content of the surrounding air has little effect on the propagation reaction once ignition has occurred.

Sources of Ignition

The variables that affect the ignition sensitivity of a dust include the ignition temperature, the minimum energy, and the minimum concentration. The Bureau of Mines has classified thousands of samples in terms of their ignition sensitivity and explosion severity, and their index of explosivity can be established based on these values. Sources of ignition include the following: open flames, smoking materials, lightbulb filaments, welding operations, electric arcs, static electricity discharge, friction sparks, heated surfaces, and spontaneous heating.

The actual ignition temperature for most dust ranges from 600°F to 1100°F (320°C to 590°C). Minimum ignition energies (MIEs) are higher for dust than for gas or vapor fuels. The lowest MIEs reported for dust are in the 1 mJ to 10 mJ range, which is significantly higher than the MIEs for most flammable gas vapors, which range from 0.02 mJ to 0.29 mJ.

Hybrid Dust Explosions

The presence of flammable vapors and gases in combination with dusts adds to the violence of a dust–air combustion, even when the concentrations of the vapor or gas are below their lower flammable limit (LFL). This kind of mixture is referred to as a hybrid mixture. In certain cases, hybrid mixtures can be deflagrable even if the vapor is below the LFL and the dust is below the MEC. In addition, hybrid mixtures can be ignited by weak ignition sources that would typically be considered nonignitable sources for the dusts alone. An example of a hybrid mixture is methane, air, and coal dust.

Multiple Explosions

Dust explosions often occur in a series within industrial and agricultural operations. The initial explosion is usually less severe than the secondary explosion, because the first explosion puts additional dust into suspension, which then results in additional explosions. The structural vibrations in the blast front from the first explosion will propagate faster than the flame front, thereby placing the dust ahead of it into suspension. The secondary explosion may progress from one area to another or from one building to another building.

Backdraft or Smoke Explosions

It is common for fires in airtight rooms to become oxygen depleted. This results in the generation of flammable gases due to incomplete combustion. The heated fuels accumulate in areas within the structure characterized by insufficient oxygen and insufficient ventilation. If a window or door is opened, the fuels readily mix with air. The fuels can then ignite and burn sufficiently fast to produce low-order damage (less than 2 psi [13.8 kPa]). These events are called backdrafts and smoke explosions.

Outdoor Vapor Cloud Explosions

The release of gas, vapor, or mist into the atmosphere may result in the formation of a cloud within the fuel's flammable limits and, subsequently, ignition culminating in an outdoor vapor cloud explosion. The phenomenon is referred to as unconfined vapor–air explosion or unconfined vapor cloud explosion. It is most frequently related to catastrophic failure of vessels and tankers, large amounts of fuel, and low-lying areas.

Similar to backdrafts and smoke explosions, outdoor vapor cloud explosions occur usually in partly restricted areas of a natural or humanmade structure. Unconfined environs, which were previously thought to be prone to these types of explosions, actually result only in a flash fire. Outdoor vapor cloud explosions may occur in chemical processing plants because significant amounts of fuel are involved; however, it is possible for congestion from natural sources (e.g., trees, vegetation) to accelerate flames and cause significant overpressures.

Explosives

Explosives are categorized into two main types—low explosives and high explosives—which should not be confused with low-order and high-order damage. The fire investigator must seek assistance from explosives specialists in any scene involving explosives.

Low explosives are characterized by deflagration (subsonic blast pressure wave) and a slow rate of reaction with the development of low pressure **FIGURE 16-10A**. Examples of low explosives include smokeless gunpowder, flash powders, and black powder. Low explosives are designed to work by pushing or heaving effects.

High explosives are designed to produce shattering effects because of their high rate of pressure rise and extremely high detonation pressure **FIGURE 16-10B**. Thus, high explosives are associated with a detonation propagation mechanism. The high, localized pressures are responsible for creating damage near the center of the explosion.

FIGURE 16-10 A. Flash powders are one type of low explosive. **B.** Dynamite is one type of high explosive.
A: © Imageman/Shutterstock; B: © Fer Gregory/Shutterstock.

TIP

If explosive devices are still present on scene, it is not yet time for the fire investigation to begin. Trained explosive disposal specialists should be called to render any explosives safe, or safely remove them, before you examine the scene.

The effects seen from a diffuse-phase (fuel–air) explosion are typically very different from those seen from solid explosives. Such a fuel–air explosion, most often a slow deflagration, will usually result in structural damage that is uniform and omnidirectional, with relatively widespread evidence of burning, scorching, and blistering. The rate of combustion of a solid explosive is extremely fast in comparison to the speed of sound. Thus, pressure does not equalize throughout the explosion volume, and high pressures are generated near the explosive. The location of the explosion should be evidenced by the crushing, splintering, and shattering that are produced by the higher pressure; however, at the greatest distances from the source, the explosion may leave little evidence of

TABLE 16-3 Fuel–Air Explosions Versus Solid Explosive Explosions

Damage	Fuel–Air	Solid Explosive
Structural	Uniform and omnidirectional	Nonuniform Crushing and shattering near the location of the explosion
Fire	Widespread burning, scorching, and blistering	Localized around the source of the explosion

intense burning or scorching except where shrapnel has landed on combustible materials. Common examples of high explosives include dynamite, water gel, TNT, ANFO, RDX, and PETN. The extremely high detonation pressures noted with these explosives may reach 1 million psi (6.9 million kPa).

TABLE 16-3 provides some tips for conducting an explosion investigation. When on the scene of an explosion of unknown origin, the investigator should always be on the lookout for evidence that explosives were involved, such as signs of high-order explosion and high detonation pressure, and should seek the assistance of explosives specialists when such evidence is present. Such a situation could include the presence of homemade explosives and/or improvised devices of varying levels of sophistication. Fire investigators should also recognize that secondary devices and explosions are possible.

Only investigators with appropriate training should conduct an explosion investigation. Investigators who lack this training should coordinate their investigation with the appropriate experts.

Investigating the Explosion Scene

The objectives of an explosion scene investigation are no different from those of a fire investigation: determine the origin, identify the fuel and ignition source, determine the cause, and establish responsibility for the incident. As with a fire investigation, an explosion investigation should be done in a systematic fashion. The following procedures can be scaled to match the scope of an event. In an extensive event, the investigator may need to obtain the assistance of structural and explosion experts.

Securing the Scene

First responders to the explosion should establish and maintain physical control of the structure and surrounding areas for safety and to minimize the potential for loss or contamination of evidence and artifacts. No unauthorized person should enter the scene or have contact with any blast debris, regardless of how remote it is from the scene. Evidence may be small and easily disturbed or removed by traffic in and around the scene. Caution must be used to prevent cross-contamination of the scene by investigators or authorized personnel wearing clothing or footwear that may contain explosive residue from other scenes or sources, such as an explosives firing range.

Establishing the Scene

The outer perimeter of the incident scene (or blast zone) should be established at one and one-half times the distance of the farthest piece of debris found. Blast debris may be propelled for great distances, and if debris is found farther than expected during the investigation, the scene perimeter should be widened accordingly.

TIP

Keep safety uppermost in mind during an explosion investigation. This includes using proper equipment; looking for additional hidden devices; and assessing and reassessing the stability of the structure (including investigator footing), status of utilities, standing water, weather conditions, air quality, and biological and chemical exposure risks. Any hazards that are identified during the investigation should be clearly marked to warn others of their presence.

Obtaining Background Information

The investigator should develop as much background information as possible to facilitate the creation of a timeline for the purposes of analysis.

First, all information should be obtained relating to the incident location itself. This would include a full pre-explosion description of the incident site, systems and operations involved, conditions, and sequence of events that led to the incident, including material safety data sheets. Importantly, the investigator should identify not only the locations of any combustibles and oxidants that were present and the conditions that existed at the time of the incident, but also information about which combustibles, oxidants, and hazardous conditions currently exist at the site.

Investigators should examine witness accounts, maintenance records, operational logs, manuals, weather reports, previous incident reports, and other relevant records. As in any failure analysis, any recent changes in equipment, procedures, and operating conditions may be especially significant. Blueprints of the building and drawings and prints of the process can assist in the proper documentation of the scene.

Explosion Scene Safety

Structures that have been involved in an explosion often experience significant structural damage. Accordingly, the possibility of floor, wall, ceiling, roof, or entire building collapse is great. Involvement of a structural engineer or construction engineer should be considered. Construction equipment may provide a temporary support mechanism, lowering the risk to the investigators. All toxic materials in the air need to be neutralized, and material safety data sheets should be consulted to identify the appropriate personal protective equipment to be used.

A thorough search of the scene should be conducted for any secondary devices before the investigation is initiated. If undetonated devices and explosives are found, the area should be evacuated and isolated, and explosives disposal personnel notified.

All fire investigation safety recommendations listed in NFPA 921, Chapter 13, also apply to investigations of explosions.

Establishing a Search Pattern

The scene should be searched from the outer perimeter inward toward the area of greatest damage. The final determination of the explosion's epicenter should be made only after the entire scene has been examined. The search pattern may be spiral, circular, or grid shaped. The scene itself, with its particular circumstances, often dictates the nature of the pattern used. The search pattern should overlap so that no evidence is lost at the edge of any search area, regardless of the methodology employed. The number of actual searchers needed depends on the size and complexity of the scene. Consistent procedures for identifying, logging, photographing, marking, and mapping of evidence must be maintained. The location of evidence may be marked with chalk, spray paint, flags, stakes, or any other marking means. All evidence should be photographed and may be secured or collected once spoliation issues have been addressed.

Initial Scene Assessment

An investigator should make an initial assessment of the type of incident that occurred. Safety is paramount, however, and the investigation must not proceed if safety has not been established. If the explosion is determined to have been fueled by explosives or an explosive device, the investigator should discontinue the scene investigation, secure the area, and contact the appropriate entities, such as bomb technicians, to check for hazards and secondary and unexploded devices. The area should also be checked with a gas detector to ensure that explosive or hazardous atmospheres no longer exist.

Once safety has been addressed, the following tasks are performed in sequence to assist in the initial scene assessment:

1. Identify whether the incident was an explosion or fire.
2. Determine whether high- or low-order damage occurred.
3. Identify whether the incident was a seated or nonseated explosion.
4. Identify the type of explosion.
5. Identify the potential general fuel type.
6. Establish the origin.
7. Establish the ignition source.

Identifying Explosion Versus Fire

The first task is to determine whether the incident was a fire, explosion, or both and, if both, which came first. Look for signs of overpressure such as displacement or bulging of walls, floors, ceilings, doors and windows, roofs, structural members, nails, screws, utility service lines, panels, and boxes. Also analyze the extent of heat damage to the structure and its components to determine whether the damage can be attributed to fire alone.

Determining High- Versus Low-Order Damage

Determine whether the damage consists of high-order or low-order damage. This determination will help the investigator classify the type, quantity, and mixture of fuel involved, as discussed earlier in this chapter and in NFPA 921, Section 22.3.

Identifying Seated Versus Nonseated Explosion

Determine whether the explosion was seated or nonseated. This determination will also be useful in the classification and discovery of the fuel that may have been involved, as discussed earlier in this chapter and in NFPA 921, Sections 22.6–22.7.

Identifying the Type of Explosion

Identify the type of explosion involved: mechanical (such as BLEVE), chemical, other chemical reaction (such as a combustible explosion or other chemical explosion), or electrical. Often, the presence of damaged equipment or fuel sources can provide evidence of the explosion type.

Identifying Potential General Fuel Type

Identify the types of fuels that were potentially available at the explosion scene by determining the condition and location of utility services (fuel gases) and sources of ignitable dusts or liquids. Analyze the nature of the damage in comparison to the damage patterns consistent with the following: lighter-than-air gases, heavier-than-air gases, liquid vapors, dust, explosives, backdrafts, and BLEVEs.

Establishing the Origin

Attempt to establish the origin of the explosion. The origin is usually identified as the area of most damage—a crater or localized area of severe damage in the case of a seated explosion. If the incident involved a diffused fuel–air explosion, the origin will be consistent with the confining volume or room of origin and likely cannot be narrowed down any further.

Establishing the Ignition Source

Attempt to identify the ignition source by looking at the various possibilities, including hot surfaces, electrical arcing, static electricity, open flames, sparks, and chemicals in which fuel–air mixtures are involved. If explosives are involved, the ignition source may be a blasting cap or pyrotechnic device. Be sure to note artifacts from the ignition sources that may have survived the explosion.

Vapors—including natural gas, propane, and gasoline vapors—can be ignited by a variety of ignition sources, including electrical arcs from equipment and appliances in normal operation, pilot lights, and sparks. When a gas explosion is under consideration, locate and inventory all such ignition sources, keeping in mind that they may not be found in the area with the most explosion damage. Gaseous fuel released in a structure can migrate among rooms and be ignited remotely from the heaviest fuel concentration, with the flame front then traveling back to areas with greater fuel loads.

Detailed Scene Assessment and Documentation

After obtaining the general information from the initial scene assessment, a more detailed study and documentation of the blast damage is recommended using the components described in this section.

Effects of Explosion

A detailed analysis of the explosion overpressure damage should be made. The items that are damaged should be identified as having been affected by one or more of the following forces: blast pressure wave—positive phase, blast pressure wave—negative phase, shrapnel impact, thermal energy, or seismic energy.

An investigator should examine the type of damage as to whether the debris was shattered, bent, broken, pushed, or flattened and determine whether this pattern changed over distance or over the length of the article involved. At a distance from a detonation explosion center, the pressure rise is moderate, and the artifacts resemble those of a deflagration explosion. Items in the immediate vicinity of the detonation center exhibit splintering and shattering.

Examine the scene carefully to identify any fragments of foreign material. Estimating the damage from an explosion takes into account the maximum pressure of the explosion compared with the construction of the structure. A light-framed structure can be damaged with much less overpressure than a reinforced structure can, for example.

Damage to human victims from explosion blast pressures usually results from acceleration of the body in the high-velocity air stream with subsequent impact against a rigid surface, rather than from compression in the airwave itself. More information on blast injuries can be found in Chapter 12, *Complex Investigations*.

Preblast and Postblast Damage

Debris that has been burned and propelled away from the point of origin may indicate that a fire preceded the explosion. Glass fragments with smoke residue and soot found some distance away from the structure may indicate a fire of some duration followed by an explosion. Glass fragments that are clean and debris that is not burned, but that are found some distance from the structure, may indicate an explosion occurred prior to the fire.

Items of Evidence

The methods used to document scene artifacts may include locating, identifying, noting, logging, photographing, and mapping physical evidence.

The probability of physical evidence being propelled both inside and outside of the structure may result in the evidence being found embedded in walls, resting in adjacent vegetation, inside adjacent structures, and within the body and clothing of victims. Take photographs of the injuries to the victims as well as any materials removed from them during medical treatment. Hardhats, gloves, boots, and respirators, as well as clothing and materials removed from the victims, should be preserved for further examination.

The condition and position of damaged structural components—walls, ceilings, floors, roofs, foundations, support columns, doors, windows, sidewalks, driveways, and patios—should be noted. The condition and position of building contents such as furnishings, appliances, heating or cooking equipment, manufacturing equipment, clothing, and personal effects should also be noted. The condition and position of the utility equipment, such as fuel gas meters, regulators, fuel gas piping and tanks, electrical boxes and meters, electrical conduits and conductors, heating oil tanks, parts of explosive devices, and fuel vessels, should be examined as well.

Force Vectors

Document the debris that has been propelled away from the area of origin as well as its direction of travel; the distance of travel; the material propelled; and the material's size, weight, and configuration. Items in the flame front path, including victims' clothing and skin, may show both damage and directional patterns. Dust explosions may also leave burned and unburned particles on the affected items. This process assists in identifying the trajectories of the artifacts involved.

Analyze the Origin (Epicenter)

In **explosion dynamics analysis**, the general path of the explosion force vectors is followed from the least to the most damaged area. This process may require more than one explosion dynamics diagram to identify the debris movement: A large-scale analysis indicates the general area or room for further analysis as to the origin, and a small-scale diagram analyzes the explosion dynamics within the area of origin itself **FIGURE 16-11**.

Plot the directions of debris movement and the relative force necessary for the movement of each major piece of debris. The explosion dynamics analysis may be complicated by secondary explosions. Secondary explosions, especially when they involve dust explosions, can be greater than the primary explosion and, therefore, cause more damage. The analysis of the explosion dynamics is based on debris movement away

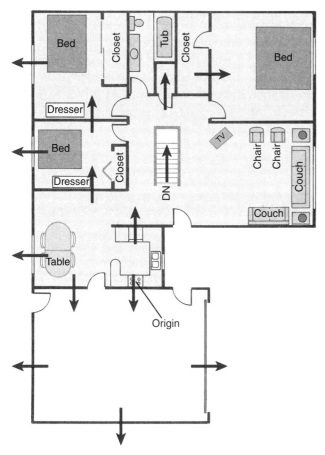

FIGURE 16-11 Diagram showing displacement of walls, doors, and windows as a result of explosion.

from the epicenter of explosion in a roughly spherical pattern. The farther an object is located from the epicenter, the less force that object will be subjected to.

Analyze the Fuel Source

All available fuel sources should be considered, with this list of possibilities then being narrowed down, if possible, to the one fuel source that meets all the physical damage criteria. Clinical analysis of debris, soot, soil, and air, including gas chromatography, spectrography, and other chemical tests of samples, may identify the fuel. Once the fuel has been identified, an investigator can determine the source.

All gas piping should be examined, and leak testing should be performed. Furthermore, odor verification should be part of any explosion investigation involving flammable gas. Although stain tubes may be used for this purpose, the collection of a sample for submission to a lab is the most accurate means of odor verification.

Analyze the Ignition Source

A careful evaluation of every possible ignition source should be made. Factors to consider include the minimum ignition energy of the fuel, ignition energy of the potential ignition source, ignition temperature of the fuel, temperature of the ignition source, location of the ignition source in reference to the fuel, presence of both fuel and ignition source at the time of ignition, and witness accounts of conditions prior to and at the time of the explosion.

Analyze to Establish Cause

An analysis to determine the simultaneous presence of the fuel and the ignition source can include timeline analysis, damage pattern analysis, debris analysis, relative structural damage analysis, correlation of blast yield with damage, analysis of damaged items in the structure(s), and the correlation of thermal effects.

Timeline Analysis

After gathering information, create a timeline of events prior to and during the explosion using hard and soft times. Consistencies and inconsistencies with cause theories can be inferred to establish a "best fit" theory.

TIP

When releasing the scene, be sure that you document this hand-off, including the proper authority taking custody of the scene, with time and date; that you have identified and disclosed all existing health and safety hazards; and that you have identified any other steps that may be needed, such as a structural report, utility company work, and cleanup/safety and salvage steps.

Damage Pattern and Debris Analysis

Damage pattern analysis is the documentation of damage patterns, primarily of debris and structural damage, for further analysis. Debris analysis includes the identification, diagramming (location found), photographing, and noting of debris pieces that indicate the direction and force of the explosion, which may allow for reconstruction of components.

Correlation of Blast Yield with Damage Incurred

Correlation of damage and projectile distance with the type and amount of fuel used should be done to see whether they agree—that is, whether the yield of the hypothesized fuel could have done the existing damage.

Analysis of Damaged Items and Structures

Determining the relationship between the type of fuel used and the damage caused may require specialized experts to examine the damage to items and structures.

Correlation of Thermal Effects

Heat damage on a collection of articles from an explosion may be evidence of a fireball or fire during the event and may assist in the investigator's identification of the type of explosion and/or the fuel. This also may require a specialist in the field to conduct the analysis of the articles.

After-Action REVIEW

IN SUMMARY

- Blast overpressure can damage containers, structures, and equipment, and cause injury to persons.
- The primary criterion defining an explosion is production and release of gas; a loud noise, although common, is not required to classify an event as an explosion.
- The failure and bursting of a container caused by purely hydrostatic pressure, such as water pressure, is not an explosion.
- Stay alert to the risk of secondary explosions and/or devices.
- The two major types of explosions are mechanical and chemical, which are distinguished by the source and the mechanism that produces the explosion.
- A BLEVE can occur in a vessel of any size and does not require ignitable liquid contents.
- If the vessel contents are ignitable, a fire of the contents is likely to be ignited upon BLEVE either from the original heat source that caused the BLEVE or by another electrical or mechanical source.
- The velocity of the flame front's propagation through the fuel determines whether the combustion reaction is classified as a deflagration (slower than the speed of sound) or a detonation (faster than the speed of sound).
- A significant difference between deflagration and detonation is the time over which the event occurred: A detonation is faster and cannot be vented because of the speed of the reaction.
- Electrical explosions can occur when high-energy arcing generates sufficient heat to cause an explosion.
- The effects of explosions can be classified into four major groups: blast overpressure and wave effect, projected fragments effect, thermal effect, and seismic effect (ground shock).
- The blast overpressure and wave effect creates a positive-pressure phase, which is characterized by the outward movement of gases and displaced air, followed by a negative-pressure phase in which the air rushes back toward the area of origin.
- The shape of the blast front from an idealized explosion is spherical, expanding evenly in all directions from the epicenter.
- Factors such as confinement, obstruction, ignition position, cloud shape, or concentration distribution at the source of the blast pressure wave can change and modify the direction, shape, and force of the front.
- Confinement changes the shape and force of the flame front.
- Slow rates of pressure rise result in pushing or bulging types of damage, with weaker parts of the structure weakening first.
- Rapid rates of pressure rise result in shattering of the confining container, with debris being launched over great distances.
- The rate of energy release, and rate of pressure rise that results, is often more significant than the total amount of energy in a blast.
- Fireballs and firebrands can result from explosions, particularly from BLEVEs of containers with ignitable liquids.
- Explosion effects are influenced by the type and configuration of the fuel; the nature, size, volume, and shape of the containment vessel; congestion and obstacles within the vessel; the location and magnitude of the ignition source; venting of the vessel; the maximum pressure; and the rate of pressure rise.

- The blast pressure front may focus in a particular area because of reflection from objects in its path, and because of refraction through air at different temperatures.

- The more ventilation that occurs in a diffuse fuel explosion, and the closer that ventilation is to the ignition source, the weaker the explosion will be.

- The seat of an explosion may be of any size, depending on the amount and strength of the explosive material.

- Fuel gases, such as natural gas and liquefied petroleum, usually produce nonseated explosions because the gases involved are found in large containers such as rooms.

- Explosion damage to structures depends on several factors, including the fuel–air ratio, vapor density, turbulence effects, volume of the confining space, location and magnitude of the ignition source, venting, and strength of the structure.

- Mixtures at or near the lower explosive limit or the upper explosive limit of a gas or vapor usually produce less-violent explosions than do those near the optimal concentration.

- A mixture slightly richer than the stoichiometric ratio produces the most efficient combustion, resulting in the highest flame speeds, quickest rates of pressure rise, and maximum pressures.

- An explosion can occur even if the fuel has not spread throughout the entire container.

- Heavier-than-air gases and vapors tend to flow toward lower areas, while lighter-than-air gases tend to rise and flow toward upper areas.

- The nature of the explosion damage can be affected by the size, shape, construction, volume, materials, and design of the confining space.

- Secondary explosions or cascade explosions may occur when gas and vapors have migrated to adjacent stories or rooms, resulting in the formation of pockets.

- Violent explosions can be fueled by dust that is dispersed within the air.

- As in a fire investigation, the objectives of an explosion investigation are to determine the origin, identify the fuel and ignition source, determine the cause, and establish responsibility.

- Responders to an explosion scene should establish and maintain control of the structure and surrounding areas and prevent entry by unauthorized persons.

- Safety should be the paramount concern in every explosion investigation.

KEY TERMS

Blast overpressure Large quantities of gas produced by the explosion of the material.

Boiling liquid/expanding vapor explosion (BLEVE) An explosion that occurs when pressurized liquefied materials (e.g., propane or butane) inside a closed vessel are exposed to a source of high heat.

Burning velocity The rate of flame propagation relative to the velocity of the unburned gas ahead of it. (NFPA 68)

Chemical explosion Explosion in which a chemical reaction is the source of the high-pressure fuel gas. The fundamental nature of the fuel is changed.

Combustion explosion Explosion caused by the burning of combustible hydrocarbon fuels and characterized by the presence of a fuel with air as an oxidizer.

Deflagration A reaction that propagates at a subsonic velocity through an unreacted medium, less than 1100 ft/sec, and can be successfully vented.

Detonation A combustion reaction that propagates at supersonic velocities, greater than 1100 ft/sec (340 m/sec), and cannot be vented because of its speed.

Explosion The sudden conversion of potential energy (chemical or mechanical) into kinetic energy, with the production and release of gases under pressure. These high-pressure gases then do mechanical work, such as moving, changing, or shattering nearby materials. (NFPA 921)

Explosion dynamics analysis The process of using force vectors to trace backward from the least to the most damaged area following the general path of the explosion force vectors.

Explosives Any chemical compounds, mixtures, or devices that function by explosion.

Fireballs Momentary balls of flame observed during or after an explosion that may present high-intensity, short-term radiation.

Firebrands Hot or burning fragments propelled from an explosion.

Flame speed The local velocity of a freely propagating flame relative to a fixed point.

Flash fire A fire that spreads rapidly by means of a flame front through a diffuse fuel such as dust, gas, or ignitable liquid vapor, but without production of damaging pressure.

High-order damage A rapid pressure rise or high-force explosion characterized by a shattering effect on the confining structure or container and long missile distances. (NFPA 921)

Kinetic energy The energy possessed by a system or object as a result of its motion.

Low-order damage A slow rate of pressure rise or low-force explosion characterized by a pushing or dislodging effect on the confining structure or container and by short missile distances. (NFPA 921)

Mechanical explosion Rupture of a vessel or container such as a cylinder, tank, or boiler, resulting in the release of pressurized gas or vapor. The pressure leading to the mechanical explosion is not due to a chemical reaction or a change in the chemical composition of the substances involved.

Nonseated explosion Explosion in which the fuels in the explosion are dispersed or diffused, characterized by moderate rates of pressure rise and subsonic explosive velocities.

Seat of the explosion A craterlike indentation created at the point of origin of an explosion. (NFPA 921)

Seismic effect The transmission of tremors through the ground as a result of the blast wave expansion causing structures to be knocked down.

Stoichiometric mixture The optimal ratio at which point the combustion will be most efficient.

REFERENCES

National Fire Protection Association. 2018. *NFPA 68: Standard on Explosion Protection by Deflagration Venting ed.* Quincy, MA: National Fire Protection Association.

National Fire Protection Association. 2020. *NFPA 921: Guide for Fire and Explosion Investigations*, 2021 ed. Quincy, MA: National Fire Protection Association.

On Scene

1. What distinguishes an explosion investigation from a fire investigation? How are they similar?

2. What type of explosion would be characterized by shattering effects, and what type by pushing or heaving effects? Why might the investigator see both types of effects in one event?

3. If a significant fire happens in gaseous fuels subsequent to an explosion, what might this mean in terms of the fuel–air mixture that preceded the explosion?

4. Why are the locations and sizes of vent openings less important in detonation explosions?

Chapter Opener: © Jones & Bartlett Learning. Photographed by Glen E. Ellman; On Scene siren: © Bildgigant/Shutterstock.

CHAPTER 17

Automobile, Marine, and Equipment Fires

KNOWLEDGE OBJECTIVES

After studying this chapter, you should be able to:

- Describe safety issues surrounding vehicle fire investigation.
- Identify the potential fire fuels present in vehicles.
- Describe the various vehicle systems and components as related to fire cause.
- Discuss the body systems in a vehicle and how they may affect a vehicle fire.
- Identify potential ignition sources present in vehicles.
- Describe the investigative techniques used to analyze a vehicle fire. (**NFPA 1033: 4.2.6**; **4.4.2**, pp. 423–425)
- Describe the process used in the examination of a vehicle fire. (**NFPA 1033: 4.2.6**; **4.4.2**, pp. 423–429)
- Explain the investigation of hybrid vehicle fires.
- Describe special considerations for vehicle fires.
- Explain how to document a vehicle fire scene. (**NFPA 1033: 4.3.1**; **4.3.2**; **4.3.3**, pp. 429–430)
- Describe the various types of propulsion systems for marine vessels and their uses.
- Identify and describe systems used on marine vessels.
- Explain design and construction features associated with marine vessels.
- List potential ignition sources on marine vessels.
- Identify issues related to cargo on marine vessels.
- List safety considerations related to marine fire investigations.
- Describe how to document marine vessel fire scenes. (**NFPA 1033: 4.3.1**; **4.3.2**; **4.3.3**, pp. 441–445)

- Describe the process for examining a marine vessel fire scene. (**NFPA 1033: 4.2.6**; **4.4.2**, pp. 441–445)
- Identify considerations for marine vessel fires in structures.
- Explain legal considerations related to marine fire investigations.
- Discuss considerations for various types of specialized vehicles and equipment.
- Describe considerations for investigating other types of vehicle fires.

SKILLS OBJECTIVES

After studying this chapter, you should be able to:

- Conduct a vehicle or marine fire investigation. (**NFPA 1033: 4.3.1**; **4.3.2**; **4.3.3**, pp. 410–445)
- Conduct a fire scene examination of a vehicle or marine fire. (**NFPA 1033: 4.3.1**; **4.3.2**; **4.3.3**, pp. 410–445)
- Ensure the safety of the fire investigator when conducting both shore- and water-based investigations.
- Properly document vehicle and marine fires. (**NFPA 1033: 4.4.2**, pp. 429–430, 441–445)
- Locate and identify the area of origin associated with a vehicle or marine fire. (**NFPA 1033: 4.2.5**, pp. 411, 424–425, 439, 444)

You are the Fire Investigator

You have been requested to investigate a fire involving an older sedan. Fire personnel report to you that upon their arrival at the scene, they encountered visible fire in the engine compartment and a small amount of fire in the front passenger compartment.

The vehicle's owner states they had not driven the vehicle for several months and needed a jump-start to start the vehicle. The owner stated that they began driving on a country road, and within five minutes noticed smoke coming from the engine compartment. You notice severe burning and melting of plastic components under the vehicle's hood, which has been pried open by the fire department. The passenger compartment has fire damage in the firewall area and dashboard.

1. What safety concerns should you consider while investigating a vehicle fire?

2. What fuel sources exist within the engine compartment that may have been ignited?

3. Why is the interview with the owner and/or driver important when investigating a vehicle fire?

4. What should be documented during your vehicle examination?

Access Navigate for more practice activities.

Introduction

This chapter discusses the investigation of fires in automobiles, marine vessels, and other types of equipment. Although investigation of vehicle fires is often considered complex and difficult, fires in automobiles, trucks, heavy equipment, farm and logging equipment, and recreational vehicles (RVs) or motor homes have the same basic requirements for successful ignition and propagation as do structural or other fires. A fire investigator should be able to recognize the burn and char indicators, which may vary from vehicle to vehicle, although patterns may be hard to identify given the tight spaces and excessive fuel loads found in vehicle compartments. The investigator should also have a good understanding of various fuel characteristics within the vehicle, fire behavior associated with those fuels, and a general knowledge of the working mechanical components.

Marine fires present unique settings and challenges for the fire investigator and are discussed at length in this chapter. These fire scenes present specific safety challenges as well as unique ignition sources and scenarios to consider. Similarly, a wide variety of other equipment exists that may present unusual investigative challenges to the investigator, although it may share similarities with other vehicles. The investigator must understand some of the systems and potential fire scenarios in marine vessels and equipment and know where to find further information or resources to support a successful investigation.

Vehicle Fire Investigation

Safety

Motor vehicle fires present unique safety challenges. When conducting an inspection of the vehicle's undercarriage, take care to support and stabilize the vehicle properly with blocking or stands to prevent movement, which may cause injuries. Lifting devices such as jacks, forklifts, tow trucks, or hydraulic/pneumatic devices must never be solely relied upon for support during the inspection.

Undeployed air bags (supplemental restraint systems) in vehicles damaged by fire or an accident may pose a serious safety hazard because they may deploy if sensors are further disturbed. Disconnecting the battery or batteries prior to the inspection may prevent actuation of these systems and circumvent inadvertent electrical faulting (and possible initiation of another fire) during the examination.

Electrical systems in hybrid or electric vehicles may present an increased shock hazard because of the higher voltages used in these systems. A **hybrid vehicle** can have any of many different and proprietary designs, some of which employ potentially dangerously high voltages, ranging from 100 V in one brand of small automotive hybrid to 800 V in hybrid commuter buses. Investigators and firefighters should exercise extra care when examining these vehicles due to the high potential for serious, if not lethal, electric shock. Hybrids feature safety systems to isolate their high battery-pack voltage, including a manual disconnect near the battery pack (often located in the trunk or cargo area, behind or beneath a small flap or door). These high-voltage power supply systems may be compromised if subjected to fire or the trauma of a motor vehicle accident. Use caution when approaching a fire- or accident-damaged hybrid vehicle.

Recreational vehicles (RVs) may be connected to 120-V AC or 240-V AC power through a power cord or **shore power** connection.

Other hazards to consider include the possibility of fuel spillage from tanks or lines; contamination with coolants, lubricants, or other fluids; cut and puncture hazards from contact with broken glass or sharp metals; and release of energy from damaged spring or hydraulic devices, such as truck parking brake assemblies and hatchback lift supports.

Vehicle Fire Fuels

Fuels in vehicle fires fall into three categories: liquid fuels, gaseous fuels, and solid fuels. Once a fire has begun to develop, any of these fuels may become involved secondarily, creating damage patterns that may be difficult to separate from those stemming from burning of the initially ignited materials. The relatively small size of typically noncombustible compartment shells may blur fire patterns, complicating the task of separating primary damage patterns from secondary ones.

Liquid Fuels

Liquid fuels present in vehicles may include engine fuels (gasoline, diesel), engine lubricants, transmission fluid, power steering fluid, coolant, and brake fluid, and sometimes hydraulic fluids or cargo fluids **TABLE 17-1**. Many of these fuels will be found in most vehicles. The ignition potential of these fuels depends on the fuel's specific properties, its physical state (liquid, atomized, or spray form), and the nature of the ignition source. Gasoline is often blended with ethanol for use as motor fuel. This combination changes the properties of the blended fuel somewhat, raising the autoignition temperature and slightly lowering the hot surface ignition temperature.

Gaseous Fuels

A significant number of motor vehicles of all sizes run on compressed natural gas (CNG), liquefied natural gas (LNG), or propane. These gas systems can be found in a range of vehicles, including fleet vehicles such as buses and shuttles. Dedicated vehicles run only on gaseous fuel, while bifuel vehicles are capable of running on either gaseous fuel or gasoline. Some heavy-duty vehicles are dual-fuel, which means that they use diesel as a pilot fuel to aid in ignition.

In addition to their use as alternative motor fuels, propane and natural gas are used in RVs for heating, cooking, and refrigeration. Hydrogen gas is often present and may be released in the area of batteries during charging. **TABLE 17-2** outlines the properties of gaseous fuels in motor vehicles.

Solid Fuels

Solid fuels encompass any fuel within or on a vehicle that is not liquid or gaseous in its normal state. Although it is less common in accidental fires, solid fuels may be the first material ignited. Examples include wiring insulation or plastic conduit reaching its ignition temperature due to short circuits or resistive heating of the conductors and heat from friction causing drive belts, bearings, and tires to ignite.

Solid fuels may contribute significantly to the fire spread. Solid fuels in vehicles include vehicle interiors and finishes such as plastics that have heat release rates similar to those of ignitable liquids. Plastics can melt, ignite, and drop flaming pieces, which may spread the fire downward.

Aluminum and magnesium or their alloys, which are often found in engine and vehicle components, can ignite and provide an additional fuel load. Most metals need to be in powder or melted form to burn; however, solid magnesium, which is present in some vehicles, burns vigorously once it is ignited by a competent external heat source.

The presence of melted metals in a vehicle is not necessarily indicative of the presence of an ignitable liquid accelerant. Many of the alloys or metals found in vehicles are not in a pure form or state, so they have melting temperatures lower than those given on charts and graphs for the pure forms. When one type of metal falls or melts onto another, the process of alloying can occur. For instance, melted aluminum or zinc could run or drip onto another metal, causing it to melt at a lower-than-normal temperature. Molten metals may also spread the fire as they contact other combustible materials or may cause short circuits between energized cables and grounded components.

A pattern of melted metals may provide additional information that can assist the fire investigator in determining the area of origin for a vehicle fire, much as a char pattern analysis is frequently used in structural fire investigations. Melted metals used as indicators should be identified, and their actual melting temperatures should be determined. Many common metals may be identified from their visible characteristics. A precise determination of the composition of metals may require laboratory analysis using technical processes such as energy-dispersive spectroscopy.

TABLE 17-1 Properties of Ignitable Liquids (NFPA 921, Table 26.3.1)

Liquid	Flash Point[a] °C	Flash Point[a] °F	Autoignition Temperature[b] °C	Autoignition Temperature[b] °F	Flammability Limits[c] LFL %	Flammability Limits[c] UFL %	Boiling Point[d] IBP °C	Boiling Point[d] IBP °F	Boiling Point[d] FBP °C	Boiling Point[d] FBP °F	Vapor Density (Air = 1)
Gasoline	−45 to −40	−49 to −40	350–460	660–860	1.4	7.6	26–49	78–120	171–233	339–452	3–4
Diesel fuel (fuel oil #2)	38–62	100–145	254–260	489–500	0.4	7	127–232	260–450	357–404	675–760	5–6
Brake fluid	110–171	230–340	300–319	572–606	1.2	8.5	232–288	111–142	460–550	238–288	5–6
Power steering fluid	175–180	347–356	360–382	680–720	1	7	309–348	588–658	507–523	945–973	>1
Motor oil	200–280	392–536	340–360	644–680	1	7	299–333	570–631	472–513	882–955	>1
Gear oil	150–270	302–510	>382	>716	1	7	316–371	601–700	>525	>977	>1
Automatic transmission fluid	150–280	302–536	330–382	626–716	1	7	239–242	462–468	507–523	945–973	>1
Ethylene glycol (antifreeze)	110–127	230–261	398–410	748–770	3.2	15.3	196–198	385–388			2.1

| Propylene glycol (antifreeze) | 93–107 | 199–225 | 371–421 | 700–790 | 2.6 | 12.5 | 187–188 | 369–370 | 2.6 |
| Methanol (washer fluid) | 11–15 | 52–55 | 464–484 | 867–903 | 6 | 36 | 65 | 149 | 1.1 |

aFlash point data were obtained from technical data sheets and safety data sheets from manufacturers and suppliers of the major brands of each type of fluid available in the United States. The flash points of gasolines reported in these sources were determined by ASTM D56, *Standard Test Method for Flash Point by Tag Closed Tester*. The flash points for diesel fuels, brake fluids, power steering fluids, motor oils, transmission fluids, gear oils, ethylene glycol (antifreeze), propylene glycol (antifreeze), and methanol were determined by ASTM D56; ASTM D92, *Standard Test Method for Flash and Fire Points by Cleveland Open Cup Tester*; or ASTM D93, *Standard Test Method for Flash Point by Pensky–Martens Closed Cup Tester*.

bAutoignition temperature data for gasoline, diesel fuel, brake fluid, ethylene glycol, propylene glycol, and methanol were obtained from technical data sheets and safety data sheets from manufacturers and suppliers of the major brands of each type of fluid available in the United States. These sources generally did not report the test method used to determine autoignition temperature; however, ASTM E659, *Standard Test Method for Autoignition Temperature of Liquid Chemicals*, is the laboratory test method typically used to determine autoignition temperature. Autoignition temperature data for power steering fluid, motor oil, gear oil, and automatic transmission fluid were obtained using ASTM E659.

cFlammability limit data were obtained from technical data sheets and safety data sheets from manufacturers and suppliers of the major brands of each in the United States. These sources generally did not specify the laboratory test method used to determine the reported flammability limits; however, ASTM E681 is a laboratory test method typically used to determine the lower flammability limit (LFL) and the upper flammability limit (UFL).

dBoiling range data for gasolines were obtained from the Alliance of Automobile Manufacturers' annual North American survey of gasoline properties for 2003. The boiling ranges reported in this survey were determined by ASTM D86, *Standard Test Method for Distillation of Petroleum*. Boiling range data for diesel fuel were obtained from technical data sheets and safety data sheets from manufacturers and suppliers of the major brands of diesel fuel in the United States. These sources generally did not report the laboratory test method used to determine the boiling range of diesel fuel. Boiling range data for brake fluid, power steering fluid, motor oil, gear oil, and automatic transmission fluid were determined by ASTM D2887, *Standard Test Method for Boiling Range Distribution of Petroleum Fractions by Gas Chromatography*. Boiling point data for ethylene glycol, propylene glycol, and methanol were obtained from technical data sheets and safety data sheets from manufacturers and suppliers of these chemicals. These sources did not report the laboratory test method used to determine boiling point. In the table, IBP and FBP are initial boiling point and final boiling point, respectively.

Specific gravity data was obtained from safety data sheets from manufacturers and suppliers of these materials.

*Studies include the following:
1. API PUBL 2216, *Ignition Risk of Hydrocarbon Vapors by Hot Surfaces in the Open Air*.
2. Arndt, S. M., Stevens, D. C., and Arndt, M. W. "The Motor Vehicle in the Post-Crash Environment: An Understanding of Ignition Properties of Spilled Fuels," SAE 1999-01-0086, International Congress and Exposition, Detroit, MI, March 1–4, 1999.
3. Colwell, J. D., and Reze, A. "Hot Surface Ignition of Automotive and Aviation Fluids," *Fire Technology*, Second Quarter 2005, pp. 105–123.
4. LaPointe, N. R., Adams, C. T., and Washington, J. "Autoignition of Gasoline on Hot Surfaces," *Fire and Arson Investigator*, October 2005, pp. 18–21.

Reproduced from NFPA 921-2021, *Guide for Fire and Explosion Investigations*, Copyright © 2020, National Fire Protection Association. This reprinted material is not the complete and official position of the NFPA on the referenced subject, which is represented only by the standard in its entirety.

TABLE 17-2 Properties of Gaseous Fuels in Motor Vehicles (NFPA 921, Table 26.3.2)

| Gas | Autoignition Temperature | | Flammability Limits (Vol. % fuel in air) | | Boiling Point | | Specific Gravity (Air) | |
	°C	°F	LFL	UFL	°C	°F	Vapor Density (air = 1)	Min. Ignition Energy (mJ)
Hydrogen	400–572	753–1061	4.0	75.0	−253	−422	0.07	0.018
Natural gas	632–650	1169–1202	5.3	15.0	−162	−259	0.60	0.280
Propane	450–493	842–919	2.2	9.5	−42	−44	1.56	0.250

Note: The data provided in this table are for generic or typical products and may not represent the values for a specific product. When possible, values specific to the product involved should be obtained from a safety data sheet, product specifications, or by standard test methods.

Reproduced from NFPA 921-2021, *Guide for Fire and Explosion Investigations*, Copyright © 2020, National Fire Protection Association. This reprinted material is not the complete and official position of the NFPA on the referenced subject, which is represented only by the standard in its entirety.

Systems and Their Function in a Vehicle

Knowing how the various systems in vehicles operate is critical, and the investigator's knowledge in this area can be enhanced by viewing various technical manuals available online or at a library, car dealership, or a parts supplier. Quite often, familiarity with a particular vehicle system can be gained by visiting a dealership or automotive repair shop and asking for specific information about the vehicle from a service manager or mechanic. The parts department may also have schematics and views of particular systems.

Fuel Systems

Two basic fuel systems are used in gasoline-powered vehicles: vacuum/low-pressure carbureted systems and high-pressure fuel-injection systems. Both take fuel from the gas tank through the use of a fuel pump and deliver that fuel through a fuel line and filter system to the engine, where it is atomized, compressed, and burned.

Vacuum/Low-Pressure Carbureted Systems

Vacuum/low-pressure carbureted systems are most often seen on older automobiles, gasoline-powered farm equipment, lawn maintenance vehicles, and small gasoline engines powering stationary equipment. These systems usually draw fuel from the fuel tank by means of a mechanical pump attached to the engine and operate only when the engine is operating. Electric fuel pumps are sometimes used for this purpose. The fuel is then pumped into the carburetor at a pressure of 3 to 5 psi (20 to 35 kPa), where it mixes with air in the carburetor (usually at a ratio of 15:1). From there, it is drawn into the engine via the intake manifold and into the cylinder, where combustion takes place. **FIGURE 17-1** shows a generic view of a vehicle fuel system and its associated components.

Potential problems with vacuum/low-pressure carbureted systems are outlined in **TABLE 17-3**.

High-Pressure Fuel-Injection Systems

High-pressure fuel-injection systems also retrieve the fuel from the fuel tank and pump the fuel to the engine via an electric fuel pump, which is energized whenever there is a key in the ignition in the "on" or "run" position. Some fuel pumps on modern cars are integrated with an onboard computer, which can trigger an inertia switch in a crash to disable the fuel pump. The fuel is pumped at 35 to 70 psi (240 to 480 kPa) in most vehicles with standard electronic or mechanical fuel-injection systems. However, this pressure may be more than 1700 psi (11.72 mPa) in newer gasoline direct-injection systems, and may reach spike pressures of several thousands of pounds per square inch in some modern heavy-duty diesel systems. **FIGURE 17-2** depicts the type of electric fuel pump used in most new vehicles with fuel-injection systems. The fuel is pumped to either a single Venturi-mounted fuel injector (throttle body) or, more commonly, a **fuel rail** assembly on the engine. In both systems, the excess fuel is pumped back to the fuel tank.

Fuel System Components from the Pump to the Injector

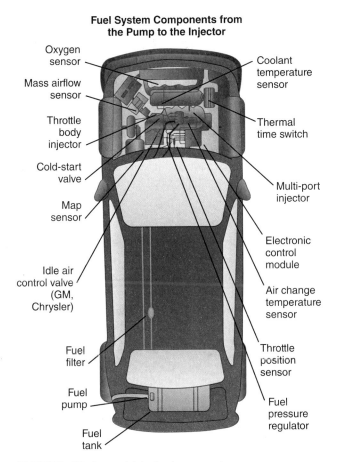

FIGURE 17-1 A vehicle fuel system has many components.

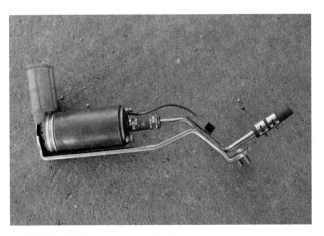

FIGURE 17-2 An electric fuel pump.
© Jones & Bartlett Learning. Photographed by Glen E. Ellman.

TABLE 17-3 Vacuum/Low-Pressure Carbureted Systems	
Location of a Leak	**Problem**
Vacuum side	Air is drawn into the system, and the engine will not run.
Pressure side	Fluid can leak as a fine mist or a heavy stream. Ignition is possible if there is an ignition source. If the fuel line leaks as a result of a fire, the leaking fuel vapors can be ignited.

Potential problems with high-pressure fuel-injected systems may include leaks under pressure at various fittings, lines, or fuel pressure regulators. In some vehicles, the pressure developed from a leak may be sufficient to propel the fuel over several feet, resulting in potentially confusing patterns when that fuel burns.

The horizontal aluminum piping in **FIGURE 17-3A** depicts the fuel rail on this vehicle. The fuel injector shown in **FIGURE 17-3B** is attached to the fuel rail. Failures can and do occur because of age and the types of materials used for the connection between the rail and the supply line **FIGURE 17-3C**.

When being interviewed, the vehicle owner may report fuel-related problems that can help the fire investigator determine whether an event led to a fire. A leak on the supply side of the system or problems with the operation of the vehicle should be noted by the operator. Other problems that may have been noticed include starting difficulty, erratic operation, and stalling. A leak on the return side of the system can go undetected, however, with no operational problems being noted by the operator. Fuel lines that enter the tank on the bottom can feed fuel to a fire by gravity, and lines that enter at the top of the tank can feed fuel by siphon effect if they are compromised. Residual pressure in the tank or pressure that is created by the fire could force fuel out of a compromised fuel line. **TABLE 17-4** can assist in testing a hypothesis in a suspected fuel-related fire.

Be aware that fuel lines may be constructed of plastic (typically Nylon 12) and may easily be breached by exposure to the fire. Trauma from a motor vehicle accident can compromise either plastic or steel fuel lines. Fuel transported or delivered in such a manner may result in the creation of very confusing patterns of intense fire damage; therefore, an investigator should carefully identify and document the condition of these lines.

Diesel Fuel Systems

Diesel fuel systems often use two pumps to deliver the fuel from the tank to the engine with lift pumps that operate under high volume but with low pressure. The diesel fuel is pumped from the tank to an engine-driven fuel-injector pump, and the higher-pressure fuel-injector pump delivers the fuel to

A

B

C

FIGURE 17-3 A. A fuel rail. **B.** A fuel injector. **C.** The connection between rail and supply line.
© Jones & Bartlett Learning. Photographed by Glen E. Ellman.

TABLE 17-4 Fuel-Related Problems	
Location of a Leak	Problem Noted
Supply side	Operating problems will be noted. Unintended ignition may occur.
Return side	No operating problems will be noted. Unintended ignition may occur.

Fuel leaks can occur over time because of the vibrations of a diesel engine. If any diesel fuel comes into contact with a hot surface, it may ignite. Another potential, albeit rare, problem associated with diesel engines is a "runaway" engine in which fuel or oil is drawn in through the air intake system, causing the engine to maintain a very high RPM (revolutions per minute) level. This high RPM level may cause the engine to fail catastrophically. Many diesel vehicles have large-capacity fuel tanks, which can contribute to a total burn if the fire suppression is delayed.

Natural Gas and Propane Fuel Systems

Natural gas and propane fuels may be stored as compressed gas or as liquids under very high pressures (about 3000 psi [21 mPa]), which are then reduced for use by a regulator at or near the engine. In such a system, the fuel flows through a regulator and into the carburetor, where it mixes with air before entering the engine. Postfire leaks found in vehicles of this type may not be indicative of prefire leaks, partly because of the different coefficients of expansion for the fittings and piping materials. If a leak does occur, fugitive gases can find an ignition source and flash back to that source. In some cases, the burning fuel or heat from other burning components can cause the fuel tank to rupture during a fire, although these tanks employ pressure relief valves designed to release excess pressure in the event of exposure to outside sources of heat.

RVs often have natural gas or liquefied petroleum gas (LPG) systems for heating, refrigeration, and cooking. Additional information about natural gas and LPG systems can be found in NFPA 54, *National Fuel Gas Code*, and NFPA 58, *Liquefied Petroleum Gas Code*.

Turbochargers

Turbochargers can be found on both gasoline- and diesel-powered engines **FIGURE 17-4**. A turbocharger increases the power of the engine by forcing

the individual injectors. Combustion air is sometimes provided by natural aspiration, or more commonly through a **turbocharger**. Diesel fuel systems ignite the fuels in the cylinder through compression. Unlike gasoline engines, they do not require a spark for ignition.

FIGURE 17-4 A turbocharger.
© Jones & Bartlett Learning. Photographed by Glen E. Ellman.

pressurized air into the cylinders. This device uses exhaust gases to rotate the impeller up to 100,000 RPM. Turbochargers and the exhaust manifolds that supply them are the hottest points on the surface of the engine, with their temperatures approaching 1500°F (815°C) under load. If a fuel leak or oil leak occurs and the fluid contacts these surfaces, the fuel can ignite readily. The center bearing of most turbochargers is also lubricated by engine oil under pressure. If the turbocharger suffers a mechanical failure during operation, engine oil may be delivered into the hot exhaust stream, potentially resulting in a fire.

TIP

It is unlikely that you will know everything about the systems in all vehicles you could possibly encounter. The key is to know where to find the needed information when investigating a fire involving a motor vehicle. Technical manuals can be obtained online or from a library, car dealership, or parts supplier. The parts department of an automotive repair shop may also have schematics and views of particular systems.

Emission Control Systems

Emission control systems are found on most vehicles today. These automotive systems are designed to reduce or control exhaust gas emissions and to collect gasoline vapors while the engine is operating. The emission control system regulates the fuel input, timing adjustments, and recirculation of exhaust gases through the exhaust gas recirculation (EGR) valve. A charcoal canister collects excess fuel vapors.

Problems that can occur in the emission control system include vapor leakage from the hoses or charcoal canister or overfilling of the fuel tank, which forces liquid into the charcoal canister. This could result in rough idling, stalls, backfires, and overheating of the catalytic converter, which can operate with an internal temperature approaching 1300°F (700°C) and a surface temperature around 600°F (315°C). If vapors leak from the charcoal canister or associated hoses, a fire may result.

Diesel trucks, under Environmental Protection Agency (EPA) guidelines, incorporate systems to reduce oxides of nitrogen (NO_x), carbon dioxide (CO_2), carbon monoxide (CO), and particulate emissions. Trucks built after 2007 typically incorporate software-controlled cooled EGR systems coupled with diesel particulate filters (DPFs), which use various methods of "regeneration," or burning of trapped carbon in the DPF. Fuel delivery and, in gasoline engines, ignition timing are computer controlled in modern engines to optimize both power delivery and emissions control. Other systems unique to the diesel exhaust system include diesel exhaust fluid (DEF) dosing, in which valves provide a fine mist of DEF to spray into the hot exhaust stream, and selective catalytic reduction, which reduces nitrous oxide by converting it into nitrogen gas and water vapor.

Exhaust Systems

Starting at the engine, exhaust system components include an exhaust manifold that is connected directly to the engine, exhaust piping, a catalytic converter, and a muffler. During normal operation of a properly maintained and operating vehicle, the temperature at the entrance of the catalytic converter can measure 650°F (343°C), and it is normally the hottest part of the exhaust system. An improperly operating engine can raise the temperature of the exhaust system at the catalytic converter enough to ignite the vehicle's undercoating and interior carpeting. A hot exhaust system can also ignite grass and other external combustibles when the vehicle is parked on them.

Internal combustion engines, particularly diesel-fueled engines, tend to eject carbon particles from their exhaust. Under load, some of these particles may ignite or glow and present an ignition hazard if they contact dry grass or duff (decomposing organic materials such as needles, bark, and leaves). Many states require construction tractors and agriculture vehicles to be equipped with spark arrestors.

Motor Vehicle Electrical Systems

The storage battery is the primary energy source in a vehicle. When the ignition key is turned on, the starter draws energy from the battery to start the engine.

Once the engine is running, the alternator provides electrical power to the vehicle. The vehicle battery is normally a 12-V DC negative-ground system. The conductor size determines the amount of electrical energy the electrical system can safely carry. The largest conductors extend from the battery to the starter and alternator, and then to a grounding location. They are rarely protected by overcurrent protection equipment. Arcing on these conductors may often be found following a fire in the vehicle because of damage to insulation and cable supports. The battery can also be a fuel source in a vehicle fire.

Small amounts of hydrogen and oxygen gases may be found in sealed "no maintenance" batteries, and these gases could be released if the battery case is damaged. Such a battery can generate hydrogen gas and ignite easily if an ignition source is present. An internal battery failure can release both hydrogen gas and significant heat, resulting in the ignition of the hydrogen as well as nearby combustible components such as the battery housing.

12-Volt Electrical Systems

Electrical energy throughout the vehicle comes through the positive side of the electrical system; from there, it is routed through a fuse block (sometimes called a load center or power distribution module), with overcurrent protection being provided in the form of circuit breakers or fuses. The electrical system for the vehicle is most often grounded through the vehicle's frame. Most electrical devices in a vehicle have only one conductor (positive); the rest of the circuit is completed through the car frame or body. The exception to this design is found in trucks and diesel-powered heavy equipment, which often employ positive and negative cables between the batteries, starter, and alternator to reduce power losses in their systems.

Only a few of the electrical circuits retain power when the engine is shut off. These may include circuits from the battery to the starter, from the battery to the ignition switch and fuse block, and to power seats. Circuits from the ignition switch to the accessories, such as the clock and cigarette lighter, onboard computers, and aftermarket accessories, may also remain powered with the ignition off. In addition, battery power may be supplied to relays for other accessories.

Wiring diagrams typically show the color of the conductor insulation, the device location, and its function. Most vehicle wiring is stranded and can become brittle when heated. Document the condition and identity of wiring before moving it.

Arc mapping has been used in structure fires but may not be effective in automobiles, trucks, and other mobile equipment that use DC with a common ground. These power circuits can consist of a number of different wire gauges and have circuit protection of different types and various locations, and sometimes no overcurrent protection at all. Since the vehicle chassis provides the ground for the electrical system, all conductive vehicle components can provide a potential path for electrical current if the insulation becomes compromised. Currents can run between adjacent conductors or find alternative electrical paths, and inadvertent paths can be created that cause arcs and shorts unrelated to the original circuit involved with the fire origin. Single-conductor primary circuits with little or no overcurrent protection may reveal potentially useful evidence of electrical activity.

Other Electrical Systems

Some older vehicles have 6-V DC systems. Some vehicles can have positive grounding systems. Some trucks, buses, and heavy equipment may have 24-V systems.

Event Data Recorders

Many modern vehicles are equipped with an **event data recorder (EDR)** as part of the air-bag actuation system. In the EDR, vehicle parameters recorded for seconds before and after a crash may include vehicle speed, engine RPM, brake actuation, air-bag deployment, and so forth. Retrieval of this information requires special skills and/or equipment and in some cases may be done only by the manufacturer. Damage to the EDR because of exposure to the fire or trauma of a crash, however, may make information retrieval impossible.

A newer passenger vehicle will commonly contain a computer connected to vehicle systems and sensors, known as an **engine control module (ECM)** or powertrain control module (PCM). If this system has not been damaged by the fire, the data can be downloaded in a readable form by an engine technician. Many truck and bus fleets use electronic communications to transmit information wirelessly regarding the vehicles' systems and operations, and data on their location as well as their operation may be available from monitoring agencies after a fire. Systems such as General Motors' OnStar have the capability of recording and reporting on many parameters, including location, operation, systems diagnosis, crash data, and communications.

Many of these systems record and report information that is considered proprietary, either to a manufacturer, a company, or a vehicle owner. As a result, legal concerns may arise regarding data ownership and/or release.

FIGURE 17-5 Mechanical failures associated with internal combustion engines are frequent fire causes.
© AlexKalashnikov/Shutterstock.

Mechanical Power Systems

Mechanical failures associated with internal combustion engines can, and frequently do, cause fires **FIGURE 17-5**. Failure of bearings, rings, or pistons can generate internal heat that then leads to catastrophic engine failure. Such failures can project fragments of connecting rods through the engine block, resulting in the release of a spray of heated engine oil, which ignites on contact with sufficiently heated surfaces. A loss of engine oil from incompletely sealed gaskets can allow the combustible fluid to come in contact with the exhaust manifolds and ignite. Coolant loss may allow the glycol component of the coolant to reach its ignition temperature atop manifolds or turbochargers. Leaks from extremely high-pressure fuel-injection systems (like those found in modern gasoline or diesel direct-injection systems) may send readily ignitable, atomized fuel into the engine compartment.

Lubrication Systems

Most vehicle engines use hydrocarbon-based oils for lubrication. The oil is pumped through the engine at 35 to 60 psi (240 to 415 kPa). An oil leak can occur in a number of locations within the engine compartment. If one does occur, the oil vapors are susceptible to igniting given the proper circumstances. A lack of oil in the engine can also cause a fire if the engine fails. In diesel-powered vehicles, a turbocharger failure may release lubricating oil into either the exhaust system, the intake system, or both. This may result in a runaway engine or a fire.

Liquid Cooling Systems

Most vehicle engines contain a liquid coolant, which is circulated through the engine by a water pump.

Generally, the coolant is 50 percent ethylene glycol and 50 percent water. Operating temperatures of the coolant generally range from 180°F to 195°F (80°C to 90°C), with an operating pressure of 16 psi (110 kPa). As the water component evaporates and the remaining ethylene glycol falls below its boiling temperature, a loss of coolant that allows contact with heated exhaust components may result in initiation of a fire.

Air-Cooled Systems

Air-cooled systems are uncommon in most modern vehicles. An air-cooled system circulates the air through the engine by a fan and ductwork. Normally, the fans are belt driven, so if the belt breaks, engine failure can occur.

Mechanical Power Distribution

Mechanical power produced by the engine must be delivered to the drive wheels for the vehicle to be put in motion. This power distribution or transfer is accomplished through a transmission and drive axles, both of which require lubrication.

Mechanically Geared Transmissions

Mechanically geared transmissions (manual transmissions) transfer power from the engine through the clutch assembly (located between the engine and the transmission) to the vehicle wheels. Most transmissions contain a transmission fluid for lubrication and the movement of internal valves. If a leak occurs in the vicinity of heated exhaust system components or another suitable ignition source, a fire can occur.

Hydraulically Actuated Transmissions

Hydraulically actuated transmissions (automatic transmissions) contain a transmission fluid that is typically cooled by routing through a cooler in the radiator or by auxiliary heat exchangers that may be installed on an aftermarket basis. The fluid can be added to the transmission through the tube provided for the transmission fluid level indicator (dipstick). Internal heat developed in the transmission or torque converter (often due to slippage of friction elements) may vaporize transmission fluid, resulting in expulsion of this fluid through vents, seals, or the dipstick tube. If transmission fluid contacts heated exhaust components, a fire may result. Transmission fluid that has been overheated may have a brown color and distinctive burned aroma and may be cloudy because of particulates suspended in the fluid.

Laboratory examination of a fluid sample may assist in identifying the nature of any transmission or torque converter failure. A disassembly inspection of both the transmission and torque converter by a qualified expert may be necessary to pinpoint the precise failure.

Accessories and Braking Systems

Other accessories attached to the engine that can fail include alternators, air-conditioning compressors, power steering pumps, air pumps, and vacuum pumps. Mechanical failures that produce friction may present viable ignition sources.

Hydraulic brake systems operate under high pressure. If a leak occurs in this system, it can generate a spray of brake fluid in the form of ignitable vapors. Fluid spilled from a master cylinder reservoir (which sometimes occurs when the cap is left ajar on the reservoir) may contact heated exhaust components, resulting in a fire. Braking systems in all vehicles convert kinetic energy to heat energy, dissipating the heat to the atmosphere. Malfunctions in the braking system may produce more heat than can be safely transferred to the air, resulting in overheating of drums, wheels, tires, or fluids and possibly a fire.

Windshield Washer Systems

Windshield washer fluid mixed with water typically contains methyl alcohol or methanol; it is listed in NFPA 921 as a fuel that could serve as a first material ignited in a vehicle fire. Some washer systems use nozzles that include heaters to keep the fluid from freezing; these heaters, if they malfunction, can provide a heat source to cause a fire. Additionally, the fluid is typically kept in a plastic container, which may be consumed in a fire. It is important to note whether another part of the vehicle has penetrated the area typically occupied by the washer fluid container.

Body Systems

Many panels in modern vehicles are made of combustible materials. These panels can contribute to the fuel load once a fire has started. Combustible metals, which are used in some panels, can burn intensely. The engine compartment bulkhead can have a number of penetrations for plastic ductwork; this ductwork may burn in a fire and provides an avenue of fire spread from one compartment to another.

Interior Finishes and Accessories

Interior seats and padding can provide a significant fuel load in a vehicle fire **FIGURE 17-6**. During

FIGURE 17-6 Interior seats and padding provide a significant fuel load in a vehicle fire.
Courtesy of Jeff Spaulding.

investigation of a vehicle fire, document any existing or missing accessories within the vehicle passenger compartment, and inventory and photograph the contents in the passenger areas.

Cargo Areas

It is important to determine whether the fire started in a cargo area or spread to it from another location. Take an inventory of the contents of the cargo area. Examine the spare tire and the depth of the tire tread.

Ignition Sources

Many sources of ignition energy in vehicles resemble those associated with structure fires. Open flames, mechanical and electrical failures, and discarded smoking materials can often lead to a vehicle fire. Vehicles do have some unique ignition sources that should be considered and evaluated, including heated exhaust components, various types of bearings, and braking systems.

Open Flames

Although open flames were a common source of fires in a carbureted vehicle if the engine backfired through the carburetor, fuel-injection systems have replaced carburetors in most of today's modern vehicles. The throttle body in most modern vehicles does not handle fuel, which is commonly injected directly into intake manifolds or combustion chambers. An after-fire from an exhaust system can also produce open flames, but fuel-injection systems reduce the likelihood of such a condition occurring.

Open flames from smoking materials can occur in ashtrays in the vehicle and could ignite combustibles. Exposure to open flames from outside sources, such as wildland or ground-cover fires, sometimes results in vehicle fires whose origin may be very difficult to determine, particularly if the vehicle is examined in a location away from the fire scene.

Electrical Sources

The primary source of electrical energy in a vehicle that is not running is the battery. During the investigation, it is important to remember that some vehicles have multiple batteries. Some of the many components that can remain energized when the vehicle is not operating include headlights, power seats, interior lights, the alternator, the starter solenoid, and relays serving various vehicle systems. Information specific to the vehicle may be obtained from service manuals and/or wiring diagrams.

Determine whether the vehicle was running at the time of the fire to determine which potential electrical ignition sources were energized.

Fuses, circuit breakers, or fusible links are designed to provide protection for electrical circuits in most vehicles **FIGURE 17-7**. Most vehicle electrical systems use direct current (DC), with the negative (ground) side of the system being connected to the body, frame, and engine. The positive side of the electrical system provides current via the electrical wiring to devices such as starters, alternators, and radios. Anytime a positively charged lead goes to ground, a short circuit will occur, with the potential to result in a fire. This may occur if battery cables or other conductors chafe against grounded portions of the vehicle, such as the frame or engine, wearing through their insulation.

The inside of the cover of most fuse panels lists the equipment protected and the size (amperage) of the fuse or breaker. If the label is illegible, this information may often be obtained from the owner's manual or a shop service manual for the vehicle. It is important to view the fuse panel to determine whether a problem exists with an electrical component or circuit and to verify that proper fuses or circuit breakers are in place.

The electrical systems in RVs may rely on a combination of 12 V DC and 120 V AC. Converters are used to change the voltage from 120 V AC to 12 V DC; inverters change the power from 12 V DC to 120 V AC. Many motor homes may have onboard generators that produce 120 V AC. Two or more batteries or as many as four auxiliary batteries as a secondary power source for "coach" systems are common **FIGURE 17-8**. During the investigation, determine whether the RV was connected to an outside power source through a shore power connection.

Another means by which electrical energy may serve as an ignition source is through resistance heating, which may be associated with overloaded wiring. Improperly installed or utilized aftermarket equipment can overload the vehicle's electrical system. Aftermarket items include audio or video systems and associated hardware, as well as enhanced vehicle lighting (running lights, spotlights, fog lights), navigation systems, and many other types of accessories. Improperly installed aftermarket equipment may lack overcurrent protection, or the protection may be inadequate for the load. Wiring can be incorrectly sized, leading to an overload condition.

On commercial vehicles, aftermarket equipment may include electrically operated body systems and communication systems. A high-resistance fault in their electrical wiring can raise the temperature of the wiring insulation to its ignition point without activating

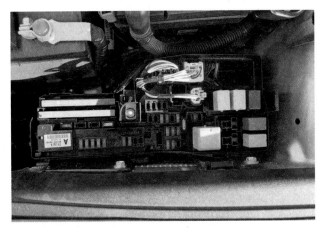

FIGURE 17-7 Fuse and relay panel.
© Jones & Bartlett Learning. Photographed by Glen E. Ellman.

FIGURE 17-8 This RV has more than one battery.
© Jones & Bartlett Learning. Photographed by Glen E. Ellman.

the circuit protection. Faults in high-current devices can ignite readily available combustible materials.

Damaged insulation can result in electrical arcing when a charged wire comes in contact with a grounded surface. Crushed, stretched, cut, or poorly secured wires can be a source of electrical arcing, as can damaged batteries. It is important to remember that the battery and starter cables are unlikely to be electrically protected (by fuses or overcurrent protection) and that they can carry a large amount of current. If these lines are severed during a crash, arcing can occur and result in a fire. Wiring that is subject to chafing may lose its insulation, allowing unintended contact with other metals, which then creates sparks or resistive heating.

Lamp filaments of broken bulbs are a potential ignition source if the fuel is present in a gas, vapor, or liquid spray form. Most lamp filaments are designed to function in a vacuum. The operating temperature of the exposed filament can be as high as 2550°F (1400°C), but when a filament is broken and exposed to air, it operates for only a few seconds and then burns away, removing the ignition source. The glass envelope of a modern headlight, while lit, operates at a temperature well above the ignition temperature of many plastics. If the bulb is displaced and rests against plastic trim components, a fire may result. Check whether the correct size, type, and wattage bulbs are installed in a headlamp assembly; bulbs with higher wattage than the original equipment manufacturer (OEM) bulb can create more heat than the assembly was designed for. An improper bulb type can produce resistive heating where the bulb connects to the vehicle wiring.

Some vehicles are equipped with high-intensity discharge (HID) headlights, which provide high levels of light output from xenon bulbs. These lights require high voltages to ignite (up to 25,000 V), although once ignited they typically operate at 40 to 90 V. HID systems include an AC inverter module and ballast that step up, and control, the power from the vehicle's 12-V electrical system. HID lights can come as OEM equipment or from aftermarket kits; if an aftermarket HID kit is used, its power draw could potentially exceed the capacity of the OEM wiring, resulting in an overload condition. As with any other aftermarket installation, consider the possibility of errors or defects in the installation of HID kits.

During the investigation of a vehicle fire, look for evidence of external electrical ignition sources. This could include ports for recreational connections, engine heaters that heat engine blocks or coolant, and battery chargers. Engine heaters, as well as battery and transmission heaters, usually require power cords

to connect them to a building's electricity supply, and these cords should be inspected for overload or damage. Battery chargers can represent an additional electrical source to consider or eliminate during the investigation.

Hot Surfaces

Exhaust system components can reach temperatures sufficient to ignite diesel spray and to vaporize gasoline. Slippage of transmission or torque converter components may result in sufficient heat being generated to vaporize fluid, causing a discharge of fluid that can ignite on contact with the heated exhaust system components.

Engine oil, power steering fluid, and brake fluid can also ignite when in contact with the heated exhaust system components. These liquids may ignite during operation or soon after the engine is shut off, when the temperature of some exhaust system components is still sufficiently elevated. Note that the ignition temperatures for fluids on hot surfaces have been reported to be substantially higher than the established autoignition temperatures.

The following factors all affect ignition of liquids by a hot surface:

- Ventilation
- Environmental factors
- Autoignition point
- Liquid flash point
- Liquid boiling point
- Liquid vaporization rate
- Atomization of the liquid (effective fuel surface area)
- Liquid temperature
- Surface temperature
- Length of exposure of the liquid to the heated surface and configuration of the hot surface

All of these factors may influence the air–fuel ratio at the location of the potential ignition source, creating mixtures that are either too rich or too lean for ignition and/or propagation of the fire.

Other combustible materials, such as plastics, papers, and vegetation, can reach their ignition temperature when in contact with heated vehicle components.

Catalytic converters normally operate with an internal temperature approaching 1300°F (700°C) and a surface temperature in the vicinity of 600°F (315°C). However, they may reach higher temperatures under heavy loads if air circulation is restricted or if unburned fuel enters the converter because of overfueling or engine misfiring. Under such extreme circumstances, the ceramic matrix material inside the converter may melt and be ejected from the tailpipe in an incandescent spray.

Mechanical Sparks

Mechanical sparks occur from metal-to-metal contact and can be associated with rotating equipment (alternators, pumps, pulleys, etc.) or bearings. Metal-to-pavement contact (e.g., a broken drive shaft, tire rim, or exhaust pipe) can also produce sparks. The vehicle must be running or moving to produce such an ignition source. Vehicle speeds of 5 mph (8 kph) have been determined to create sparks with a temperature of 1470°F (800°C), while greater speeds may create sparks with a temperature of 2190°F (1200°C). Aluminum-to-pavement sparks are not considered a competent ignition source for most materials. Usually, these types of sparks will not ignite solids. Thus, the investigator must consider the fuels present near the area from which the sparks emanated and the amount of time the sparks may have been in contact with these fuels.

Farm, logging, construction, or highway maintenance equipment may have mechanical attachments that come in contact with cellulosic crop or plant materials. If this equipment is damaged or not properly maintained, friction may produce heat capable of igniting those materials. When humidity is low and fuels are dry, mechanical sparks can occur when metal blades or implements hit a rock. The mechanically produced sparks may have enough potential energy to ignite surrounding fuels under these conditions. The ability of such sparks to ignite other materials depends on their mass–surface area ratio: Smaller sparks will cool more rapidly with exposure to ambient air.

Smoking Materials

Another ignition source identified in vehicle fires is improperly discarded or misused smoking materials. In this ignition scenario, the ignition sequence for discarded smoking materials would center on the vehicle's ability to confine and retain the heat of the smoking material and the ventilation aspects of the vehicle. In assessing the possibility of such an ignition sequence, be aware of federal regulations such as Federal Motor Vehicle Safety Standard (FMVSS) #302 (47 CFR 571.302), which provides the burn resistance requirements for materials used in motor vehicle interiors. Consider the possibility that other fuels, such as papers or clothing in the vehicle, might have provided the first fuel for a cigarette ignition. If ignition of these materials occurs, the fire may spread to other fuels, including foam and other materials in the vehicle interior.

Investigative Techniques

Before conducting any vehicle fire investigation, the investigator should learn about the general systems and characteristics of powered vehicles and equipment. Most mobile equipment uses similar types of mechanical systems. For example, cooling systems or fuel systems in automobiles share many design aspects of those employed in trucks and heavy equipment.

Be aware of the vehicle's safety features; have a general knowledge of the type and use of the vehicle, including its motor and components; and have a plan for conducting the investigation. It is helpful to gather, when possible, information regarding the use and maintenance of the vehicle and circumstances surrounding the fire prior to examination of the site and vehicle. Internet sites may provide information regarding failures and/or recalls for specific vehicles. **TABLE 17-5** provides a suggested method for planning the investigation.

During the initial phase of a vehicle examination, an investigator should look for fire or damage patterns inside or outside the vehicle. These patterns can then direct the investigator toward the area, and then point, of origin.

Motor Vehicle Examinations

The examination of a vehicle fire is a tedious and often complex task that can take longer than the investigation of a simple house fire. A more complete examination of the vehicle can often be better accomplished after its removal from the fire scene. As is the case with all fire evidence examinations, it is important to recognize that other entities (such as insurers, manufacturers, and other injured parties) may have a legitimate and legal interest in the investigation and that their interest may be compromised by any alteration or removal of evidence. Prior to conducting any destructive examination, those parties should be identified and notified to preserve their access to the examination and to prevent a later claim of spoliation of evidence. Refer to Chapter 5, *Legal Considerations for Fire Investigators*, for specific coverage of spoliation of evidence and NFPA 921; ASTM E860, *Standard Practice for Examining and Preparing Items That Are or May Become Involved in Criminal or Civil Litigation*; and ASTM E1188-11, *Standard Practice for Collection and Preservation of Information and Physical Items by a Technical Investigator*.

TABLE 17-5 Planning the Investigation of a Motor Vehicle Fire

Safety Concerns	Qualifications	Vehicle Examination
Before viewing the vehicle, in any condition: ■ Ensure its stability. ■ Check that the electrical system on the vehicle is disconnected. ■ Be aware of undeployed air bags and bumpers. (The expelling agent for the air bag is a serious safety hazard.) ■ Check for fuel and other fluid leaks that may pose a fire hazard. ■ Take the necessary steps to neutralize hazards. ■ Use proper protective clothing and equipment.	Perform an analysis of the vehicle systems: ■ Know the specific function of each system. ■ Determine whether a system malfunctioned. ■ Determine whether a system has been altered. ■ Determine whether a malfunction or alteration could be responsible for the fire. ■ Preserve evidence for future laboratory examination. ■ Seek guidance from a qualified individual when needed.	Differentiate five major compartments: ■ Engine compartment ■ Passenger compartment or interior ■ Cargo compartment ■ Exterior ■ Underbody Differentiate ignition scenarios: ■ Electrical ■ Mechanical ■ Human intervention (actions or inactions) ■ Determine an area of origin

FIGURE 17-9 Vehicle examinations should begin with documentation of exterior damage patterns.
© D Russell 78/Shutterstock.

A

B

FIGURE 17-10 Radial fire patterns.
Reproduced from NFPA 921-2017, *Guide for Fire and Explosion Investigations*, Copyright © 2017, National Fire Protection Association. This reprinted material is not the complete and official position of the NFPA on the referenced subject, which is represented only by the standard in its entirety.

An early step in the investigation is to determine an area of fire origin. As with structural fire investigations, the investigator should work from the area of least damage to the area of greatest damage. Note and document exterior damage patterns as indicators of the most intensely damaged compartments **FIGURE 17-9**. These patterns will assist in working toward the area of origin.

It is common to see discoloration and changes in surface coatings, as well as radial color and pattern variations on metal surfaces after coatings have been consumed. On bare metal surfaces, metal oxides can form during a fire, and patterns may be created by different metal oxides or even variations in a single oxide **FIGURE 17-10**. Such patterns can be useful in tracing the origin of a vehicle fire, though the investigator should be careful not to rely exclusively on them without other substantiating evidence.

During the investigation process, vehicles can be generally divided into five compartments: the engine compartment, the passenger and driver areas, the cargo space, the exterior, and the underbody (underchassis). On the exterior, paint damage, sheet metal distortion, and glass damage often provide directional patterns to lead an investigator toward areas of early fire development **FIGURES 17-11** and **17-12**. **TABLE 17-6** provides an overview of windshield indicators.

TABLE 17-6 Windshield Indicators of First Involvement

Compartment	Indicators
Passenger	Top of the windshield fails Radial burn patterns on the hood
Engine	Bottom of the windshield fails Radial patterns on the doors

Examination of Vehicle Systems

The inspection of vehicle fuel systems should be systematic, proceeding from outside to inside, starting

FIGURE 17-11 Incipient windshield failure caused by an engine compartment fire.
Reproduced from NFPA 921-2017, *Guide for Fire and Explosion Investigations*, Copyright © 2017, National Fire Protection Association. This reprinted material is not the complete and official position of the NFPA on the referenced subject, which is represented only by the standard in its entirety.

FIGURE 17-12 Radial pattern on the passenger's-side door caused by an engine compartment fire.
Reproduced from NFPA 921-2017, *Guide for Fire and Explosion Investigations*, Copyright © 2017, National Fire Protection Association. This reprinted material is not the complete and official position of the NFPA on the referenced subject, which is represented only by the standard in its entirety.

with the fuel tank and working toward the engine. Fuel supply lines and vapor return lines should be documented in the process. Note and photograph all signs of fire-related damage or rupture, if present. In the case of larger diesel engines, the fuel lines may consist of hoses that are reinforced with steel braid or may be composed of nylon tubing. Carefully examine these hoses for signs of breaching, mid-hose, due either to contact with energized electrical cables or to chafing, and examine the ends of hoses for evidence of failed connections. Modern fuel lines that are constructed or reinforced with nylon or other plastics are vulnerable when exposed to fire. Remaining fuel-line materials in those vehicles may be important in defining those areas least affected by fire exposure. Be aware that many gasoline injection systems utilize a fuel pressure regulator near the fuel rail that, due to diaphragm failure, may release gasoline under pressure. When examining areas of suspected failure of pressurized fuel components, notice any acute patterns of damage, which may be either indicators of the area of origin or examples of secondary damage.

Other hoses to examine for breaching or crimp failures include coolant hoses, power steering hoses (particularly on the high-pressure side of the system), and oil lines to turbochargers or other engine systems.

Switches, Handles, and Levers

During inspection of the interior compartment, note and document the positions of switches, handles, and levers, if possible. The ignition switch should also be checked to see whether the key is present or absent. If this area is severely damaged, check the floor below for any signs of keys or the ignition lock assembly, which may be found sufficiently intact to support an examination for evidence of key use or prefire damage. Ascertain and document the positions of the windows prior to the fire and determine the position of the transmission gearshift lever.

TIP

Interview owners and operators regarding equipment operation, cleaning, and maintenance schedules. If repairs have recently been performed, interview the service providers and mechanics involved. Inspect and document maintenance on the equipment if records are available.

Hybrid Vehicles

Hybrid Vehicle Technology

A hybrid vehicle is any vehicle that uses two or more types of power. Hybrid cars are increasingly popular and usually consist of a combination of a gasoline engine and an electric motor, with the goal of maximizing the benefits of each. As the hybrid designs are relatively new, there is little uniformity in their technical features and specifications. Their general features include a combination of electric power and an internal combustion engine, which is electronically controlled to maximize efficiency. Electric motors typically operate on DC, although variable-frequency AC motors are also produced.

Although the drive system may use up to 600 V DC, supplied by a battery pack, peripheral systems such as lighting, sound, and other convenience systems use a traditional automotive-style 12-V DC system, supplied from a standard 12-V DC battery. Electric motors may also provide power for steering assistance instead of hydraulic power steering. High-voltage circuits and harnesses are colored orange for identification. The systems are typically designed to disconnect the battery pack electrically when the key is in the OFF position, in the event of an electrical fault, or upon sensing a crash. A manual disconnect is usually provided as well **FIGURE 17-13**.

Currently, hybrid technology is being applied to more and larger vehicles, including mid-sized trucks and delivery vehicles and Class 8 or over-the-road (OTR) trucks.

Investigation of Hybrid Vehicle Fires

Hybrid systems differ from previous automotive designs because of their battery packs, which can provide potentially lethal electrical shocks and which may, if compromised, provide a unique set of potential ignition sources, as compared with internal combustion-engine vehicles.

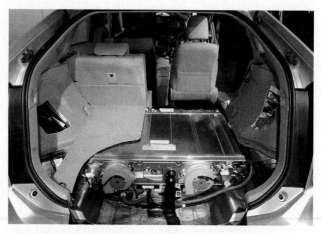

FIGURE 17-13 Hybrid vehicles feature safety systems to isolate high battery-pack voltage, including a manual disconnect near the battery pack.
© Takehiko Suzuki/The Yomiuri Shimbun/AP Images.

The automotive wiring that carries more than 60 V DC or 30 V AC will be orange in color. This color is intended as a warning that the wiring may contain sufficient voltage to be a lethal hazard. The majority of these leads are contained in protective sheathing beneath the vehicle. In addition to being present in the traction motor, higher voltages may be found inside the inverter/converter, which contains the computer control for the system. Because the hybrid's engine does not run at all times, the air-conditioning compressor must be electrically driven. Power is drawn from the high-voltage battery pack to run the 300-V DC motor; therefore, the supply wiring harness to the AC compressor motor is colored orange. The steering system requires a pump but cannot be driven as in a conventional automobile; instead, the pump is electrically driven, albeit by a lower-voltage motor (less than 60 V DC).

Hybrid systems and vehicles are undergoing rapid developmental changes as their technology continues to evolve, so be sure to obtain product-specific information regarding the high-voltage system prior to beginning an examination. It may be prudent to consult with technicians trained in maintenance and repair of the specific vehicle prior to pursuing the vehicle examination. Exercise caution with postcrash vehicles or postfire suppression environments that are contaminated with water. Keep in mind that the internal combustion equipment of the hybrid vehicle will be much the same as that found on a gasoline-engine automobile. After positive isolation of the high-voltage system, the investigation may proceed as for other vehicles.

///////////////

TIP

Automotive wiring containing more than 60 V DC or 30 V AC will be orange in color. This color indicates that the wiring may contain sufficient voltage to be a lethal hazard.

Additional Vehicle Considerations

Total Burns

A **total burn** is a fire that has consumed all, or nearly all, of the combustible materials present at the scene. Total burns are challenging fires to investigate, whether the fire is structural or vehicular. However, even though much of the evidence an investigator is seeking may have been altered or destroyed by the fire, some remaining information may still be retrievable. Enter the process with a goal of maximizing the collection and documentation of that information.

Determine the condition of the vehicle prior to the fire and whether any components were missing. Examine the floorboards for the presence of ignitable liquids or their trace remnants. It is advisable to analyze the fluid levels in the various systems and to take samples. An engine oil analysis or laboratory analysis of the engine oil filter can indicate the condition of the engine prior to the fire, even if the vehicle is severely damaged by fire. Similarly, the prefire condition of the transmission can frequently be determined by analysis of the transmission fluid.

Stolen Vehicles

History has shown that the chances of an accidental fire after a vehicle is stolen are low. Often, vehicles are stolen for parts and are burned to cover the crime. Parts that may have been removed include the wheels, major body panels, engines, transmissions, air bags, stereos, and passenger seats. Sometimes the vehicle was stolen to conceal another crime. Vehicles have also been burned by or on behalf of their owners, often to support a fraudulent insurance claim, and reported stolen in an attempt to deflect suspicion from the owner.

Through debris sifting and inspecting, it may be possible to recover the ignition switch locking tumblers—the portion of the ignition switch that holds the key. Forensic laboratory inspection of the tumblers can often indicate whether the ignition switch was defeated and sometimes whether a specific key was used to last operate the vehicle. This examination must be done by a qualified technician, such as a forensic locksmith. An investigator should secure the ignition lock assembly and transport it to the technician, documenting its discovery, removal, and transportation or shipping to maintain the evidence chain properly. If the ignition lock is still in place in the column, removal and transportation of the entire column will allow the forensic technician to document the tumbler position and steering lock engagement as well.

Vehicle Ignition Components

When investigating vehicles that were reported stolen and recovered in a burned condition, physical evidence of the prefire condition of the ignition lock may be discoverable if the vehicle contains a cut key-style ignition lock. A common location for the ignition lock is on the right side of the steering column; alternatively, it may be mounted in the dash or center console. Vehicle manufacturers use only a few styles of ignition lock wafers; you should be familiar with the construction of the wafer associated with the vehicle that is being investigated **FIGURE 17-14**.

Typical ignition lock cylinders are constructed of die-cast zinc mounted in a die-cast zinc steering column housing, whereas the lock wafers are typically made of brass. The lower melting temperature of the die-cast zinc steering column housing will typically release the ignition lock from the housing during a fire, and the ignition lock will fall to the floor or into the floorboard debris, often remaining at least partially intact. The recovery of the ignition lock artifacts can provide information regarding the prefire condition of the ignition lock and the time during the fire when the lock collapsed.

Layer the debris on the driver's-side floorboard to recover the ignition lock artifacts. If the lock was not forcibly removed prior to the fire, some or all of the lock artifacts should be identifiable and recovered from the debris for further analysis. The recovery of the lock remains from within the debris just below the original location of installation would indicate that the lock was likely in the proper position at the time of the fire. The lock remains should show thermal damage but may also show mechanical damage that may have occurred prior to the fire.

The discovery of an ignition lock on the floor beneath the fire debris could indicate that the lock was on the floor prior to the fire. This could be an indicator

FIGURE 17-15 Remains of ignition lock wafer within the debris.

Reproduced from NFPA 921-2017, *Guide for Fire and Explosion Investigations*, Copyright © 2017, National Fire Protection Association. This reprinted material is not the complete and official position of the NFPA on the referenced subject, which is represented only by the standard in its entirety.

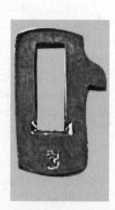

FIGURE 17-16 Remains of ignition lock wafer on top of the debris.

Reproduced from NFPA 921-2017, *Guide for Fire and Explosion Investigations*, Copyright © 2017, National Fire Protection Association. This reprinted material is not the complete and official position of the NFPA on the referenced subject, which is represented only by the standard in its entirety.

FIGURE 17-14 Examples of commonly used ignition lock wafers.

Reproduced from NFPA 921-2017, *Guide for Fire and Explosion Investigations*, Copyright © 2017, National Fire Protection Association. This reprinted material is not the complete and official position of the NFPA on the referenced subject, which is represented only by the standard in its entirety.

that the ignition lock was removed from its installed position prior to the fire **FIGURES 17-15** and **17-16**.

When inspecting the ignition lock for evidence of mechanical damage, magnification may be necessary to evaluate the wafer "key way." A detailed examination may show whether the lock was forcibly rotated, or "picked." It may also reveal a rotational deformation of the wafer. Sharp gouges or scratches across the inside of the key way may indicate that the lock was picked.

If the remains of the lock are not readily identifiable during the examination of the debris, it may be concealed within component materials. In such a case, radiographic images may be necessary to identify the ignition lock artifacts.

Some manufacturers use non-brass wafers in ignition locks, which may be melted or destroyed by fire. Use other ignition lock system components, such as locking lugs, armored lock cylinder caps, tumblers, detent pins, springs, ring antenna, lock core retainer, and other case-hardened parts found in the floor-level debris, to determine whether the ignition lock was present in the vehicle at the time of the fire, although these components may not be useful in establishing whether the ignition lock was defeated prior to the fire.

Vehicles in Structures

When a vehicle is located in a structure where a fire has occurred, the vehicle should be considered as a potential ignition source until it is properly eliminated. The vehicle might be covered by structural debris that has to be removed before inspection of the vehicle can take place. An external fire could have caused the fuel in the fuel tank or other vehicle fluids to be released and to become involved in the fire. A fully involved automobile burning inside a garage or other building presents a dramatic fuel load and, as the liquid fuels contained in the vehicle become involved, can burn with a high heat release rate. This may play a role in the location and extent of structural damage.

Documenting Motor Vehicle Fires and Fire Scenes

Vehicle fire scenes can be documented using the same procedures that are used for documenting a structural fire scene. Start with the documentation of the area surrounding the vehicle, and then focus on the vehicle itself. When possible, the position and location of the vehicle and its relationship to buildings, streets, and other site features should be documented photographically and by measuring and diagramming the site.

Vehicle Identification

Vehicle identification begins with the make, model, and year as well as other identifying features, including the vehicle identification number (VIN) **FIGURE 17-17**. The VIN is commonly found on the dashboard in front of the driver's position; in some vehicles, it may be stamped on the engine side of the bulkhead (also known as the firewall). Sometimes the VIN will be located on a label affixed to the driver's

FIGURE 17-17 The vehicle identification number provides important information about the production of a vehicle.
Courtesy of the NJ State Fire Marshal's Office, Arson/K-9 Unit.

door pillar or on the engine block. Using the VIN, you can obtain information about the vehicle's manufacturer, country of origin, body style, engine type, model year, assembly plant, and production number.

If standard VIN locations are obscured or destroyed, other derivative VINs may be retrievable in the form of "confidential" VINs. You may need to contact your local auto theft unit, the National Insurance Crime Bureau, or the Canadian Police Information Centre for assistance in locating these hidden VINs.

Vehicle Fire Details

It is advisable to determine the actions and events that led up to the fire by conducting interviews with the person(s) who first observed the fire. The owner, passengers, and operator of the vehicle at the time of the fire or the last person to drive the vehicle; nearby residents; fire department personnel; and police officers might also provide pertinent information. Information that may be helpful to obtain includes the following:

- Last use of the vehicle and by whom
- Mileage at the time of the fire
- Operation or problems
- Service or maintenance history
- Fuel level and type, and when last fueled and where
- Equipment, including aftermarket equipment
- Personal effects in each area of the vehicle
- Photos or videos prior to, during, or after the fire

Vehicle Particulars

For assistance in identifying systems and component configurations, it is often worthwhile to inspect a similar (exemplar) vehicle. During an inspection of the vehicle, the checklists in Annex A of NFPA 921, *Guide for Fire and Explosion Investigations*, can be used to gather information. Notes and diagrams taken during the inspection are important to document details observed that might otherwise be overlooked and may help later to clarify details of photographs taken during the examination.

Information about fire causes in vehicles of the same make, model, and year could be important to the investigation. This information can be obtained from the National Highway Traffic Safety Administration, the Insurance Institute for Highway Safety, the Center for Auto Safety, and, in Canada, Transport Canada. Online automotive or fire investigation forums may also serve as a resource.

Documenting the Scene

As discussed earlier, when documenting the scene, define the entire fire scene, including the vehicle and the area surrounding the vehicle. The scene itself should be considered evidence that needs to be protected and preserved until it can be thoroughly documented. When the vehicle has been removed from the scene, document both the vehicle and the scene, making every effort to understand where the vehicle was located and in what position or orientation it sat. Record the fire scene by creating a scene diagram showing reference points and distances. Photograph the scene, as well as surrounding buildings, highway structures, vegetation, other vehicles, tire and foot impressions, fire damage, signs of fuel discharge, and any parts or debris.

The vehicle should be photographed in a systematic manner similar to that employed with a structure fire. Begin on the outside, documenting all surfaces (including the top and underside, if possible), damaged and undamaged areas, tires, and tire tread depth. Document the engine compartment, taking overview photographs of all sides and focusing on specific engine areas and components. Determine and document the positions of windows and doors, inspecting window lift mechanisms for their positions and side-window channels for evidence of remaining glass. Document the paths of fire that spread into or out of compartments and cargo spaces. For the cargo space, obtain photographs of the spare tire and any special equipment found there, such as stereo gear or add-on devices, as appropriate. Photograph any changes or alterations made to the vehicle during the fire suppression effort and/or the initial inspection. Finally, when possible, document the vehicle removal process.

Towing Considerations

As noted earlier, the vehicle is often best examined in detail in a controlled setting, away from the fire scene. Removal of the vehicle from the scene, however, may dislodge and potentially lose valuable evidence. Make sure you carefully document the scene surrounding and beneath the vehicle both prior to moving the vehicle and as the vehicle is being moved. It may be helpful to wrap the damaged portions of the vehicle prior to transport, preventing evidence from dropping off it en route.

Documenting Away from the Scene

Documentation should identify any parts that might be missing or damaged when the vehicle is inspected after it has been moved from the scene. If possible, the vehicle should be protected from the elements.

Fires in Marine Vessels

The field of marine fire investigation is one of the most challenging areas facing investigators. A shipboard fire can encompass many facets of fire investigative knowledge, from compartment fires to vehicle fires to industrial fires. A **vessel** can vary in size from a small recreational boat less than 20 ft (6 m) long to ultralarge crude carriers thousands of times larger. Vessel fires can occur on land, in a stable drydock, or thousands of miles offshore in hurricane-force winds and high seas. The myriad possibilities facing a marine fire investigator require not only a solid understanding of fire dynamics but also familiarity with marine terminology, ship construction, vessel operations, maritime safety and fire protection regulations, and international admiralty law.

Types of Marine Vessels

There are hundreds of different types of marine vessels, each specifically designed for an intended use. Recreational or personal vessels include jet skis, wave runners, runabouts, catamarans, sailboats, and high-performance speedboats. Commercial vessels include fishing boats, container ships, ferries, recovery vessels, and oil supertankers. Military vessels include destroyers, aircraft carriers, hovercrafts, and submarines.

Because of the wide range of vessels in which an investigator may be called to examine a fire scene and render an opinion as to origin and cause, NFPA 921, *Guide for Fire and Explosion Investigations*, focuses on boats that are less than about 65 ft (20 m) long **FIGURE 17-18**. Vessels in this category number well into the hundreds of thousands and experience the majority of marine fires in the United States. They include most pleasure powerboats, sailboats, and larger luxury yachts. Many of the smaller commercial boats, such as fishing boats and passenger craft, also fit into this size range.

Marine vessels can be constructed using a variety of techniques and materials. Wood has been used for thousands of years as a primary construction material; however, as technology has advanced, steel, aluminum, fiberglass, and other materials have become more popular choices. Vessels may be created using multiple pieces or modules or may be formed as one solid component. Fixtures and furnishings are also made from a variety of materials that have properties similar to those found in a residence or motor vehicle.

Knowing the many terms specific to marine vessels will be helpful during marine fire investigations. **TABLE 17-7** defines several of those terms; additional terms are discussed further in this chapter.

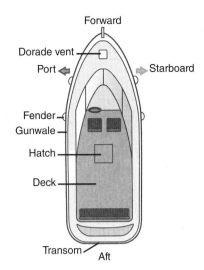

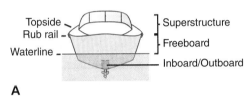

A

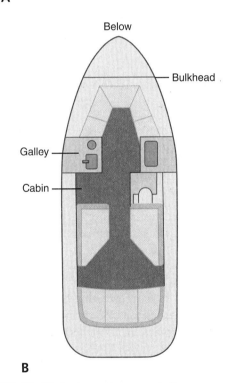

B

FIGURE 17-18 Basic marine vessel diagram. **A.** Exterior. **B.** Interior.

TABLE 17-7 Common Marine Vessel Terms	
Term	**Definition**
Aft	Toward the rear of the vessel
Aground	Touching the bottom
Beam	The widest part of a vessel
Below	Beneath the deck
Cabin	A compartment for passengers or crew
Capsize	Turn over
Chain plate	A metal plate used to fasten a shroud or stay to the hull of a sailboat
Dock	A protected water area in which vessels are moored (usually a pier or wharf)
Dorade vent	Deck box ventilation that uses a baffle to keep water out while letting air in below decks
Fender	A cushion placed between marine vessels or between a marine vessel and a pier to prevent damage to the vessel(s)
Forward	Toward the bow of the marine vessel
Freeboard	The vertical distance between the waterline and the gunwale
Galley	The kitchen area of a marine vessel
Gear	Ropes, blocks, tackle, and other equipment
Gunwale	The upper edge of a marine vessel's side
Port	The left side of the marine vessel when looking forward
Rub rail	The rubberized plastic or metal bumper that extends along both sides of the vessel, usually immediately below the gunwales
Sole	The cabin floor, timber extensions under the rudder, or the molded fiberglass deck of a cockpit
Starboard	The right side of the marine vessel when looking forward
Waterline	A line painted on a hull showing the point to which a marine vessel sinks when it is properly trimmed

System Identification and Function

Like any fire investigation, the investigation of a marine vessel fire can be a challenging endeavor. An investigator must be able to properly recognize and identify the various systems and components to determine their role, if any, in the cause of the fire. The extensive use of combustible materials such as foam, plastics, and wood may make the identification of components and systems difficult; however, noncombustible items will often survive and can be examined.

Fuel Systems: Propulsion and Auxiliary

The fuels used in propulsion systems for marine vessels vary depending on the type of vessel. For example, most personal watercraft and small marine vessels are powered by gasoline or diesel engines that use the combustion energy to turn the propeller shaft. In contrast, military naval vessels may develop propulsion from direct-drive diesel engines or nuclear power plants that generate heat to operate steam boilers.

The most common types of propulsion systems that you will encounter will be gasoline or diesel engine power plants.

Gasoline-Based Systems

Vacuum/Low-Pressure Carbureted. Unlike the engines in automobiles, the vacuum/low-pressure carbureted **inboard** and **outboard** engines on marine vessels must have backfire flame arrestors and a specialized gasket to prevent the escape of fuel from the **Venturi** opening of the carburetor if flooding occurs.

High-Pressure/Marine Fuel-Injection Systems. Fuel-injection systems, which operate at high pressures, vary among manufacturers. In general, the fuel-injection system of all makes and models of marine vessel engines will include a throttle body, plenum and fuel rail assembly, knock sensor, and engine control module **FIGURE 17-19**. The fuel storage and delivery systems, however, will still be low-pressure systems with vented fuel tanks.

EPA regulations require gasoline-powered vessels built after August 1, 2012, to meet specified emissions standards. Vessels meeting these standards may be equipped with specialized equipment including fuel caps, vents, carbon canisters, relief valves, and diurnal controls.

Occasionally, multiple versions of a specific engine model may exist to allow the manufacturer to meet individual needs and requirements. To determine the specific system design, an investigator must record the serial number of the engine, as well as the make and model.

FIGURE 17-19 Marine fuel-injection system.
Courtesy of Jeff Spaulding.

Diesel

A fuel-injection system is used for diesel-fueled engines and is specific to each manufacturer. Diesel engines have similar fuel system requirements as gasoline engines, and require large air exchanges in the engine room or compartment to maintain adequate combustion air and appropriate ambient room temperatures.

Fuel Systems: Cooking and Heating

It is not uncommon to find both cooking and heating systems on a marine vessel, especially those designed for continuous habitation. These appliances are very similar in size and use to those found in RVs. If these devices are found during the examination, the investigator should determine whether they were potentially involved with the fire's development and/or spread.

Liquefied Petroleum Gases

LPG systems may be used for either heating or cooking fuel on a marine vessel; the LPG cylinders will be stored in compartments that vent at the bottom directly to the outside of the vessel. When LPG is used for cooking, it is necessary to control other potential ignition sources. Appliances that have integrated fuel cylinders of less than 16 ounces are governed by the American Boat and Yacht Council (ABYC) A-30, *Cooking Appliances with Integral LPG Cylinders*. LPG may be used for either heating or cooking if the appliance meets ABYC A-26, *LPG and CNG Fueled Appliances*.

Compressed Natural Gas

Recently, the use of CNG for cooking has declined in popularity; however, an investigator may still encounter these systems. The CNG cylinders will often be stored in an **accommodation space** of less than 100 ft³ (2.8 m³).

CNG may be used for either heating or cooking if the appliance meets ABYC A-26, *LPG and CNG Fueled Appliances.*

Alcohol

Often used in galley ranges, alcohol may be stored either in tanks integrated with the appliance or in an independent container and pressurized by a hand pump.

TIP

If an explosion has occurred, it is important to identify all potential flammable or combustible gases and liquids that may have been present on the vessel.

Solid Fuels

NFPA 302, *Fire Protection Standard for Pleasure and Commercial Motor Craft*, provides design and installation requirements for solid fuels used on marine vessels. Such fuels may include gels, charcoal, and wood.

Diesel and Kerosene

NFPA 302; ABYC A-3, *Galley Stoves*; and ABYC A-7, *Boat Heating Systems*, provide design and installation requirements for the use of diesel fuels for heating and cooking on marine vessels. Diesel and kerosene stoves are often used in place of propane, reducing the risk of migrating vapors. Combustion air is drawn from inside the compartment and is exhausted outside the vessel. Improper installation or use may cause heated gases to enter the compartment, creating a risk of carbon monoxide poisoning or ignition of nearby combustible items.

Turbochargers/Superchargers

Both diesel and gasoline engines may have turbochargers or superchargers installed to improve horsepower for the engine. Although both turbochargers and superchargers send pressurized air into the intake manifold, a turbocharger is powered by hot exhaust gases, whereas a supercharger is powered by the engine via a drive belt.

Turbochargers and superchargers may use the engine oil to lubricate or cool the devices because they operate at very high temperatures. To protect the surrounding materials, water jackets and/or heat blankets or shields may be fitted to these units.

Exhaust Systems

When combustion engines are used for propulsion, the hot gases produced must be safely exhausted. This can be achieved with either wet or dry exhaust systems. A wet system draws water surrounding the marine vessel into the exhaust system, where it travels through the muffler (if provided) and hoses until it exits the marine vessel. The components of these systems must be designed in accordance with ABYC P-1, *Installation of Exhaust Systems for Propulsion and Auxiliary Engines*, and NFPA 302. If a de-watered exhaust system is used, the exhaust gases are separated from the water at the muffler. The water is discharged from the **transom**, the stern cross-section of a square-sterned vessel, while the exhaust is released through the bottom of the marine vessel. These systems are also designed to meet the specifications in ABYC P-1.

Dry exhaust systems vent combustion gases through vertical exhaust pipes that are covered with insulating materials. These systems are seldom present on recreational boats.

Electrical Systems

In saltwater marine environments, a common difficulty facing vessel owners and operators is the need to protect electrical and mechanical components from corrosion. Not only is the metal structure of the ship in danger of corroding, but wiring, connectors, and electrical distribution equipment are also at heightened risk of experiencing this problem **FIGURE 17-20**. Once copper conductors and parts begin to corrode

FIGURE 17-20 Corrosion.
© Gari Wyn Williams/Alamy Stock Photo.

after exposure to saltwater or other corrosive environments, resistance can increase in the electrical circuits. Furthermore, if saltwater infiltrates or wicks under wire or cable insulation, such problems can occur at locations in the electrical system other than just at connections. As a result, areas of high resistance can lead to overheating, even in conductors well away from connections where such failures tend to occur most often in electrical systems on land.

Corrosion can be a particular problem for AC power supplied to vessels through shore-tie connections. Conductors exposed to saltwater and not adequately maintained can result in high-resistance connections. In the case of partially shorted systems, overcurrent protection devices may not trip, and electrical power may continue to flow. Fires can then occur at the plastic or rubberized boots of the shore-to-vessel connection.

Anywhere above deck where electrical or electro-mechanical components are present, corrosion poses a concern. Although **topside** (above deck) electrical items are usually protected from sea spray contact and saltwater intrusion, at times they may still occur. Such conditions may produce high resistance points in the circuitry, causing heating and sometimes fires.

Both AC and DC electric systems may be found on marine vessels and are powered by a variety of sources. A shore-line is often used to provide AC power to marine vessels; however, generators and inverters may also be used as an onboard source. Direct current is provided by onboard battery systems, which power lights and equipment.

Engine Cooling Systems

Marine vessel engines are cooled by either drawing surrounding water into the engine and then discharging the water, or through the use of closed-coolant systems. A seawater system draws the water into the engine via a pump, which then circulates the water through the engine and discharges it through the muffler. A closed system uses a 50/50 blend of glycol antifreeze, similar to an automobile, contained in a heat exchanger. Seawater is circulated into the heat exchanger and then discharged through the exhaust system.

TIP

Statistics on both large ship and recreational vessel fires suggest that between one-half and three-fourths of marine fires occur in engine spaces. Many of those involve electrical failures and/or fuel systems.

Ventilation

Many marine vessels use a permanently mounted fuel tank that is often concealed within an accommodation space or specifically designed compartment. These tanks are required to have flame arrestors installed and are vented via the top of the tank through a hose, which is then directed **overboard** (out of the vessel). Gasoline fuel tanks are naturally ventilated (not forced), whereas compartments and associated bilges in these compartments are required to be power ventilated.

Many marine vessels have **hatches** or openings in the deck that are fitted with watertight covers, as well as portholes that allow for natural ventilation of the accommodation spaces when they are open. Dorade and cowl vents may be present on a marine vessel and allow for continuous ventilation. Gasoline-powered vessels dated from 2012 and later may have specialized equipment to comply with applicable EPA emissions standards.

When interpreting fire patterns, an investigator must consider the presence of these ventilation points and their effects on fire movement patterns.

Transmissions

A marine vessel may have either a mechanical or hydraulic-geared transmission. Generally, inboards and some high-performance inboard/outboard (I/O) engines will use a hydraulic-geared transmission, whereas outboards and most I/O engines will use a mechanical transmission. When a marine vessel uses a hydraulic-geared transmission, it will have its own lubricating oil (SAE 90) and cooler.

Accessories

Like other vehicles, marine vessels may contain various accessories for convenience or comfort. Air conditioners use a heat exchanger that circulates seawater through the coils. Air-conditioning compressors will often be found in lower accommodation spaces or the engine room.

A power steering pump may also be present in the marine vessel. It operates similarly to the corresponding equipment in a motor vehicle, where a belt-driven pump is present on the propulsion engine.

Refrigeration units such as refrigerators and freezers may be powered by AC or DC current, LPG, or a combination of both. If an LPG system is used in an accommodation space, it must meet the requirements of UL 1500, *Standard for Safety Ignition-Protection Test for Marine Products*.

Hydraulic Systems

Hydraulic systems may be used either to steer or to adjust the trim settings on a marine vessel. Trim tabs are powered by a pump motor that is mounted next to a hydraulic tank near the transom. Hydraulic steering systems may be either pump driven or mechanical, and will use either rigid or flexible hoses to steer the marine vessel. Although both are closed systems and are relatively safe from ignition, a release of hydraulic fluid on electrical components nearby or a hot surface may pose a potential ignition risk. Hydraulic thruster systems are available and generally used in larger vessels.

Construction and Design of Marine Vessels

It is imperative that marine fire investigators be familiar with the basics of ship and marine vessel construction. Maintaining adequate strength through well-designed, robust construction allows vessels to operate in heavy weather conditions without failure.

Knowing how these factors apply will assist in the proper application of fire dynamics principles to explain fire patterns.

Exterior Construction

High-tensile steel is used almost exclusively in the construction of large, seagoing vessels. The **hull** and primary construction of smaller commercial vessels or recreational boats may be steel but may also consist of wood, fiberglass-reinforced plastic (FRP), aluminum, or ferro-cement. The use of FRP and wood in the construction of marine vessels should be taken into account when determining the total fuel load. In certain instances, marine vessels may be constructed of FRP that used balsa or high-density foam as a core, which can act as a heat insulator during a fire.

Additionally, an investigator should consider the accessory items either attached to or within the marine vessel, which may be combustible and add to the total fuel load. These items include communication equipment; low- and medium-density plastic fixtures such as marker and navigation lights, spotlights, and railings; wooden components such as decking, masts, and booms; interior finishes and fixtures; and other combustible items such as sails, personal flotation devices (PFDs), life rafts, and seat cushions.

Generally, **superstructures** (cabins and other structures above the main deck) and decks are made of the same material as the hull. Horizontal partitions that form the tops and bottoms of compartments are known as **decks**. Decks are supported on the underside by beams and longitudinals; they provide strength to the hull by helping resist transverse compressive loads.

Bulkheads are the vertical partitions that extend perpendicularly to the long axis of the vessel and establish the basic compartmentation of the interior of the vessel. Bulkheads can be either structural or nonstructural. Structural bulkheads connect decks, frames, and hull plating and can withstand severe fluid dynamic pressures of the sea and liquid cargos. Nonstructural bulkheads serve as compartment separations, as in living or accommodation spaces. Structural bulkheads and decks can also function in steel ships to provide various degrees of passive fire protection.

Interior Construction

Depending on the design of the marine vessel, various materials may be used to achieve a lightweight, yet durable structure similar to those found in automobiles. Generally, interiors of recreational boats are made with FRP, plywood, and veneers. Steel and aluminum are noncombustible and may be used for structural components such as bulkheads and support structures.

Depending on the size and use of a marine vessel, the interior may resemble that of an RV, an ordinary home, or a luxury hotel. The interior of accommodation spaces may be finished with wood paneling or painted walls; have carpeting; and contain items such as televisions, lamps, candles, and fans. These spaces often have bedding and furniture constructed of fabric-covered foam padding. Wood items may be coated with organic oils and varnishes as a finish. These items, as well as the potential for human actions in these occupied spaces, must be considered when determining ignition sources, fuel loads, and fire spread issues.

Engine and machinery compartments contain the engine and accessory devices and components needed to propel and operate the marine vessel. These include batteries; generators; inverters; and storage units for waste water, hydraulic fluids, and fuel. Both DC and AC power may be present, with various electrical circuits available to accommodate the different electrical needs of the vessel. Fuel tanks may be constructed of aluminum, steel, polyethylene, or fiberglass. Fittings and hoses may be constructed of neoprene, synthetic rubber, metal, or nylon. The tanks, fittings, and hoses must meet strict design requirements to prevent the release of fuel and should be examined to determine their suitability for marine use. Leakage may be the result of damage caused by deterioration, age, heat, vibration, mechanical damage, or corrosion from contact with water.

Because fuel is often stored or transferred in the vessel's interior spaces, flammable/explosive vapor detectors may be present to detect fugitive vapors from either a spill or a leak. The investigator should search for these devices and examine any found to determine whether the device was operating properly at the time of the fire.

In some instances, a fire protection and suppression system may be present in the vessel's interior spaces. These systems can displace oxygen and pose a dangerous hazard should they discharge. An investigator must ensure these systems are disabled while conducting the investigation.

Storage tanks and **holds** are often found in the lower portions of the marine vessel, below the accommodation spaces. The materials stored in these tanks may vary greatly and are often combustible. When conducting a fire investigation, determine which items were present in the vessel's storage or holding tanks. Items stored improperly may either come in contact with an ignition source or, in the case of volatile organics used for maintenance or cleaning, spontaneously ignite.

Propulsion Systems

Propulsion systems are often determined by the use and operation of a particular vessel. Types found in marine vessels include electric, liquid fuel, and fuel gas systems, and combinations of two or more of these systems. After determining the type of propulsion system(s) present, the fire investigator can begin to identify components that may have contributed to the fire's cause or spread.

Electric Systems

The most common electric propulsion unit an investigator will encounter is a **trolling motor**: a small electric motor connected to a battery and used primarily for slow-speed maneuvering of recreational boats **FIGURE 17-21**. Main propulsion from electric motors is not generally used in such vessels.

Bow and stern thrusters may be found in horizontal tunnels forward and aft in a marine vessel. These electrically driven units may be driven by either a 12- or 24-V DC motor and generally involve high-amperage circuitry to operate the thruster. They should be protected by appropriately rated circuit protection, often a 200-amp fuse installed near the thruster unit. Thermal circuit protection may also be provided for these units.

Fuels for Marine Vessels with Motorized Propulsion Systems

As discussed earlier, fuel systems may be present for propulsion, appliances, or the generation of electricity.

FIGURE 17-21 The most common electric propulsion unit you will encounter is a trolling motor.
Courtesy of Rodney J. Pevytoe.

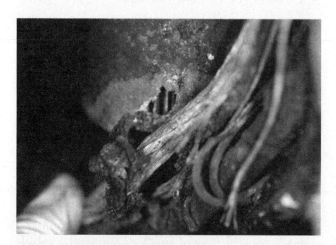

FIGURE 17-22 Outboard fuel filter fire.
Reproduced from NFPA 921-2021, *Guide for Fire and Explosion Investigations*, Copyright © 2020, National Fire Protection Association. This reprinted material is not the complete and official position of the NFPA on the referenced subject, which is represented only by the standard in its entirety.

Engines will fall into one of three general categories: outboard engines with self-contained or portable fuel tanks, inboard gasoline engines, or inboard diesel engines.

Outboard motors may be either two- or four-cycle gasoline engines with a carburetor or fuel-injected fuel delivery system. Two-cycle engines use a mixture of gasoline and oil to lubricate the motor, whereas four-cycle engines operate similarly to automobile engines, with the oil being contained in a separate reservoir. Fuel is delivered by a low-pressure delivery pump or, in fuel-injected systems, by a high-pressure, engine-mounted fuel pump **FIGURE 17-22**.

Inboard motors are mostly four-cycle engines that use built-in fuel tanks, fuel lines, fuel filters, and fuel pumps for fuel delivery. The marine vessel may contain more than one tank, which will be connected by an equalizing line known as a manifold to ensure equal distribution of weight. The tanks themselves are vented, and all connections, fittings, and valves are located on the top of the tank to prevent draining or leakage. A fill plate is located on top of the fill port and connected to the marine vessel's ground; it prevents spilled fuel from entering the hull and reduces the risk of a static discharge while fueling. The fill nozzle must be in contact with the fill plate prior to the pump being activated. Marine vessels constructed after 1977 are required to have antisiphon devices installed on the tanks if the engine is below the level of the tank.

Inboard diesel engines operate similarly to an automotive engine, with a high-pressure system delivering the fuel. However, a low-pressure return line to the tank will likely be present.

Fuels may also be used for cooking, heating, or generators. LPG is the most common heating, refrigeration, and cooking fuel that you will encounter, and it may collect in the lower portions of the marine vessel if vapors leak from the system. Generators are often supplied by a separate fuel line from the marine vessel's fuel tank.

When examining either the propulsion units or accessory items, obtain samples of all engine fluids, including fuel, hydraulic, and lubricating fluids, for analysis of their prefire conditions, just as with a motor vehicle fire.

Other Fuel Systems Used for Propulsion

On rare occasions, an investigator may encounter a marine vessel that is propelled by a steam-driven engine in which a boiler is fired by wood, charcoal, coal, or paraffin. Older ships such as paddle-wheels often used the steam power created by the boilers to drive the pistons that turned the large rear- or side-mounted wheels; they were popular in the mid- to late 1800s.

Marine Vessel Ignition Sources

A marine fire investigator will likely find sources of ignition similar to those observed in both structural and vehicular fires; however, unique sources of ignition such as turbochargers and exhaust manifolds may also be present in these vessels. Take the time to understand thoroughly the components of the marine vessel's mechanical and electrical systems during the investigation to ensure that all potential sources of ignition are identified and examined.

Open Flames

Backfire events through unprotected carburetors are a common source of open flames. To prevent such fires, all inboard and I/O marine vessels are required to have a backfire flame arrestor attached to the air intake with a flame-tight connection. These flame arrestors must be approved for marine use by the U.S. Coast Guard (USCG) or comply with the Society of Automotive Engineers' Standard SAE J-1928. Even with flame arrestors in place, however, fires are still possible if the owner does not keep the arrestor clean and in good condition.

Other sources of open-flame ignition include fuel-fired burners used for cooking. Since 1977, however, appliances with pilot flames have been prohibited on gasoline-fueled marine vessels.

Electrical Sources

Most marine vessels will have some type of electrical system used to either start the engine, provide power to running lights, operate auxiliary lighting, or provide other conveniences like those found in automobiles and buildings. As described earlier, high-amperage thrusters may be found on these vessels as well. These systems and devices can all be potential sources of ignition and must be examined and evaluated.

Unless installed with a main disconnect switch, marine vessels equipped with batteries may have several electrical circuits that remain energized despite the fact that the engine is not running and the ignition switch is in the "off" position. These circuits, which are generally always energized, may include the bilge pump and the primary wiring circuits to the alternator, ignition switch, and battery cables, among others. When the marine vessel is operating, the accessory items and fixtures become potential sources of ignition.

An investigator may also find components present that were not rated for watercraft use. Overcurrent devices may have been replaced with oversized fuses and breakers, and accessory items may have been added that overloaded the electrical system.

Marine vessels will most often be found to have both AC and DC electrical systems present that require examination. Components located below the waterline are grounded or bonded to the marine vessel's electrical system to prevent electrical shock as well as to serve as a form of lightning protection. As in an automobile, the negative or ground side of the battery is bonded to the engine block, whereas the positive or hot side supplies power to the fuses, accessories, and equipment that operate on the DC electric system. This two-wire system can be damaged or fail, and may serve as a potential ignition source if the positive side (wiring, connections, or components) becomes grounded to another object. AC systems that generate power from either a generator or inverter may also represent potential ignition sources within the circuit, connections, and/or connected appliances.

Overloaded Wiring

When circuits are overloaded by excessive use of accessory items, heating of the conductor may degrade and ignite the protective insulation. Because most wiring is routed through bulkheads and in the concealed structures of the marine vessel in large bundles and harnesses, such heat is not easily dissipated, and a significant fire can develop without activating the overcurrent devices. Be sure to interview the appropriate person(s) to determine whether any prefire deficiencies existed and what role they may have played in the ignition of the fire.

Electrical Short Circuiting and Arcs

If wiring becomes worn, brittle, or damaged, an arc or short circuit may occur if exposed conductors come in contact with a grounded surface. Inspect the condition of all electrical wiring to determine whether insulation degradation or damage as a result of charring or other mechanical forces exists. If large current circuits such as the battery cables short or arc, they can easily ignite engine oil, ignitable vapors, or the insulation on the wiring itself.

Electrical Connections

Just as in automobiles, connections in a marine vessel may become corroded because of a high-resistance connection in addition to exposure to water and salt. A high-resistance connection can easily ignite surrounding combustible items. Examine all connections for evidence of resistance heating.

Lightning

If lightning strikes near the marine vessel or strikes the vessel itself, this high-energy discharge can conduct through the marine vessel's structure and electrical system. In turn, they can become heated and easily ignite combustible items located nearby.

Static Electricity and Incendive Arcs

When loading or unloading cargo, attention must be paid to reducing the possibility of static build-up and discharge. If piping and other mechanical devices used in transferring cargo to or from a ship are not properly grounded, then a static arc could, depending on the type of cargo, result in either a fire or an explosion.

If ignitable vapors are present, static electricity and incendive arcs can pose a significant hazard of fire or explosion. As in automobiles, refueling operations may produce both ignitable vapors and static electricity and require proper grounding between the fuel tank and fuel nozzle. Gasoline vapors, which are heavier than air, can collect in the lower portions of the marine vessel, including the accommodation spaces and bilge areas where the generation of static electricity may occur as a result of normal operations.

Care must be taken to ensure that dissimilar cargos, particularly ignitable liquids, are not accidentally mixed. When two liquids with dissimilar volatilities are mixed, the overall volatility may potentially increase above that of the liquid with the higher boiling point. If this occurs, precautions that would normally prevent ignition of the less volatile liquid may no longer be sufficient to avoid a fire or explosion.

Hot Surfaces

Exhaust systems and cooking and heating appliances also can ignite combustibles located too close to these objects. Exhaust manifolds generate temperatures that can easily ignite engine fluids, including diesel sprays, vaporized gasoline, oils, and hydraulic fluids. Similarly to automobiles, the temperatures of exhaust components may actually increase shortly after the engine is stopped because the cooling system is no longer circulating fluid through the engine. Regardless of the type of fuel used to generate heat, cooking and heating appliances can readily ignite combustibles located near the exposed hot surfaces.

Adequate clearances from combustible items should be maintained. Be sure to conduct interviews with the appropriate person(s) to determine whether any combustible items may have been stored in compartments or locations near one of these hot surfaces.

Mechanical Failures

Main bearing failures generally do not result in a fire; however, if a bearing in the alternator, the motor, a pulley, or a pump fails, combustible items nearby may ignite because of either excessive heating or hot metal being expelled from the bearing. If suspicion arises that a bearing may have failed in one of these devices, inspect the inner and outer **races** of the bearing for physical damage.

If a pulley, alternator, or water pump seizes, the drive belt can heat as it passes over these components, generating sufficient heat to ignite the belt and start a fire. An inspection of the engine components can reveal whether a pulley wheel is seized or locked up.

Smoking Materials

Lit cigarettes are capable of starting a fire, particularly when a lit cigarette contacts upholstery and other fabrics. Urethane foam and other materials often used in seating burn easily and can promote spread and intensity of the fire.

Cargo Issues

A concern related to the shipment of cargo is ensuring that incompatible materials are not stored near each other or shipped in the same container or location in a hold. Thousands of tons of oxidizers and peroxides, for instance, are transported annually aboard ships. It is imperative that such materials are isolated from fuels that might combust in their presence. Even though materials are packaged separately, incompatible cargos must not be placed close to each other in the event that rough seas cause breakage of containers and/or accidental mixing. Once fires do involve spaces containing oxidizers, they can easily grow out of control with little hope of extinguishment by firefighters.

Marine investigators examining the origin and cause of fires that may be cargo related should review the vessel manifest to identify what materials were being transported at the time of the fire. The International Maritime Dangerous Goods Code provides information on the classification of various hazardous products and cargo compatibility issues.

In break bulk cargo ships and container vessels, if a fire starts deep in the middle of stacks of cargo, it may be a long time before it is detected. In some cases, ships have unloaded cargo containers only to discover that a fire burned in their cargo unnoticed for the entire voyage. On many ships, cargo containers are mounted with only about 18 inches (45.7 cm) of space between them laterally and several containers high. Should a fire occur near the middle of one of these stacks or in a below-deck hold near the center of packaged cargo, the seat of the fire can be extremely difficult to locate and can easily spread before being identified **FIGURE 17-23**. The job of the investigator in locating origin and cause correspondingly becomes more difficult.

FIGURE 17-23 A fire occurring near the middle of a stack of cargo can be extremely difficult to locate and can easily spread before being identified.
© Malcolm Fife/iStock/Thinkstock.

Marine Vessel Investigation Safety

Prior to inspecting a marine vessel, either on land or afloat, determine the general safety and stability of the craft and provide for the appropriate level of personal protective equipment (PPE) to conduct the investigation safely. Ensure that the craft is stable and will not fall over or capsize during the investigation. Vessels still afloat pose a significant hazard because they may be filled with water, which can make them very unstable and prone to capsizing. In these cases, the water should be removed (de-watering) and the craft stabilized prior to continuing the investigation. If an investigator is unable to remove the craft from the water, wear a PFD. In cases of submerged or sunken marine vessels, skilled divers may be required to ensure safety. If electrical service—either shore based or from onboard battery systems—is present, be sure to de-energize these systems or risk being shocked.

The threat of taking on too much water, resulting in flooding or instability, is always a concern with marine vessels and demands the attention of all involved. Ships' crews are generally well trained in such matters and are familiar with de-watering as well as firefighting.

FIGURE 17-24 Conduct air monitoring to ensure that the air in the confined environment is not hazardous.
Courtesy of Jeff Spaulding.

Confined Spaces

Marine vessels are often compartmentalized to allow for containment of water should flooding occur. This often creates small, single-access, confined spaces that pose a significant risk to an investigator during an investigation. Prior to entering any confined space, conduct air monitoring to ensure that hazardous levels of ignitable gases, toxic vapors, or oxygen-enriched/deficient environments are not present **FIGURE 17-24**. Use an appropriate level of PPE and ensure that any lighting or powered equipment used in these spaces is intrinsically safe.

Airborne Particulates

Because many marine vessels are constructed of fiberglass, consider the use of appropriate respiratory protection while conducting the investigation. When fiberglass burns, the resin is consumed, leaving small particulates of fiberglass, which can cause severe irritation to the respiratory system. Cancer is also a high concern when someone breathes in fiberglass particulates. To mitigate these risks, the investigator should wear self-contained breathing apparatus or a particulate filter mask when conducting the examination. If arriving at the scene soon after the fire is extinguished, air monitoring should be conducted prior to and during the examination.

Energy Sources and Hazardous Materials

As noted earlier, electrical power for a marine vessel may come from a variety of onboard sources. Locate and disable these systems to protect everyone from personal injury and to eliminate their potential for creating another fire. Photograph these electrical sources and create diagrams of their locations prior to manipulating and disabling them.

Batteries are often used in marine vessels, and more than one may be present in different locations. Atmospheric monitoring must be conducted to ensure that an explosive environment is not present prior to disconnecting the battery cables on each battery. Hydrogen gas may collect in battery compartments, and care should be taken to reduce the potential for static discharge or electrical arcs when removing the battery cables. An arc may be produced when the cables are disconnected, creating a risk of ignition if such an environment is present. If an inverter is present, the DC input should be disconnected.

Shore lines may also be in use, especially if the marine vessel is docked. The shore line should be de-energized and then disconnected from the marine vessel.

Various types of liquid and gaseous fuels may be found on a marine vessel, including both fuels used to propel the marine vessel and fuels for cooking and heating purposes. Be aware of the potential presence of these fuels, because a leak may pose a significant risk of fire or explosion should the gas migrate from an uninspected area of the marine vessel.

> **TIP**
>
> Marine vessels contain materials similar to those found in both structural and vehicle fires and will produce similar products of combustion.

An investigator may encounter other fluids, lubricants, and oils used on the marine vessel that not only pose a risk of fire but are also potentially harmful to the environment. In addition, hydraulic fluids, antifreeze, engine oil, and other lubricants could have leaked from the marine vessel's mechanics, creating a slippery walking surface.

Sewage Holding Tank

On marine vessels that contain sewage holding tanks, methane gas may be present. During the fire, these tanks may have been damaged, allowing these gases to escape. Ensure that proper venting occurs to reduce the risk of fire or explosion. Besides the risk of fire and explosion, the sewage may present biological hazards.

Structural Concerns

Either during the fire suppression operations or as a result of damage sustained by the fire, water may enter

FIGURE 17-25 The structural components and finish of the vessel can be damaged during the fire, creating additional hazards.
© Jones & Bartlett Learning. Photographed by Glen E. Ellman.

the bilges and lower sections of the marine vessel, causing it to become very unstable. This poses the greatest hazard to the fire investigator while conducting an investigation on a marine vessel that is still floating in the water. Ensure that the craft is stable prior to conducting the investigation, because if the vessel lists or capsizes, a chance exists of becoming trapped within the vessel.

As in a structure fire, the structural components and finish of the marine vessel can be damaged during the fire, creating additional hazards **FIGURE 17-25**. Thoroughly inspect and sound the decking and other walking sources to reduce the risk of collapse. To sound the decking, an investigator should use techniques similar to those applied by firefighters in sounding floors, using a tool to test the integrity of the decking in front of them.

TIP

Flammable atmospheres on vessels are common because of the presence of not only hydrocarbon fuels for propulsion but also chemicals such as paints, solvents, lubricants, and volatile cargos. Vapors from many of these products are often present in compartments such as fuel tanks, fuel day-tank pump rooms, paint lockers, **lazarettes**, workshops, head spaces of cargo tanks, engine spaces, and void spaces. Most are heavier than air and tend to settle to lower levels. As with ignitable vapors on land, if these vapors collect in pockets within the flammable range of a particular chemical or product and a sufficient ignition source is present, then a fire or explosion may result.

Both vertical and horizontal openings may be present on a marine vessel to allow access below decks and to other portions of the marine vessel.

Although some openings may be visible, hatches, doors, and covers may conceal them after a fire. If flooding is present, unprotected openings may be concealed and pose a significant risk while examining the marine vessel.

Wharves, Docks, and Jetties

Wharves, docks, and jetties are often slippery as a result of their presence near the water. They may also be structurally unstable, especially if they were in contact with a vessel subjected to fire. Take care while on these surfaces, especially while transferring to or from a marine vessel.

Submerged Marine Vessels

Ideally, marine vessels that are submerged will be thoroughly documented and photographed prior to being raised and removed from the water. Although the initial examination should be done where the vessel is, a complete examination of the vessel should be conducted once it is on dry land.

Specialized personnel trained to conduct such unique investigations are needed for submerged vessels. During the recovery operation, monitor and manage the operation from the surface. Investigators should dive only if they are qualified and certified in underwater diving; otherwise, raising the vessel may be the best alternative. Take care to reduce the potential for damaging or altering the marine vessel and its appurtenances during these operations. Also, be aware of the potential hazards from fuels and other harmful products escaping into the water and contaminating the environment.

Visual Distress Signals and Pyrotechnics

Pyrotechnic signal flares may be present on marine vessels and should be handled with care if encountered. If such a device is discovered, secure it to prevent its accidental activation.

Documenting Marine Vessel Fire Scenes

Approach the documentation of marine vessel fires in the same manner as the documentation of a structure or automobile fire. Ideally, the boat should be inspected at the loss location prior to being moved, although this often proves impractical. Most inspections will be conducted out of the water in a salvage yard, repair facility, or storage lot.

Marine Vessel Fire Patterns

The techniques used by ship- and land-based firefighters may differ. Consequently, an investigator must be cognizant of the potential differences in the ways in which a fire may be attacked at sea as compared with in port, and must pay particular attention to how suppression efforts may have altered fire pattern development.

Although some might argue that "fire patterns are fire patterns," certain conditions on a vessel may substantially alter fire development and the resulting patterns. The very nature of vessel construction and operation is the source of significant differences in fire behavior and pattern generation. Other factors that can affect the interpretation of fire ignition, growth, and pattern development on a marine vessel include the following:

- Common instances of flammable atmospheres
- Effects on ventilation of maintaining watertight integrity at or after a fire's ignition
- High conductivity through steel bulkheads
- Six-way heat transfer to surrounding spaces
- Hidden/rarely accessible spaces
- Subsurface effects of boundary cooling
- Movement of the shipboard platform in heavy seas
- A wide variety of hazardous materials in both the ship's stores and its cargo
- Effects of wind
- Availability of air in the fire compartment
- Use of saltwater for firefighting

Differences in Shipboard Versus Land-Based Firefighting

On Land

First establish the location of the fire. If the marine vessel was damaged at its current location, determine whether it was connected to any shore-based power sources (i.e., shore line). The power sources, shore line, and connection to the marine vessel should be photographed and documented. In addition, all potential ignition sources and fuels either in or near the marine vessel should be identified, photographed, and documented. The surrounding area, including the manner of storage of the marine vessel, should be documented as well. Also, look for evidence of prefire vandalism or mechanical problems with the marine vessel. Vessels that are improperly resting on their trailers, covered with debris, and located in obscure locations with large growing foliage around the vessel may have been abandoned.

TIP

Marine vessels should be removed from the water and secured to allow the fire investigator an opportunity to examine the entire vessel. Evidence of prefire damage or poor maintenance may be apparent below the waterline.

In Water

Similar to land-based investigation, determine whether the vessel is floating at the location where the fire occurred. If the vessel has been towed away from the original location, examine this area as well to locate any possible items of evidentiary value. Interviews with witnesses should reveal whether the vessel was **underway**, moored, or anchored at the time of the fire. Underwater examinations may also be necessary to locate items with potential evidentiary value.

Moored

If a marine vessel is moored at a dock, slip, or seawall, check that area for potential ignition sources. The construction materials of piers or docks may have contributed to the intensity of the fire or fire spread from another vessel **FIGURE 17-26**.

Anchored and Underway

Ignition sources may vary between vessels that are underway, **adrift**, or at anchor. The investigator should

FIGURE 17-26 A moored boat fire.

Reproduced from NFPA 921-2021, *Guide for Fire and Explosion Investigations*, Copyright © 2020, National Fire Protection Association. This reprinted material is not the complete and official position of the NFPA on the referenced subject, which is represented only by the standard in its entirety.

determine which systems would be expected to be in use depending on the status of the vessel.

Underwater

As noted earlier, only specialized personnel should conduct underwater surveys of submerged vessels. The vessel and surrounding area should be examined for items with potential evidentiary value, including components of the vessel, dock, or pier or items that might potentially be ignition sources. Prior to raising the vessel, thoroughly document the orientation and condition of the vessel, as items may be damaged, displaced, or lost during the raising operation.

Marine Vessel Identification

An investigator can locate various types of registrations and markings for the boat after a fire. Generally, a unique identifier is given to each vessel, though it will vary depending on applicable regulations and the vessel's manufacturer. Vessels manufactured or imported into the United States, including homemade vessels, are required to have a hull identification number (HIN) permanently attached (e.g., stamped, engraved) in two separate places on the vessel. The primary location is generally at the **starboard** (right side) side of the transom, below the rubrail **FIGURE 17-27**. The secondary location will vary and can be obtained by contacting the manufacturer. If the HIN can be located, the first three letters will indicate the manufacturer code, which can provide further information. Registration numbers are normally located underneath the bow of the vessel and can be used to establish ownership. USCG numbers may also be found inside some

FIGURE 17-27 Boats in the United States are required to have an HIN permanently attached in two separate places on the boat.
Courtesy of David Ritter.

vessels. Sometimes a hailing port and name may be painted on the vessel, usually on the transom; however, this is not usually regulated.

Witness Interviews

As with any fire, establish the scenario in regard to the circumstances of the fire. Seek out and interview anyone who may have information about the fire, including the owner, the operator at the time of the fire, bystanders, and the police and fire department personnel who responded to the incident. Obtain information about the vessel's history and any specific activities related to the fire, including activities just prior to, during, and after the fire occurred. The questions to ask will be very similar to questions posed during other types of fire investigation. Questions include, but are not limited to, the following topics:

- Operational condition of the vessel
- Accessory use and locations
- Recent repairs or modifications
- Water and weather conditions
- Activities and operations on the vessel prior to the fire
- How the fire was first discovered
- Whether the boat was working correctly at the time the fire was discovered
- Actions taken after the fire's discovery
- How the fire was extinguished
- An estimated timeline for fire discovery, summoning of help, and extinguishment
- Salvage operations
- Actions taken by the EPA or similar agencies
- Locations of occupants and fire victims
- Actions taken by occupants and any public officials
- Details about any previous investigations involving the vessel
- Any other questions that may arise as a result of the interview itself

Marine Vessel Particulars

After identifying the vessel and obtaining all relevant information, begin to determine the construction and design features that would affect the fire's development and growth. This can be achieved by examining an exemplar and reviewing the owner/operator's manual and sales literature associated with the vessel. A recall database search may also prove useful in determining any preexisting safety issues. The Consumer Product

TABLE 17-8 Examination Outline

Exterior	Hull, outdrive, vents, windows, prefire damage
Top side	Mast, decking, accessory items, application patterns, and evidence of materials being removed or replaced prior to the fire
Interior	Cockpit, sleeping berths, storage areas, galley, and state rooms
Mechanical	Propulsion system, bilge rooms/compartments, and other accessory locations

FIGURE 17-28 Begin by examining the exterior of the boat.
Courtesy of Jeff Spaulding.

FIGURE 17-29 Burned boats inside a marina building.
Reproduced from NFPA 921-2021, *Guide for Fire and Explosion Investigations*, Copyright © 2020, National Fire Protection Association. This reprinted material is not the complete and official position of the NFPA on the referenced subject, which is represented only by the standard in its entirety.

Safety Commission and USCG both maintain an extensive database on product recalls related to vessels based on their make and model information.

Marine Vessel Examination

As with any fire, the area of origin must be determined. It may be sited in one of four major areas:

- Cockpit/topside
- Engine/fuel compartment
- Accommodation compartment or cabin
- Bilge areas

Establish a systematic process to evaluate and examine the vessel **TABLE 17-8**. This may include processes similar to those conducted on structural and vehicular examinations.

Examination of Marine Vessel Systems

Begin by examining the exterior of the vessel, identifying fire patterns that may help to determine the area of origin **FIGURES 17-28** and **17-29**. Consider the potential roles of vents and openings around the engine and fuel compartments, as these can affect fire patterns. If a fire occurs in an accommodation space, it may consume the topside or cabin without affecting the other spaces. Because these spaces are usually airtight when closed, the fire will often self-extinguish. Items within these spaces must be examined to determine their role in either the ignition or the spread of the fire. Some spaces may have been used as sleeping areas, lounges, cooking, or storage, all of which pose unique ignition and fire spread issues.

The construction of the hull causes accommodation spaces to have sloped surfaces, limited vertical spaces, and offsets that may alter normal fire development and, therefore, limit or alter the patterns that investigators typically find. By examining the fire patterns present, an investigator can determine whether the fire occurred within or outside an accommodation space. If an engine or fuel compartment fire occurs, it will likely involve fuel vapors and cause extensive damage, spreading to the accommodation spaces. Thoroughly check the carburetor; fuel-injection systems; and fuel delivery, exhaust, and ignition systems to determine their roles in the ignition of the fire. **Bilge** areas are usually located in remote and isolated areas of the vessel and are often sites where heavier-than-air fuel gases, such as those from gasoline, diesel, or oil, collect; these gases may still be present after the fire. Bilge pumps themselves are usually intrinsically safe and not a likely source of ignition if operating properly.

After the compartment of origin is located, conduct a detailed examination of the area. Look at the individual systems to locate evidence of their role in the fire. Examine fuel tanks for failures around the edges or bottom or for corrosion that might have allowed fugitive gases to escape. Inspect the fuel fill and vent hoses to determine whether chafing, corrosion, or other damage is present. In addition, test the ground between the tank fill plate and the tank for electrical continuity. If a fuel tank is exposed to heat, a demarcation line may be present where the fuel level inside has cooled the exterior surface. Plastic tanks may still be intact and contain fuel. The plastic container often softens and fails at the level of the fuel within.

Examine switches and handles; document and photograph their positions. Inspect port lights and hatches and determine their positions at the time of the fire. Also look at generators, battery, and shore line power transfer switches. If the vessel's ignition device can be identified, determine whether the key is present and whether any damage or tampering occurred. Although many of these switches are made of materials that are easily consumed, components may still be present and identifiable after the fire.

Marine Vessels in Structures

Marine vessels stored within fire-affected garages or carports should be examined as though they are potentially the source of ignition; in such cases, however, the vessel may simply be an additional combustible fuel package. Other potential sources of ignition within the storage space must also be examined because a fire originating within the structure can easily spread to the vessel.

Vessels afloat that are moored under shelters should be examined and the shelter itself should be considered as potentially containing heat and fire gases that could have influenced the fire development and patterns present on the vessel. The electrical service to and on floating docks must also be checked, as damage from corrosion or movement of the dock itself may be present.

Specialized Vehicles and Equipment

Recreational Vehicles

RVs combine elements of both vehicle and residential structures. Their living spaces are similar to those in houses and mobile homes in terms of contents and materials. Some have plywood flooring, carpets with foam padding, wooden wall paneling, and polyurethane foam furniture. The RV may also be a motorized vehicle, similar to a truck, bus, or automobile, so it may

TABLE 17-9 Types of Recreational Vehicles

Type	Description
Fifth-wheel travel trailer	Mounted on wheels and towed via a towing mechanism mounted above or forward of a motor vehicle's rear axle
Folding camping trailer	Mounted on wheels and towed by a motor vehicle; features collapsible side walls and a roof that can be unfolded for use and refolded for travel
Travel trailer	Mounted on wheels and towed by a motor vehicle, with a roof and sidewalls made of rigid material

incorporate many of the same systems seen in those types of vehicles **TABLE 17-9**. RVs also include trailers such as fifth-wheel trailers and camping trailers.

Resources that may help in determining the origin and cause of an RV fire include sales brochures, owner's manuals, and websites maintained by the manufacturers of the chassis, coach, and other components.

Safety Concerns

The RV may, through its design, be a confined space, so be aware of entry/egress and atmospheric issues. Ensure that such spaces do not contain hazardous levels of explosive or toxic vapors or gases prior to entering them. Toxic, carcinogenic, or irritating materials such as ammonia, sodium chromate, plastic residues, and fiberglass particles may be present in the thermally damaged RV. Some such materials may become airborne. Evaluate their presence prior to entry, and use appropriate PPE.

Also be aware of electrical hazards in the form of 12 V DC from battery-supplied circuits and 120- and 240-V AC power supplied by generators, inverters, or shore power connections. Liquid (gasoline, diesel) or gaseous (propane or hydrogen) fuels may be onboard. Perform a proper evaluation and mitigate such potential hazards before undertaking the onboard examination.

Prior to beginning the examination, evaluate the RV for stability hazards. Tires, plywood flooring, sidewalls, and roof structures compromised by the fire may fail during the exam, causing the RV or its components to drop. These elements must be blocked or shored prior to allowing any personnel within or beneath the RV.

Systems and Components of the RV

RVs may be equipped with equipment and systems unlike those found in other vehicles, including the following items:

- Shore power connections: These consist of a cord, sometimes with a collection of adapters, for connection to a nearby receptacle, as may be available at an RV park or at the owner's home.

- Auxiliary power generator systems: These systems, which may be either fixed or portable, are powered by an internal combustion engine fueled by gasoline, diesel, or propane. An automatic generator starting system that automatically starts and stops the generator under preset conditions of load, demand, battery condition, or time may be incorporated into the system.

- Electrical converters and inverters: These devices allow power to be interchanged between 12-V DC and 120-V AC systems. The converter transforms 120-V AC power to 12-V DC, while the inverter performs the reverse function. Often, these functions are combined in one electronically controlled unit, which also may be connected to the coach and chassis batteries as an automatic charger.

- Battery systems: Most often, the self-propelled RV will use two separate battery systems. The chassis will have one or more 12-V DC batteries for starting the engine and operating other vehicle systems, such as exterior lights and heating systems. The coach will typically employ a separate set of two to four deep-cycle batteries. The sets of batteries are often connected through an electronic isolator system to allow the batteries to be evenly charged by the engine generator, the auxiliary generator, or the converter system.

- Holding tanks: The RV may be equipped with multiple holding tanks for toilet waste (black water), sink and shower waste (gray water), and fresh or potable water.

- Propane systems: Compressed propane stored in one or more tanks on the RV supplies fuel to power the range or cooktop, refrigerator, water heater, and furnace, if the RV is so equipped. The pressure is modified by a two-stage regulator, and fuel is delivered through pipes or tubing to the individual appliances. The propane system is subject to conditions as specified in

NFPA 1192, *Standard on Recreational Vehicles*; NFPA 58, Title 49 of the Code of Federal Regulations (CFR), relating to transportation; and the American Society of Mechanical Engineers' (ASME's) *Boiler and Pressure Vessel Code*, Section VIII.

- Stove: A stove may be either a cooktop, which drops into a countertop, or a range, which includes an attached oven. In either case, the stove is usually fueled by propane and is equipped with a nonadjustable propane regulator.

- Water heaters: Powered by propane, 120 V AC, or a combination of both. Typically, 12-V DC power is used for control circuits, employing an internal thermostat to regulate water temperature.

- Furnaces: Typically forced-air, using propane as the fuel, with controls and blowers operating on 12 V DC.

- Refrigerators: RV refrigerators do not use a compressor like that found in home refrigerators; rather, their design involves a closed system of tubes containing water, anhydrous ammonia, sodium chromate, and hydrogen gas under pressure. The ammonia–water mixture is heated by a small propane burner, a 120-V AC heating element, or a 12-V DC heating element. Automatic controls may switch between fuel sources, or the operator may manually select one mode of operation. Historically, fires involving RV refrigerators have involved loss of coolant containment (ammonia and/or hydrogen gas), electrical problems with control circuits, or venting restrictions.

Investigation of Recreational Vehicle Fires

Investigation of recreational vehicle fires is similar to investigation of fires in other motor vehicles. The RV's systems should be examined to determine whether they played a role in ignition or as fuel, and how they may have been involved in the spread of the fire. If appliances are involved in the fire development, determine their types, brands, models, and conditions. Take care in removing components because of close installation tolerances. Assistance from an electrical or mechanical engineer could be indicated in the evaluation of an RV's heating and electrical systems if a particular system is suspected of having been the ignition source or is otherwise under consideration as a cause hypothesis.

As with the vehicle itself, if individual appliances are potentially involved in the fire cause, research their service and recall history, and interview the RV's users for any history of malfunction or replacement.

Much information regarding a particular RV's layout and specifications may be found on the manufacturers' websites or through local RV dealers. Most new RVs come from the manufacturer with a packet of owner's manuals, including the RV's specific information as well as that pertaining to installed appliances or equipment. These packets, if available, will help identify originally installed equipment and appliances. Many manufacturers maintain specific records of installed parts and equipment for their products by serial number. Some manufacturers also maintain websites or telephone "help lines" to provide technical assistance to owners and RV technicians.

Additional Investigation Considerations

Investigation of RV fires should include examination and analysis of all systems to determine their potential for involvement in the fire, as a source of ignition or fire spread. A failure analysis of these systems includes checking for poor installation or function, prefire damage, fuel leaks, and function of safety and control features. Ignition sources may include the same types as are involved in structural fires, those common to motor vehicles, and those related to movement, towing, and mechanical damage.

Like other motor vehicles, RVs are required by the National Highway Traffic Safety Administration to have a unique VIN. In the case of RVs, the manufacturer also applies a VIN, which can be found in the owner's documents, on a label on the exterior of the coach, and often on a placard on a galley cabinet. Numbers stamped on the chassis may not be directly related to the VIN.

When investigating an RV fire within a structure, such as a garage, care must be taken to differentiate between the structure's systems and combustibles and the RV's systems and combustibles. When the fire occurs in a park or campground, all services provided by the park or campground need to be examined, in addition to the RV's systems. NFPA 1194, *Standard for Recreational Vehicle Parks and Campgrounds*, is a valuable resource in these cases.

Heavy Equipment

Heavy equipment includes earth-moving equipment as well as construction, mining, forestry, landfill, and agricultural equipment. These types of equipment normally serve one or more specific materials-handling functions. Large equipment is often diesel powered with a hydraulic transmission. These systems are sometimes susceptible to failure due to overloading the engine or transmission, a failure of the hydraulic or electrical systems, bearing or engine failures, or the unintended ignition of the materials being handled. Some heavy equipment is equipped with fire suppression systems, which should be inspected for function and maintenance history.

Medium- and Heavy-Duty Trucks and Buses

With the exception of over-the-road (OTR) tractors, most medium- and heavy-duty trucks are custom built for specific applications. The original manufacturer typically delivers these trucks as a bare cab and chassis, and then beds, bodies, and other equipment are built and installed by one or more secondary shops, manufacturers, or "body companies." Buses frequently are delivered from the original manufacturer as a bare chassis, sometimes with minimal front-end body parts. The complete body is then built and installed by a second manufacturer. Often, a separate company will install other body equipment and systems, such as wheelchair lifts, specialized seating, and so forth. Thus, after a fire in relatively new equipment, several companies may have a warranty or liability interest in the investigation and should be notified prior to any disassembly or other alteration of evidence.

Trucks and buses are often designed for a specific purpose and, accordingly, may have unique equipment and design. For instance, they can be constructed from a number of different materials. They frequently have various hydraulic components that, when they fail, can spray fluid over hot exhaust components, resulting in potential ignition. As a result, extensive involvement of plastics, fuels, and oils can create misleading patterns and evidence. Because diesel vehicles often have large-capacity fuel tanks, a fire in a truck where suppression is delayed may involve all the fuel and cause severe damage.

In addition to unique equipment and design features, electrical systems in trucks and buses often are more complex than those in cars. Electrical systems can include 12-V and 24-V systems, low-voltage systems, and even AC to some components. High currents can be found with intake air heaters and glow plugs. Often these vehicles have fusible links in circuits with high-amperage components; the links will open in the event of a fault involving the components.

Locate and document these links (if they appear normal, they may have failed internally and need to be tested with an ohmmeter), and examine the relevant connections such as battery and starter posts.

Mass Transit Vehicles

Many transit districts are converting their fleets to use alternative fuels. Although diesel engines are still standard in most buses, many districts are now using CNG, LNG, diesel–electric hybrid, or pure electric drive systems, with the last often seen in light rail or streetcar services. Most transit districts perform their own maintenance in-house. The transit district shop may be a valuable reference or technical resource.

Earth-Moving, Forestry/Logging, Landfill, and Construction Equipment

Earth-moving, forestry/logging, landfill, and agricultural equipment typically all employ hydraulic systems, providing a reservoir of additional fuel as well as hoses, fittings, and valves that may wear, chafe, or rupture. Chafing between electrical conductors and hydraulic hoses is a frequent cause of fires in these types of equipment. If a conductor chafes through harness coverings and insulation, contact with a grounded steel-reinforced hydraulic hose may provide not only a source of fuel under pressure but also a competent ignition source.

Logging equipment may accumulate large quantities of duff, consisting of needles, bark, and other organic material. This material may absorb hydraulic fluid, oil, or diesel fuel and is easily ignited by contact with heated exhaust components, electrical events, or friction-related heat sources. With all equipment operating in an environment in which loose particulate matter is present, radiators and oil coolers may clog, causing overheating of coolant, oils, or other fluids.

Agricultural Equipment

Agricultural equipment is nearly always subject to accumulations of cellulosic materials such as grain dust, wheat or corn chaff, cotton fibers, hay, or straw. Also, harvesting operations are most frequently performed during periods of high ambient temperatures and low humidity. Cellulosic materials are predictably subject to ignition when exposed to high levels of friction, sparks, or contact with sufficiently heated surfaces.

Fires in agricultural equipment may also result from failure of electrical, fuel, or hydraulic systems; friction developed from bearing or other mechanical component failures; hot surfaces within the equipment; sparks from foreign metals picked up in the field; or friction and sparks from filed mechanical parts that are transported through the crop processing areas of combines, balers, or pickers.

TIP

Rodents and birds can build nests in agricultural equipment that is not in use. If the nest is located near the exhaust components, it can be ignited by the heat of the exhaust system when the equipment is used.

Agricultural equipment requires routine maintenance on a schedule specified in the operator's manual supplied by the manufacturer. Interview the owners and/or operators with regard to equipment operation and cleaning/maintenance schedules, and inspect and document maintenance if records are available.

Aftermarket accessories are often installed on agricultural equipment and are sometimes a factor in the initiation of fires. Such accessories should be identified through interviews and/or records and documented.

TIP

The following hazards are common with agricultural equipment; take care to ensure safety with these hazards in mind:
- Particulate from burned composite material and vegetation
- Sharp edges from glass or metals
- Overhead hazards due to the size of equipment
- Slippery or uneven surfaces on the equipment or unstable equipment
- Fuels, herbicides, pesticides, and other agricultural chemicals
- Failure of mechanical parts, leading to collapse or spills
- In agricultural fields, presence of mud, snow, plants, snakes, rodents, and insects

The fuels used in agricultural equipment are similar to those used in other motor vehicles, including diesel and biodiesel. The body of the equipment can include plastics, composite materials, and rubber. The materials used will greatly impact the investigation.

	TABLE 17-10 Agricultural Equipment

Type	Description
Farm tractors	■ Used for pulling or pushing machinery or trailers, lifting, loading, plowing, tilling, and similar tasks. ■ Engines are usually forward of the operator and run on gasoline or diesel fuel. ■ Transmission and brake types vary, but the transmission includes a power take-off (PTO). ■ They often include hydraulic components.
Combines	■ Used to harvest grain crops: the crop is cut by the header, carried through a threshing mechanism, cleaned in a hopper or tank, and then offloaded. ■ Most have turbocharged diesel engines mounted to the rear of the grain tank. ■ Most have hydrostatic drive transmissions.
Forage harvesters	■ Used to harvest field crops to make silage. ■ The cutting heads cut plants and feed them into processing drums, then to an accelerator device and through a chute to a trailer. ■ Most have turbocharged diesel engines mounted in the rear. ■ They are usually equipped with electronically assisted hydraulic transmissions.
Cotton pickers	■ Used to harvest cotton. ■ Barbed spindles separate the cotton from the plant, which is then blown into a large basket. ■ Most have turbocharged diesel engines mounted near the center of the equipment, below and to the rear of the operator. ■ They are usually equipped with hydraulic drive transmissions. ■ Additional components include row units attached to a header, a moistener system, a hydraulic system, and a lubrication system.
Sprayers	■ Similar to tractors but fitted with solution tanks containing chemicals to treat growing crops. ■ Most do not have PTO attachments. ■ Similar equipment includes windrowers, floaters, spreaders, and fertilizer applicators.
Baling equipment	■ Used to compress harvested crops into round or square bales that can then be wrapped. ■ In round balers, the crop is rolled inside the equipment and is bound and released through a hydraulic gate when it has reached a certain size. ■ In square balers, a mechanical plunger compresses the crop into thin sheets. When it has reached a certain size, it is bound and released. ■ Additional considerations include the pick-ups used to pick up crop material, the chain drive systems used to drive the pick-up and belt and roller systems, the hydraulic system, and the electrical system.

For example, at high enough temperatures, plastic components may melt to the point where any fire patterns or other evidence is lost.

Many kinds of agricultural equipment exist, and within these groupings are several makes and models. When investigating a fire involving such equipment, refer to the manufacturer's information for specific details on the particular equipment involved. **TABLE 17-10** lists common types of agricultural equipment.

TIP

In agricultural equipment, it is important to check for a build-up of combustible materials such as vegetation. This debris can collect in all parts of the machinery, leading to ignition through friction. For example, the practice of scrapping involves harvesting scrap cotton from dead plants. The conditions in which scrapping is performed are very dry, and crop residue can build up in the machinery, leading to malfunction and possible ignition.

Safety Concerns

Although investigating fires in agricultural equipment involves many of the same safety concerns as investigating fires in other motor vehicles, some additional safety considerations should be kept in mind:

- Due to the size of the equipment, there is the potential for more fluid (e.g., fuel) to be stored in the equipment. This can affect the fire spread and intensity and can eliminate early fire patterns.
- Special tools or equipment may be necessary to access internal areas of agricultural equipment. In these cases, it is best to confer with someone who has experience with the particular type of equipment to avoid spoliation issues.

- The height of the equipment may interfere with overhead electrical wires, either at the scene or during equipment removal from the scene. If the equipment is potentially affected by high voltage, investigators and others should stay at least 10 ft (3.1 m) away from the equipment until it has been de-energized by qualified personnel.
- A fire in an elevated cotton picker basket may cause instability and accessibility issues if the operator was unable to release the basket and drive away from it.
- The gate on round balers can create severe crushing and entrapment hazards if the gate's hydraulic system fails.

After-Action REVIEW

IN SUMMARY

- Although investigation of vehicle fires is often considered complex and difficult, fires in vehicles and machinery have the same basic requirements for successful ignition and propagation as structural, marine, or wildland fires.
- Well before you are called upon to conduct any vehicle fire investigation, prepare yourself by learning the general systems and characteristics of powered vehicles and equipment.
- When conducting an inspection of the vehicle's undercarriage, take care to support and stabilize the vehicle properly with blocking or stands to prevent the vehicle's movement, which may cause injuries.
- Fuels in vehicle fires fall into three categories: liquid fuels, gaseous fuels, and solid fuels.
- Vehicles have some unique ignition sources, including heated exhaust components, various types of bearings, and braking systems.
- Knowing how the various systems in vehicles operate is critical and can be enhanced by viewing technical manuals available at the library, car dealerships, and parts suppliers.
- Vehicle components, including panels and bulkheads, can influence the spread of fire.
- Document the fire-involved vehicle at the fire scene using the same procedures used for documenting a structural fire scene.
- A more complete examination of the fire-involved vehicle can often be better accomplished after its removal from the fire scene.
- Although much of the evidence investigators are seeking may have been altered or destroyed by a total burn fire, there is always remaining information that is retrievable.
- Solid fuels in vehicles include plastics that have heat release rates similar to those of ignitable liquids and can contribute significantly to the spread of fire.
- Residual pressure in the vehicle's tank or pressure that is created by the fire could force fuel out of a compromised fuel line.
- Vehicle fuel systems can include systems that run on gasoline, diesel, and other fuels.
- An improperly operating engine can raise the temperature of the exhaust system at the catalytic converter enough to ignite undercoating and interior carpeting, grass, and other combustibles.
- In automobiles, various circuits may continue to be energized even when the engine is shut off.

- In most cases, the sources of ignition energy in vehicles are the same as those associated with structure fires, including open flames, mechanical and electrical failures, and discarded smoking materials and others.

- Exhaust system components can reach a sufficient temperature to ignite diesel spray and to vaporize gasoline.

- Gasoline, engine oil, power steering fluid, and brake fluid can ignite when in contact with heated exhaust system components.

- Mechanical sparks occur from metal-to-metal contact and can be associated with rotating equipment, bearings, and contact with the pavement by parts of a moving vehicle.

- Hybrid vehicles use many differing and proprietary designs, some of which employ potentially dangerously high voltages, ranging from 100 V in one brand of small automotive hybrid to 800 V in hybrid commuter buses.

- Marine vessel fire investigations pose unique safety issues for the fire investigator.

- Hundreds of different types of marine vessels exist, each specifically designed for an intended use.

- You must be cognizant of the potential differences in the ways in which a fire may be attacked at sea as compared with in port, and must pay particular attention to how fire suppression efforts may have altered fire pattern development.

- You must be able to recognize and properly identify the various systems and components to determine their role, if any, in the cause of the fire in a marine vessel.

- It is imperative that marine fire investigators be familiar with the basics of ship and marine vessel construction.

- Once you have determined the type of propulsion system(s) present in the marine vessel, you can begin to identify components that may have contributed to the fire's cause or spread.

- Take the time to understand thoroughly the components of the marine vessel's mechanical and electrical systems during the investigation to ensure that all potential sources of ignition are identified and examined.

- Marine investigators examining the origin and cause of fires that may be cargo related should review the vessel manifest to identify what materials were being transported at the time of the fire.

- Approach the documentation of marine vessel fires in the same manner as you would a structure or automobile fire.

- Marine vessels are usually divided into four areas: topside/cockpit, accommodation space, engine/fuel compartment, and bilge area.

- Marine vessels stored within garages or carports should be examined as though they are potentially the source of ignition, while recognizing that the vessel may simply be an additional combustible fuel package.

- Large equipment is often diesel powered with a hydraulic transmission. These systems are sometimes susceptible to failure due to overloading the engine or transmission; a failure of the fuel, hydraulic, or electrical systems; bearing or engine failures; or the unintended ignition of the materials being handled.

- Agricultural equipment is nearly always subject to accumulations of cellulosic materials such as grain dust, wheat or corn chaff, cotton fibers, hay, or straw. These cellulosic materials are subject to ignition when exposed to high levels of friction, sparks, or contact with sufficiently heated surfaces.

KEY TERMS

Accommodation space A space on a vessel designed for people to live in.

Adrift Loose; not on moorings or towline.

Bilge The lowest, interior part of a vessel's hull; the area where spilled water and oil can collect.

Bulkheads The vertical separations in a vessel that form compartments and that correspond to walls in a building. (This term does not refer to the vertical sections of a vessel's hull.)

Catalytic converter A device in the exhaust system that exposes exhaust gases to a catalyst metal to promote oxidation of hydrocarbon materials in the exhaust gas.

Decks Usually continuous, horizontal divisions running the length of a vessel and extending athwart ships; they correspond to floors in a building on land.

Engine control module (ECM) An electronic device that controls engine operation parameters, including fuel delivery, throttle control, and safety systems operations.

Event data recorder (EDR) A device to record data before and after a crash event.

Fuel rail An internal passage or external tube connecting a pressurized fuel line to individual fuel injectors.

Hatches Openings in a vessel's deck that are fitted with a watertight cover.

Holds Compartments used for carrying cargo below deck in a large vessel.

Hull The outer skin of a vessel, including the bottom, sides, and main deck, but not including the superstructure, masts, rigging, and other fittings.

Hybrid vehicle A vehicle that uses a combination of internal combustion and electric motors for propulsion.

Inboard Toward the interior of a vessel from the outer hull; a vessel engine that is enclosed within the hull of the boat.

Lazarettes Stowage compartments, often in the aft end of a vessel, sometimes used as workshops.

Outboard Lying outward from the centerline of a vessel toward or beyond its sides; a type of engine that is attached to and lying aft of a vessel's stern.

Overboard Over the side or out of the vessel.

Races A component of a rolling element bearing that contains the elements and transfers the load to the bearing; generally used in pairs (inner and outer races).

Shore power Electrical power supplied from shore via a cord set.

Starboard The right side of a vessel when looking forward.

Superstructures All vessel structures above the main deck or weather deck.

Topside When below decks, the areas on or above the main deck.

Total burn A fire that has consumed all, or nearly all, combustible materials.

Transom The stern cross section of a square-sterned vessel.

Trolling motor A small electric motor connected to a battery and used primarily for slow-speed maneuvering of recreational boats.

Turbocharger An exhaust-driven device that compresses intake air to increase engine power.

Underway Not attached to land or mooring by tie, anchor, or grounding.

Venturi A narrowed area within a carburetor that causes air to accelerate and create a low-pressure area that draws fuel into the intake manifold of the engine.

Vessel A broad grouping of every description of watercraft, other than a seaplane on the water, used or capable of being used as a means of transportation on the water.

On Scene

1. Summarize some of the ways in which the investigation of a vehicle fire may be similar to that of a structural fire. In what ways are these investigations different?

2. Unlike in a structural fire, a vehicle can often be moved from its fire location, after proper documentation, for closer inspection and examination. What are the benefits of doing so? Are there drawbacks?

3. Of the list of fire effects and fire patterns provided in this text and NFPA 921, which are you more likely to see in an automotive context?

4. Although many jurisdictions and areas do not have oceans, most have waterways and some boats or marine vessels. Which such vessels are likely to be found in your area or jurisdiction?

5. Identify some of the heavy equipment and/or agricultural equipment commonly found in your area or jurisdiction. What issues might arise in the investigation of fires of this equipment, and where can you turn for product information if assigned to investigate a fire in such equipment?

Chapter Opener: © Jones & Bartlett Learning. Photographed by Glen E. Ellman; On Scene siren: © Bildgigant/Shutterstock.

CHAPTER 18

Wildland Fires

KNOWLEDGE OBJECTIVES

After studying this chapter, you should be able to:

- List special safety considerations associated with wildland fire investigation. (**NFPA 1033: 4.2.1**, pp. 454–455)
- Describe the effects of wind, fuels, topography, weather, fire suppression, and other natural mechanisms on fire spread. (**NFPA 1033: 4.2.4**; **4.2.5**, pp. 455–456)
- Describe the differences among ground fuels, surface fuels, and aerial fuels and their effects on fire spread. (**NFPA 1033: 4.2.4**; **4.2.5**, pp. 455–456)
- Describe indicators a fire leaves behind that can lead the investigator to the origin of the fire. (**NFPA 1033: 4.2.4**; **4.2.5**, pp. 456–459)
- Describe methods of conducting a wildland investigation. (**NFPA 1033: 4.6.5**, pp. 462–468)
- Describe fire cause categories used in wildland fire investigation. (**NFPA 1033: 4.6.5**, pp. 465–466)
- Identify methods of evidence collection for wildland investigations. (**NFPA 1033: 4.4.2**, p. 468)

SKILLS OBJECTIVES

After studying this chapter, you should be able to:

- Apply safety considerations associated with wildland fire investigation. (**NFPA 1033: 4.2.1**, pp. 454–455)
- Locate and protect the general area of origin of a wildland fire.
- Conduct an investigation of a wildland fire. (**NFPA 1033: 4.6.5**, pp. 462–468)

You are the Fire Investigator

You are at the scene of a fire that has consumed several acres of grassland and timber. The fire was reported during moderate winds, and a witness tells you she heard popping noises that she associated with fireworks or gunshots a few minutes before she spotted the fire. You begin your examination of the scene and immediately notice footprints and other signs of human presence. The scene is located approximately 150 ft (45.7 m) from a footpath.

1. What types of fuels are present, and how will they influence the development of this fire?

2. What fire movement indicators will be present that can assist with determining the origin of the fire?

3. What hazards need to be considered during your investigation?

 *Access Navigate for more practice activities.*

Introduction

Wildland fire investigations involve special techniques, practices, equipment, and terminology, and differ significantly from structural fire investigations. Any fire investigator performing wildland fire investigations should know these particular skills and information and must have a thorough understanding of wildfire behavior. This chapter describes topography, fuel arrangement, wildfire fuel types, factors that affect fire spread, indicators of directional patterns, conducting an origin and cause investigation, and special safety considerations in wildland fire investigations.

As in other specialized investigation situations, consider enlisting the assistance of personnel who specialize in wildland firefighting and/or wildfire investigations. The investigator is also encouraged to obtain training through wildland-specific fire investigation courses such as those offered in conjunction with the National Wildfire Coordinating Group (NWCG), including the courses denominated FI-210 and FI-310. The NWCG publishes a guide for wildland fire investigation, available online, called *Guide to Wildland Fire Origin and Cause Determination* (2016).

Safety Considerations

As with other fire investigations, safety is of paramount importance in wildland fire incidents. These incidents present different safety considerations than those associated with structural incidents, and the fire investigator must know about wildland fire safety if operating in the wildland arena.

Many wildfire investigations take place while the fire is still being extinguished. Investigators should check in with the incident commander, ensure that firefighting personnel are always aware of their location, and ensure that a clear escape route is always available. Investigators should also participate in any daily safety messaging or meetings, and make sure they have the ability to contact incident command in the event that resources are needed or fire conditions are spotted. When examining the scene, it is preferable to be accompanied by another fire investigator or firefighter.

Wildfire safety includes following the ten standard firefighting orders:

1. Keep informed about the fire weather conditions and forecasts.
2. Know what the fire is doing at all times.
3. Base all actions on the current and expected behaviors of the fire.
4. Identify escape routes and safety zones and make them known.
5. Post lookouts when there is possible danger.
6. Be alert. Keep calm. Think clearly. Act decisively.
7. Maintain prompt communications with your forces, your supervisor, and adjoining forces.
8. Give clear instructions and ensure that they are understood.
9. Maintain control of your forces at all times.
10. Fight fire aggressively, having provided for safety first.

All appropriate personal protective equipment (PPE) for wildland firefighting should be worn as required. Information on the proper type of protective equipment can be found in NFPA 1977, *Standard on Protective Clothing and Equipment for Wildland Fire Fighting*, and NFPA 1500, *Standard on Fire Department Occupational Safety and Health Program*.

Safety is a major concern on all fire scenes, but investigating wildfires involves some unique hazards.

▰▰▰▰▰▰▰▰▰▰▰▰▰▰

TIP

The NWCG has published a list of 18 "Watch Out Situations"—conditions that pose potential hazards to the firefighter and investigator, and that require heightened attention and care:

- Fire not scouted and sized up
- In country not seen in daylight
- Safety zones and escape routes not identified
- Unfamiliarity with weather and local factors influencing fire behavior
- Uninformed on strategy, tactics, and hazards
- Instructions and assignments not clear
- No communication link between crew members and supervisors
- Constructing line without a safe anchor point
- Building line downhill with fire below
- Attempting a frontal assault on fire
- Unburned fuel between you and the fire
- Cannot see the main fire and not in contact with anyone who can
- On a hillside where rolling material can ignite fuel below
- Weather gets hotter and drier
- Wind increases and/or changes direction
- Frequent spot fires crossing line
- Terrain or fuels make escape to safety zones difficult
- Feel like taking a nap near the fire line

Accountability for all personnel at the scene must be maintained at all times. Other challenges and considerations that must be addressed on a wildland fire scene include underground burning hazards, ongoing suppression efforts (including air operations and equipment use), hazardous materials, and downed power lines.

Wildfire Fuels

Wildfire fuels are classified as **ground fuels**, which include all flammable materials lying on or in the ground; **surface fuels**, which include all flammable materials just above the ground; and **aerial fuels**, which include all green and dead materials located in the upper forest canopy. Although aerial fuels may be important indicators that can help the fire investigator narrow the fire's start to a general area of origin, ground fuels are critical in determining the point of origin.

Fuel Condition Analysis

The physical characteristics of the fuel are classified based on both the burning characteristics of individual materials and the combined effects of the various types of flammable materials present. This classification then allows for predictions of the rate of spread and general fire behavior.

Ground Fuels

Ground fuels include all flammable materials located between the mineral soil layer and the ground surface. **Duff** (decomposing organic material above the soil) is not a major influence on the fire spread rate because it is typically moist and tightly compressed with little surface exposure. Roots restrict air supply, but they may also provide an avenue for a fire to spread.

Surface Fuels

Surface fuels are those flammable materials located from the ground surface to approximately 6 ft (2 m) above the ground. Dead leaves and **coniferous litter** are highly flammable materials and should be considered separately when evaluating surface fuels **FIGURE 18-1**. Examples include needles dropped from trees. Dead needles attached to trees are especially flammable because they are exposed to air and do not touch the ground, which is often moist.

Grass, weeds, and other small plants are surface fuels that influence the rate of fire spread, based primarily on their degree of curing. Cured grass is dry and extremely flammable.

Fine fuels and dead wood, which consist of twigs, small limbs, needles, leaves, bark, and rotting material with a diameter of less than 1 inch (2.5 cm), ignite easily and frequently carry fire from one area to another. They are often found in areas where logging has taken place. Dry, rotten wood lying on the ground may be ignited by the main fire, and it will burn hotter and longer than fine fuels.

FIGURE 18-1 Dry leaves and needle litter are highly flammable surface fuels.
© MrIncredible/iStock/Thinkstock.

Downed logs, stumps, and large limbs require long periods of hot, dry weather before they become highly flammable. These heavy fuels are more of a factor in determining fire duration than in aiding fire spread. Low brush and reproduction vegetation may either accelerate or slow the spread rate of a fire because **understory vegetation** can shade fuels, preventing them from drying rapidly. The understory vegetation often provides the link between ground and aerial fuels.

Aerial Fuels

Aerial fuels are those located from approximately 6 ft (2 m) above the ground surface to the crowns of the canopy. **Crowns** (twigs and needles or leaves of a tree) are a highly flammable fuel whose arrangements allow free circulation of air and exposure to wind and sun. A **snag** (standing dead tree) is an important aerial fuel that influences fire behavior. Fires often start in snags because they are drier and much easier to ignite.

Moss hanging on trees is the most apt to ignite of all aerial fuels and provides a means of spreading fires from ground fuels to other aerial fuels. Moss reacts quickly to changes in relative humidity.

Crowns of high brush are considered aerial fuels because they are separated by distance from ground fuels. The key factors in evaluating fires in high brush are the fuel volume, live fuel moisture, the fuel arrangement, the general condition of the ground fuels, and the presence of fine, dead aerial fuels.

Fuel Matrix

Wildfire fuels are classified into four major groups: grass, shrub, timber litter, and logging debris. Each group is further divided into fuel-type models based on the predictable behavior under specified weather conditions. These models group fuel elements based on species, fuel form, size, and arrangement.

Factors Affecting Fire Spread

The primary factors that affect fire spread are heat transfer, lateral confinement, weather influence, fuel influence, suppression efforts, and topography. A variety of other factors, both natural and humanmade, may influence fire spread as well.

Heat Transfer

In a wildfire, fire spread is influenced by convective and radiant heat. Convective heat allows the fire to spread from the lower-level grasses and duff to the upper-level branches and crowns. Radiant heat then becomes the dominant primary heat transfer method as the fire progresses laterally. Radiant heat is likewise the dominant heat transfer method in lower-level fuels, which consist of brush and grass on level surfaces.

Lateral Confinement

Lateral confinement occurs when the fire cannot spread laterally because of inflammable terrain features such as roads, rocky areas, or streams. These confinements can alter the speed, intensity, and direction of spread. Burn patterns may be deceptive when the fire flanks these obstacles.

Wind Influence

Wind influences a fire by pushing the flames ahead and preheating fuel; providing more oxygen; drying out vegetation; and blowing embers, sparks, and airborne firebrands (pieces of burning material) ahead of the fire **FIGURE 18-2**.

Four classifications of winds are distinguished:

1. Meteorological winds are caused by pressure differences in the atmosphere that create weather patterns. They are greatly impacted by the rotation and topography of the Earth.
2. Diurnal winds result from air being heated by the sun during the day and then cooled at night. As the temperature of the air increases, the air rises; as the temperature of the air decreases, the air sinks. The fluctuation between the two creates the diurnal winds.
3. Fire winds are created when air is pulled into the bottom of a powerful convection column to replace the rising warmer air. This movement pulls oxygen into the rising plume and encourages fire spread.
4. Foehn winds are the result of air flowing between pressure gradients. As gravity increases the air's speed and the air is compressed, the temperature of the air increases and its humidity decreases, improving the conditions for fire spread.

Fire Head, Flanks, and Heel (Rear)

The direction in which the local wind is blowing determines the primary route of the fire's advance. That advance is led by the **fire head**, the area of greatest fire intensity. The **fire flanks** are located on either side of the head. Fire progression on the flanks of the fire is characterized by less intense fire behavior than

Wind Speed and Shape of Fires

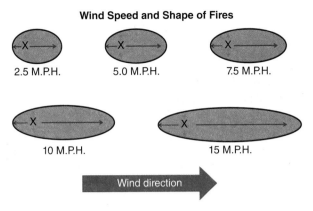

FIGURE 18-2 Wind speed and wind direction are important factors to consider.

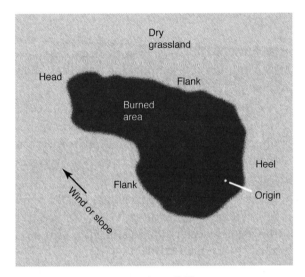

FIGURE 18-3 Anatomy of a wildfire.

TABLE 18-1 Fuel Size		
Size	**Burning Characteristics**	**Examples**
Small (fine) fuels	■ Easiest to ignite ■ Rapidly consumed	■ Seedlings and small trees ■ Twigs and leaves ■ Dry grass ■ Brush ■ Dry field crops ■ Pine needles and cones
Large (heavy) fuels	■ Harder to ignite ■ Burn slower and longer	■ Large-diameter trees and brush ■ Large limbs ■ Logs ■ Stumps

observed with the head of the fire. The **fire heel** is located on the opposite end of the fire from the fire head and is less intense. It will generally be backing or burning slowly against the wind or downhill **FIGURE 18-3**.

Fuel Influence

Ignitibility, rate of burning, and fire spread are influenced by several fuel characteristics in the wildland setting. Vegetation species vary in terms of their moisture content, shape, and density. The moisture present in the fuel is further influenced by the condition of the vegetation, solar exposure, weather, and geographic location. Oil or resin content within the vegetation is another influence.

The smaller the fuel, the more easily the fuel will ignite and the faster it will be consumed **TABLE 18-1**. Ground fuels, such as roots, can burn along their entire length underground and can ignite a surface fire in a different location much later. Surface fuels, including grasses, twigs, needles, or low brush, are

commonly involved with the spread of fire. Crown fuels include tall brush, hanging mosses, limbs, leaves, and tree crowns. If fire spreads in crown fuels, the results can often be devastating because all flammables are normally consumed.

Dead fuels are often classified in terms of the time it takes for them to reach a percentage of equilibrium with their environment. Specifically, they are referred to as 1-hour, 10-hour, 100-hour, and 1000-hour fuels, depending on their composition, thickness, and location.

Weather

Weather plays an important role in wildfire behavior because it influences the atmospheric stability, temperature, relative humidity, wind velocity, cloud cover, and precipitation.

Weather history is a description of atmospheric conditions over the preceding days or weeks; it should be analyzed to determine how it might have influenced the fire's ignition and burning characteristics. This information can be obtained through wildland fire dispatch centers, the National Weather Service website, and a variety of other Internet sites.

The **ambient temperature** (air temperature of the surrounding environment) influences the temperature of the fuel. As the radiant energy of the sun heats the atmosphere, the ambient temperature increases, in turn increasing the fuel's temperature and reducing its moisture content.

Humidity, a measure of water vapor suspended in air, is usually expressed as **relative humidity**. The moisture in the air directly affects the amount of moisture in the fuel. Warm air can hold more water compared to cool air. If the air is dry, it will try to pull moisture out of the surrounding vegetation, thereby drying out the vegetation and making it more susceptible to fire. If the air is more humid, however, moisture will be transferred to the surrounding vegetation, reducing the chance of fire and slowing any fire spread.

Topography

Topography relates to the form of natural and human-made earth surfaces. It affects the intensity and spread of the fire and can greatly influence winds.

The **slope** is the change in elevation over a given distance, measured by the rise over the run (the differential in height divided by the total horizontal distance, normally measured in feet, then multiplied by 100, equals the percent slope). It allows fuel on the uphill side of the geographic area to be preheated more rapidly than if it were on level ground and is an important factor in fire spread. Wind currents moving uphill during the day also accelerate fire spread. Burning fuels can roll downhill, igniting other fuels.

The **aspect** is the direction the slope faces. If a slope faces the sun, it is typically drier and may have a more combustible fuel type than slopes that do not experience such solar heating. This results in a greater ease of ignition and faster spread within the sun-exposed area. **TABLE 18-2** summarizes these relationships.

Fire Suppression

Fire suppression comprises all activities that lead to the extinguishment of a fire. Fire crews should attempt to protect the potential areas of origin so that

FIGURE 18-4 An air drop is the application of water, foam, or retardant from the air.
© Photocdn6/iStock/Thinkstock.

the fire's origin and cause can be accurately determined. Common fire suppression activities include the following:

- **Fire breaks:** Any natural or humanmade barriers that are used to stop the spread or reroute the direction of the fire by separating the fuel from the fire. They are also called fire lines or control lines.
- **Air drop:** The aerial application of water, foam, or retardant mixture directly onto the fire or threatened area or along a strategic position ahead of the fire **FIGURE 18-4**.
- **Firing out:** The process of burning the fuel between a fire break and the approaching fire to extend the width of the fire barrier.

TIP

Suppression crews may be requested to carefully place wet lines or construct fire lines a short distance from the fire's edge in the area of origin. This will allow the fire to burn undisturbed to the lines, resulting in protection of the fuels, burn patterns, and the origin.

TIP

Wildland–urban interface fires may result from fire starting in a structure and spreading to the wildlands or starting in the wildlands and spreading to a structure. A wildland fire investigator must determine whether and how such spread may have happened.

TABLE 18-2 Effect of Slope Aspect on Fire		
Direction	**Conditions**	**Effect**
Facing toward the sun	- Has less vegetation - Drier	- Greater ease of ignition and fire spread
Facing away from the sun	- Has more vegetation - More moist	- Less ease of ignition and fire spread

Class A Foam

Air mixed with water and Class A foam produces a sudsy coating on combustibles. On wood, it reduces the surface tension of water, allowing it to penetrate much more easily rather than shedding off the surface. This foam also suspends water for a period of time, allowing for a longer cooling effect, a reduction of oxygen to the fuel source, and prolonged protection to unburned fuels.

Other Natural Mechanisms of Fire Spread

Windborne firebrands are hot embers picked up by wind and blown into unburned fuel great distances from the original fire. They can start **spot fires** remotely from the main fire and could be mistakenly interpreted as deliberately set fires.

A **fire storm** is a natural phenomenon in which a fire attains such intensity that it creates and sustains its own wind system. This system can be so strong that it is capable of producing small dust devil–type circulations called fire whirls, which dart around in unpredictable directions, damage homes and other structures, and quickly spread the fire.

Animals and birds can also cause fire to spread if their fur or feathers catch fire (e.g., from a power line). Once the animal is on fire, it can spread the fire wherever it goes.

Indicators of Directional Pattern

Analysis of the directional pattern, when shown by multiple indicators in a specific area, can identify the path of fire spread through the area. By using a systematic approach of backtracking the progress of the fire, an investigator can retrace the path of the fire to its point of origin. Visual indicators include differential damage; char patterns; discoloration; carbon staining; and the shape, location, and condition of residual, unburned fuel.

Wildfire V-Shaped Patterns

Wildfire V-shaped patterns are ground-surface burn patterns created by the fire spread. When viewed from above, they are generally shaped like the letter "V." These patterns should not be confused with the traditional plume-generated vertical V patterns associated with structural fire investigations.

Wildfire V-shaped patterns have the following characteristics:

- Horizontal, not vertical, patterns
- Affected by wind direction and slope

- Fire flanks (perimeter of the fire parallel to the direction of spread) that widen up the slope in the direction of the wind
- Origin that is normally near the base of the V

Degree of Damage

The degree and type of damage to fuels indicate the intensity, duration, and direction of fire passage **TABLE 18-3**. Leaves, branches, and large woody material on the ground may sustain greater damage on the side from which the fire approached. Grass that grows in clumps may not be completely consumed, resulting in protected areas on the side opposite the fire's advance.

TABLE 18-3 Damage Indicators of Fire Passage

Fuel	Appearance	Interpretation
General	Fuel is more damaged on one side than another	Shows direction of spread due to protection
Grass stems	Unburned grass stalks or seed heads lying on the ground	Grass heads typically fall in the opposite direction of the fire's travel
Brush	More upper foliage is burned	Exit side
	Some upper branch tips fall unburned to the ground	Entry side
	Ash deposits not found on fuels	Fuels still burning when the ash was deposited
	Brush cupping Tips of burned stubs are blunt or cupped	Upwind side of fire
	Tips of burned stubs are sharp	Downwind side of fire
	Brush die-out pattern Decreasing fire intensity, charring, and burned branch size	Fire died out after entering brush growth

//////////////////////////////////////

TIP

Within the fire zone, place a stem of low-growing brush, 1 to 3 ft (0.3 to 0.9 m) in height, between two of your fingers near the midpoint of the stem. Squeeze the stem slightly and move your fingers upward. Sooting often appears heavier on one finger than the other, showing the direction from which the fire advanced. Several samples should be taken from various stems for confirmation. This technique can also be useful on wire fencing.

Grass Stems

As a low-intensity fire burns the bottoms of grass stems, the stems fall back into the fire. If they fall into a burned area behind the fire head, they may remain unburned. Unburned stalks of grass on the ground generally point in the direction of the fire's approach. The investigator should examine several grass stems to increase the reliability of this method. The investigator should also view the stems outside the burned area to determine whether any other factors, such as wind or heavy rains, had a prefire effect on their vertical arrangement.

Brush

Brush can be very valuable as a fire direction indicator. It is often damaged by the fire but not fully consumed, leaving a reliable indicator that has not been moved or destroyed, even during light mop-up. Brush can display several types of indicators, such as freezing, degree of damage, depth of char, angle of char, curling of leaves, sooting, white ash deposits, **cupping** (a concave or cup-shaped char pattern on grass stem ends, small stumps, and the terminal ends of brush and tree limbs), and die-out patterns. There are few other objects that can display as many indicators as brush.

Trees

Trees are significant indicators of fire direction, particularly in areas of frontal fire damage **FIGURE 18-5**.

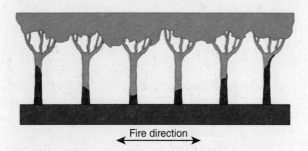

<— Fire direction —>

FIGURE 18-5 Trees indicate fire direction.

Fire movement is recorded at ground level around the root base and tree trunk, and at flame height by the lower foliage and crown canopy. The char created on the trunk surface of a tree is affected by the topography of the land surface. A fire burning uphill or with the wind creates a char pattern that slopes to a greater degree than the ground slope.

Crown damage can also be used to interpret fire direction. Convection and radiant heat will ignite lower limbs and then spread upward into the rest of the tree. This action progresses in intensity as the wind action drives the fire away from the windward foliage and branches **TABLE 18-4**. As an example, if an investigator observes a triangular unburned area on the side of the crown, the interpretation would be that the unburned area is on the windward, or approaching, side of the fire.

Angle of Char

Angle of char indicators are divided into two groups based on the types of fuels in which they occur: in pole-type fuels (e.g., tree trunks, utility poles, fence posts) **FIGURE 18-6** or in the foliage crowns of brush or timber-type fuels **FIGURE 18-7**. Angles of char on pole-type fuels are described in more detail in Table 18-4. Angles in foliage crowns are similar but are often not as obvious.

White Ash Deposit

White ash can be the by-product of combustion. More white ash will be created on the sides of objects

TABLE 18-4 Trunk Char		
Fire Direction	**Wind Direction**	**Observation**
Upslope	Upslope	The angle of char on the tree will be greater than the slope angle of the hill.
Downslope	Upslope	The char line on the tree is nearly parallel to the slope of the hill.
Downslope	Downslope	The angle of tree char will be higher on the downhill side.
Upslope	Downslope	The char line is level with the slope of the hill, with only light upslope damage to the tree trunk.

exposed to greater amounts of heat and flame. Ash is often dispersed downwind and deposited on the windward sides of objects. Thus, fuels facing the advancing fire will appear lighter on the side facing the oncoming fire and darker on the side opposite.

Cupping

Cupping is a concave or cup-shaped char pattern found on grass stem ends, small stumps, and the terminal ends of brush and tree limbs. The exposure of the surface to the windward side of the fire leaves a

FIGURE 18-6 Char patterns on a tree and fence post, indicating fire movement.

Reproduced from NFPA 921-2021, *Guide for Fire and Explosion Investigations*, Copyright © 2020, National Fire Protection Association. This reprinted material is not the complete and official position of the NFPA on the referenced subject, which is represented only by the standard in its entirety.

charred surface on the fuel. This side is charred more deeply than the opposite side, which is protected from exposure.

Die-Out Pattern

As a fire progresses into different types of fuels, its spread may slow. In areas where there is increased fuel moisture, lack of fuels, or other conditions causing a decrease in rate of spread and intensity, the fire may become completely extinguished.

Exposed and Protected Fuels

A noncombustible object or a fuel itself shields the unexposed side of a fuel from heat damage. Fuels will be unburned or show less damage on the side shielded from the advancing fire.

Staining and Sooting

Staining is caused by hot gases, resins, and oils condensing on the surface of objects. This phenomenon occurs most commonly with noncombustible objects such as metal cans, glass bottles, or rocks. Stains will appear on the side of the object exposed to the flames.

Soot is caused by incomplete combustion and the natural fatty oil content in some vegetation. Carbon soot is typically more heavily deposited on the side facing the approaching fire. Soot deposits can be noticed by rubbing a hand or finger across the surface, comparing opposite sides. Sampling soot on small brush or weed stems in multiple locations can be an effective technique. Staining and sooting on one side of the object indicate that the fire direction was from the stained side **FIGURE 18-8**.

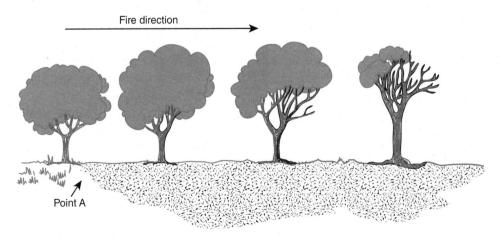

Fire direction

Point A

FIGURE 18-7 Progressive crown burning from the point of origin.

Reproduced from NFPA 921-2021, *Guide for Fire and Explosion Investigations*, Copyright © 2020, National Fire Protection Association. This reprinted material is not the complete and official position of the NFPA on the referenced subject, which is represented only by the standard in its entirety.

FIGURE 18-8 Look for sooting and/or staining to determine the direction of fire spread.
Courtesy of Michael R. Denney.

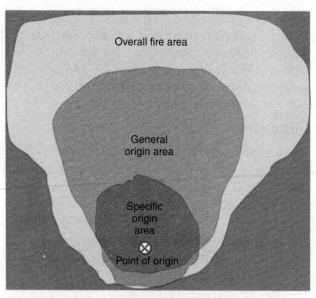

FIGURE 18-9 Anatomy of the origin of a wildland fire.
Reproduced from NFPA 921-2021, *Guide for Fire and Explosion Investigations*, Copyright © 2020, National Fire Protection Association. This reprinted material is not the complete and official position of the NFPA on the referenced subject, which is represented only by the standard in its entirety.

Depth of Char

Char on branches or logs exhibits a scale-like appearance. Wood materials lose mass and shrink as they burn, forming this scale-like surface. The deepest charring will typically appear on the side facing the advancing fire.

Spalling

Spalling is caused by a breakdown in the tensile strength of a rock's surface when the rock is exposed to heat; it appears as shallow, light-colored craters or chips in the surface of rocks within the fire area. Spalling is usually accompanied by slabs or flakes exfoliated from the surface of the rock. It is generally associated with advancing fire areas and will appear on the side of the rock exposed to the flames.

Curling

Curling occurs when green leaves curl inward toward the heat source. They fold in the direction from which the fire is coming. This effect usually occurs with slower-moving, lighter burns associated with backing and lateral fire movement.

Foliage Freezing

Freezing can occur as fire passes through the needles or leaves of trees. Especially in an advancing fire, the foliage tends to become soft and pliable and is easily bent in the direction of prevailing winds. As it cools, the foliage will often remain pointing in the direction of the prevailing wind.

Noncombustibles

A protected area in fuels immediately adjacent to a noncombustible object indicates that the fire direction was from the opposite side of the protected area.

Loss of Material

As materials are heated, they undergo physical changes. Typically, when wood or other combustible surfaces burn, they lose material and mass. The shapes and quantities of the remaining combustibles can themselves produce lines of demarcation and ultimately fire patterns for the investigator to analyze.

Conducting an Origin Investigation

In identifying the area of origin, the investigator must consider the factors of wind, topography, and fuels. In most cases, the origin is located close to the heel, or rear, of the wildland fire **FIGURE 18-9**.

Interviews with various parties can help an investigator narrow down the area where the fire first started. Observations of reporting parties can help because they may have observed the fire at a relatively small stage.

Observations of the initial attack crew can also prove very useful. Crew members may have observed people or vehicles, can report weather conditions, and can detail initial fire conditions and location. Sometimes they may have taken photos. Airborne personnel can also make observations and have a broader perspective on the location of the fire.

Vital information for the investigation of a wildfire can often be provided by other witnesses as well. Local witnesses tend to be familiar with the area, can give information about the area of origin, and may be able to direct an investigator to a possible cause by describing the condition and movement of the fire (e.g., smoke conditions, intensity, rate of spread, weather).

Satellite or imaging tools that are used primarily to establish fire suppression tactics can also be used to assist in determining the area of origin. Based on the direction of fire spread and data indicating the fire location when first detected, these tools can provide a broader perspective on the fire and fire spread. If vehicles are seen leaving the area, consider a vehicle line-up when appropriate. A vehicle line-up can be used as effectively as a suspect line-up, as witnesses may be able to identify a vehicle seen at a fire location. If a vehicle of interest is identified, it can be confirmed or eliminated by this process. The witness should sign and date the photo of the suspect vehicle if one is selected from the line-up, and the witness is absolutely sure of their choice.

Origins are often located in what is the coolest part of the fire by the time the investigator arrives. Identify and secure any potential evidence as soon as possible so that it is protected during any additional firefighting and investigation.

Security of the Area or Point of Origin

As in any fire investigation, the integrity of the wildland fire scene needs to be preserved. The area of origin should be secured by flagging off the area and by posting personnel at the scene to restrict access and to protect evidence such as tire tracks, footprints, and potential ignition sources. If the area has been tampered with, it could mislead the investigation and affect evidence credibility.

Search Techniques

When completing a search for a wildland fire's origin, apply the following principles:

- Walk the exterior of the fire.
- Look for burn patterns and identify advancing fire and transition zone areas.
- Identify the general area of origin.
- Enter the general area of origin from the advancing side if possible, identifying and flagging advancing and lateral fire patterns, and working towards the specific origin area.
- Identify the specific area of origin.
- Grid the specific area of origin.
- Identify the origin.

The investigator should fully walk the search area in two directions to view any burn patterns from different angles. While examining burn patterns, identify clusters or groups of indicators to confirm findings on the direction of fire spread.

Once an investigator is sure of the general area of origin, the flagged-off area can be reduced in size to allow suppression crews to work the area. The general area of origin is the area of the fire that can be narrowed down based on macroscale indicators, witness statements, and analysis of fire behavior. It may be a limited area for a small fire or several acres for a large fire. The general area of origin can then be entered from the advancing side and worked to the specific area of origin.

The specific origin area is a smaller area, within the general origin area, where the fire's direction of spread was first influenced by wind, fuel, or slope. Generally, this area is characterized by subtle and microscale fire direction indicators. It will usually be no smaller than about 5 × 5 ft (1.5 × 1.5 m) but may be substantially larger, depending on fire spread indicators and other factors. It is typically characterized by less-intense burning. If the identified specific area of origin is small, the area can be searched in its entirety from its perimeter. If the area is large, the site should be broken into smaller areas for systematic close-up examination **FIGURE 18-10**.

Use metal flagging stakes or labeled flags to identify evidence **FIGURE 18-11**. Labeled flags can also be used to mark the original position of evidence that has been removed from the scene. Some flag staking may take place widely in the general area of origin but will be used much more extensively in the specific area of origin as the true origin is approached. Red-colored flagging stakes are often used to indicate advancing fire, whereas yellow stakes are used for flanking fire and blue stakes for backing fire. White stakes are reserved for evidence and the point of origin.

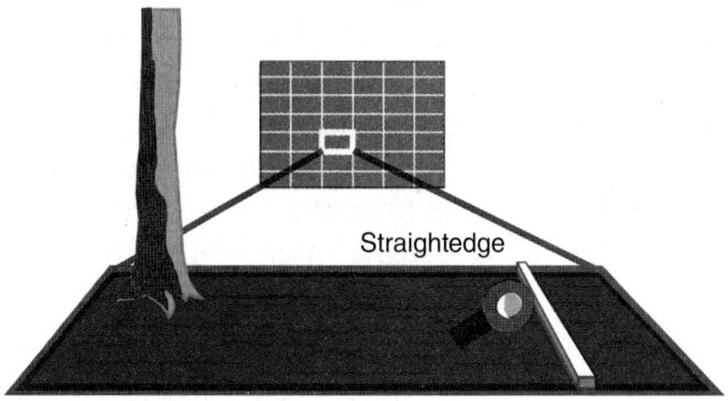

FIGURE 18-10 Searching each segment is necessary for large areas.

FIGURE 18-11 Mark off the general area of origin to protect evidence.
Courtesy of Michael R. Denney.

Analyzing fire spread may involve many steps to establish and plot the locations of the parts of the fire when it was discovered. Wind direction and available weather information should be documented and plotted based on the observations of those who discovered the fire. Weather information involves plotting the fire head, fire flank, fire heel, and wind direction and speed.

TIP

Have security (firefighters, investigators, or law enforcement) on scene at all times during the investigation. This may require multiple shifts of security personnel if the investigation is lengthy. Ensure that each shift is documented.

Care must be exercised continually not to destroy evidence or other signs, such as footprints or vehicle tracks. If the area is large, it will be helpful to break the area into segments. This "segment division" allows an investigator to look more closely at specific areas of the fire to determine which may be eliminated as areas of interest. This prevents the need for reentry and reduces potential for destruction of evidence.

The loop or spiral technique is effective for examining small areas. When using this method, the investigator approaches from the head of the fire and works from flank to flank, back and forth, closing on the origin. One risk with this technique is that if the loops are too large, indicators and evidence can be overlooked or destroyed. The distance between the loop lines should get tighter as the investigator nears the area of origin.

The grid, or strip, technique is used when multiple investigators are inspecting a large area. This method can narrow the origin search area very quickly.

The lane technique is one of the best strategies for inspecting an area in great detail. In this technique, the investigator places string lines, creating lanes to be individually inspected. This method is relatively quick and simple to implement, and is most commonly used when indicators show the origin is near.

Conduct the examination of the origin area with the minimum amount of disturbance, seeking to expose the evidence of fire spread at the site of the origin rather than destroying or removing it during the investigation. A photographic record sketch with commonly used symbols, along with field notes, should be maintained throughout the investigative process.

Search Equipment

The following list identifies some of the equipment that can be used in a wildfire investigation:

- Magnifying glass or reading glasses to enhance small details or identify evidence that was not visible without it
- Magnets to locate ferrous metal fragments or particles
- Straight edge to segment the origin area
- Probe to uncover small pieces of evidence from the surrounding vegetation
- Comb to separate evidence from debris or pick up evidence to avoid damage
- Hand-held lights to locate items in low-light areas
- Air blower (nasal aspirator) to separate light ash from items of interest
- Metal detector to locate metals that may be of evidentiary value and that could not be easily located with a magnet
- Sifting screen to separate a suspected item of evidence from the surrounding dirt and vegetation
- Global positioning satellite (GPS) recorder to obtain the accurate longitudinal and latitudinal position of the fire origin, location of evidence, mapping purposes, and reference points

GPS data can be cross-referenced against on-site survey information, lightning strike data, aerial photography, or satellite imagery for fire cause determination.

Fire Cause Determination

The objective of every origin and cause investigation is to establish and confirm the cause of the fire. In doing so, an investigator must also determine what did *not* cause the fire. All findings must be well documented. The heat source or ignition device should be recovered, if possible, and processed properly as evidence.

In addition, the investigator should test the hypothesis of how the fire started by referring to the literature or by conducting live testing under controlled conditions. To perform a live test of a hypothesis, collect fuels similar to those at the origin and re-create the ignition under controlled conditions, at a safe site away from the fire area. Investigators must follow agency policies and local, state, and federal laws when doing so. Simulate the same weather and topographic conditions, if possible, and record a video of the process. This method can be very useful in court in explaining how the fire started and generally how fire burns in wildland fuels.

Many wildfires are ignited by natural causes such as lightning, while many others are caused by people. People can cause fires either through their actions or by failing to act. An **accidental fire** is one in which the cause does not involve a deliberate human act to ignite or spread fire into an area where the fire should not be. An incendiary fire is one that is intentionally ignited in an area or under circumstances where and when there should not be a fire.

The following causes are organized consistent with the various causes described in the NWCG's *Guide to Wildland Fire Origin and Cause Determination* (2016). Discussions are under way that may result in future changes to the causes listed in the NWCG Guide.

Lightning

The action of a lightning strike may create **fulgurites**—slender, usually tubular bodies of glassy rock produced by electrical current striking and then fusing dry, sandy soil—by melting sand in the soil near the base of a tree or area of contact with the ground. Fresh scars may be noticeable on the vegetation. Other physical evidence at the origin may include pieces of wood and bark shredded from the lightning-struck tree or shrub, multiple strikes, holes in the ground, or strange burn patterns on the ground.

Lightning maps are available from several sources and may be used for verification or elimination purposes **FIGURE 18-12**.

Campfires

If a campfire is suspected as the cause of a wildfire, look for a ring of rocks and/or soil, along with mounds with the presence of ash, coals, wood, and garbage. Disturbed grasses and soil may indicate sleeping areas.

FIGURE 18-12 Lightning data maps are often available from wildland fire agencies.
Courtesy of the World Meteorological Organization (WMO)/NOAA.

Wildfires caused by campfires are normally accidental and are a result of improper extinguishment.

Smoking

Discarded ignited smoking materials, such as cigarettes, cigars, pipe tobacco, and matches, can start wildfires. Certain conditions must exist for a fire to be started by smoking materials:

- The relative humidity must be less than 22 percent and fine dead fuel moisture must be dry, with a moisture level normally less than about 14 percent (although other environmental conditions can override the humidity).

- The fuel source must be fine or powdery (litter or punky wood) or otherwise easily ignited.

- The cigarette or other material must be in contact with the fuel bed and at a particular angle and orientation, depending on the weather and fuel conditions.

Debris Burning

Fires may occur at dumpsites, timber harvesting operations, and land-clearing operations as well as at residences from garbage and other debris set on fire. These fires can spread to the neighboring vegetation. Burn barrels or incinerators may be a consideration as a fire cause. In windy conditions, hot ash and debris can blow away from a debris burn and start a fire some distance away. Large, woody debris piles, especially if mixed with soil, have been known to hold long-term thermal energy for many months, including over winter, before escaping into adjacent wildland. Witnesses are often useful in determining whether debris burning was the cause of the fire. Check for burn permits, when applicable, to determine whether a permit was issued to the occupant in question, and whether its conditions were complied with.

Incendiary Wildfires

Incendiary wildfires are deliberately and/or maliciously set with the intent to cause damage. These fires are often set in frequently traveled routes and may be set in multiple locations. Evidence such as matches, paper, cigarettes, rubber bands, or candles may be present. It is possible that a fire was started with a device that was then taken from the scene (e.g., a butane lighter in which case evidence of origin may be missing). Such a method is referred to as a **hot set**.

A **prescribed fire** is the result of an intentional ignition by a person or a naturally caused fire that is allowed to continue to burn according to approved plans to achieve resource management objectives. Sometimes these fires may escape the control of the responsible party because of weather conditions or negligence. The investigator should be aware of local laws and prosecution authorities to determine whether escaped fires are subject to criminal or other penalties.

Equipment Use (Including Automobiles)

Any power or motorized equipment that uses electricity or flammable products in its operation is capable of starting a fire when in proximity to combustible vegetation. Fires can be started by exhaust systems, faulty spark arresters, vehicle fires, and friction. The potential for fire increases when the machinery or vehicle has defective or failed parts. For example, friction heating of bearings, worn brakes, "frozen" shafts, and abrasion may lead to a fire.

Railroad

Along a railroad, fires can be caused by a train, maintenance equipment, or personnel. Common examples include wheel wear on corners, carbon emissions from the exhaust, hot brakes, rail grinding, and personnel smoking. Witnesses are important to the investigation in such a case. Trains can sometimes be stopped through dispatch for inspection and to review maintenance reports to determine fire cause, to provide safety to firefighters in the fire area, and to prevent additional fires.

Children

Some fires are started by children engaged in fire play or otherwise. With these incidents, matches, lighters, cigarettes, or other ignition devices may be found in the area of origin. Other evidence may include bicycle tracks, snack wrappers, toys, or drink containers. Investigators should determine whether the fires were started out of curiosity or whether the child has an underlying pathological issue. In the latter case, the child may benefit from referral to a juvenile fire-setter intervention program.

Miscellaneous

The NWCG Guide collects nine cause types into a "miscellaneous" category. The fire investigator should determine the categories and classifications that are required by their jurisdiction and agency for fire cause determination.

Power Lines

Fires can result from wind-damaged trees coming in contact with power line equipment or from the equipment experiencing hardware failure; bird, small animal, or balloon interference; or heavy dust build-up. Sparks or molten metal can ignite flammables, and fulgurites or indentations in the ground are often found if an electric wire contacts the ground **FIGURE 18-13**. Pits are often found where the power line comes in contact with a tree. Electric fences may also start fires by transferring rapid electric pulse charges through wires. These pulses ground against dry fuels, causing ignition **FIGURE 18-14**.

Fireworks

Fireworks are classified as being ground-based, hand-held, aerial, or explosive devices. Charred cardboard, paper, colored sticks, plastics, or packaging may be found at or near the origin of a fire caused by fireworks, and witnesses to the fireworks can be very helpful in providing additional information. Fireworks can produce hot sparks that reach surface temperatures as high as 1200°F (648°C), which provides sufficient temperature to ignite wildland fuels.

Cutting, Welding, and Grinding

Hot metal fragments and sparks can emanate from cutting, welding, and grinding operations and can start fires when they land in finely particulated fuel. They can also burrow into materials and cause a smoldering fire that may not be noticed by the personnel engaged in the operation.

FIGURE 18-13 This fulgurite was created by a downed power line contacting the ground in sandy soil.
Courtesy of Michael R. Denney.

FIGURE 18-14 This electric fence was the source of the fire.
Courtesy of Michael R. Denney.

Firearms Use

Using black powder firearms and firearms discharging tracer, incendiary, and steel core ammunition can cause wildfires. With black powder weapons, normally the burning patch material rather than the burning black powder has ignited the wildland fuels. The burning chemical compounds within tracer and incendiary ammunition can ignite wildland fuels, too. Ammunition with a steel core can strike a rock or other material hard enough to cause sparks and ignite wildland fuels as well.

TIP

Several reference materials give more details that may be helpful in determining the cause of a wildland fire. If they are used in the investigation, ensure that they do not conflict with, but instead enhance, NFPA standards.

Glass Refraction and Magnification

Clear glass or reflective objects found at or near the point of origin may be the cause of the fire if environmental conditions exist to support ignition. Such conditions may include prolonged direct sunlight exposure on the reflective object, which focuses the sun's rays into small, dry fuels such as grass or duff. Fires started by these items are rare, however.

Spontaneous Combustion

Spontaneous heating can be caused by a biological or chemical action. It is more likely to occur on warm, humid days where green or moist grass, hay, wood chips, and/or oily cloths are piled.

Flare Stack/Pit Fires

Ignition scenarios relating to oil and gas drilling may be similar to ignitions discussed in earlier sections on smoking and equipment use. Further, a well blow-out, flare pit, or stack operation is designed to dispose of excess or unwanted petroleum by-products. These operations can start fires by direct flame contact or by emitting sparks into flammable fuels.

Other Miscellaneous Fires

Other potential causes of wildland fires include blasting, fires or activities related to structures, flares and fuses, coal seam fires, electric fences, flying lanterns, failures involving wind turbines, and home outdoor wood burning furnaces.

Evidence

Evidence collection procedures for wildfire investigations are the same as those for other fires and include the following steps:

- Identifying and gathering evidence
- Preventing contamination
- Collecting evidence
- Detailed identification of evidence
- Transporting and storage of evidence
- Testing

Documentation

As with other types of investigations, the purpose of documentation is to preserve and memorialize the facts so that the investigator can describe the scene to others and recall the scene at a later date. This documentation may be relied upon later by third parties or in legal proceedings.

Written field notes should include preliminary and dispatch information; fire location and movement and observations about the fire; witness interviews and statements; weather; GPS data; and the factors, steps, data, and analysis that led to a determination of origin and cause. Sketches can reflect fire movement and behavior, origin evidence, and photo points (separate sketches may be required for all of this information). The locations of items of evidence should be documented.

Photos should be taken and carefully downloaded to storage media without allowing changes or enhancements to those photos. A photo log should be created. The photos should include general areas in the scene, photos of comparison, photos demonstrating the relationship of important objects to others and to their location within a scene, the locations and identification of evidence and comparison samples, pattern and progression indicators, and origin areas.

The investigator should prepare a comprehensive report that includes all pertinent information received, conditions encountered, indicators and evidence found and retrieved, and witness statements. Of course, the report should also include a section detailing the investigator's opinions on the fire's origin and cause, if appropriate, and the bases for those opinions.

After-Action REVIEW

IN SUMMARY

- Fire investigators operating in the wildland arena should understand the special techniques, practices, equipment, and terminology used in this type of investigation and must have a thorough understanding of wildfire behavior.
- Wildland incidents present different safety considerations than do structural incidents, and the fire investigator must know about wildland fire safety if operating in a wildland setting.
- Topography affects the intensity and spread of a wildland fire and can greatly influence winds.
- If a slope faces the sun, it is typically drier and may contain a more combustible fuel type compared to slopes not exposed to such solar heating, resulting in a greater ease of ignition and faster fire spread.
- Wildfire fuels are classified into four major fuel groups: grass, shrub, timber litter, and logging debris.
- The degree and type of damage to fuels indicate the intensity, duration, and direction of fire passage.
- The deepest charring on wood materials will typically appear on the side facing the advancing fire.
- Curling leaves fold in the direction from which the fire is coming.
- A protected area in fuels immediately adjacent to a noncombustible object indicates that the fire direction was from the opposite side of the protected area.
- To secure the area of origin in a wildland fire, flag off the area and post personnel at the scene to restrict access and to protect evidence such as tire tracks, footprints, and potential ignition sources.
- Analyzing fire spread may involve many steps to establish and plot the location of the head, flanks, and heel of the fire when it was discovered.
- As with other investigations, the objective of every wildland origin and cause investigation is to establish and confirm the cause of the fire.
- Any power or motorized equipment that uses electricity or flammable products in its operation is capable of starting a fire when in proximity to combustible vegetation.
- Along a railroad, fires can be caused by a train, maintenance equipment, or personnel.
- Evidence collection procedures for wildfire investigations are the same as those for other fires.
- As with other types of investigations, the purpose of documentation in a wildland fire investigation is to preserve and memorialize the facts so that the investigator can describe the scene to others and recall the scene at a later date.

KEY TERMS

Accidental fire A fire for which the proven cause does not involve an intentional human act to ignite or spread fire into an area where the fire should not be.

Aerial fuels All green or dead materials located in the forest canopy.

Ambient temperature The air temperature of the surrounding environment.

Aspect The direction the slope faces (north, east, south, west).

Coniferous litter Primarily needles dropped from coniferous trees, but may also include branches, bark, and cones.

Crowns Twigs, needles, or leaves of a tree or bush.

Cupping A charred surface on a fuel that is caused by exposure of the surface to the windward side of the fire.

Duff The layer of decomposing organic materials lying below the litter layer of freshly fallen twigs, needles, and leaves and immediately above the mineral soil (USFA, 2002).

Fire breaks Natural or humanmade barriers used to stop the spread or reroute the direction of a wildland fire by separating the fuel from the fire.

Fire flanks The parts of a fire's perimeter that are roughly parallel to the main direction of spread.

Fire head The portion of a fire that is moving most rapidly, subject to influences of slope and other topographic features. Large fires burning in more than one fuel type can develop additional heads.

Fire heel The portion of a fire located at the opposite side of the fire head; it is less intense and is easier to control.

Fire storm A natural phenomenon in which a fire attains such intensity that it creates and sustains its own wind system.

Freezing The phenomenon in which leaves, needles, and small branches are heated by a passing fire front, and become very soft and pliable. Prevailing winds or drafts created by the fire bend them in the direction the fire advanced. Once cooled, they remain pointing in this direction.

Fulgurites Slender, usually tubular, bodies of glassy rock produced by electrical current striking and then fusing dry sandy soil.

Ground fuels All combustible materials below the surface litter, including duff, tree or shrub roots, punky wood, peat, and sawdust, that normally support a glowing combustion without flame.

Hot set A deliberately set fire in which the ignition device is removed by the individual starting the fire.

Prescribed fire A fire that resulted from intentional ignition by a person or a naturally caused fire that is allowed to continue to burn according to approved plans to achieve resource management objectives. (NFPA 921)

Relative humidity The amount of moisture in a given volume of air compared with how much moisture the air could hold at that same temperature. (NFPA 921)

Slope The steepness of land or a geographic feature, which can greatly influence fire behavior. (NFPA 921)

Snag A standing dead tree, or part of a dead tree, from which at least the smaller branches have fallen.

Spot fires Projections of flaming or burning particles (firebrands) that are found ahead of the flame front.

Surface fuels All combustible materials from the surface of the ground up to about 6 ft (2 m).

Understory vegetation The area under a forest or brush canopy that grows at the lowest height level. Plants in the understory consist of a mixture of seedlings, saplings, shrubs, grasses, and herbs.

Weather history A description of atmospheric conditions over the preceding few days or several weeks.

Wildfire V-shaped patterns Horizontal ground surface burn patterns generated by the fire spread.

REFERENCES

Introduction to Wildland/Urban Interface Firefighting for the Structural Company Officer. U.S. Fire Administration, National Fire Academy, 2002.

National Fire Protection Association. 2020. *NFPA 921: Guide for Fire and Explosion Investigations*, 2021 ed. Quincy, MA: National Fire Protection Association.

National Wildfire Coordinating Group. n.d. *10 Standard Firefighting Orders*. https://www.nwcg.gov/committee/6mfs/10 -standard-firefighting-orders

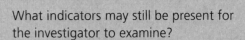

On Scene

1. What are some of the safety questions a fire investigator should ask upon their initial arrival at a wildland fire scene? What questions should be asked as the investigation progresses and the investigator checks in with incident command?

2. Not all wildland fires involve trees and tall vegetation. Can a wildland fire investigation be successfully conducted in a field of high or low grass?

What indicators may still be present for the investigator to examine?

3. What sources are available in your jurisdiction for important data relevant to the investigation of a wildland fire, such as temperature, wind speeds, relative humidity, and fuel moisture?

4. What are some advantages and disadvantages of the various search patterns that an investigator can use to examine an origin area?

Chapter Opener: © Jones & Bartlett Learning. Photographed by Glen E. Ellman; On Scene siren: © Bildgigant/Shutterstock.

Appendix A

Correlation Grid

NFPA 1033, 2022 Edition

Objectives	Corresponding Chapter(s)	Corresponding Page(s)
4.1	1, 5, 8, 12	1–18, 123–148, 219–237, 291–312
4.1.1	1	3, 8
4.1.2	1	10–14
4.1.3	6	150–153, 159–165
4.1.4	5, 8, 12	132, 222, 220–223, 292–299
4.1.5	5	124–139, 140–145
4.1.6	6, 8, 12	162–164, 219–224, 293–294
4.1.7	1	3
4.1.7.1	1	3
4.1.7.2	1	4
4.1.7.3	1	4
4.2	2, 3, 5, 6, 8, 9, 12, 18	19–39, 41–94, 123–148, 149–167, 219–238, 239–253, 291–312, 453–470
4.2.1	5, 6, 8, 9, 12, 18	132–135, 150–152, 162–164, 224–227, 241, 293–297, 454–455
4.2.1(A)	5, 6, 8, 9, 12, 18	132–135, 150–152, 162–164, 224–227, 241, 293–297, 454–455
4.2.1(B)	5, 6, 8, 18	130–131, 151–153, 226, 454–455
4.2.2	3, 6, 8, 9	43–52, 151–164, 227, 241
4.2.2(A)	3, 6, 8, 9	43–52, 151–164, 227, 241
4.2.2(B)	5, 6, 8	130–131, 151–153, 156–159, 227

Objectives	Corresponding Chapter(s)	Corresponding Page(s)
4.2.3	3, 6, 15	43–58, 152–164, 370–380
4.2.3(A)	3, 6, 8, 15	43–58, 152–164, 227–228, 370–380
4.2.3(B)	3, 6, 8	43–58, 152–153, 156–159, 227–228
4.2.4	2, 3, 4	20–36, 43–46, 96–118
4.2.4(A)	2, 3, 4	20–36, 43–46, 96–118
4.2.4(B)	4, 18	115–118, 455–456
4.2.5	2, 3, 4, 7	20–36, 43–46, 96–118, 177–197
4.2.5(A)	2, 3, 4, 7	20–36, 43–46, 96–118, 177–197
4.2.5(B)	4, 8, 17, 18	115–118, 232–235, 411, 424–425, 439, 444, 455–456
4.2.6	7, 8, 9, 13, 15, 17	177–197, 232–235, 241, 314–318, 328–330, 334–339, 370–380, 423–429
4.2.6(A)	7, 8, 9, 14, 15, 17	177–197, 232–235, 241, 314–318, 328–330, 334–339, 370–380, 423–429
4.2.6(B)	7, 14	173, 179–181, 182–184, 197, 330–333, 336–354
4.2.7	8	233–234
4.2.7(A)	8	233–234
4.2.7(B)	8	232–235
4.2.8	2, 7, 8, 14	33–36, 54–56, 58–66, 69–84, 177–197, 199–212, 228, 334–339
4.2.8(A)	2, 3, 7, 8, 14	33–36, 54–56, 58–66, 69–84, 177–197, 199–212, 228, 334–339
4.2.8(B)	7, 14	173, 182–184, 197, 209–212, 336–354
4.2.9	16	390–406
4.2.9(A)	16	390–406
4.2.9(B)	16	401–406
4.3	11, 14, 17	273–289, 327–358, 409–452
4.3.1	11, 17	283–286, 429–430, 441–445
4.3.1(A)	11, 17	283–286, 429–430, 441–445
4.3.1(B)	11	274–283

Objectives	Corresponding Chapter(s)	Corresponding Page(s)
4.3.2	11, 17	274–283, 429–430, 441–445
4.3.2(A)	11, 17	274–283, 429–430, 441–445
4.3.2(B)	11	274–283
4.3.3	11, 14, 17	286–287, 354–356, 429–430, 441–445
4.3.3(A)	11, 14, 17	286–287, 354–356, 429–430, 441–445
4.3.3(B)	11	274–283
4.4	8, 9, 12, 17	219–238, 239–253, 291–312, 409–452
4.4.1	12	308–310
4.4.1(A)	12	308–310
4.4.1(B)	12	300–306
4.4.2	8, 9, 17	227–228, 240–248, 423–429
4.4.2(A)	8, 9, 17	227–228, 240–248, 423–429
4.4.2(B)	5, 9, 17	130–131, 243–250, 29–430, 441–445
4.4.3	9, 14, 15	249–250, 334–339, 360–362
4.4.3(A)	9, 14, 15	249–250, 334–339, 360–362
4.4.3(B)	9, 14, 15	243–250, 336–354, 364–365
4.4.4	9	243–249
4.4.4(A)	9	243–249
4.4.4(B)	9	243–250
4.4.5	9	250–251
4.4.5(A)	9	250–251
4.4.5(B)	9	243–250
4.5	8, 10	219–237, 255–272
4.5.1	10	256–259
4.5.1(A)	10	256–259
4.5.1(B)	10	256–259

Objectives	Corresponding Chapter(s)	Corresponding Page(s)
4.5.2	10	256–259, 265
4.5.2(A)	10	256–259, 265
4.5.2(B)	10	256–259
4.5.3	10	256–259
4.5.3(A)	10	256–259
4.5.3(B)	10	256–259
4.6	8, 10	219–237, 255–272
4.6.1	10	256, 259–269
4.6.1(A)	10	256, 259–269
4.6.1(B)	10	256
4.6.2	10, 15	256, 265–269, 271, 360–368
4.6.2(A)	10, 15	256, 265–269, 271, 360–368
4.6.2(B)	10, 15	256–259, 364–365
4.6.3	8	222–223
4.6.3(A)	8	222–223
4.6.3(B)	8	220–223
4.6.4	15	378–385
4.6.4(A)	15	378–385
4.6.4(B)	15	375–380, 382–385
4.6.5	1, 13, 14, 15, 18	10–14, 314–324, 228–339, 360–368, 462–468
4.6.5(A)	1, 13, 14, 15, 18	10–14, 314–324, 228–339, 360–368, 462–468
4.6.5(B)	1, 13, 14, 15, 18	10–14, 314–316, 318–324, 228–333, 336–354, 360–368, 462–468
4.7	5, 11	123–148, 273–289
4.7.1	11	286–287
4.7.1(A)	11	286–287
4.7.1(B)	11	274–283

Objectives	Corresponding Chapter(s)	Corresponding Page(s)
4.7.2	5	132–135, 140–145
4.7.2(A)	5	132–135, 140–145
4.7.2(B)	5	144–145
4.7.3	5	132–139, 140–145
4.7.3(A)	5	132–139, 140–145
4.7.3(B)	5	145

Header image: © Jones & Bartlett Learning. Photographed by Glen E. Ellman.

Glossary

A

Absorption method Collection method for liquid evidence in which the liquid is absorbed into an absorbent material prior to packaging.

Accelerant Any fuel or oxidizer, often an ignitable liquid, used to initiate a fire or increase the rate of growth or spread of fire. (NFPA 921)

Accidental fire A fire for which the proven cause does not involve an intentional human act to ignite or spread fire into an area where the fire should not be.

Accommodation space A space on a vessel designed for people to live in.

Active fire protection system A system that uses moving mechanical or electrical parts to achieve a fire protection goal. (NFPA 72)

Addressable technology Fire alarm and detection systems in which devices and system components are assigned a data address, are capable of two-way communication, and can perform certain activities based on signal input.

Administrative review Review of an investigator's work carried out within the investigator's organization to ensure that policy has been followed and quality standards have been met.

Adrift Loose; not on moorings or towline.

Aerial fuels All green or dead materials located in the forest canopy.

Affidavit A voluntary written statement of fact or opinion made under oath and signed by the author.

Air entrainment The process of air or gases being drawn into a fire, plume, or jet. (NFPA 921)

Air sampling smoke detection A method of smoke detection designed to draw air from the protected area into a detection chamber for analysis.

Alarm signal A warning signal that alerts occupants of a fire emergency.

Alloying Mixing of two or more metals when one or more are in a liquefied state, forming an alloy.

Alternating current (AC) An electric current that changes its direction many times per second at regular intervals.

Ambient temperature The air temperature of the surrounding environment.

Ampacity The current, in amperes, that a conductor can carry continuously under the conditions of use without exceeding its temperature rating. (NFPA 70)

Amperage The strength of an electric current, measured in amperes.

Annunciation panel A device that uses indicating lamps, alphanumeric displays, or other means to provide first responders with status, condition, and location information concerning fire alarm system components and devices.

Arc A high-temperature luminous electric discharge across a gap or through a semiconductive medium such as charred insulation. (NFPA 921)

Arc-fault circuit interrupter (AFCI) A device designed to protect against fires caused by arcing faults in home electrical wiring. The circuitry continuously monitors current flow for the electrical waveforms characteristic of arcing.

Arc mapping Identification and documentation of a fire pattern derived from the identification of arc sites; aids in the determination of fire origin and spread. (NFPA 921)

Arc severed The result of an arcing event in which the conductors are severed owing to the arc.

Arc survey diagram A diagram of the affected area of the structure that identifies and plots the locations of electrical arcing. The spatial relationship of the arc sites can create a pattern and help establish the sequence of damage. This analysis can be used on building circuits and electrical devices within a compartment to help with origin analysis.

Arc tube A cylinder of fused silica/quartz with electrodes at either end that is used in most mercury and metal halide lamps.

Area of origin A structure, part of a structure, or general geographic location within a fire scene, in which the "point of origin" of a fire or explosion is reasonably believed to be located. (NFPA 921 and NFPA 1033)

Areas of demarcation Areas in which the fire damage or effects are similar in magnitude or characteristics.

Arson The crime of maliciously and intentionally, or recklessly, starting a fire or causing an explosion. Legal definitions of "arson" are provided by statutes and judicial decisions that vary among jurisdictions. (NFPA 921)

Arson reporting/immunity statute State law requiring insurance companies to provide information on a fire investigation when requested by state authorities working on a potential arson or other crime.

Artifact evidence Objects or items on which fire patterns are present, the remains of the first fuel ignited, the competent source of ignition that ignited the first fuel, and materials that influenced the fire's growth and development.

Aspect The direction the slope faces (north, east, south, west).

Assemblies Manufactured parts put together to make a completed product.

Atoms The smallest unit of an element that can take part in a chemical reaction.

Authority having jurisdiction (AHJ) An organization, office, or individual responsible for enforcing the requirements of a code or standard, or for approving equipment, materials, an installation, or a procedure. (NFPA 921)

Autoignition temperature (AIT) The lowest temperature at which a combustible material ignites in air without a spark or flame. (NFPA 921)

B

Backflow valve A valve that prevents gas from reentering a container or distribution system.

Balanced pressure proportioner A proportioner that uses an atmospheric tank and pump to introduce the foam concentrate into the system with the water supply.

Balloon-frame construction A construction type in which the exterior wall studs go from the foundation wall to the roofline. The floor joists are attached to the walls by the use of a ribbon board, which creates an open stud channel between floors, including the basement and attic.

Bar holes Holes driven into the surface of the ground or pavement with either weighted metal bars or drills. Gas detectors are then inserted into the holes to detect the location of a gas leak.

Benchmark events Events that are particularly valuable as a foundation for the timeline or that may have significant relation to the cause, spread, detection, or extinguishment of a fire.

Beveling A fire pattern that indicates fire direction on wood wall studs. The bevel (angled side) points toward the general source of the heat.

Bilge The lowest, interior part of a vessel's hull; the area where spilled water and oil can collect.

Blast overpressure Large quantities of gas produced by the explosion of the material.

Boiler A closed vessel in which water is heated, steam is generated, steam is superheated, or any combination thereof by the application of heat from combustible fuels in a self-contained or attached furnace. (NFPA 85)

Boiling liquid/expanding vapor explosion (BLEVE) An explosion that occurs when pressurized liquefied materials (e.g., propane or butane) inside a closed vessel are exposed to a source of high heat.

Brady materials Information in the possession of the government in a criminal case that is favorable to the accused person and that is either material to the person's guilt or to their punishment if they should be convicted of the crime charged.

Branch circuits The individual circuits that feed lighting, receptacles, and various fixed appliances. (NFPA 921)

Bulkheads The vertical separations in a vessel that form compartments and that correspond to walls in a building. (This term does not refer to the vertical sections of a vessel's hull.)

Burning velocity The rate of flame propagation relative to the velocity of the unburned gas ahead of it. (NFPA 68)

C

Calcination A fire effect realized in gypsum products, including wallboard, when exposure to heat drives off free and chemically bound water. (NFPA 921)

Canine–handler team A team composed of a handler and an ignitable liquid detection canine that has been properly trained and certified through a documented and recognized program.

Carboxyhemoglobin (COHb) The carbon monoxide saturation in the blood.

Catalytic converter A device in the exhaust system that exposes exhaust gases to a catalyst metal to promote oxidation of hydrocarbon materials in the exhaust gas.

Cause The circumstances, conditions, or agencies that bring together a fuel, the ignition source, and the oxidizer, resulting in a fire or explosion. (NFPA 921 and NFPA 1033)

Ceiling jet A relatively thin layer of flowing hot gases that develops under a horizontal surface (e.g., ceiling) as a result of plume impingement and the flowing gas being forced to move horizontally. (NFPA 921)

Central supervising station A facility that receives fire alarm signals from properties that is staffed with trained, qualified, and proficient personnel who take the appropriate action to process, record, respond, and maintain monitored fire alarm systems.

CFD models Analytical models for fire behavior based on computational fluid dynamics.

Chain of custody The trail of accountability that documents the possession of evidence in an investigation.

Char Carbonaceous material that has been burned or pyrolyzed and has a blackened appearance. (NFPA 921)

Chemical explosion Explosion in which a chemical reaction is the source of the high-pressure fuel gas. The fundamental nature of the fuel is changed.

Circumstantial evidence Proof of a fact indirectly based on logical inference rather than personal knowledge.

Classification of cause An effort to categorize the fire cause for purposes of assigning responsibility, reporting purposes, or statistics.

Clean burn A fire effect that appears on noncombustible surfaces after any combustible layers (such as soot, paint, and paper) have been burned away. The effect may also appear where soot was not deposited owing to high surface temperatures.

Code A standard that is an extensive compilation of provisions covering broad subject matter or that is suitable for adoption into law independently of other codes and standards. (NFPA 921)

Coded signal A signal that generates a predetermined visual or audible pattern to identify the location of the initiating device that is operating.

Cognitive testing The use of a person's thinking skills and judgment to evaluate the empirical data and challenge the conclusions of the final hypothesis.

Collapse zones Distances around a structure that may be affected by a structural collapse. The area should be identified by markers or specialized scene tape that indicates there is a potential for structural collapse.

Combustible gas indicator An instrument that samples air and indicates whether combustible vapors are present. (NFPA 921)

Combustion A chemical process of oxidation that occurs at a rate fast enough to produce heat and usually light, in the form of either a glow or flames. (NFPA 921)

Combustion explosion Explosion caused by the burning of combustible hydrocarbon fuels and characterized by the presence of a fuel with air as an oxidizer.

Commercial propane Derived from the refining of petroleum, a liquefied gas comprising 95 percent propane and propylene and 5 percent other gases.

Compartmentation A concept in which fire is kept confined in its room of origin, minimizing smoke movement to other areas of a building.

Complex investigation An investigation that generally includes multiple simultaneous investigations and involves a number of interested parties.

Compounds Combinations of elements.

Concurrent Flame spread that is occurring in the same direction as the gas flow from the fire or in the wind direction.

Conduction Heat transfer to another body or within a body by direct contact. (NFPA 921)

Confirmation bias The attempt to prove a hypothesis instead of disproving a hypothesis, by relying only on data that support the hypothesis and ignoring or dismissing contradictory information, resulting in a failure to consider alternative hypotheses or the premature discounting of an alternative hypothesis.

Coniferous litter Primarily needles dropped from coniferous trees but may also include branches, bark, and cones.

Consent Permission from the owner or occupant to examine property.

Container appurtenances The devices that are connected to the openings in tanks and other containers; they include pressure relief devices, liquid level gauges, and pressure gauges.

Control of Hazardous Energy (Lockout/Tagout) standard The federal OSHA regulation that governs work around equipment or wiring where an unexpected energization, start-up of machines, or release of stored energy could result in the injury of the investigators. This standard utilizes a lockout/tagout device that disables the electrical equipment. Specifics can be found in 29 CFR 1910.147.

Control thermostat A device that automatically regulates temperature or that activates a device when the temperature reaches a certain point.

Convection Heat transfer by circulation within a medium such as a gas or a liquid. (NFPA 921)

Convective heater A heater that relies on either natural or forced ventilation to move air across a hot surface, heating the air and dispersing it throughout the room.

Conventional technology The technology used in fire alarm and detection systems that provides limited communication between the fire alarm control panel and system devices and, on activation of an alarm, provides only general information concerning the device and its location.

Counterflow Flame spread that is counter to, or opposed to, the gas flow from the fire.

Crazing A complicated pattern of short cracks in glass that can be either straight or crescent shaped and can extend through the entire thickness of the glass.

Credibility The likelihood that a person is telling the truth.

Cross-contamination The unintentional transfer of a substance contaminated with a residue from one fire scene or location to an evidence collection site.

Crowns Twigs, needles, or leaves of a tree or bush.

Cupping A charred surface on a fuel that is caused by exposure of the surface to the windward side of the fire.

D

Data analytics The process of analyzing raw data from various sources in an effort to find trends and answer questions, including questions about criminal activity.

Dead load The weight of materials that are part of a building, such as the structural components, roof coverings, and mechanical equipment.

Decks Usually continuous, horizontal divisions running the length of a vessel and extending athwart ships; they correspond to floors in a building on land.

Deductive reasoning The process by which conclusions are drawn by logical inference from given premises. (NFPA 921)

Defendant The entity against whom a claim or prosecution is brought in a court.

Deflagration A reaction that propagates at a subsonic velocity through an unreacted medium, less than 1100 ft/sec, and can be successfully vented.

Deformation Change in the shape characteristics of an object.

Deluge sprinkler system A type of automatic fire sprinkler system equipped with open fire sprinkler heads, which simultaneously discharges water from all sprinkler heads.

Demonstrative evidence Any type of tangible evidence relevant to a case—for example, diagrams and photographs.

Density/area curves Graphs that establish the relationship between the required amount of water flow from a sprinkler head (density) and the area that must be covered by the water for different hazard classifications.

Deposition Collection and adherence of smoke particulates and liquid aerosols on surfaces during a fire.

Deposition (legal) One type of pretrial oral testimony made under oath.

Depth of calcination surveys Measurements of the relative depth of calcination (observable physical changes in gypsum wallboard) plotted on a detailed scene diagram to determine the locations within a structure that were exposed to a heat source for the longest durations.

Depth of char analysis Measurements of the relative depth of char on identical fuels plotted on a detailed scene diagram to determine the locations within a structure that were exposed to a heat source for the longest durations.

Design density The minimum predetermined amount of water that must flow from the fire sprinkler heads in the most hydraulically demanding part of the fire sprinkler system (remote area) to control or extinguish a fire.

Determination of cause The process of identifying the factors that allowed a fire to occur, including the type and form of the first fuel, the competent ignition source, and the circumstances and human actions that brought the ignition and the fuel together.

Detonation A combustion reaction that propagates at supersonic velocities, greater than 1100 ft/sec (340 m/sec), and cannot be vented because of its speed.

Diagram A formal drawing that is completed after the scene investigation is completed.

Diffusion flame A flame in which the fuel and air mix or diffuse together at the region of combustion. (NFPA 921)

Direct current (DC) An electrical system that has current flow from the source to the load and back via the circuit return path with one polarity only.

Discoloration Change of color on the surface of a material affected by fire or heat.

Discovery Pretrial procedure in a civil or criminal legal case in which evidence and information about the case is provided to the opposing party(ies) according to the rules of procedure.

Documentary evidence Any type of written record or document that is relevant to the case.

Dry chemical extinguishing agents Dry powder suppression agents made from sodium bicarbonate–based, potassium-based, or ammonium phosphate chemicals that cover and smother the burning material.

Dry-pipe sprinkler system A type of automatic fire sprinkler system equipped with automatic sprinkler heads; the pipes contain pressurized air or nitrogen. When a sprinkler head activates to reduce the air or nitrogen pressure, the dry pipe valve clapper opens, flooding the system piping with water.

Duct smoke detector A smoke detector that samples the air moving through the air distribution system ductwork or plenum; on detecting smoke, the detector sends a signal to shut down the air distribution unit, close any associated smoke dampers, or initiate smoke control system operation.

Duff The layer of decomposing organic materials lying below the litter layer of freshly fallen twigs, needles, and leaves and immediately above the mineral soil.

E

Electrical power The time rate at which work is finished or energy emitted or transferred. The unit of power is the watt.

Engine control module (ECM) An electronic device that controls engine operation parameters, including fuel delivery, throttle control, and safety systems operations.

Electronic valve supervisory devices Devices or switches that integrate with or attach to a water control valve; they detect movement or changes in the position of the valve and then send a signal that the normal open position has changed.

Empirical data Factual data that are based on actual measurements, observations, or direct sensory experiences, which can be verified or are known to be true. (NFPA 921)

Endothermic reaction A reaction or process that absorbs or uses energy.

Energy A property of matter manifested as an ability to perform work, either by moving an object against a force or by transferring heat. (NFPA 921)

Entity in control The interested party who has or represents ownership of the scene or is in effective management of the site, scene, or evidence and is organizing, directing, or controlling the joint actions of the other interested parties.

Estimated time An approximation based on information or calculations that may or may not be relative to other events or activities.

Eutectic melting Any combination of metals with a melting point lower than that of any of the individual metals of which it consists.

Event data recorder (EDR) A device to record data before and after a crash event.

Evidence custodian Person who is responsible for managing all aspects of evidence control.

Evidence technicians Individuals who are specially trained to document, collect, preserve, and transport evidence.

Exemplar An exact duplicate of an appliance.

Exigent circumstance A doctrine in criminal law in which law enforcement or other government agents may be permitted to enter property without a search warrant to respond to an emergency, dangerous condition, or other extraordinary circumstance.

Exothermic reaction A reaction or process that releases energy in the form of heat.

Expectation bias Preconceived determination or premature conclusions as to what the cause of the fire was without having examined or considered all of the relevant data.

Expert witness Someone who is recognized by a court as qualified by specialized knowledge, skills, experience, or education on an issue.

Explosion The sudden conversion of potential energy (chemical or mechanical) into kinetic energy, with the production and release of gases under pressure. These high-pressure gases then do mechanical work, such as moving, changing, or shattering nearby materials. (NFPA 921)

Explosion dynamics analysis The process of using force vectors to trace backward from the least to the most damaged area following the general path of the explosion force vectors.

Explosives Any chemical compounds, mixtures, or devices that function by explosion.

Exterior survey A detailed examination of the exterior and exterior surfaces of a structure involved in fire, which is performed to collect data and information, scan for hazards, and gather information for comparison with interior patterns.

Eyedropper collection method Collection method for liquid evidence in which the liquid is collected into a glass vial through the use of an eyedropper.

F

Fact witness Someone who testifies about facts from their personal knowledge or experiences; also called a lay witness.

Failure mode and effects analysis (FMEA) A technique used to identify basic sources of failure within a system and to follow the consequences of these failures in a systematic fashion.

Fault trees Logic diagrams that can be used to analyze a fire or explosion; also known as decision trees.

Fifth Amendment An amendment to the U.S. Constitution, part of the Bill of Rights, that protects people from being compelled to be witnesses against themselves in criminal proceedings. It also specifies other rights, including the right to due process and protection against double jeopardy (prosecution more than once for the same offense).

Fire A rapid oxidation process; an exothermic chemical reaction resulting in the evolution of light and heat in varying intensities. (NFPA 921)

Fire alarm control units (FACUs) Equipment that monitors the integrity of the fire alarm system's circuits and devices, processes manual and automatic input signals from the initiating devices, drives the notification appliances, provides an interface to control or activate other fire protection and building systems in a fire emergency, and provides the power to support all of the system devices.

Fire alarm system A group of components assembled to monitor and annunciate the status of automatic and manual initiating devices so building occupants may respond appropriately.

Fire area The boundary of fire effects in the scene, characterized by the border between undamaged areas and areas that are damaged by flame, heat, and smoke.

Fire barrier A wall, other than a fire wall, that has a fire-resistance rating. Fire walls and fire barrier walls do not need to meet the same requirements as smoke barriers. (NFPA 101)

Fire breaks Natural or humanmade barriers used to stop the spread or reroute the direction of a wildland fire by separating the fuel from the fire.

Fire dynamics The detailed study of how chemistry, fire science, and the engineering disciplines of fluid mechanics and heat transfer interact to influence fire behavior. (NFPA 921)

Fire effects The observable or measurable changes in or on a material as a result of the fire. (NFPA 921)

Fire flanks The parts of a fire's perimeter that are roughly parallel to the main direction of spread.

Fire head The portion of a fire that is moving most rapidly, subject to influences of slope and other topographic features. Large fires burning in more than one fuel type can develop additional heads.

Fire heel The portion of a fire located at the opposite side of the fire head; it is less intense and is easier to control.

Fire investigation The process of determining the origin, cause, and development of a fire or explosion. (NFPA 921)

Fire investigator An individual who has demonstrated the skills and knowledge necessary to conduct, coordinate, and complete a fire investigation. (NFPA 1033)

Fire pattern The visible or measurable physical changes or identifiable shapes formed by a fire effect or group of fire effects. (NFPA 921)

Fire pattern analysis The process of identifying and interpreting fire patterns to determine how the patterns were created and their significance.

Fire point The lowest temperature at which a volatile combustible substance continues to burn in air after its vapors have been ignited (as when heating is continued after the flash point has been reached).

Fire scene reconstruction The process of removing debris and replacing contents or structural elements in their prefire positions. Reconstruction can also include re-creating the physical scene at a different location to more clearly assess damage and patterns. (NFPA 921)

Fire spread analysis The process of identifying fire patterns related to the movement of fire from one place to another, along with the sequence in which the patterns were produced, so as to trace the fire back to its origin.

Fire storm A natural phenomenon in which a fire attains such intensity that it creates and sustains its own wind system.

Fire wall A wall separating buildings or subdividing a building to prevent the spread of fire, with a fire-resistance rating and structural stability. (NFPA 221)

Fireballs Momentary balls of flame observed during or after an explosion that may present high-intensity, short-term radiation.

Firebrands Hot or burning fragments propelled from an explosion.

Fixed level gauges Devices that are used primarily to indicate when the filling of a tank or cylinder has reached its maximum allowable fill volume. They do not indicate liquid levels above or below their fixed lengths. (NFPA 921)

Flame detector A radiant energy detector that senses specific portions of the visible and invisible light spectrum.

Flame speed The local velocity of a freely propagating flame relative to a fixed point.

Flammable range The range of concentration of a gas or vapor to air between the upper explosive limit (UEL) and the lower explosive limit (LEL), at which combustion can occur in air.

Flash fire A fire that spreads rapidly by means of a flame front through a diffuse fuel such as dust, gas, or ignitable liquid vapor, but without production of damaging pressure.

Flash point The lowest temperature of a liquid, as determined by specific laboratory tests, at which it gives off vapors at a sufficient rate to support a momentary flame across its surface. (NFPA 921)

Flashover A transition phase in the development of a compartment fire in which surfaces exposed to thermal radiation reach ignition temperature more or less simultaneously and, given sufficient availability of oxygen, fire spreads rapidly throughout the space, resulting in full room involvement or total involvement of the compartment or enclosed space. (NFPA 921)

Foam concentrate A condensed form of the foam product.

Foam solution A solution created by mixing foam concentrate and water in the correct proportions.

Freezing The phenomenon in which leaves, needles, and small branches are heated by a passing fire front, and become very soft and pliable. Prevailing winds or drafts created by the fire bend them in the direction the fire advanced. Once cooled, they remain pointing in this direction.

Friction loss As water flows away from the source through a water system, a progressive pressure drop occurs caused by changes in the direction of the pipe, length of pipe, type of pipe, size of the pipe, types and sizes of the fittings, and other components in line with the piping.

Fuel A material that will maintain combustion under specified environmental conditions. (NFPA 921)

Fuel analysis Identification of the first fuel item or package that sustained combustion beyond the ignition source as well as subsequent target fuels beyond the initial fuel.

Fuel-controlled fire A fire in which the heat release rate and growth rate are controlled by the characteristics of the fuel, such as quantity and geometry, and in which adequate air for combustion is available. (NFPA 921)

Fuel gas Includes natural gas, liquefied petroleum gas in the vapor phase only, liquefied petroleum gas–air mixtures, manufactured gases, and mixtures of these gases, plus gas–air mixtures within the flammable range, with the fuel gas or the flammable component of a mixture being a commercially distributed product. (NFPA 921)

Fuel items Any articles that are capable of burning.

Fuel load The total quantity of combustible contents of a building, space, or fire area, including interior finish and trim, expressed in heat units or the equivalent weight in wood. (NFPA 921)

Fuel package A collection or array of fuel items in close proximity with one another such that flames can spread throughout the array.

Fuel rail An internal passage or external tube connecting a pressurized fuel line to individual fuel injectors.

Fugitive gas Fuel gases that escape from their piping, storage, or utilization systems and serve as easily ignited fuels for fires and explosions. (NFPA 921)

Fulgurites Slender, usually tubular, bodies of glassy rock produced by electrical current striking and then fusing dry sandy soil.

Furnace A device used to heat air for residential homes and commercial buildings.

Fusible plugs Thermally activated devices that open and vent the contents of a container.

G

Gas burner A device that allows fuel gases and air to properly mix and produce a flame.

Gas-sensing fire detection A type of detector designed to sense a specific gas, toxic gases, or a variety of gases and vapors produced when processing hydrocarbons.

Ground fault An inadvertent connection of current-carrying equipment or wiring to ground, causing the current to run to ground instead of returning through the neutral cable.

Ground-fault circuit interrupter (GFCI) A device intended for the protection of personnel, which functions to de-energize a circuit or portion thereof within an established period of time when a current is detected to be flowing outside of the intended current path.

Ground fuels All combustible materials below the surface litter, including duff, tree or shrub roots, punky wood, peat, and sawdust, that normally support a glowing combustion without flame.

Grounding A connection, whether intentional or accidental, between an electrical circuit or equipment and earth or to some conducting body that serves in place of the earth. (NFPA 921)

Guide A document that is advisory or informative in nature and that contains only nonmandatory provisions. A guide may contain mandatory statements, such as when a guide can be used, but the document as a whole is not suitable for adoption into law. (NFPA 921)

H

Halocarbons (clean agents) Chemical compounds made up of carbon and hydrogen, chlorine, fluorine, bromine, or iodine.

Halogenated hydrocarbons (halons) Chemical compounds made up of carbon and one or more elements from the halogen series of elements (fluorine, chlorine, bromine, or iodine).

Hard times Specific points in time that are directly or indirectly linked to a reliable clock or timing device with known accuracy.

Hatches Openings in a vessel's deck that are fitted with a watertight cover.

HAZWOPER (Hazardous Waste Operations and Emergency Response) standard The federal OSHA regulation that governs hazardous materials waste site and response training. Specifics can be found in 29 CFR 1910.120.

Headspace The zone inside a sealed evidence can between the top of fire debris and the bottom of the lid.

Heat A form of energy characterized by vibration of molecules that is capable of initiating and supporting chemical changes and changes of state. (NFPA 921)

Heat and flame vector analysis A method used to assist with fire spread analysis and origin determination, in which arrows representing the investigator's assessment of the direction of heat/flame spread are placed on a detailed diagram of the fire scene; the resulting patterns can be traced back to the fire origin.

Heat capacity The amount of heat necessary to raise the temperature of a unit mass by one degree, under specified conditions; it is measured in units such as J/kg-K or Btu/lb-°F.

Heat detector An initiating device that operates after detecting a predetermined fixed temperature or sensing a specified rate of temperature change.

Heat flux The measure of the rate of heat transfer to a surface or an area, typically expressed in kW/m^2 or W/cm^2. (NFPA 921)

Heat release rate (HRR) The rate at which heat energy is generated by burning. (NFPA 921)

Heat shadowing A pattern that results from an object blocking the travel of radiant heat from its source to a target material on which the pattern is produced.

Heat transfer The exchange of thermal energy between materials through conduction, convection, and/or radiation. (NFPA 921)

Heat transfer models Models that allow the investigator to determine how heat was transferred from a source to a target by one or more of the common heat transfer modes: conduction, convection, or radiation.

Heavy timber construction A construction type in which structural members (i.e., columns, beams, arches, floors, and roofs) are made of unprotected, solid, or laminated wood, with large cross-sectional areas (6 or 8 inches [15 or 20 cm] in the smallest dimension, depending on the reference).

High-order damage A rapid pressure rise or high-force explosion characterized by a shattering effect on the confining structure or container and long missile distances. (NFPA 921)

High-resistance fault An unintended path for electricity that allows sufficient current for heating to occur but not enough to activate the overcurrent protection device.

High-temperature accelerants (HTAs) Mixtures of fuels with Class 3 or Class 4 oxidizers and thermite mixtures.

Holds Compartments used for carrying cargo below deck in a large vessel.

Horsepower (hp) A unit of power that is used to express mechanical energy use or production. One kilowatt is equal to 1.34 horsepower.

Hot legs The energized conductors in a circuit.

Hot set A deliberately set fire in which the ignition device is removed by the individual starting the fire.

Housing The outer shell of an appliance.

Hull The outer skin of a vessel, including the bottom, sides, and main deck, but not including the superstructure, masts, rigging, and other fittings.

Hybrid vehicle A vehicle that uses a combination of internal combustion and electric motors for propulsion.

Hydraulic data nameplate A permanent and rigid sign posted on or near a fire sprinkler system riser that provides information about the hazard, design density, size of the remote area, required gallons and pressure for the water supply, and location that the system is serving.

Hydraulic design A mathematical method of determining flow and pressure at any point along the sprinkler system piping for the purpose of determining pipe size throughout the system.

Hypothesis Theory supported by the empirical data that the investigator has collected through observation and then developed into explanations for the event, which are based on the investigator's knowledge, training, experience, and expertise. (NFPA 921)

Hypoxia Condition caused by a victim breathing in a reduced-oxygen environment.

I

Ignition sequence The sequence of events and circumstances that allowed the initial fuel and the ignition source to come together and result in a fire.

Ignition source analysis Process through which all potential ignition sources in the area of origin are identified and then considered in light of the physical properties of the first fuel ignited, fundamental scientific principles, and other available data.

Impact load A sudden added load to a structure.

Inboard Toward the interior of a vessel from the outer hull; a vessel engine that is enclosed within the hull of the boat.

Incendiary devices A wide range of mechanisms used to initiate an incendiary fire.

Incendiary fire A fire that is intentionally ignited in an area or under circumstances where and when there should not be a fire. (NFPA 921)

Incident command system (ICS) The combination of facilities, equipment, personnel, procedures, and communications under a standard organizational structure to manage assigned resources effectively to accomplish stated objectives for an incident.

Inductive reasoning The process by which a person starts from a particular experience and proceeds to generalizations; the process by which hypotheses are developed based on observable or known facts and the training, experience, knowledge, and expertise of the observer. (NFPA 921)

Inert gas A gas that could contain a mixture of helium, neon, argon, nitrogen, and small amounts of carbon dioxide and that does not become contaminated, react, or become flammable.

Initial scene assessment A preliminary phase of the fire investigation that seeks to ensure preservation of the scene, identify safety hazards and personnel and equipment needs, and determine areas that require further inspection.

Initiating devices Systems of components and devices that provide a manual or automatic means to activate fire alarm and supervisory signals.

Intent Normally an element of the proof of a crime; the state of mind that exists at the time a person acts or fails to act.

Interested party Any person, entity, or organization, including their representatives, with statutory obligations or whose legal rights or interests may be affected by the investigation of a specific incident. (NFPA 921)

Interior survey A systematic inspection of building interior areas to collect data, narrow down inspection needs, and identify hazard and stability issues.

Interrogatories A set of questions served by one party involved in litigation to another involved party.

Interrupting current rating The maximum amount of current a device is capable of interrupting.

Interstitial spaces The spaces between the building frame and interior walls and the exterior façade, and between ceilings and the bottom face to the floor or deck above. (NFPA 921)

Investigative file The organized collection of all of the documentary information in a fire case, including verbal, written, and visual information from the scene; the reports of investigators and other professionals; and any other investigations or research that has been conducted.

Ionization smoke detector A smoke detector that uses a small amount of a radioactive material, which electrically charges air particles to produce a measurable current in the sensing chamber; once smoke enters the sensing chamber, a current drop below a predetermined level initiates an alarm.

Isochar A line on a diagram connecting equal points of char depth.

J

Job performance requirements (JPRs) A statement that describes a specific job task, lists the items necessary to complete the task, and defines measurable or observable outcomes and evaluation areas for the specific task. (NFPA 1033)

K

K-factor A number assigned to represent the discharge coefficient for the orifice of a sprinkler head that is used when calculating the water flow or water pressure at a specific location in the fire sprinkler system.

Kinetic energy The energy possessed by a system or object as a result of its motion.

L

Laminated beams Structural elements that have the same characteristics as solid wood beams. They are composed of many wood planks that are glued or laminated together to form one solid beam and are designed for interior use; the effects of weathering decrease their load-bearing ability.

Lay witness Someone who testifies about facts from his or her personal knowledge or experiences; also called a fact witness.

Layering Systematic removal of debris within an area of origin from the top down, to identify evidence and patterns.

Lazarettes Stowage compartments, often in the aft end of a vessel, sometimes used as workshops.

Lightweight wood trusses Similar to other trusses in design; individual members are fastened using nails, staples, glue, metal gusset plates, or wooden gusset plates.

Lines of demarcation The borders defining the differences in fire effects on materials between the affected area and adjacent, less-affected areas. (NFPA 921)

Line-type detector A type of detection device that uses a tube or wires running in different directions as the sensing element to provide coverage over a wide area.

Liquefied petroleum (LP gas) Petroleum gases condensed to a liquid state with moderate pressure and normal temperatures to allow for more efficient distribution; also called propane.

Live load The weight of temporary loads that needs to be factored into the weight-carrying capacity of the structure, such as furniture, furnishings, equipment, machinery, snow, and rainwater.

Lividity Pooling of the blood in the lower elevations of the body after death, which is caused by the effects of gravity.

Lower explosive limit (LEL) The minimum concentration of a gas or vapor to air at which the gas or vapor will burn in air.

Low-order damage A slow rate of pressure rise or low-force explosion characterized by a pushing or dislodging effect on the confining structure or container and by short missile distances. (NFPA 921)

M

Main disconnect A component that provides a means to shut off power to the entire electrical system. It may incorporate a circuit protection device such as a fuse or circuit breaker.

Manual fire alarm box A type of initiating device that requires a person to pull a handle on the device to send an alarm signal.

Manual initiating device A type of initiating device that requires a person to make physical contact with the device to operate the device.

Manufactured housing A construction technique whereby the structure is built in one or more sections; these sections are then transported to, and assembled at, the building site.

Mass arson The setting of three or more fires at the same site or location during a limited period of time.

Mass loss Loss of mass of a material attributable to consumption by fire or heat.

Mechanical explosion Rupture of a vessel or container such as a cylinder, tank, or boiler, resulting in the release of pressurized gas or vapor. The pressure leading to the mechanical explosion is not attributable to a chemical reaction or a change in the chemical composition of the substances involved.

Melting A physical change caused by exposure to heat. (NFPA 921)

Meter A device that plugs into the meter base to measure the amount of electricity consumed at a site.

Meter base The enclosure that contains the connections for the utility's electrical meter and for the service entrance conductors.

Mill construction An early form of heavy timber construction influenced and developed largely by insurance companies that recognized a need to reduce large fire losses in factories.

Modular home A dwelling constructed in a factory and placed on a site-built foundation, all or in part, in accordance with a standard adopted, administered, and enforced by the regulatory agency, or under reciprocal agreement with the regulatory agency, for conventional site-built dwellings.

Modus operandi (MO) The method of operation used by the offender.

Mosaic photographs A series of photographs taken when a sufficiently wide-angle lens is not available and a panoramic view is desired. The mosaic is created by assembling a number of photographs in overlay form to give a more-than-peripheral view of an area.

Motion A request for court action regarding facts, documents, and evidence that are identified during the discovery phase of a legal proceeding. Motions may argue the admissibility of the involved facts, documents, or evidence.

Motive An inner drive or impulse that is the cause, reason, or incentive that induces or prompts a specific behavior. (NFPA 921)

N

Natural gas A naturally occurring, largely hydrocarbon gas product recovered by drilling wells into underground pockets, often found in association with crude petroleum.

Negligence Failure to provide the same care that a reasonably prudent person would take under similar circumstances.

Neutral plane The line where the flow of the hot gas and cooler air changes.

Noncoded signal A constant visual or audible signal that remains activated until reset.

Noncombustible construction A construction type used primarily in commercial and industrial storage and in high-rise construction. The major structural components (e.g., brick, stone, steel, masonry block, cast iron, or nonreinforced concrete) are noncombustible; the structure itself will not add fuel to the fire.

Noncontact voltage tester A device that will emit a visual and/or audio signal in the presence of an electromagnetic field produced by an AC voltage.

Nonseated explosion Explosion in which the fuels in the explosion are dispersed or diffused, characterized by moderate rates of pressure rise and subsonic explosive velocities.

Non-water–based fire suppression A type of fire suppression system that uses gas- or chemical-based agents.

Notification appliances Constant visual or audible signals that remain activated until reset.

O

Ohm's law A basic law of electricity that defines the relationship among voltage, current, and resistance. If two of these three values are known, it is possible to determine the third.

Ordinary construction A construction type in which exterior walls are masonry or other noncombustible material and in which the roof, floor, and wall assemblies are wood. (NFPA 921)

Origin matrix analysis A cognitive testing process used in the analysis of fully involved compartment fires to determine their origin.

Outboard Lying outward from the centerline of a vessel toward or beyond its sides; a type of engine that is attached to and lying aft of a vessel's stern.

Overboard Over the side or out of the vessel.

Overcurrent Any current in gross excess of the rated current of equipment or the ampacity of a conductor; it may result from an overload, short circuit, or ground fault. (NFPA 921)

Overfusing A dangerous condition that occurs when the circuit protection (fuse or circuit breaker) rating significantly exceeds the ampacity of the conductor, leading to increased heat in the conductors.

Overhaul The final stages of extinguishment by the fire department after knockdown of the main body of fire, during which any pockets of fire are found and extinguished, often through cutting or pulling down wall and ceiling segments. (NFPA 921)

Overload Operation of equipment in excess of normal, full-load rating or of a conductor in excess of rated ampacity; when it persists for a sufficient length of time, it could cause damage or dangerous overheating. (NFPA 921)

Oxidation A chemical reaction in which an element combines with oxygen, resulting in loss of electrons.

Oxidative reaction A reaction in which the atoms in a material are broken down and rearranged, producing heat energy.

Oxidizing agent A substance that promotes oxidation during the combustion process.

P

Paddle-type water flow device A type of initiating device installed on a wet-pipe sprinkler system; a paddle inserted into the pipe moves to initiate an alarm signal when there is sustained water flow through the system piping.

Parting arc An arc that occurs when conductors separate while current is flowing, either in normal conditions (e.g., when a switch is activated or a plug is pulled from an outlet) or in abnormal ones (e.g., when an energized conductor contacts a grounded conductor or other item with near-zero circuit resistance, causing a surge of current that melts the points of contact and creates a gap).

Passive fire protection system Any portion of a building or structure that provides protection from fire or smoke without any type of system activation or movement. (NFPA 921)

Peer review Formal review of a professional's work performed by other qualified professionals not selected by the author of the work, and often conducted anonymously.

Permit-Required Confined Space standard The federal OSHA regulation that governs any space that an employee can bodily enter and perform assigned work, with a limited or restricted means of entrance or exit, and that is not designed for continuous employee occupancy. Specifics can be found in 29 CFR 1910.146.

Personation Unusual behavior by an offender, beyond that necessary to commit the crime.

Phase change The conversion of a material from one state of matter to another; it is reversible and does not change the chemical composition of the material.

Photo diagram A diagram indicating the directions from which photographs were taken.

Photoelectric detector A smoke detector that uses a light source and receiver within the sensing chamber to initiate an alarm when the light source reflects or is obscured by smoke particles.

Photo log A written chart or list of all photos taken at a fire scene, including photo numbers or other identifiers and a brief description of the item or area photographed.

Physical evidence Any physical or tangible item that tends to prove or disprove a particular fact or issue. (NFPA 921)

Pipe schedule A list of pipe sizes and the number of fire sprinkler heads that the pipe can support based on the hazard classification.

Plank-and-beam construction A construction type in which a few large members replace the many small wood members used in typical wood framing; that is, large dimension beams, more widely spaced, replace the standard floor and/or roof framing of smaller dimensioned members.

Platform-frame construction The most common construction method currently used for residential and lightweight commercial construction. In this method of construction, separate platforms or floors are developed as the structure is built.

Plume The column of hot gases, flames, and smoke rising above a fire; also called *convection column, thermal updraft,* or *thermal column.* (NFPA 921)

Point of origin The exact physical location within the area of origin where a heat source and a fuel first interacted, resulting in a fire or explosion. (NFPA 921 and NFPA 1033)

Polarized plug A plug made after 1987 for 20 A or less, which has a neutral prong that is wider than the hot prong.

Polyvinyl chloride (PVC) High-strength plastic that is used in many applications, including pipes and wire insulation.

Post-and-frame construction Similar to plank-and-beam construction, but the structure uses larger elements. An example is a typical barn construction in which the posts provide a majority of the support and the frame provides a place for the exterior finish to be applied.

Power A property of a process, such as fire, which describes the amount of energy that is emitted, transferred, or received per unit of time. (NFPA 921)

Preaction sprinkler system A type of automatic fire sprinkler system equipped with automatic fire sprinkler heads that interface with fire detection equipment; it requires a fire detector to activate the system and an automatic fire sprinkler head to flow water.

Preinvestigation team meeting A meeting that takes place prior to the on-scene investigation. The team leader or investigator addresses questions of jurisdictional boundaries and assigns specific responsibilities to the team members. Personnel are advised of the condition of the scene and the safety precautions required.

Premixed burning Burning in which the fuel and oxidizer are mixed prior to combustion, as in a laboratory Bunsen burner or a gas cooking range; propagation of the flame is governed by the interaction between flow rate, transport processes, and chemical reaction. (NFPA 921)

Prescribed fire A fire that resulted from intentional ignition by a person or a naturally caused fire that is allowed to continue to burn according to approved plans to achieve resource management objectives. (NFPA 921)

Pressure gauge A type of container appurtenance depicting the internal pressure of a tank. The gauge is connected directly to the tank or sometimes through the valve. Pressure gauges do not indicate the quantity of liquid propane in the tank.

Pressure proportioner A proportioner that redirects some of the water supply into the foam concentrate tank to either exert pressure on a collapsible bladder or push the concentrate out of the tank to the proportioner for mixing.

Pressure relief valve A valve designed to open at a specific pressure, usually around 250 psi (1700 kPa). It is generally placed in the container, where it releases the vapor.

Pressure switch A type of initiating device installed on dry, deluge, preaction, and older wet-pipe sprinkler systems; the device initiates an alarm signal once a water pressure threshold is met or a supervisory signal once an air pressure threshold is met.

Private mode notification An alarm signal that is sent to a location within a facility so that trained individuals can interpret and implement the appropriate response procedures.

Productive sampling area An area for fire debris or liquid sampling where the potential for laboratory identification of ignitable materials is maximized owing to its location or contents.

Projected beam detector A fire detector that projects a light beam to a receiver over a hazard; once the beam is obscured or scattered, it activates an alarm.

Proprietary supervising station A group of fire alarm systems at one location or multiple locations that are under constant supervision and monitoring by the property owner's trained personnel.

Protocol A description of the specific procedures and methods by which one or more tasks are to be accomplished. (NFPA 921)

Public mode notification An alarm signal that propagates throughout a building to audibly or visually alert the building occupants so they can take appropriate action.

Pugilistic attitude A crouching stance with flexed arms, legs, and fingers.

Pyrolysis Process in which material is decomposed, or broken down, into simpler molecular compounds by the effects of heat alone; pyrolysis often precedes combustion. (NFPA 921)

R

Races A component of a rolling element bearing that contains the elements and transfers the load to the bearing; generally used in pairs (inner and outer races).

Radiant energy–sensing detector A type of fire detector that is not dependent on smoke and heat plumes to operate; instead, the detector looks for specific portions of the visible and invisible light spectrum produced by flames, sparks, or embers.

Radiant heater A portable electric heater that utilizes a glowing heating element.

Radiation Heat transfer by way of electromagnetic energy.

Rainbow effect A diffraction pattern formed when hydrocarbons float on a surface.

Recommended practice A document that is similar in content and structure to a code or standard but that contains only nonmandatory provisions using the word "should" to indicate recommendations in the body of the text. (NFPA 921)

Regular current rating The level of current above which the protective device will open, such as 15 A, 20 A, or 50 A.

Regular or ordinary dry chemicals Dry chemical agents rated only for Class B and C fires.

Relative humidity The amount of moisture in a given volume of air compared with how much moisture the air could hold at that same temperature. (NFPA 921)

Relative time The chronological order of events or activities that can be identified in relation to other events or activities.

Relaxation time The amount of time for a charge to dissipate.

Remote area The minimum square footage of the most hydraulically remote pipe in a fire sprinkler system; the minimum design density must be available from all heads in that area.

Remote supervising station A supervising station to which alarm, supervisory, or trouble signals, or any combination of those signals, emanating from protected premises' fire alarm systems are received and where personnel are in attendance at all times to respond. (NFPA 72)

Residential fire sprinkler system A type of automatic fire sprinkler system equipped with fast response automatic sprinkler heads specifically made for low heat release and low water pressures.

Resistance heating Heating that occurs when current flows through a path that provides resistance to current flow, such as a heating element (intentional) or a resistive connection (unintentional).

Resistive circuit A circuit that does not contain inductive or capacitive elements.

Response time index (RTI) The amount of time a head takes to activate once exposed to temperatures above the predetermined activation temperature.

Ring flash A type of flash that reduces glare and gives adequate lighting for the subject matter in close-up photography.

Root mean square (RMS) A mathematical computation that equates the voltage level of an alternating current (AC) system to that of a direct current (DC) system.

S

Sabotage Intentional damage or destruction.

Scaled timeline Timeline in which the spacing between the time events being scaled is depicted in a manner that would show the elapsed time between each event.

Schedule A list that details the types of equipment in a structure, which is typically used on larger projects.

Scientific method The systematic pursuit of knowledge involving the recognition and definition of a problem; the collection of data through observation and experimentation; analysis of the data; the formulation, evaluation, and testing of a hypothesis; and, when possible, the selection of a final hypothesis. (NFPA 921)

Search and seizure A legal term encompassing the government's search or examination of a person, their property, or their land; or interference with the person's right to the use of the same.

Seat of the explosion A craterlike indentation created at the point of origin of an explosion. (NFPA 921)

Seismic effect The transmission of tremors through the ground as a result of the blast wave expansion, causing structures to be knocked down.

Self-heating A result of exothermic reactions that occur spontaneously in some materials, whereby heat is generated at a sufficient rate to raise the temperature of the material. (NFPA 921)

Sequential pattern analysis Application of principles of fire science to the analysis of fire pattern data (including fuel packages and geometry, compartment geometry, ventilation, fire suppression operations, and witness information) to determine the origin, growth, and spread of a fire.

Sequential photography A series of photographs that shows the relationship of a small subject to its relative position in a known area. The small subject is first photographed from a distant position, where it is shown in context with its surroundings. Additional photographs are then taken increasingly closer until the subject is the focus of the entire frame.

Serial arson Multiple arsons by an offender who sets three or more fires, with a cooling-off period between fires.

Serial arsonist A serial fire-setter.

Serial fire-setter An individual or group involved in three or more fire-setting incidents.

Service drop The overhead service conductors between the power pole and the structure, up to and including any splices, which connect to the service entrance conductors.

Service entrance The area where the electrical service enters a building.

Service equipment The equipment used to shut off, distribute, and protect the wiring to a facility. It includes the main disconnect, overcurrent protection devices, and the main distribution panelboard.

Service lateral The service entrance wiring entering the structure from underground.

Service mast The conduit between the weatherhead and the meter base.

Sever arc An arc site where one or more of the circuit conductors were physically severed by the arcing event at that location. (NFPA 921)

Shore power Electrical power supplied from shore via a cord set.

Short circuit An abnormal connection of low resistance between normal circuit conductors where the resistance would normally be much greater. (NFPA 921)

Shutoff valve A valve located in the piping system and readily accessible and operable by the consumer; it is used to shut off individual appliances or equipment.

Sine wave The waveform that an alternating current follows.

Site safety assessment A thorough assessment of all hazards and safety factors performed by the fire investigator prior to entering a fire scene to determine whether it is safe to enter and which personal protective equipment will be required.

Sixth Amendment An amendment to the U.S. Constitution, part of the Bill of Rights, that provides that an accused person shall have the assistance of counsel in their defense, a speedy and public trial, the right to "confront" opposing witnesses, and other rights in criminal cases.

Sketch Freehand diagram or diagram drawn with minimal tools that is completed at the scene; it can be either three- or two-dimensional.

Slope The steepness of land or a geographic feature, which can greatly influence fire behavior. (NFPA 921)

Smoke alarm A single- or multistation smoke detector with an integrated power source, sensing device, and alarm device.

Smoke barrier A continuous membrane, either vertical or horizontal, designed and constructed to restrict the movement of smoke. (NFPA 92)

Smoke deposits Hot products of combustion that may adhere on collision with a surface.

Smoke detector A fire detector that senses visible and invisible particles of combustion. (NFPA 72)

Snag A standing dead tree, or part of a dead tree, from which at least the smaller branches have fallen.

Soffits The horizontal undersides of the eaves or cornice.

Soft time Estimated or relative point in time.

Spalling The chipping or pitting of concrete or masonry surfaces. (NFPA 921)

Spark/ember detector A radiant energy detector that senses specific portions of the visible and invisible light spectrum.

Spoliation of evidence Loss, destruction, or material alteration of an object or document that is evidence or potential evidence in a legal proceeding by the person who has the responsibility for its preservation. (NFPA 921)

Spot fires Projections of flaming or burning particles (firebrands) that are found ahead of the flame front.

Spot-type detector A type of detection device that provides coverage in a specific area where the sensing element is in a fixed location.

Spree arson The setting of three or more fires at separate locations with no emotional cooling-off period between fires.

Staging Purposeful alteration of the crime scene prior to the arrival of police.

Standard An NFPA standard, whose main text contains only mandatory provisions using the word "shall" to indicate requirements, and which is in a form generally suitable for mandatory reference by another standard or code or for adoption into law. Nonmandatory provisions are not to be considered a part of the requirements of a standard and shall be located in an appendix, annex, footnote, or other means as permitted in the *NFPA Manual of Style*. (NFPA 921)

Starboard The right side of a vessel when looking forward.

Static electricity The electrical charging of materials through physical contact and separation and the various effects that result from the electrical charges formed by this process.

Steel-framed residential construction A construction type with characteristics similar to those of wood-frame construction but that is noncombustible. Steel framing can lose its structural capacity during extreme exposure to heat.

Step-down transformer Device that reduces the 120 V provided at the receptacle to the required voltage.

Stoichiometric mixture The optimal ratio at which the combustion will be most efficient.

Stoichiometric ratio The optimal ratio in a fuel–air mixture at which combustion will be most efficient (above the lower explosive limit and below the upper explosive limit).

Superstructures All vessel structures above the main deck or weather deck.

Supervisory alarm signal A type of fire alarm system alert signal that indicates when the normal ready status of other fire protection systems or devices connected to or integrated with the fire alarm panel has changed.

Surface fuels All combustible materials from the surface of the ground up to about 6 ft (2 m).

Surface-to-mass ratio Ratio of the surface area of a solid or gas to its mass. The higher the surface-to-mass ratio, the more surface area of the material that is exposed to air.

Switch Device used to turn an appliance on or off or to change its operating conditions.

Switch loading Loading of a product into a tank or compartment that previously held a product with a different vapor pressure and flash point.

System analysis An analytical approach that takes into account characteristics, behavior, and performance of a variety of elements.

T

Tamper switch A device that detects changes in the normal position of a fire protection system valve and sends a supervisory signal when the valve is moved from this position.

Technical review Review of a professional's work by other professionals qualified in all aspects of the author's profession, with access to all relevant documentation.

Temperature The degree of sensible heat of a body as measured by a thermometer or similar instrument. (NFPA 921)

Temporary Interim Amendment (TIA) An amendment to an NFPA standard processed according to NFPA regulations on an urgent basis after the publication of the standard, without the opportunity to publish in the first and second draft reports for review and comment.

Testimonial evidence Verbal testimony of a witness given under oath or affirmation and subject to cross-examination by the opposing party.

Textual signal A messaging signal that provides constant and specific information via voice, pictorial, or alphanumeric means.

Thermal conductivity (k) The measure of the amount of heat that will flow across a unit area with a temperature gradient of one degree per unit of length; it is measured in units such as W/m-K and Btu/hr-ft-°F.

Thermal decomposition An irreversible change in chemical composition as a result of pyrolysis.

Thermal inertia The properties of a material that characterize its rate of surface temperature rise when exposed to heat; related to the product of the material's thermal conductivity ($\downarrow$), density ($\uparrow$), and heat capacity (c). (NFPA 921)

Thermal runaway A condition in which the heat generated exceeds the amount of heat lost from a material.

Thermistor A sensor whose resistance changes in response to temperature; used in temperature measurement.

Thermometry The study of the science, methodology, and practice of temperature measurement. (NFPA 921)

Time–current curve A graph, provided by the manufacturer, used to determine the activation time of an overcurrent protection device when it is exposed to various currents.

Timeline A graphic or narrative representation of events related to the fire incident, arranged in chronological order.

Topside When below decks, the areas on or above the main deck.

Total burn A fire that has consumed all, or nearly all, combustible materials.

Traditional forensic physical evidence Evidence that includes, but is not limited to, finger and palm prints, bodily fluids such as blood and saliva, hair and fibers, footwear impressions, tool marks, soils and sand, woods and sawdust, glass, paint, metals, handwriting, questioned documents, and general types of trace evidence. (NFPA 921)

Trailer Solid or liquid fuel used to intentionally spread or accelerate the spread of a fire from one area to another. (NFPA 921)

Transformers Devices that can increase or decrease an AC voltage.

Transom The stern cross section of a square-sterned vessel.

Trolling motor A small electric motor connected to a battery and used primarily for slow-speed maneuvering of recreational boats.

Trouble alarm signal A type of fire alarm system alert signal that indicates a problem with the system's integrity, such as a power or component failure, device removal, communication fault or failure, ground fault, or a break in the system wiring.

Turbocharger An exhaust-driven device that compresses intake air to increase engine power.

U

Understory vegetation The area under a forest or brush canopy that grows at the lowest height level. Plants in the understory consist of a mixture of seedlings, saplings, shrubs, grasses, and herbs.

Underway Not attached to land or mooring by tie, anchor, or grounding.

Uninhibited chemical chain reaction One of the elements of the fire tetrahedron; it provides for the combination and interaction of the other elements.

Upper explosive limit (UEL) The maximum concentration of a gas or vapor to air at which the gas or vapor will burn in air.

V

Vandalism Mischievous or malicious fire setting that results in damage to property.

Vaporization A phase transition from a liquid or a solid phase to a gas phase.

Vaporizer A heater used to heat and vaporize propane where larger quantities of propane are required, such as for industrial applications.

Variable gauge A gauge that gives readings of the liquid contents of containers, primarily tanks or large cylinders. It gives readings at virtually any level of liquid volume.

Venting The escape of smoke and heat through openings in a building. (NFPA 921)

Venturi A narrowed area within a carburetor that causes air to accelerate and create a low-pressure area that draws fuel into the intake manifold of the engine.

Venturi proportioner A proportioner that uses water moving over an open orifice to create a lower pressure at the opening, which draws the foam into the water stream.

Vessel A broad grouping of every description of watercraft, other than a seaplane on the water, used or capable of being used as a means of transportation on the water.

Video detector A type of fire detector that uses high-definition video cameras to analyze digital images for changes in the pixels caused by smoke and flame generation.

W

Water control valve A device used to control the flow of water.

Water flow detection device An attachment to a sprinkler system that detects water flow within the system piping and initiates an alarm signal; also called a water flow alarm device.

Water mist system A fixed fire protection system that uses specialized nozzles to discharge a very fine spray mist of water droplets that extinguish by cooling, displacing oxygen, or blocking radiant heat.

Weatherhead A conduit body at the service entrance used to prevent water from entering the service mast and meter base.

Weather history A description of atmospheric conditions over the preceding few days or several weeks.

Wet chemical extinguishing agents Suppression agents that mix water with potassium acetate, potassium carbonate, potassium citrate, and, in some instances, a mixture of these agents and other additives; used primarily to suppress Class K fires.

Wet-pipe sprinkler system A type of automatic fire sprinkler system equipped with automatic fire sprinkler heads that has water in the pipes at all times; when a sprinkler head activates, water flow is immediate.

Wildfire V-shaped patterns Horizontal ground surface burn patterns generated by the fire spread.

Wood-frame construction A construction type in which exterior walls and load-bearing components are wood. This type of construction is often associated with residential construction and contemporary lightweight commercial construction.

Wood I-beams Constructed with small-dimension or engineered lumber, as the top and bottom chord, with oriented strand board or plywood as the web of the beam.

Work plan An outline of the tasks to be completed as part of the investigation, including the order or timeline for completion. (NFPA 921)

Z

Zone models Computer fire models that divide a compartment into two zones: a hot upper zone and a cooler lower zone.

Index

Note: Page numbers followed by "*f*" and "*t*" indicate figures and tables, respectively.

Header image: © Jones & Bartlett Learning. Photographed by Glen E. Ellman.